Particles and Fields

Particles and Fields

DAVID LURIÉ

Technion, Israel Institute of Technology,
Haifa, Israel

1968

INTERSCIENCE PUBLISHERS
a division of John Wiley & Sons
New York · London · Sydney

Library of Congress catalog card number 68-22312

SBN 470 55642 0

Set on Monophoto Filmsetter and Printed by
J. W. Arrowsmith Ltd., Bristol, England

Foreword

The quantum theory of fields was invented about forty years ago to give expression to the relativistic quantum mechanics of the radiation field. Within a decade it had established itself as the theoretical framework for the description not only of electromagnetic interaction, but also of nuclear interactions. Within another two decades it was extended to cover the weak interactions as well.

There have been several occasions in the past when it appeared that the quantum theory of fields was inconsistent and had to be abandoned. The difficulties with quantum electrodynamics were circumvented by the development of renormalization theory towards the middle of this century and it was shown that meaningful and accurate predictions about the properties of electrons in electromagnetic fields could be made. The basic difficulties themselves were not resolved; instead one succeeded in reformulating the theory in such a fashion that the embarrassing infinities of the quantum field theory did not appear explicitly.

The successful exploitation of analytic properties to develop physically satisfactory approximation methods for strong interactions and the population explosion among hadrons diverted attention from the basic problems of the quantum field theory of interacting systems. Group-theoretic and dispersion-theoretic methods of calculation had a moderate amount of success. On the basis of this, the possibility was seriously entertained that the quantum theory of fields was irrelevant for the dynamical description of subnuclear phenomena.

Recent years have seen a resurgence of interest in the quantum field theory. Dynamical calculations based on suitable Lagrangians for interacting fields have had a certain amount of success. Considerable attention is now being devoted to an examination of basic field theory; and some of the older problems are being analyzed anew. Field variables like currents and fields are being allotted their rightful place as fundamental dynamical variables.

It is appropriate that this elegant presentation of orthodox quantum field theory by my friend Dr. David Lurié is appearing at this juncture. It contains an account of the elements of field quantization and covariant perturbation theory in the first half of the book. One cannot fail to note the graceful simplicity of the treatment of this traditional domain. The expert would also note the explicit discussion of the spin 3/2 and spin 1 fields. The treatment of the renormalization theory, reduction formulae and spectral representations brings out the essential ideas without getting lost in mathematical details. The discussion of coupling constants and sum rules is quite attractive.

But it is most pleasing to see the excellent treatment of the Bethe–Salpeter equation, of bound states and of the functional method; in most existing books on field theory these topics are dismissed without any ceremony.

On the whole the book does a very creditable job of presenting the quantum theory of fields in its essentials. I have enjoyed this book. It is my hope that many students will enjoy this book and gain from it an understanding of the accomplishments and promise of quantum field theory.

New Delhi, E. C. G. SUDARSHAN
January 1968

Preface

This book is primarily designed to serve as a text for a graduate course on quantum field theory. As a prerequisite it is assumed that the student is familiar with ordinary non-relativistic quantum mechanics, including the formal theory of scattering. In addition, a knowledge of the basic phenomenology of elementary particle physics has been assumed.

In addition to the material designed to serve as a graduate text, the book includes discussion of advanced topics of interest to research workers in quantum field theory. Throughout the book, with the exception of Section 3-4 and part of Section 9-1, the emphasis is on *relativistic* quantum field theory, with a consequent orientation toward elementary particle physics. On the other hand, the successful adaptation, during the past decade, of quantum field theoretic methods to problems in statistical and solid state physics encourages one in the hope that research workers in these fields will also find this book useful.

The first six chapters constitute the basic subject matter of elementary quantum field theory. The opening chapter deals with the relativistic one-particle theories for spin 0, $\frac{1}{2}$, 1 and $\frac{3}{2}$. The inadequacy of the one-particle formulation of relativistic quantum mechanics is marked by the appearance, in each theory, of negative energy solutions. To clear the way for a new departure, emphasis is placed in Chapter 2 on the field aspect of these theories and the Lagrangian formalism is set up. Field and quantum particle aspects are then unified by means of field quantization in Chapters 3 and 4. Chapter 3 presents the canonical quantization of the spin 0 and spin $\frac{1}{2}$ fields and includes, among other features, an introductory discussion of Schwinger's quantum action principle. Chapter 4 is devoted to the quantization of spin 1 and spin $\frac{3}{2}$ fields, with special attention given to the electromagnetic field which is quantized in both the radiation and Lorentz gauges.

Interacting quantum fields are introduced in Chapter 5. Coupling schemes for elementary particle interactions are constructed and, in the case of strong and weak interactions, considerable weight is given to the symmetry principles and conservation laws which play such an important role in modern applications. Chapter 6 is concerned with perturbation theory. The first half of the chapter develops the Feynman rules for calculating transition matrix elements and presents simple applications to electromagnetic and weak processes. The second half of the chapter is devoted to renormalization theory.

The last four chapters of the book constitute an introduction to advanced quantum field theory. Chapter 7 is concerned with the link between transition amplitudes and vacuum expectation values of products of field operators. Particular stress is laid on the abstract, nonperturbative reformulation of field theory, dating from the mid-1950's and based on the use of the asymptotic condition. Chapter 8 presents a variety of applications of field-theoretic techniques to particle physics. Topics include the Goldberger–Treiman relation, the Adler–Weisberger sum rule, and the universality of the vector coupling constant in the theory of weak interactions. Chapter 9 deals with a number of topics relating to bound states. These include the Bethe–Salpeter equation and the question of assigning field operators to composite particles. The final chapter develops the powerful functional approach to quantum field theory pioneered by Schwinger. Applications of the functional technique to the Goldstone theorem and to one-dimensional quantum electrodynamics are presented. Two major topics—dispersion relations and axiomatic field theory—are omitted from discussion in the book. However, the work of Chapters 7 and 8 should equip the student with the necessary background for the study of these subjects in more specialized texts.

Acknowledgements

I was fortunate to be able to write the bulk of this book at the School of Theoretical Physics of the Dublin Institute for Advanced Studies. I could have found no more congenial atmosphere for the task and I wish to express my gratitude to the Governing Board of the School and to its Director, Professor J. L. Synge, for having made it possible for me to come to Dublin. I wish to thank my colleagues at the Institute, and especially Professor Y. Takahashi, for many helpful remarks and discussions. I am particularly grateful to the Rev. Dr. C. Ryan for a critical reading of the entire manuscript and for scores of constructive comments and suggestions. I am also indebted to Professor R. E. Marshak and to Dr. A. J. Macfarlane for encouragement and advice in the initial stages of writing at the University of Rochester.

The assistance of Dr. E. Pechlaner in the preparation of the manuscript for the printer is most gratefully acknowledged. Above all, I am especially grateful to my wife, who typed the manuscript, and whose constant encouragement has been in no small way responsible for the completion of this book.

Dublin, D. LURIÉ
February 1967

Notation

We denote 3-vectors by means of bold-faced type, 4-vectors by light-faced type. Components of 3-vectors are labelled by Roman indices, while components of 4-vectors carry Greek indices.

Our spacetime metric is $\delta_{\mu\nu}$, 4-vectors being assigned imaginary fourth components. Thus the momentum 4-vector is $k = (\mathbf{k}, ik_0)$ with norm squared

$$k^2 = k_\mu k_\mu = \mathbf{k}^2 - k_0^2 = -m^2$$

Summation over repeated indices will always be understood, unless otherwise indicated. The $\delta_{\mu\nu}$ metric is convenient, in that there is no need to distinguish between covariant and contravariant 4-vectors. Also, the Dirac γ's, $\gamma_\mu (\mu = 1 \ldots 4)$, are all hermitian with square equal to one. Complex-conjugation, (denoted by an asterisk *), and hermitian conjugation, (denoted by a dagger †), pose no problem; the conjugation is performed on *all* imaginary units, including the metric i. Thus, for example, the complex conjugate of the gradient 4-vector

$$\partial_\mu = \frac{\partial}{\partial x_\mu} = (\boldsymbol{\nabla}, -i\partial/\partial x_0)$$

is $\partial^*{}_\mu = (\boldsymbol{\nabla}, i\partial/\partial x_0)$. However, in dealing with *complex* 4-vectors, $A_\mu = (\mathbf{A}, iA_0)$ where $\mathbf{A} \neq \mathbf{A}^*$, $A_0 \neq A_0^*$, it is convenient to work with the conjugate vector $A^*{}_\mu = (\mathbf{A}^*, iA_0^*)$ in which the metric i is left unconjugated, as distinct from $A_\mu{}^* = (\mathbf{A}^*, -A_0{}^*)$.

Time derivatives will frequently be denoted by a dot, i.e. $\partial_t A = \dot{A}$.

We shall consistently work with natural units in which $\hbar = c = 1$. Plane waves will almost always be normalized in a large box of volume V with periodic boundary conditions.

Contents

1

Relativistic One-Particle Equations

1-1 Introduction

This first chapter presents a survey of relativistic wave mechanics for non-interacting single particles of spin 0, $\frac{1}{2}$, 1 and $\frac{3}{2}$. In each case the theory will be characterized by a relativistic wave equation for a classical field. We shall not attempt to cover the subject completely*; our purpose will be to:

a. Trace the historical motivation for the development of relativistic quantum field theory by exhibiting the limitations of the one-particle theories, and
b. Assemble those formulae which we shall need for quantum field theory.

We shall begin by reviewing the spin 0 and spin $\frac{1}{2}$ theories. From there we shall pass to the general equations for arbitrary spin due to Bargmann and Wigner, and then to the particular cases of spin 1 and spin $\frac{3}{2}$. For reasons of space we shall forego detailed treatment of spins greater than $\frac{3}{2}$.

1-2 Klein–Gordon field

Formulation The simplest relativistic wave equation is the Klein–Gordon equation (Klein, 1926; Gordon, 1926),

$$(\Box - \mu^2)\phi(\mathbf{x}, t) = 0$$
$$\Box = \nabla^2 - \partial_t^2 \qquad 1(1)$$

obtained by applying the well-known quantization prescription $\mathbf{k} \to -i\nabla$ and $E \to i\partial_t$ to the relativistic relation between energy and momentum for a free particle of rest mass μ.

* For more detailed treatment the reader is referred to E. Corinaldesi and F. Strocchi, *Relativistic Wave Mechanics*, Interscience, New York, 1963; J. D. Bjorken and S. Drell, *Relativistic Quantum Mechanics*, McGraw-Hill, New York, 1964.

It was originally hoped that the Klein–Gordon equation might provide the basis for a general relativistic quantum theory, just as the Schrödinger equations serves as the basis for non-relativistic quantum mechanics. We now know that the Klein–Gordon field can only describe particles of spin 0.

Negative Energies A major difficulty of the theory is encountered from the outset. The plane wave solutions of 1(1) are of the form

$$e^{(i\mathbf{k}.\mathbf{x}-iEt)}$$

with $\mathbf{k}^2-E^2 = -\mu^2$ or $E = \pm(\mathbf{k}^2+\mu^2)^{1/2}$. We see that solutions exist for both positive *and negative* values of the energy E. To form a complete set of solutions we may take the set of all exponentials

$$\frac{1}{\sqrt{V}}\frac{1}{\sqrt{2\omega_{\mathbf{k}}}}e^{i\mathbf{k}.\mathbf{x}-i\omega_{\mathbf{k}}t} \qquad 1(2a)$$

and

$$\frac{1}{\sqrt{V}}\frac{1}{\sqrt{2\omega_{\mathbf{k}}}}e^{-i\mathbf{k}.\mathbf{x}+i\omega_{\mathbf{k}}t} \qquad 1(2b)$$

with $\omega_{\mathbf{k}} = (\mathbf{k}^2+\mu^2)^{1/2}$. The normalization factor $(2\omega_{\mathbf{k}}V)^{-1/2}$ is chosen for future convenience, and periodic boundary conditions on the surface of the volume V are assumed.

Strictly speaking this is not really a difficulty, so long as one restricts attention to the *free*-particle theory; one can simply legislate that only positive energy states are to be regarded as physically realizable states. The energy of a free particle being a constant, there are no transitions from positive to negative energy states and a particle in a positive energy state will remain in that state. Nevertheless, the free-particle theory in itself is of little physical relevance, and as soon as interactions are taken into account the negative energy solutions give rise to serious difficulties of interpretation. Consider for example the interaction of a charged Klein–Gordon particle with an electromagnetic field described by the four-potential $A_\mu = (\mathbf{A}, iV)$. We introduce the interaction by the substitution $k_\mu \to k_\mu - eA_\mu$ or

$$\partial_\mu \to \partial_\mu - ieA_\mu \qquad 1(3)$$

in the Klein–Gordon equation 1(1), e ($e < 0$) being the electron charge*. This yields the equation

$$(\partial_\mu - ieA_\mu)(\partial_\mu - ieA_\mu)\phi - \mu^2\phi = 0 \qquad 1(4)$$

* Throughout the book we shall use Heaviside–Lorentz units in which $\alpha = e^2/4\pi = \frac{1}{137}$.

The substitution 1(3) is suggested by both classical and non-relativistic quantum mechanics and produces what is known as *minimal electromagnetic coupling*. Now, if the electromagnetic potentials $A_\mu(\mathbf{x}, t)$ vary sufficiently rapidly in time—or, more precisely, if they have sizeable Fourier components $A_\mu(\mathbf{k}, k_0)$ for $k_0 > 2\mu$—then transitions between positive and negative energy states may occur. This is a direct consequence of standard quantum mechanical perturbation theory.

This is a fundamental drawback of the theory. It is common to all relativistic one-particle theories and can only be resolved within the framework of quantum field theory.

Charge–Current Vector It is well known that for non-relativistic Schrödinger theory one can construct a probability density—namely $\psi^\dagger(x)\psi(x)$—which is positive-definite and satisfies a continuity equation; the probabilistic interpretation of the theory is based on this fact. If one attempts to follow the same path in the Klein–Gordon theory, however, one meets with a basic difficulty. Proceeding in analogy with the Schrödinger case, let us multiply 1(1) by ϕ^*:

$$\phi^*(\Box - \mu^2)\phi = 0$$

and multiply the conjugate equation by ϕ:

$$\phi(\Box - \mu^2)\phi^* = 0$$

Subtracting, we get

$$\phi^*(\Box - \mu^2)\phi - \phi(\Box - \mu^2)\phi^* = 0$$

which can be written as the conservation law

$$\partial_\mu j_\mu(x) = 0 \tag{1(5a)}$$

with

$$j_\mu(x) = i[(\partial_\mu \phi^*(x))\phi(x) - \phi^*(x)\partial_\mu \phi(x)] \tag{1(5b)}$$

where we have multiplied by i to insure the reality of the current $\mathbf{j}$ and the density

$$\rho(x) = -ij_4(x) = -i(\dot{\phi}^*(x)\phi(x) - \phi^*(x)\dot{\phi}(x)) \tag{1(6)}$$

As in Schrödinger theory one is tempted to identify 1(6) as the probability density for the relativistic particle, but this interpretation breaks down upon noting that 1(6) is not positive-definite. If, for example, the particle is in an eigenstate of energy E, we have $i\dot{\phi} = E\phi$ and

$$\rho(x) = 2E\phi^*(x)\phi(x)$$

which can be either positive or negative, depending on the sign of E. Again, for free particles, this difficulty may be ignored along with the more general problem of negative energy states, but the real difficulty lies in the interacting case. For the minimal electromagnetic interaction it is a simple matter to check that the conserved current is given by

$$j_\mu = i[(\partial_\mu + ieA_\mu)\phi^* . \phi - \phi^*(\partial_\mu - ieA_\mu)\phi]$$
$$= i[\partial_\mu \phi^* . \phi - \phi^* \partial_\mu \phi + 2ie\phi^* \phi A_\mu] \qquad 1(7)$$

involving the electromagnetic field explicitly. The corresponding density

$$\rho = -i[\dot{\phi}^* \phi - \phi^* \dot{\phi} - 2ie\phi^* \phi A_0] \qquad 1(8)$$

is again non-positive-definite.

The failure to support a positive-definite probability density led to the abandonment of the Klein–Gordon theory after it was first proposed, in favour of the Dirac equation (Dirac, 1928). However, the Klein–Gordon and Dirac theories apply to particles of different spin and, from our present viewpoint, both are equally good examples of relativistic quantum theories. Shortly after its abandonment the Klein–Gordon theory was revived by Pauli and Weisskopf (Pauli, 1934) who reinterpreted it as a *quantized field theory* (cf. Chapter 3). In this interpretation ρ is no longer expected to represent the probability density of a single particle; instead it represents the *charge** density of an assembly of positively charged particles and negatively charged antiparticles.

Hilbert Space of Positive-Energy Solutions As a consequence of the continuity equation 1(50), the total ' charge '

$$\int_V \rho(x)\, d^3x = i \int_V (\phi^* \partial_t \phi - \phi^* \overleftarrow{\partial_t} \phi)\, d^3x \qquad 1(9)$$

is conserved in time. Furthermore, it is positive so long as ϕ is a positive-energy solution to 1(1). Now for free particles or for weakly varying external fields we can safely restrict our attention to the manifold of positive-energy solutions. It is therefore convenient to regard this manifold as a *Hilbert space* where 1(9) represents the *norm* of ϕ and

$$(\phi, \psi) = i \int (\phi^* \partial_t \psi - \phi^* \overleftarrow{\partial_t} \psi)\, d^3x$$
$$= i \int \phi^* \overleftrightarrow{\partial_t} \psi\, d^3x \qquad 1(10)$$

* Or some other quantum number (see Section 3-2).

represents the *scalar product* of ϕ and ψ. Here we have introduced the useful symbol $\overleftrightarrow{\partial}_t = \overrightarrow{\partial}_t - \overleftarrow{\partial}_t$. It is a simple matter to check that according to this convention the positive energy solutions 1(2a)

$$f_{\mathbf{k}}(\mathbf{x}, t) = \frac{1}{\sqrt{V}} \frac{1}{\sqrt{2\omega_{\mathbf{k}}}} e^{i\mathbf{k}.\mathbf{x} - i\omega_{\mathbf{k}} t}$$

form an orthonormal set

$$\int f^*_{\mathbf{k}}(\mathbf{x}, t) i\overleftrightarrow{\partial}_t f_{\mathbf{k}'}(\mathbf{x}, t)\, d^3x = \delta_{\mathbf{k}\mathbf{k}'} \qquad 1(11)$$

The definition 1(10) can be shown to satisfy all the usual requirements for a scalar product, namely

$$(\phi, \psi) = (\psi, \phi)^* \qquad 1(12a)$$

$$(\phi_1 + \phi_2, \psi) = (\phi_1, \psi) + (\phi_2, \psi) \qquad 1(12b)$$

$$(\phi, \phi) \geqslant 0 \qquad 1(12c)$$

where the equality sign in 1(12c) holds if, and only if, ϕ vanishes identically. Furthermore, the scalar product 1(10) is conserved in time, since

$$\partial_t(\phi, \psi) = i \int d^3x[\phi^* \partial_t^2 \psi - (\partial_t^2 \phi^*)\psi]$$

$$= i \int d^3x[\phi^*(\nabla^2 - \mu^2)\psi - (\nabla^2 - \mu^2)\phi^* . \psi]$$

$$= 0$$

where we have used 1(1) and integrated by parts.

Lorentz Invariance The Klein–Gordon equation 1(1) will remain invariant under a Lorentz transformation $x' = ax$, i.e.

$$x'_\mu = a_{\mu\nu} x_\nu \quad \begin{cases} a_{\mu\lambda} a_{\nu\lambda} = \delta_{\mu\nu} \\ a_{\lambda\mu} a_{\lambda\nu} = \delta_{\mu\nu} \end{cases} \qquad 1(13)$$

if the field ϕ transforms as a scalar, i.e.

$$\phi'(x') = \phi(x) \qquad 1(14)$$

$\phi'(x')$ represents the transformed field in the new frame of reference. The simple law 1(14) is characteristic of a *spinless* field; spin-carrying fields obey more complicated transformation laws, as we shall see below.

Borrowing the terminology of linear vector space theory, we shall refer to 1(13) and 1(14) as the *passive* transformation. The statement of invariance can also be rephrased in terms of an *active* transformation; the emphasis is then entirely on the functional change in the field $\phi(x)$. Thus the Klein–Gordon equation is invariant under the active transformation

$$\phi(x) \rightarrow \phi'(x)$$

with

$$\phi'(x) = \phi(a^{-1}x) \tag{1(15)}$$

The *infinitesimal* form of 1(15) is often useful. Setting

$$a_{\mu\nu} = \delta_{\mu\nu} + \omega_{\mu\nu} \tag{1(16a)}$$

with $\omega_{\mu\nu}$ infinitesimal and antisymmetric, we write 1(13) in the form

$$\delta x_\mu = x'_\mu - x_\mu = \omega_{\mu\nu} x_\nu \tag{1(16b)}$$

Then $a^{-1}x = x - \delta x$ and 1(15) takes the form

$$\begin{aligned} \delta\phi(x) &= \phi(x - \delta x) - \phi(x) \\ &= -\delta x_\mu \partial_\mu \phi(x) \end{aligned} \tag{1(17)}$$

Hamiltonian Form The Klein–Gordon equation can be cast into Hamiltonian form by observing that ϕ and $\dot{\phi} = \partial_t \phi$ represent two independent degrees of freedom, as a result of the appearance in 1(1) of the second time derivative of ϕ. Rewriting 1(1) as a set of two coupled *first order* differential equations for ϕ and $\theta = \partial_t \phi$ (Feshbach, 1958), we obtain the Hamiltonian form

$$\begin{aligned} i\partial_t \phi &= i\theta \\ i\partial_t \theta &= i\nabla^2 \phi - i\mu^2 \phi \end{aligned}$$

It is more convenient to work with the linear combinations

$$\Psi_1 = \frac{1}{\sqrt{2}}\left(\phi + \frac{i}{\mu}\partial_t \phi\right) = \frac{1}{\sqrt{2}}\left(\phi + \frac{i}{\mu}\theta\right) \tag{1(18a)}$$

$$\Psi_2 = \frac{1}{\sqrt{2}}\left(\phi - \frac{i}{\mu}\partial_t \phi\right) = \frac{1}{\sqrt{2}}\left(\phi - \frac{i}{\mu}\theta\right) \tag{1(18b)}$$

which satisfy the equations

$$i\partial_t\Psi_1 = -\frac{\nabla^2}{2\mu}(\Psi_1+\Psi_2)+\mu\Psi_1 \tag{1(19a)}$$

$$i\partial_t\Psi_2 = \frac{\nabla^2}{2\mu}(\Psi_1+\Psi_2)-\mu\Psi_2 \tag{1(19b)}$$

The linear combinations 1(18a) and 1(18b) have simple non-relativistic limits. For a positive-energy particle at rest we have $i\partial_t\phi = \mu\phi$ and therefore

$$\Psi_1 = \sqrt{2}\phi \qquad \Psi_2 = 0 \qquad (E = \mu)$$

For negative-energy particles, on the contrary

$$\Psi_1 = 0 \qquad \Psi_2 = \sqrt{2}\phi \qquad (E = -\mu)$$

Equations 1(19a) and 1(19b) can be rewritten in more compact form by defining the two-component column vector

$$\Psi = \begin{pmatrix}\Psi_1\\ \Psi_2\end{pmatrix}$$

and the associated Pauli matrices

$$\tau_1 = \begin{pmatrix}0 & 1\\ 1 & 0\end{pmatrix} \qquad \tau_2 = \begin{pmatrix}0 & -i\\ i & 0\end{pmatrix} \qquad \tau_3 = \begin{pmatrix}1 & 0\\ 0 & -1\end{pmatrix}$$

Then 1(19a) and 1(19b) take the form

$$i\partial_t\Psi = \mathbf{H}\Psi \tag{1(20a)}$$

with the Hamiltonian operator **H** given by*

$$\mathbf{H} = (\tau_3+i\tau_2)\frac{\mathbf{k}^2}{2\mu}+\mu\tau_3 \tag{1(20b)}$$

where $\mathbf{k} = -i\boldsymbol{\nabla}$.

The total ‘charge’, 1(9), is easily expressed in this notation. We find

$$\begin{aligned}\int_V \rho\, d^3x &= \mu\int_V (\Psi_1^*\Psi_1-\Psi_2^*\Psi_2)\, d^3x\\ &= \mu\int_V \Psi^*\tau_3\Psi\, d^3x\end{aligned} \tag{1(21)}$$

* We use bold-faced notation for the Hamiltonian of the one-particle theory, to distinguish it from the field theoretic Hamiltonian introduced in Chapter 2.

Another important quantity is the integral

$$\int_V \Psi^* \tau_3 \mathbf{H} \Psi \, d^3x = \int_V \Psi^* \left[(1+\tau_1) \frac{\mathbf{k}^2}{2\mu} + \mu \right] \Psi \, d^3x$$

$$= \int_V \left\{ (\Psi_1^* + \Psi_2^*) \frac{\mathbf{k}^2}{2\mu} (\Psi_1 + \Psi_2) + \mu(\Psi_1^* \Psi_1 + \Psi_2^* \Psi_2) \right\} d^3x \qquad 1(22)$$

which is *positive-definite*; in contrast note that 1(21) can take on both positive and negative values, depending on the sign of the eigenvalue of **H**. Evaluating 1(22) by straightforward substitution of 1(18a) and 1(18b), we find

$$\int_V \Psi^* \tau_3 \mathbf{H} \Psi \, d^3x = \frac{1}{\mu} \int_V (-\phi^* \mathbf{\nabla}^2 \phi + \dot{\phi}^* \dot{\phi} + \mu^2 \phi^* \phi) \, d^3x \qquad 1(23)$$

We shall reencounter the right-hand of 1(23) in Chapter 2, where it will be identified as the *field theoretic* Hamiltonian of the Klein–Gordon field (to within a factor μ^{-1}).

1-3 Dirac field

Formulation The relativistic wave equation describing particles of spin $\frac{1}{2}$ is known as the Dirac equation (Dirac, 1928). It has the form

$$(\gamma_\mu \partial_\mu + m)\psi = 0 \qquad 1(24)$$

where m is the rest mass. The wave function $\psi(x)$ is a four-component object known as a *spinor*, and the γ_μ are a set of four 4×4 hermitian matrices satisfying

$$\gamma_\mu \gamma_\nu + \gamma_\nu \gamma_\mu = 2\delta_{\mu\nu} \qquad 1(25)$$

$$(\mu, \nu = 1 \ldots 4)$$

In more explicit form, the Dirac equation may be written as

$$\sum_{\beta=1}^{4} \left(\gamma_\mu^{\alpha\beta} \frac{\partial}{\partial x_\mu} + m\delta^{\alpha\beta} \right) \psi_\beta(x) = 0 \qquad (\alpha = 1 \ldots 4)$$

with all matrix indices displayed.

Historically, Dirac discovered the above equation in attempting to construct a relativistic wave equation which, unlike the Klein–Gordon equation, would support a positive probability density. This requirement led him to consider only those equations which are of first order in the time derivative. For the equation to be relativistic, it should then be of first order in the space derivatives as well. From this point, the line of argument which leads to 1(24) is given in many texts* and will not be repeated here. In any case, the historical development tends to obscure the fact that the Dirac and Klein–Gordon are equally good examples of relativistic field theories, as mentioned earlier.

As a result of the commutation relations 1(25), all four components of ψ can be shown to satisfy the Klein–Gordon equation. Multiplying 1(24) by $\gamma_\lambda \partial_\lambda - m$ we have

$$(\gamma_\lambda \partial_\lambda - m)(\gamma_\mu \partial_\mu + m)\psi = (\gamma_\lambda \gamma_\mu \partial_\lambda \partial_\mu - m^2)\psi = 0$$

and since $\partial_\lambda \partial_\mu$ is symmetric, we get, by 1(25)

$$\tfrac{1}{2}(\gamma_\lambda \gamma_\mu + \gamma_\mu \gamma_\lambda)\partial_\lambda \partial_\mu \psi - m^2 \psi = (\partial^2 - m^2)\psi = 0$$

This guarantees that the relativistic relation

$$\mathbf{k}^2 - E^2 = -m^2 \tag{1(26)}$$

between energy and momentum will be satisfied.

Note that all four γ-matrices satisfy

$$\gamma_\mu^2 = 1 \qquad (\mu = 1 \ldots 4) \tag{1(27)}$$

as is seen by setting $\mu = \nu$ in 1(25). A convenient representation of the γ-matrices is

$$\gamma_i = \begin{pmatrix} 0 & -i\sigma_i \\ i\sigma_i & 0 \end{pmatrix} \qquad \gamma_4 = \begin{pmatrix} I & 0 \\ 0 & -I \end{pmatrix} \tag{1(28)}$$

where I denotes the 2×2 unit matrix and the σ_i ($i = 1, 2, 3$) are the set of three Pauli spin matrices

$$\sigma_1 = \begin{pmatrix} 0 & 1 \\ 1 & 0 \end{pmatrix} \qquad \sigma_2 = \begin{pmatrix} 0 & -i \\ i & 0 \end{pmatrix} \qquad \sigma_3 = \begin{pmatrix} 1 & 0 \\ 0 & -1 \end{pmatrix} \tag{1(29)}$$

The representation 1(28) is clearly hermitian and it is straightforward to check that the anticommutation relations 1(25) are satisfied.

Charge–Current Vector Let us check that 1(24) does indeed give rise to a positive-definite probability density, as Dirac originally required. Taking the hermitian adjoint of 1(24), i.e.

$$\psi^\dagger(\gamma_\mu \overleftarrow{\partial}_\mu^* + m) = 0 \tag{1(30)}$$

* See for example, L. I. Schiff, *Quantum Mechanics*, 2nd ed., McGraw-Hill, New York, 1955.

with $\partial_\mu^* = (\boldsymbol{\partial}, -\partial_{it})$, we define the *adjoint spinor*

$$\bar{\psi} = \psi^\dagger \gamma_4 \tag{1(31)}$$

satisfying

$$\bar{\psi}(x)(\gamma_\mu \overleftarrow{\partial}_\mu - m) = 0 \tag{1(32)}$$

We now multiply 1(24) on the left by $\bar{\psi}$ and 1(32) on the right by ψ. Adding, we get the conservation law

$$\partial_\mu j_\mu(x) = 0 \tag{1(33a)}$$

with

$$j_\mu(x) = i\bar{\psi}(x)\gamma_\mu \psi(x) \tag{1(33b)}$$

the factor i being inserted to make $\mathbf{j}$ and $\rho = -ij_4$ real. The density

$$\rho(x) = \bar{\psi}(x)\gamma_4\psi(x) = \psi^\dagger(x)\psi(x) \tag{1(34)}$$

is thus positive-definite, and we conclude that the Dirac theory supports a probability interpretation.

Actually, the positive-definiteness of ρ does not survive the transition to quantum field theory where ρ appears as the *charge* density of an assembly of positively and negatively charged particles, just as in the Klein–Gordon theory. We shall anticipate this by referring to 1(33b) as the *charge–current* vector and to ρ as the charge density.

Hamiltonian To get the Hamiltonian form of the Dirac equation we multiply 1(24) by γ_4 and isolate the time-derivative term. Since $\gamma_4^2 = 1$, we find

$$i\,\partial_t \psi = (\gamma_4 \boldsymbol{\gamma} \,.\, \boldsymbol{\nabla} + \gamma_4 m)\psi$$

Defining

$$\boldsymbol{\alpha} = i\gamma_4 \boldsymbol{\gamma} \tag{1(35a)}$$

$$\beta = \gamma_4 \tag{1(35b)}$$

we can write the Hamiltonian as

$$\mathbf{H} = \boldsymbol{\alpha} \,.\, \mathbf{k} + \beta m \tag{1(36)}$$

In the representation 1(28), $\boldsymbol{\alpha}$ and β have the explicit matrix form

$$\boldsymbol{\alpha} = \begin{pmatrix} 0 & \boldsymbol{\sigma} \\ \boldsymbol{\sigma} & 0 \end{pmatrix} \qquad \beta = \begin{pmatrix} I & 0 \\ 0 & -I \end{pmatrix} \tag{1(37)}$$

Spin It is a simple matter to check that the commutator of the orbital angular momentum $\mathbf{L} = -i\mathbf{x} \times \nabla$ with the Hamiltonian 1(36) is non-vanishing. However the sum

$$\mathbf{J} = \mathbf{L} + \tfrac{1}{2}\boldsymbol{\Sigma}$$

with $\Sigma_i = -i\gamma_j\gamma_k$ (i, j, k cycl.) or explicitly

$$\boldsymbol{\Sigma} = \begin{pmatrix} \boldsymbol{\sigma} & 0 \\ 0 & \boldsymbol{\sigma} \end{pmatrix} \qquad 1(38)$$

does commute with $\mathbf{H}$ and so represents a constant of the motion. $\mathbf{J}$ is interpreted as the total angular momentum and $\frac{1}{2}\boldsymbol{\Sigma}$ as the spin—the residual angular momentum in the rest frame of the particle. As the eigenvalues of $\frac{1}{2}\boldsymbol{\Sigma}$ are $\pm\frac{1}{2}$ we are assured that the Dirac equation describes particles of spin $\frac{1}{2}$.

The projection of $\boldsymbol{\Sigma}$ along the direction of motion is known as the *helicity*:

$$h = \frac{\boldsymbol{\Sigma} . \mathbf{k}}{|\mathbf{k}|} \qquad (\mathbf{k} = -i\nabla) \qquad 1(39)$$

One easily verifies that

$$[h, \mathbf{H}] = 0 \qquad 1(40a)$$

$$h^2 = 1 \qquad 1(40b)$$

so that the helicity is a constant of the motion with eigenvalues ± 1.

We now define an important operator which reduces to the spin for a particle at rest and which transforms as a four-vector under Lorentz transformations. This is the Pauli–Lubanski covariant spin vector (Lubanski, 1942)

$$\omega_\alpha = -\tfrac{1}{2}\varepsilon_{\alpha\mu\nu\lambda}\Sigma_{\mu\nu}\partial_\lambda \qquad 1(41)$$

where $\varepsilon_{\alpha\mu\nu\lambda}$ is the completely antisymmetric Levi–Civita tensor and $\Sigma_{\mu\nu}$ is the antisymmetric *spin tensor*.

$$\Sigma_{\mu\nu} = \frac{1}{2i}(\gamma_\mu\gamma_\nu - \gamma_\nu\gamma_\mu) \qquad 1(42)$$

whose three space components are $\Sigma_{23} = \Sigma_1$, $\Sigma_{31} = \Sigma_2$, and $\Sigma_{12} = \Sigma_3$. We shall see below that $\Sigma_{\mu\nu}$ has the character of a second rank tensor under Lorentz transformations; this provides the justification for calling ω_α a vector. For a positive-energy particle at rest we have $\partial_\lambda = -\delta_{\lambda 4}m$, and the only non-vanishing components of ω_α are the three space components

$$\omega_i = \frac{m}{2}\varepsilon_{ijk4}\Sigma_{jk}$$

or

$$\boldsymbol{\omega} = m\boldsymbol{\Sigma}$$

Thus for positive and negative energy particles at rest we have, respectively,

$$\omega_\alpha = (m\boldsymbol{\Sigma}, 0) \qquad 1(43a)$$

and

$$\omega_\alpha = (-m\boldsymbol{\Sigma}, 0) \qquad 1(43b)$$

We shall exhibit the existence of negative energy solutions to the Dirac equation presently.

An alternative form for ω_α may be written down with the aid of the important matrix

$$\gamma_5 = \gamma_1\gamma_2\gamma_3\gamma_4 = -\begin{pmatrix} 0 & I \\ I & 0 \end{pmatrix} \qquad 1(44)$$

with the properties

$$\gamma_5^2 = 1 \qquad 1(45a)$$

$$\gamma_5\gamma_\mu + \gamma_\mu\gamma_5 = 0 \qquad 1(45b)$$

With the aid of 1(25) it is a simple matter to check that

$$\gamma_5\Sigma_{\alpha\lambda} = -\tfrac{1}{2}\varepsilon_{\alpha\mu\nu\lambda}\Sigma_{\mu\nu}$$

and, therefore

$$\omega_\alpha = \gamma_5\Sigma_{\alpha\lambda}\partial_\lambda \qquad 1(46)$$

Plane Wave Solutions Let us construct solutions to 1(24) of the form

$$w(\mathbf{k}, E)\, e^{i\mathbf{k}.\mathbf{x} - iEt}$$

The four-component spinors $w(\mathbf{k}, E)$ must satisfy the equation

$$(i\boldsymbol{\gamma}\,.\,\mathbf{k} - \gamma_4 E + m)w(\mathbf{k}, E) = 0 \qquad 1(47)$$

Using 1(28) and setting

$$w(\mathbf{k}, E) = \begin{pmatrix} \xi(\mathbf{k}, E) \\ \eta(\mathbf{k}, E) \end{pmatrix}$$

where ξ and η are two-component objects, we may write 1(47) as a set of two coupled equations for ξ and η:

$$\boldsymbol{\sigma}\,.\,\mathbf{k}\eta = (E-m)\xi \qquad 1(48a)$$

$$\boldsymbol{\sigma}\,.\,\mathbf{k}\xi = (E+m)\eta \qquad 1(48b)$$

Let us examine these equations in the rest frame, $\mathbf{k} = 0$. We have then either

$$E = m \begin{cases} \xi \neq 0 \\ \eta = 0 \end{cases} \qquad 1(49a)$$

or

$$E = -m \begin{cases} \xi = 0 \\ \eta \neq 0 \end{cases} \qquad 1(49b)$$

where the non-zero components, namely ξ for $E = m$ and η for $E = -m$, may be chosen arbitrarily. It is convenient to choose them to be eigenstates of σ_3, namely $\begin{pmatrix}1\\0\end{pmatrix}$ or $\begin{pmatrix}0\\1\end{pmatrix}$; this yields the following set of four independent solutions for $\mathbf{k} = 0$

$$\begin{array}{cccc} \multicolumn{2}{c}{E = m} & \multicolumn{2}{c}{E = -m} \\ \Sigma_3 = +1 & \Sigma_3 = -1 & \Sigma_3 = +1 & \Sigma_3 = -1 \\ \begin{pmatrix}1\\0\\0\\0\end{pmatrix} & \begin{pmatrix}0\\1\\0\\0\end{pmatrix} & \begin{pmatrix}0\\0\\1\\0\end{pmatrix} & \begin{pmatrix}0\\0\\0\\1\end{pmatrix} \end{array} \qquad 1(50)$$

Physically these states correspond to particles at rest with spin up or down along the z-axis. The fact that there exist solutions for both $E = m$ and $E = -m$ confirms that the Dirac theory is not free from the negative energy difficulty which plagued the Klein–Gordon theory. This difficulty will be resolved by field quantization which associates the negative energy solutions to *anti*-particles.

For arbitrary $\mathbf{k}$ we have $E = \pm(\mathbf{k}^2+m^2)^{1/2}$ by virtue of 1(26). The positive energy solution to 1(48a) and 1(48b) is easily seen to be

$$w(\mathbf{k}, E_{\mathbf{k}}) = \begin{pmatrix} \xi \\ \dfrac{\boldsymbol{\sigma}\,.\,\mathbf{k}}{E_{\mathbf{k}}+m}\xi \end{pmatrix}$$

with $E_{\mathbf{k}} = (\mathbf{k}^2+m^2)^{1/2}$, while the negative energy solution is

$$w(\mathbf{k}, -E_{\mathbf{k}}) = \begin{pmatrix} -\dfrac{\boldsymbol{\sigma}\,.\,\mathbf{k}}{E_{\mathbf{k}}+m}\eta \\ \eta \end{pmatrix}$$

We are again at liberty to fix the two-component spinors ξ and η arbitrarily; the most common convention is to take them to be eigenstates of σ_3, as in 1(50)*. Moreover, it is convenient to deal with $w(-\mathbf{k}, -E_{\mathbf{k}})$ rather than $w(\mathbf{k}, -E_{\mathbf{k}})$. We therefore adopt as our basic set of four linearly independent solutions for momentum $\mathbf{k}$ the spinors

$$u_{\mathbf{k}\sigma} = w_\sigma(\mathbf{k}, E_{\mathbf{k}}) = \left(\frac{E_{\mathbf{k}}+m}{2m}\right)^{1/2} \begin{pmatrix} \xi_\sigma \\ \dfrac{\boldsymbol{\sigma}\,.\,\mathbf{k}}{E_{\mathbf{k}}+m}\xi_\sigma \end{pmatrix} \qquad (\sigma = 1, 2) \qquad 1(51a)$$

and

$$v_{\mathbf{k}\sigma} = w_\sigma(-\mathbf{k}, -E_{\mathbf{k}}) = \left(\frac{E_{\mathbf{k}}+m}{2m}\right)^{1/2} \begin{pmatrix} \dfrac{\boldsymbol{\sigma}\,.\,\mathbf{k}}{E_{\mathbf{k}}+m}\xi_\sigma \\ \xi_\sigma \end{pmatrix} \qquad (\sigma = 1, 2) \qquad 1(51b)$$

with $\xi_1 = \begin{pmatrix} 1 \\ 0 \end{pmatrix}$ and $\xi_2 = \begin{pmatrix} 0 \\ 1 \end{pmatrix}$. The spinors $u_{\mathbf{k}\sigma}$ and $v_{\mathbf{k}\sigma}$ satisfy

$$(\gamma\,.\,k-im)u_{\mathbf{k}\sigma} = 0 \qquad 1(52a)$$

$$(\gamma\,.\,k+im)v_{\mathbf{k}\sigma} = 0 \qquad 1(52b)$$

as can be seen by referring to 1(47). It will be understood throughout that

$$k_0 = E_{\mathbf{k}} = +(\mathbf{k}^2+m^2)^{1/2} > 0$$

A complete set of Dirac wave functions is obtained by taking the set of all solutions

$$\frac{1}{\sqrt{V}}\sqrt{\frac{m}{E_{\mathbf{k}}}}u_{\mathbf{k}\sigma}\, e^{i\mathbf{k}.\mathbf{x}-iE_{\mathbf{k}}t} \qquad (\sigma = 1, 2) \qquad 1(53a)$$

and

$$\frac{1}{\sqrt{V}}\sqrt{\frac{m}{E_{\mathbf{k}}}}v_{\mathbf{k}\sigma}\, e^{-i\mathbf{k}.\mathbf{x}+iE_{\mathbf{k}}t} \qquad (\sigma = 1, 2) \qquad 1(53b)$$

* This convention ensures that the solutions 1(51a) and 1(51b) go over into 1(50) for $\mathbf{k} \to 0$. An alternative procedure, exploited by Jacob and Wick (Jacob, 1959), is to take the ξ_σ to be eigenstates of $\boldsymbol{\sigma}\,.\,\mathbf{k}/|\mathbf{k}|$. The solutions 1(51a) and 1(51b) are then eigenstates of the helicity operator 1(39).

Note the normalization factors which have been introduced in 1(51a), 1(51b) and 1(53a), 1(53b). The factor $(E_{\mathbf{k}}+m)^{1/2}/(2m)^{1/2}$ in 1(51a) and 1(51b) ensures that

$$\bar{u}_{\mathbf{k}\sigma}u_{\mathbf{k}\sigma'} = \delta_{\sigma\sigma'} \tag{1(54a)}$$

$$\bar{v}_{\mathbf{k}\sigma}v_{\mathbf{k}\sigma'} = -\delta_{\sigma\sigma'} \tag{1(54b)}$$

where the adjoints $\bar{u} = u^{\dagger}\gamma_4$ and $\bar{v} = v^{\dagger}\gamma_4$ satisfy

$$\bar{u}_{\mathbf{k}\sigma}(\gamma\,.\,k-im) = 0 \tag{1(55a)}$$

$$\bar{v}_{\mathbf{k}\sigma}(\gamma\,.\,k+im) = 0 \tag{1(55b)}$$

The verification of 1(54a) and 1(54b) is left as an exercise. The advantage of normalizing $\bar{u}u$ rather than $u^{\dagger}u$ to unity is that $\bar{u}u$ is a Lorentz scalar, as we shall show later; our normalization prescription is therefore Lorentz-invariant*. We also note, for future reference, the easily derived orthogonality relations

$$\bar{u}_{\mathbf{k}\sigma}v_{\mathbf{k}\sigma'} = \bar{v}_{\mathbf{k}\sigma}u_{\mathbf{k}\sigma'} = 0 \tag{1(56a)}$$

and

$$\bar{u}_{\mathbf{k}\sigma}\gamma_4 v_{-\mathbf{k}\sigma'} = \bar{v}_{\mathbf{k}\sigma}\gamma_4 u_{-\mathbf{k}\sigma'} = 0 \tag{1(56b)}$$

The factor $(m/E_{\mathbf{k}}V)^{1/2}$ in 1(53a) and 1(53b), on the other hand, ensures that the wave functions are properly normalized according to

$$1 = \int_V \bar{\psi}(x)\gamma_4\psi(x)\,d^3x = \int_V \psi^{\dagger}(x)\psi(x)\,d^3x \tag{1(57)}$$

$\psi^{\dagger}(x)\psi(x)$ being the probability density $\rho(x)$, as shown earlier. The proof that the wave functions 1(53a) and 1(53b) satisfy 1(57) is straightforward; in effect one must establish that

$$\bar{u}_{\mathbf{k}\sigma}\gamma_4 u_{\mathbf{k}\sigma'} = \frac{E_{\mathbf{k}}}{m}\bar{u}_{\mathbf{k}\sigma}u_{\mathbf{k}\sigma'} \tag{1(58a)}$$

$$\bar{v}_{\mathbf{k}\sigma}\gamma_4 v_{\mathbf{k}\sigma'} = -\frac{E_{\mathbf{k}}}{m}\bar{v}_{\mathbf{k}\sigma}v_{\mathbf{k}\sigma'} \tag{1(58b)}$$

We also note that the functions 1(53a) and 1(53b) form an orthogonal set with respect to the norm 1(57); this can be seen by using 1(54a), 1(54b) and 1(56b).

* In certain instances, however, it is more convenient to normalize according to $u^{\dagger}u = v^{\dagger}v = 1$. This is the case in neutrino theory.

To what spin orientation do the solutions 1(51a) and 1(51b) correspond? To answer this question we must specify what we mean by 'spin orientation'. Except in the special case when $k_x = k_y = 0$, the solutions 1(51a) and 1(51b) do not have sharp spin along the z-axis; only for $\mathbf{k} = 0$ do they reduce to eigenstates of Σ_z. Indeed, we cannot expect to form solutions which are eigenstates of Σ_z for $\mathbf{k} \neq 0$, since $[\mathbf{H}, \Sigma_z] \neq 0$. Nor are the solutions 1(51a) and 1(51b) eigenstates of the helicity operator $\mathbf{\Sigma}\,.\,\mathbf{k}/|\mathbf{k}|$ except, again, in the special case when $\mathbf{k}$ points along the z-axis. To get a satisfactory answer we must refer to the Pauli–Lubanski covariant spin vector 1(46), which when applied to the positive and negative energy solutions 1(53a) and 1(53b) takes the form

$$\omega_\alpha \begin{cases} u_{\mathbf{k}\sigma}\,\mathrm{e}^{ik.x} \\ v_{\mathbf{k}\sigma}\,\mathrm{e}^{-ik.x} \end{cases} = \begin{cases} i\gamma_5\Sigma_{\alpha\lambda}k_\lambda u_{\mathbf{k}\sigma}\,\mathrm{e}^{ik.x} & \text{1(59a)} \\ -i\gamma_5\Sigma_{\alpha\lambda}k_\lambda v_{\mathbf{k}\sigma}\,\mathrm{e}^{-ik.x} & \text{1(59b)} \end{cases}$$

In the rest frame, $u_{\mathbf{k}\sigma}$ and $v_{\mathbf{k}\sigma}$ reduce to eigenstates of Σ_z or equivalently to eigenstates of $\mathbf{\Sigma}\,.\,\mathbf{e}$ where $\mathbf{e}$ is the polarization vector (0, 0, 1). Let us introduce a four-vector e_α which reduces to $(\mathbf{e}, 0)$ in the rest frame $k_\alpha = (0, 0, 0, im)$. For arbitrary k_α, e_α will be given by

$$e_\alpha = \begin{cases} \mathbf{e} + \dfrac{\mathbf{k}(\mathbf{k}\,.\,\mathbf{e})}{m(E_{\mathbf{k}}+m)} & (\alpha = 1, 2, 3) \\[2ex] i\dfrac{\mathbf{k}\,.\,\mathbf{e}}{m} & (\alpha = 4) \end{cases} \qquad \text{1(60)}$$

as is easily verified by performing a Lorentz transformation from the rest frame. Note that e_α satisfies the covariant relations

$$k\,.\,e = 0 \qquad \text{1(61a)}$$

$$e^2 = 1 \qquad \text{1(61b)}$$

Since the Lorentz transformation which takes $(\mathbf{e}, 0)$ into 1(60) takes $(m\mathbf{\Sigma}, 0)$ into the vector operator $\omega_\alpha(k) = i\gamma_5\Sigma_{\alpha\lambda}k_\lambda$, [compare 1(43) and 1(59)], it follows that $\mathbf{\Sigma}\,.\,\mathbf{e}$ goes over into the operator

$$\frac{\omega(k)\,.\,e}{m} = \frac{i\gamma_5\Sigma_{\alpha\lambda}k_\lambda}{m}e_\alpha$$

Let us verify that $u_{\mathbf{k}\sigma}$ and $v_{\mathbf{k}\sigma}$ are eigenstates of this operator with eigenvalues ± 1. As a preliminary we establish the identities

$$\omega(k)\,.\,eu_{\mathbf{k}\sigma} = mi\gamma_5\gamma\,.\,eu_{\mathbf{k}\sigma} \qquad \text{1(62a)}$$

$$\omega(k)\,.\,ev_{\mathbf{k}\sigma} = -mi\gamma_5\gamma\,.\,ev_{\mathbf{k}\sigma} \qquad \text{1(62b)}$$

Indeed, by 1(46), 1(25) and 1(52a)

$$\begin{aligned}\omega_\alpha(k)e_\alpha u_{\mathbf{k}\sigma} &= \tfrac{1}{2}\gamma_5(\gamma_\alpha\gamma_\lambda-\gamma_\lambda\gamma_\alpha)k_\lambda e_\alpha u_{\mathbf{k}\sigma}\\ &= \gamma_5(\gamma_\alpha\gamma_\lambda-\delta_{\lambda\alpha})k_\lambda e_\alpha u_{\mathbf{k}\sigma}\\ &= mi\gamma_5\left(\gamma_\alpha+\frac{i}{m}k_\alpha\right)e_\alpha u_{\mathbf{k}\sigma}\end{aligned}$$

Since $k\,.\,e = 0$ we recover 1(62a). 1(62b) is derived in the same way. The operator $\gamma_5\gamma\,.\,e$ is easier to work with than $\gamma_5\Sigma_{\alpha\lambda}e_\alpha k_\lambda$ and it is now a simple matter to check that

$$\frac{\omega(k)\,.\,e}{m}u_{\mathbf{k}\sigma} = i\gamma_5\gamma\,.\,e u_{\mathbf{k}\sigma} = \begin{cases} u_{\mathbf{k}\sigma} & (\sigma = 1)\\ -u_{\mathbf{k}\sigma} & (\sigma = 2)\end{cases} \tag{1(63a)}$$

and

$$\frac{\omega(k)\,.\,e}{m}v_{\mathbf{k}\sigma} = -i\gamma_5\gamma\,.\,e v_{\mathbf{k}\sigma} = \begin{cases} v_{\mathbf{k}\sigma} & (\sigma = 1)\\ -v_{\mathbf{k}\sigma} & (\sigma = 2)\end{cases} \tag{1(63b)}$$

To establish 1(63a), for example, we use 1(28), 1(44) and 1(51a) and evaluate the matrix product

$$\begin{pmatrix} \boldsymbol{\sigma}\,.\,\mathbf{e}+\dfrac{(\boldsymbol{\sigma}\,.\,\mathbf{k})(\mathbf{k}\,.\,\mathbf{e})}{(E_{\mathbf{k}}+m)m} & \dfrac{-\mathbf{k}\,.\,\mathbf{e}}{m}\\ \dfrac{\mathbf{k}\,.\,\mathbf{e}}{m} & -\boldsymbol{\sigma}\,.\,\mathbf{e}-\dfrac{(\boldsymbol{\sigma}\,.\,\mathbf{k})(\mathbf{k}\,.\,\mathbf{e})}{(E_{\mathbf{k}}+m)m}\end{pmatrix} \times \begin{pmatrix}\xi_\sigma\\ \dfrac{\boldsymbol{\sigma}\,.\,\mathbf{k}}{E_{\mathbf{k}}+m}\xi_\sigma\end{pmatrix}$$

using the identity

$$(\boldsymbol{\sigma}\,.\,\mathbf{k})(\boldsymbol{\sigma}\,.\,\mathbf{e})+(\boldsymbol{\sigma}\,.\,\mathbf{e})(\boldsymbol{\sigma}\,.\,\mathbf{k}) = 2\mathbf{k}\,.\,\mathbf{e}$$

The result is

$$\begin{pmatrix}\boldsymbol{\sigma}\,.\,\mathbf{e} \quad \xi_\sigma\\ \dfrac{\boldsymbol{\sigma}\,.\,\mathbf{k}}{E_{\mathbf{k}}+m}\boldsymbol{\sigma}\,.\,\mathbf{e} \quad \xi_\sigma\end{pmatrix}$$

in agreement with 1(63a). We conclude that the solutions 1(51a) and 1(51b) are eigenstates of the operator formed by the projection of $\omega(k)$ on the covariant spin direction e.

Charge Conjugation An important operation in relativistic quantum mechanics is that which transforms positive energy solutions $u_{\mathbf{k}\sigma}$ into negative energy solutions $v_{\mathbf{k}\sigma}$. One can prove on general grounds* that there exists a *charge conjugation matrix* C with the properties

$$C^{-1}\gamma_\mu C = -\gamma_\mu^T \tag{1(64a)}$$

$$C^\dagger = C^{-1} \tag{1(64b)}$$

$$C^T = -C \tag{1(64c)}$$

In the representation 1(28) for example, one can easily check that the above properties are satisfied by

$$C = \gamma_2\gamma_4 = \begin{pmatrix} 0 & i\sigma_2 \\ i\sigma_2 & 0 \end{pmatrix} \tag{1(65)}$$

We also note, for future reference, the property

$$C^{-1}\gamma_5 C = \gamma_5^T \tag{1(66)}$$

which follows from 1(64a) and 1(44).

Let us now take the transpose of 1(55a)

$$(\gamma^T . k - im)\bar{u}_{\mathbf{k}\sigma}^T = 0$$

and multiply on the left by C. Using 1(64a) we get

$$(\gamma . k + im)C\bar{u}_{\mathbf{k}\sigma}^T = 0$$

which shows that $C\bar{u}_{\mathbf{k}\sigma}^T = C\gamma_4^T u_{\mathbf{k}\sigma}^*$ satisfies 1(52b) and is therefore a negative energy spinor of momentum $-\mathbf{k}$, i.e. some linear combination of $v_{\mathbf{k}1}$ and $v_{\mathbf{k}2}$. To determine the spin direction we apply the operator $\omega . e$. By 1(62a) and 1(62b) we have

$$\begin{aligned} C\gamma_4^T\left(\frac{\omega . e}{m} u_{\mathbf{k}\sigma}\right)^* &= -C\gamma_4^T i\gamma_5^T\gamma^T . e^* u_{\mathbf{k}\sigma}^* \\ &= i\gamma_5\gamma . eC\gamma_4^T u_{\mathbf{k}\sigma}^* \\ &= -\frac{\omega . e}{m} C\gamma_4^T u_{\mathbf{k}\sigma}^* \end{aligned}$$

so that $C\bar{u}_{\mathbf{k}\sigma}^T$ has the opposite spin direction to $u_{\mathbf{k}\sigma}$:

$$C\bar{u}_{\mathbf{k}\sigma}^T = v_{\mathbf{k}\bar{\sigma}} \quad \text{with} \quad \bar{\sigma} = \begin{cases} 2 \text{ for } \sigma = 1 \\ 1 \text{ for } \sigma = 2 \end{cases} \tag{1(67)}$$

Thus charge conjugation transforms a positive energy spinor into a negative energy spinor with the opposite momentum and spin. Now we

* See for example H. Umezawa, *Quantum Field Theory*, North Holland, Amsterdam, 1956, Chapter 3.

shall see in Section 3.6* that negative energy spinors with momentum $-\mathbf{k}$ and spin down represent antiparticles with momentum $+\mathbf{k}$ and spin up. Thus charge conjugation may be viewed as the operation which interchanges particles into antiparticles with the *same* momentum and spin. In this connection it is instructive to apply charge conjugations to the Dirac equation in the presence of an external electromagnetic field

$$\gamma_\mu(\partial_\mu - ieA_\mu)\psi + m\psi = 0 \tag{1(68)}$$

Here the minimal substitution 1(3) has been applied to the free Dirac equation 1(24). The equation for the charge conjugate wave function

$$\psi^C = C\bar{\psi}^T = C\gamma_4^T\psi^*$$

is easily derived from 1(68); it is

$$\gamma_\mu(\partial_\mu + ieA_\mu)\psi^C + m\psi^C = 0$$

Thus ψ and ψ^C refer to particles of opposite charge.

Lorentz Invariance In contrast to the scalar transformation law 1(14) the Dirac field undergoes a linear transformation in passing from one Lorentz frame to another. To see this one need only compare the wave functions for momentum $\mathbf{k}$ 1(51a) and 1(51b) with the corresponding wave functions 1(50) in the rest frame. To test Lorentz invariance for the Dirac theory we must check that bilinear expressions in $\bar{\psi}$ and ψ, like the probability current $j_\mu = i\bar{\psi}\gamma_\mu\psi$, transform correctly under the change in ψ. For example, j_μ should transform as a 4-vector to ensure the invariance of the conservation law 1(33a).

The linear transformation law for ψ

$$\psi'(x') = L(a)\psi(x) \tag{1(68)}$$

under the Lorentz transformation 1(13) is determined by the requirement that $\psi'(x')$ satisfy the same equation in the transformed frame as does $\psi(x)$ in the original frame, namely

$$\left(\gamma_\mu \frac{\partial}{\partial x'_\mu} + m\right)\psi'(x') = 0 \tag{1(69)}$$

We have

$$\frac{\partial}{\partial x'_\mu} = \frac{\partial x_\nu}{\partial x'_\mu}\frac{\partial}{\partial x_\nu} = (a^{-1})_{\nu\mu}\frac{\partial}{\partial x_\nu} = a_{\mu\nu}\frac{\partial}{\partial x_\nu}$$

* See in particular the paragraphs following 3(168b).

so that 1(69) may be written as

$$\left(a_{\mu\nu}\gamma_\nu\frac{\partial}{\partial x_\nu}+m\right)L(a)\psi(x) = 0$$

Multiplying by $L^{-1}(a)$ on the left, we recover the Dirac equation 1(24) in the original frame, provided that $L^{-1}a_{\mu\nu}\gamma_\mu L = \gamma_\nu$ or

$$L^{-1}\gamma_\mu L = a_{\mu\nu}\gamma_\nu \tag{1(70)}$$

Let us focus our attention on infinitesimal transformations, setting $a_{\mu\nu} = \delta_{\mu\nu}+\omega_{\mu\nu}$, as in 1(16a); correspondingly we set $L = 1+iS$ or

$$\psi'(x') = \psi(x)+iS\psi(x) \tag{1(71)}$$

where S is an infinitesimal operator. We can expand S in terms of the six coefficients $\omega_{\mu\nu}$ according to

$$S = \tfrac{1}{4}\omega_{\mu\nu}\Sigma_{\mu\nu} \tag{1(72)}$$

where the $\Sigma_{\mu\nu}(\Sigma_{\mu\nu} = -\Sigma_{\nu\mu})$ are a set of six matrices. In terms of the $\Sigma_{\mu\nu}$, the condition 1(70) takes the form

$$[\gamma_\mu, i\Sigma_{\lambda\nu}] = \delta_{\lambda\mu}\gamma_\nu-\delta_{\nu\mu}\gamma_\lambda \tag{1(73)}$$

with the solution

$$\Sigma_{\mu\nu} = \frac{1}{2i}(\gamma_\mu\gamma_\nu-\gamma_\nu\gamma_\mu) \tag{1(74)}$$

Thus the $\Sigma_{\mu\nu}$ are just the components of the spin tensor 1(42). Combining 1(71) and 1(72) we can write the transformation law for $\psi(x)$ in the form

$$\begin{aligned}\psi'(x')-\psi(x) &= \frac{i}{4}\omega_{\mu\nu}\Sigma_{\mu\nu}\psi(x)\\ &= \frac{i}{2}\sum_{\mu<\nu}\omega_{\mu\nu}\Sigma_{\mu\nu}\psi(x)\end{aligned} \tag{1(75)}$$

Equation 1(75) defines the transformation law of a *four-component spinor*. In group-theoretic language, 1(75) provides a spinor representation of the Lorentz group with generators $\frac{1}{2}\Sigma_{\mu\nu}$.

To summarize, we have shown that the Dirac equation remains invariant under the Lorentz transformation 1(13), provided that $\psi(x)$ undergoes the simultaneous transformation 1(75). We can also rephrase this as the statement that the Dirac equation is invariant under the active transformation

$$\psi(x) \to \psi'(x)$$

with

$$\psi'(x) = L(a)\psi(a^{-1}x) \tag{1(76)}$$

The infinitesimal form of 1(76) is easily found; it is

$$\begin{aligned}\delta\psi(x) &= \psi'(x)-\psi(x)\\ &= -\delta x_\mu \partial_\mu \psi(x) + \frac{i}{4}\omega_{\mu\nu}\Sigma_{\mu\nu}\psi(x)\end{aligned} \tag{1(77)}$$

which differs from 1(17) by the presence of the spin term.

The proof that the current $j_\mu(x) = i\bar{\psi}(x)\gamma_\mu\psi(x)$ transforms as a vector under 1(75) is left as an exercise for the reader. One can also show that $\bar{\psi}(x)\gamma_\mu\gamma_5\psi(x)$ transforms as a vector under the infinitesimal transformation 1(75), while $\bar{\psi}(x)\Sigma_{\mu\nu}\psi(x)$ transforms as a second rank tensor. $\bar{\psi}(x)\psi(x)$ and $\bar{\psi}(x)\gamma_5\psi(x)$ transform as scalars.

The difference between $\bar{\psi}\psi$, $\bar{\psi}\gamma_\mu\psi$ on the one hand, and $\bar{\psi}\gamma_5\psi$, $\bar{\psi}\gamma_\mu\gamma_5\psi$ on the other, lies in their behaviour under the discrete operation of space reflection. One readily verifies that the Dirac equation remains unchanged under a space reflection $\mathbf{x}' = -\mathbf{x}$, $t' = t$, provided the spinor field $\psi(x)$ undergoes the corresponding change

$$\psi'(\mathbf{x}'t') = \gamma_4\psi(\mathbf{x}, t) \tag{1(78)}$$

Applying 1(78) to $\bar{\psi}\psi$ and $\bar{\psi}\gamma_5\psi$ we see that the former remains invariant, whereas the latter changes sign. Thus $\bar{\psi}\psi$ is a true scalar, while $\bar{\psi}\gamma_5\psi$ is a pseudoscalar. Similarly $\bar{\psi}\gamma_\mu\psi$ and $\bar{\psi}\gamma_\mu\gamma_5\psi$ are vector and pseudovector respectively, as evidenced by the behaviour of their space components under 1(78).

The sixteen quantities $\bar{\psi}\psi$, $\bar{\psi}\gamma_\mu\psi$, $i\bar{\psi}\gamma_\mu\gamma_5\psi$, $\bar{\psi}\Sigma_{\mu\nu}\psi$ and $\bar{\psi}\gamma_5\psi$ are known as the bilinear covariants. They feature the sixteen matrices

$$\underset{(A=1\ldots16)}{\gamma_A =} \begin{cases} 1 \\ \gamma_1, \gamma_2, \gamma_3, \gamma_4 \\ i\gamma_2\gamma_3, i\gamma_3\gamma_1, i\gamma_1\gamma_2, i\gamma_1\gamma_4, i\gamma_2\gamma_4, i\gamma_3\gamma_4 \\ i\gamma_2\gamma_3\gamma_4, i\gamma_3\gamma_1\gamma_4, i\gamma_1\gamma_2\gamma_4, i\gamma_1\gamma_3\gamma_2 \\ \gamma_1\gamma_2\gamma_3\gamma_4 \end{cases} \tag{1(79)}$$

which have several interesting properties. They satisfy

$$\gamma_A^2 = 1 \qquad (A = 1\ldots16) \tag{1(80a)}$$

and*

$$Tr\gamma_A = 0 \qquad (A = 2, \ldots 16) \tag{1(80b)}$$

$$Tr\gamma_A\gamma_{A'} = 4\delta_{AA'} \qquad (A, A' = 1 \ldots 16) \tag{1(80c)}$$

Moreover, the γ_A are linearly independent in the sense that

$$\sum_{A=1}^{16} c_A\gamma_A = 0$$

implies $c_A = 0$ $(A = 1 \ldots 16)$. The proof is a simple application of the property 1(80c) and is left to the reader.

Spin Sums and Projection Operators In applications it is frequently necessary to evaluate spin sums of the form

$$P_+(\mathbf{k}) = \sum_{\sigma=1}^{2} u_{\mathbf{k}\sigma}\bar{u}_{\mathbf{k}\sigma} \tag{1(81a)}$$

$$P_-(\mathbf{k}) = -\sum_{\sigma=1}^{2} v_{\mathbf{k}\sigma}\bar{v}_{\mathbf{k}\sigma} \tag{1(81b)}$$

We first note that as a consequence of the normalization and orthogonality properties 1(54a), 1(54b) and 1(56a) we have

$$P_+P_+ = P_+ \qquad P_-P_- = P_- \tag{1(82a)}$$

$$P_+P_- = P_-P_+ = 0 \tag{1(82b)}$$

For example

$$\begin{aligned} P_-P_- &= \sum_{\sigma}\sum_{\sigma'} v_{\mathbf{k}\sigma}\bar{v}_{\mathbf{k}\sigma}v_{\mathbf{k}\sigma'}\bar{v}_{\mathbf{k}\sigma'} \\ &= -\sum_{\sigma} v_{\mathbf{k}\sigma}\bar{v}_{\mathbf{k}\sigma} = P_- \end{aligned}$$

Thus P_+ and P_- are projection operators; they project out the positive and negative energy solutions respectively from a given plane wave solution of momentum $\mathbf{k}$.

$P_+(\mathbf{k})$ and $P_-(\mathbf{k})$ can be expanded in terms of the sixteen matrices γ_A with coefficients depending on k $(k^2 = -m^2)$. However, as $\bar{u}P_\pm u$ and $vP_\pm v$ are Lorentz scalars, not all γ_A's will appear in the expansion. It is easy to see that the most general form is

$$P_\pm(\mathbf{k}) = a_\pm + b_\pm\gamma \,.\, k$$

* See the theorems on traces of γ-matrices proved in Section 6-4.

To determine the coefficients we use 1(82a) together with the restrictions which result from applying 1(52a) and 1(52b) to 1(81a) and 1(81b). In this way we derive the important formulae

$$P_+(\mathbf{k}) = \sum u_{\mathbf{k}\sigma}\bar{u}_{\mathbf{k}\sigma} = \frac{\gamma\,.\,k + im}{2im} \qquad 1(83a)$$

$$P_-(\mathbf{k}) = -\sum_{\sigma} v_{\mathbf{k}\sigma}\bar{v}_{\mathbf{k}\sigma} = -\frac{\gamma\,.\,k - im}{2im} \qquad 1(83b)$$

We also note the completeness relation

$$\sum_{\sigma=1}^{2} (u_{\mathbf{k}\sigma}\bar{u}_{\mathbf{k}\sigma} - v_{\mathbf{k}\sigma}\bar{v}_{\mathbf{k}\sigma}) = 1 \qquad 1(84)$$

obtained by adding 1(83a) and 1(83b).

One further remark which will be of use presently is that if, instead of 1(54a), the Dirac spinors are normalized according to

$$u_{\mathbf{k}\sigma}{}^{\dagger}u_{\mathbf{k}\sigma'} = v_{\mathbf{k}\sigma}{}^{\dagger}v_{\mathbf{k}\sigma'} = \delta_{\sigma\sigma'} \qquad 1(85)$$

then by 1(58a) and 1(58b)

$$\bar{u}_{\mathbf{k}\sigma}u_{\mathbf{k}\sigma'} = \frac{m}{E_{\mathbf{k}}}\delta_{\sigma\sigma'} \qquad 1(86a)$$

$$\bar{v}_{\mathbf{k}\sigma}v_{\mathbf{k}\sigma'} = -\frac{m}{E_{\mathbf{k}}}\delta_{\sigma\sigma'} \qquad 1(86b)$$

and we have, instead of 1(83a) and 1(83b)

$$P_+(\mathbf{k}) = \sum_{\sigma} u_{\mathbf{k}\sigma}\bar{u}_{\mathbf{k}\sigma} = \frac{\gamma\,.\,k + im}{2iE_{\mathbf{k}}} \qquad 1(87a)$$

$$P_-(\mathbf{k}) = -\sum_{\sigma} v_{\mathbf{k}\sigma}\bar{v}_{\mathbf{k}\sigma} = -\frac{\gamma\,.\,k - im}{2iE_{\mathbf{k}}} \qquad 1(87b)$$

Dirac spinors normalized according to 1(85) are obtained from 1(51a) and 1(51b) by the replacement

$$\left(\frac{E_{\mathbf{k}} + m}{2m}\right)^{1/2} \rightarrow \left(\frac{E_{\mathbf{k}} + m}{2E_{\mathbf{k}}}\right)^{1/2}$$

Neutrino Theory To describe neutrinos, for which the rest mass m is zero, certain modifications of the formalism are called for. In the first place, we can no longer choose the normalization factors as in 1(51a), 1(51b) and 1(53a), 1(53b). Instead, we shall take as our four linearly independent spinors for momentum **k** the set

$$u_{\mathbf{k}\sigma} = \frac{1}{\sqrt{2}} \begin{pmatrix} \xi_\sigma \\ \dfrac{\boldsymbol{\sigma} . \mathbf{k}}{|\mathbf{k}|} \xi_\sigma \end{pmatrix} \quad (\sigma = 1, 2) \tag{1(88a)}$$

and

$$v_{\mathbf{k}\sigma} = \frac{1}{\sqrt{2}} \begin{pmatrix} \dfrac{\boldsymbol{\sigma} . \mathbf{k}}{|\mathbf{k}|} \xi_\sigma \\ \xi_\sigma \end{pmatrix} \quad (\sigma = 1, 2) \tag{1(88b)}$$

normalized according to 1(85), i.e.

$$u_{\mathbf{k}\sigma}{}^\dagger u_{\mathbf{k}\sigma'} = v_{\mathbf{k}\sigma}{}^\dagger v_{\mathbf{k}\sigma'} = \delta_{\sigma\sigma'}$$

instead of 1(54a) and 1(54b). The positive and negative energy projection operators are then

$$P_+(\mathbf{k}) = \sum_\sigma u_{\mathbf{k}\sigma} \bar{u}_{\mathbf{k}\sigma} = \frac{\gamma . k}{2i|\mathbf{k}|} \tag{1(89a)}$$

$$P_-(\mathbf{k}) = -\sum_\sigma v_{\mathbf{k}\sigma} \bar{v}_{\mathbf{k}\sigma} = -\frac{\gamma . k}{2i|\mathbf{k}|} \tag{1(89b)}$$

and a complete set of normalized neutrino wavefunctions is constructed by taking the set of all solutions

$$\frac{1}{\sqrt{V}} u_{\mathbf{k}\sigma}\, \mathrm{e}^{ik.x} \tag{1(90a)}$$

and

$$\frac{1}{\sqrt{V}} v_{\mathbf{k}\sigma}\, \mathrm{e}^{-ik.x} \tag{1(90b)}$$

A further important modification is that we choose the spinors ξ_σ to be eigenstates of $\boldsymbol{\sigma} . \mathbf{k}/|\mathbf{k}|$ rather than σ_z, i.e.

$$\begin{aligned} \boldsymbol{\sigma} . \mathbf{n}\xi_1 &= \xi_1 \\ \boldsymbol{\sigma} . \mathbf{n}\xi_2 &= -\xi_2 \end{aligned} \quad \left(\mathbf{n} = \frac{\mathbf{k}}{|\mathbf{k}|} \right) \tag{1(91)}$$

The reason for this change will be indicated presently. The eigenstates of $\boldsymbol{\sigma}\,.\,\mathbf{n}$ are easily constructed and are given by

$$\xi_1 = \frac{1}{\sqrt{2(n_z+1)}}\begin{pmatrix} n_z+1 \\ n_x+in_y \end{pmatrix} \tag{1(92a)}$$

$$\xi_2 = \frac{1}{\sqrt{2(n_z+1)}}\begin{pmatrix} -n_x+in_y \\ n_z+1 \end{pmatrix} \tag{1(92b)}$$

with eigenvalues $+1$ and -1 respectively; the normalization factor ensures that the $\xi_{\pm 1}$ are normalized to unity. With the choice 1(92a) and 1(92b) the solutions 1(88a) and 1(88b) are eigenstates of the helicity operator 1(39). Moreover, they are eigenstates of the γ_5-matrix,

$$\gamma_5 = \begin{pmatrix} 0 & -I \\ -I & 0 \end{pmatrix}$$

It is a simple matter to check that, for positive energy solutions, the helicity eigenvalues $h = +1$ and $h = -1$ correspond to γ_5-eigenvalues -1 and $+1$ respectively; for negative energy solutions on the other hand, $h = +1$ and $h = -1$ correspond respectively to $\gamma_5 = +1$ and $\gamma_5 = -1$. This is summarized in the following table:

	$E > 0$		$E < 0$	
h	$+1$	-1	$+1$	-1
γ_5	-1	$+1$	$+1$	-1

The possibility of diagonalizing γ_5 results from the fact that for $m = 0$, the Dirac equation remains invariant under the change $\psi \to \gamma_5\psi$, or equivalently that the Dirac Hamiltonian commutes with γ_5:

$$[\mathbf{H}, \gamma_5] = 0 \tag{1(93)}$$

The choice of helicity eigenstates for the ξ_σ is connected to a profound difference in the meaning of spin for massive particles on the one hand and massless particles on the other. In the massive case one can fix ξ_σ arbitrarily, each choice of ξ_σ representing a different direction of the rest frame polarization $\mathbf{e}$ and, correspondingly, of the polarization 4-vector e_α given by 1(60); by convention $\mathbf{e}$ is usually taken to lie along the z-axis, as in 1(51a) and 1(51b). For $m = 0$, however, this freedom disappears: the particle may no longer be referred to a rest frame and the 4-vector e_α is undefined. The spinors ξ_σ must then be taken to be eigenstates of $\boldsymbol{\sigma}\,.\,\mathbf{k}/|\mathbf{k}|$ and the polarization is always in the direction of motion. To confirm this we may refer to the Pauli–Lubanski vector $\omega_\alpha(k)$. By the same calculation which led to 1(62a) we have, for $m = 0$

$$\omega_\alpha(k)u_{\mathbf{k}\sigma} = -\gamma_5 k_\alpha u_{\mathbf{k}\sigma} \tag{1(94a)}$$

$$\omega_\alpha(k)v_{\mathbf{k}\sigma} = -\gamma_5 k_\alpha v_{\mathbf{k}\sigma} \tag{1(94b)}$$

Therefore, if $u_{\mathbf{k}\sigma}$ and $v_{\mathbf{k}\sigma}$ are eigenstates of γ_5 they are also eigenstates of the covariant spin operator $\omega_\alpha(k)$ and the eigenvalue 4-vector of ω_α points in the direction of k_α. In the next section we shall see that this feature is common to all massless particles.

Since γ_5 commutes with $\mathbf{H}$ it is frequently convenient, when discussing neutrinos, to adopt a representation of the γ-matrices in which γ_5 is diagonal. Such a representation is provided by

$$\boldsymbol{\gamma} = \begin{pmatrix} 0 & -i\boldsymbol{\sigma} \\ i\boldsymbol{\sigma} & 0 \end{pmatrix} \qquad \gamma_4 = \begin{pmatrix} 0 & I \\ I & 0 \end{pmatrix} \tag{1(95a)}$$

where

$$\gamma_5 = \begin{pmatrix} I & 0 \\ 0 & -I \end{pmatrix} \tag{1(95b)}$$

This representation is particularly convenient in the *two-component* theory of the neutrino (Lee, 1957), in which the neutrino wave function is subject to the supplementary condition

$$\gamma_5 \psi = \psi \tag{1(96)}$$

In this theory, positive energy solutions necessarily have $h = -1$, while negative energy solutions (which represent antiparticles) have $h = +1$. The physical significance of the reduction 1(96) stems from the fact that only the $\gamma_5 = +1$ neutrinos are emitted and absorbed in physical interactions.

1-4 Bargmann–Wigner equations

Introduction We shall derive the field equations for spin 1 and spin $\frac{3}{2}$ as particular cases of a general system of relativistic wave equations for arbitrary spin. Such a system was first written down by Dirac (1936) (see also Fierz, 1939a, b). Here we shall follow the formulation of Bargmann and Wigner (Bargmann, 1948) in which a field of rest mass m and spin $s \geqslant \frac{1}{2}$ is represented by a *completely symmetric multispinor* of rank $2s$

$$\underbrace{\psi_{\alpha\beta\gamma\ldots\tau}}_{2s}(x)$$

satisfying Dirac-type equations in all indices:

$$(\gamma\,.\,\partial+m)_{\alpha\alpha'}\psi_{\alpha'\beta\gamma\ldots\tau}(x) = 0$$
$$(\gamma\,.\,\partial+m)_{\beta\beta'}\psi_{\alpha\beta'\gamma\ldots\tau}(x) = 0 \qquad 1(97)$$
$$\vdots$$

For $s = \frac{1}{2}$ this system reduces to the single Dirac equation 1(24).

To check that the system 1(97) describes particles of rest mass m we multiply, say, the first equation by $(\gamma\,.\,\partial-m)_{\omega\alpha}$ and sum over α. The result

$$(\partial^2-m^2)\psi_{\alpha\beta}\ldots(x) = 0$$

ensures that 1(26) is satisfied for rest mass m.

Spin To prove that a symmetric multispinor of rank $2s$ satisfying 1(97) describes a particle of spin s, we construct positive energy plane-wave solutions. Setting

$$\psi_{\alpha\beta\gamma\ldots}(x) = w_{\alpha\beta\gamma\ldots}(\mathbf{k}, E_{\mathbf{k}})\,\mathrm{e}^{i\mathbf{k}.\mathbf{x}-iE_{\mathbf{k}}t}$$

we have

$$(\gamma\,.\,k-im)_{\alpha\alpha'}w_{\alpha'\beta\gamma\ldots}(\mathbf{k}, E_{\mathbf{k}}) = 0$$
$$(\gamma\,.\,k-im)_{\beta\beta'}w_{\alpha\beta'\gamma\ldots}(\mathbf{k}, E_{\mathbf{k}}) = 0 \qquad 1(98)$$
$$\vdots$$

Let us assume that $m \neq 0$ and go to the rest frame $\mathbf{k} = 0$. Adopting the representation 1(28) in which γ_4 is diagonal with diagonal elements 1, 1, -1, -1 we see that $w_{\alpha\beta\gamma\ldots}(0, m)$ is non-zero only when all indices $\alpha, \beta\ldots$ are restricted to the values 1 and 2. This is just the generalization of the result 1(49a). Taking into account the requirement that $w_{\alpha\beta\ldots}$ must be completely symmetric in all indices, we can construct the following complete set of linearly independent solutions for $E_{\mathbf{k}} = m$.

$$w^{(0)}_{\alpha\beta\gamma\ldots\tau} = \delta_{\alpha1}\delta_{\beta1}\delta_{\gamma1}\ldots\delta_{\tau1}$$
$$\begin{aligned} w^{(1)}_{\alpha\beta\gamma\ldots\tau} = {} & \delta_{\alpha2}\delta_{\beta1}\delta_{\gamma1}\ldots\delta_{\tau1} \\ & +\delta_{\alpha1}\delta_{\beta2}\delta_{\gamma1}\ldots\delta_{\tau1} \\ & +\delta_{\alpha1}\delta_{\beta1}\delta_{\gamma2}\ldots\delta_{\tau1} \\ & +\ldots \end{aligned} \qquad 1(99)$$

$$w^{(2)}_{\alpha\beta\gamma\ldots\tau} = \delta_{\alpha 2}\delta_{\beta 2}\delta_{\gamma 1}\ldots\delta_{\tau 1}$$
$$+\ldots$$
$$\vdots$$
$$w^{(2s)}_{\alpha\beta\gamma\ldots\tau} = \delta_{\alpha 2}\delta_{\beta 2}\delta_{\gamma 2}\ldots\delta_{\tau 2}$$

generalizing the positive energy solutions of 1(50). Since $2s+1$ is the number of linearly independent states for angular momentum s, we should expect the solutions 1(99) to refer to the different polarization states of a particle of spin s at rest. To confirm this, we enlist the aid of the covariant Pauli-Lubanski spin operator

$$(\omega_\rho)_{\alpha\alpha',\beta\beta',\ldots} = -\tfrac{1}{2}\varepsilon_{\rho\mu\nu\lambda}(\Sigma_{\mu\nu})_{\alpha\alpha',\beta\beta',\ldots}\partial_\lambda \tag{1(100)}$$

which generalizes 1(41); $\Sigma_{\mu\nu}$ is the spin tensor*

$$(\Sigma_{\mu\nu})_{\alpha\alpha',\beta\beta',\ldots} = \frac{1}{2i}\{(\gamma_\mu\gamma_\nu-\gamma_\nu\gamma_\mu)_{\alpha\alpha'}+(\gamma_\mu\gamma_\nu-\gamma_\nu\gamma_\mu)_{\beta\beta'}+\ldots\} \tag{1(101)}$$

For positive energy particles at rest, ω_ρ becomes

$$\boldsymbol{\omega} = m\boldsymbol{\Sigma}$$
$$\omega_4 = 0$$

with the eigenvalues of $\frac{1}{2}(\Sigma_z)_{\alpha\alpha',\beta\beta',\ldots}$ giving the spin along the z-axis. Let us operate with Σ_z on the solutions 1(99). In the representation 1(28), $-i\gamma_1\gamma_2$ is diagonal with eigenvalues $1, -1, 1, -1$, so that when we apply $\frac{1}{2}\Sigma_z$ to a solution $w^{(i)}_{\alpha\beta\gamma\ldots}$ with i indices equal to 1 and $2s-i$ indices equal to 2, we get the eigenvalue $\frac{1}{2}i-\frac{1}{2}(2s-1) = i-s$. Since the integer i runs from 0 to $2s$, we are assured that the solutions 1(99) form a set of eigenstates of $\frac{1}{2}\Sigma_z$ with integer eigenvalues running from $-s$ to $+s$.

Case of Zero Rest Mass If the rest mass is equal to zero we cannot go to a rest frame, but can choose a frame in which $k_\mu = k^0_\mu = (0, 0, k, ik)$ corresponding to motion along the positive z-axis. The system 1(98) then reduces to

$$(\gamma_3\gamma_4)_{\alpha\alpha'}w_{\alpha'\beta\gamma\ldots}(k^0) = iw_{\alpha\beta\gamma\ldots}(k^0)$$
$$(\gamma_3\gamma_4)_{\beta\beta'}w_{\alpha\beta'\gamma\ldots}(k^0) = iw_{\alpha\beta\gamma\ldots}(k^0) \tag{1(102)}$$
$$\vdots$$

after multiplication by γ_3. It is convenient to adopt a representation in

* See Problem 7.

which $\gamma_3\gamma_4$ is diagonal, for example 1(95a) and 1(95b). The diagonal matrix elements of $\gamma_3\gamma_4$ are then $-i, +i, +i, -i$ so that $w_{\alpha\beta\gamma\ldots}(\mathbf{k}, E_{\mathbf{k}})$ can only be non-zero in this frame if all its indices $\alpha, \beta, \gamma\ldots$ are restricted to the values 2 and 3. When we take into account the symmetry requirement, we are again led to a set of $2s+1$ states of the type 1(99), the indices 1 and 2 being replaced by 2 and 3 respectively wherever they occur. In the zero-mass case, however, only two of these states, namely

$$w^{(0)}_{\alpha\beta\gamma\ldots\tau}(k^0) = \delta_{\alpha 2}\delta_{\beta 2}\ldots\delta_{\tau 2} \qquad 1(103a)$$

and

$$w^{(2s)}_{\alpha\beta\gamma\ldots\tau}(k^0) = \delta_{\alpha 3}\delta_{\beta 3}\ldots\delta_{\tau 3} \qquad 1(103b)$$

are in fact solutions. Indeed, when $m = 0$, we see from 1(98) that the solutions to the Bargmann–Wigner equations may be multiplied by any one of the operators

$$\begin{aligned}(\gamma_5)_{\alpha\alpha'} &= (\gamma_1\gamma_2\gamma_3\gamma_4)_{\alpha\alpha'} \\ (\gamma_5)_{\beta\beta'} &= (\gamma_1\gamma_2\gamma_3\gamma_4)_{\beta\beta'} \\ &\vdots\end{aligned} \qquad 1(104)$$

without modifying the system. Since γ_5 is diagonal in the representation 1(95a), with diagonal eigenvalues $+1, +1, -1, -1$, a moment's thought reveals that the only two states compatible with the above requirement are 1(103a) and 1(103b).

Hence, when $m = 0$, there are only two spin states, rather than $2s+1$ for each sign of the energy. To identify their spin value we call on the Pauli–Lubanski operator 1(100), which for $-i\partial_\mu \equiv k^0_\mu = (0, 0, k, ik)$ takes the form

$$\begin{aligned}\omega_1(k^0) &= (\Sigma_{23}+i\Sigma_{24})k \\ \omega_2(k^0) &= -(\Sigma_{13}+i\Sigma_{14})k \\ \omega_3(k^0) &= \Sigma_{12}k \\ \omega_4(k^0) &= i\Sigma_{12}k\end{aligned} \qquad 1(105)$$

The first two components yield zero eigenvalues when operating on the solution of 1(102):

$$\omega_1(k^0)_{\alpha\alpha',\beta\beta'\ldots}w^{(i)}_{\alpha'\beta'\ldots}(k^0) = 0 \qquad 1(106a)$$

$$\omega_2(k^0)_{\alpha\alpha',\beta\beta'\ldots}w^{(i)}_{\alpha'\beta'\ldots}(k^0) = 0 \qquad 1(106b)$$

1(106a), for example, is a consequence of the fact that $\gamma_2\gamma_3$ and $-i\gamma_2\gamma_4$ yield the same eigenvalue when operating on $w^{(i)}$; this follows by multiplying each equation of 1(102) by the appropriate $\gamma_2\gamma_3$. On the other hand, for ω_3 we have

$$\omega_3(k^0)_{\alpha\alpha',\beta\beta'\ldots} w^{(0)}_{\alpha'\beta'\ldots}(k^0) = -2skw^{(0)}_{\alpha\beta\ldots}(k^0)$$

$$\omega_3(k^0)_{\alpha\alpha',\beta\beta'\ldots} w^{(2s)}_{\alpha'\beta'\ldots}(k^0) = 2skw^{(2s)}_{\alpha\beta\ldots}(k^0) \qquad 1(107)$$

Indeed, multiplying each equation of 1(102) by $-i\gamma_1\gamma_2$ we find that $-i\gamma_1\gamma_2$ gives the same eigenvalue as $-\gamma_5$ when operating on w. The result 1(107) follows from the fact that all the γ_5's yield eigenvalues $+1$ and -1 when operating on $w^{(0)}$ and $w^{(2s)}$ respectively. Finally, for the fourth component ω_4 we have the same result, 1(107), multiplied by i. Denoting the eigenvalues of $\frac{1}{2}\omega_\mu(k)$ by s_μ we have found that

$$s_\mu = \pm sk_\mu$$

which generalizes the result 1(94a) for spin $\frac{1}{2}$. Thus for a massless particle of spin s, the spin points along the direction of motion and can take on only the two extreme values $\pm s$. The important difference in the meaning of spin for massive and massless particles is that when $m \neq 0$ one can refer the particle to a rest frame in which the spin can point in any direction, whereas for massless particles there is no rest frame and the spin can only point in the direction of motion. Note that along the direction of motion the orbital angular momentum is zero, so that the projection of the spin along **k** may be identified with the projection of the *total* angular momentum.

Outlook Despite their generality, the Bargmann–Wigner equations are difficult to handle, especially from the point of view of Lagrangian field theory (see Chapter 2). The wave equations for higher spin must be recast in a more manageable form and this entails treating each spin value individually. In the following sections this will be done explicitly for spin 1 and spin $\frac{3}{2}$.

1-5 Massive vector field

Field Equations For spin 1 the Bargmann–Wigner system reduces to two equations, which can be written in the form

$$(\gamma\,.\,\partial+\mu)\psi(x) = 0 \qquad 1(108a)$$

$$\psi(x)(\gamma^T\,.\,\overleftarrow{\partial}+\mu) = 0 \qquad 1(108b)$$

where we have represented the bi-spinor $\psi_{\alpha\beta}$ as a 4×4 matrix and denoted the rest mass by μ.

Let us expand ψ in terms of the set of 16 matrices given by 1(79), taking care to ensure that the assumed symmetry of $\psi_{\alpha\beta}$ is preserved in the expansion. We can construct a complete set of symmetric 4×4 matrices with the aid of the charge conjugation matrix C, satisfying 1(64a), 1(64b) and 1(64c). Such a set is given by $\gamma_\mu C$ and $\Sigma_{\mu\nu} C$; the remaining six matrices C, $i\gamma_5 C$ and $i\gamma_\mu\gamma_5 C$ are *anti*symmetric. Indeed, using 1(64a), 1(64c) and 1(66), we find

$$(\gamma_\mu C)^T = C^T \gamma_\mu^T = -C\gamma_\mu^T = \gamma_\mu C$$

$$(\gamma_\mu \gamma_5 C)^T = C^T \gamma_5^T \gamma_\mu^T = \gamma_5 \gamma_\mu C = -\gamma_\mu \gamma_5 C$$

and so on. Accordingly we set

$$\psi(x) = i\mu\gamma_\lambda C A_\lambda(x) + \tfrac{1}{2}\Sigma_{\lambda\nu} C F_{\lambda\nu}(x) \qquad 1(109)$$

where A_λ and $F_{\lambda\nu}$ are vector and antisymmetric second-rank tensor fields respectively, and where numerical factors have been introduced for later convenience. We now apply the Bargmann–Wigner equations 1(108) to 1(109). Adding 1(108a) to 1(108b) and using 1(64a), we find

$$\begin{aligned} 0 &= i\mu[\gamma_\nu, \gamma_\lambda] C \partial_\nu A_\lambda + \tfrac{1}{2}[\gamma_\mu, \Sigma_{\lambda\nu}] C \partial_\mu F_{\lambda\nu} \\ &\quad + 2i\mu^2 \gamma_\lambda C A_\lambda + \mu \Sigma_{\lambda\nu} C F_{\lambda\nu} \\ &= -2\mu \Sigma_{\lambda\nu} C \partial_\lambda A_\nu + 2i\gamma_\lambda C \partial_\nu F_{\lambda\nu} \\ &\quad + 2i\mu^2 \gamma_\lambda C A_\lambda + \mu \Sigma_{\lambda\nu} C F_{\lambda\nu} \end{aligned}$$

where we have used the easily derived identity

$$[\gamma_\mu, \Sigma_{\lambda\nu}] = 2i\delta_{\mu\nu}\gamma_\lambda - 2i\delta_{\mu\lambda}\gamma_\nu$$

Setting the coefficients of γ_λ and $\Sigma_{\lambda\nu}$ equal to zero, we get the system of coupled equations

$$F_{\lambda\nu} = \partial_\lambda A_\nu - \partial_\nu A_\lambda \qquad 1(110a)$$

$$\partial_\lambda F_{\lambda\nu} = \mu^2 A_\nu \qquad 1(110b)$$

or, in terms of $A_\nu(x)$ by itself,

$$\Box A_\nu - \partial_\nu(\partial_\lambda A_\lambda) = \mu^2 A_\nu \qquad 1(111)$$

Equations 1(110a) and 1(110b) or 1(111) are the fundamental field equations for a vector field of rest mass μ. Assuming $\mu \neq 0$, we derive from 1(110a) and 1(110b) or 1(111) the supplementary condition

$$\partial_\lambda A_\lambda(x) = 0 \qquad 1(112)$$

so that 1(111) may also be written as the set of equations

$$\Box A_\nu(x) - \mu^2 A_\nu(x) = 0 \qquad 1(113a)$$

$$\partial_\nu A_\nu(x) = 0 \qquad 1(113b)$$

As we are working with a space-time metric in which $x_4 = it$ is imaginary, we must take care in defining the conjugate 4-vector A^*_ν. With $A_\nu = (\mathbf{A}, iA_0)$ we define the conjugate vector A^*_ν by

$$A^*_\nu = (\mathbf{A}^*, iA_0^*) \qquad 1(114)$$

without taking the complex conjugate of *i*. In other words, we define

$$A^*_4 = -A_4^*$$

Similarly, we define the conjugate tensor $F^*_{\lambda\nu}$ by $F^*_{ij} = F_{ij}^*$ but $F^*_{i4} = -F_{i4}^*$. With this convention the equations of motion 1(110a) and 1(110b) are valid for both $(A_\nu, F_{\lambda\nu})$ and $(A^*_\nu, F^*_{\lambda\nu})$, i.e. we have

$$F^*_{\lambda\nu} = \partial_\lambda A^*_\nu - \partial_\nu A^*_\lambda \qquad 1(115a)$$

$$\partial_\lambda F^*_{\lambda\nu} = \mu^2 A^*_\nu \qquad 1(115b)$$

Charge Current Vector To derive a conserved current for the vector theory we multiply 1(115b) by A_ν and 1(110b) by A^*_ν. Summing over ν and subtracting the two equations, we get

$$(\partial_\lambda F^*_{\lambda\nu})A_\nu - (\partial_\lambda F_{\lambda\nu})A^*_\nu = 0$$

which can be written as the conservation law

$$\partial_\lambda j_\lambda(x) = 0 \qquad 1(116a)$$

with

$$j_\lambda(x) = i(F^*_{\lambda\nu}A_\nu - F_{\lambda\nu}A^*_\nu) \qquad 1(116b)$$

the difference $F^*_{\lambda\nu}\partial_\lambda A_\nu - F_{\lambda\nu}\partial_\lambda A^*_\nu$ vanishing by virtue of 1(110a) and 1(115a). The total ‘ charge ’

$$-i\int j_4(x)\, d^3x = \int (F^*_{4\nu}A_\nu - F_{4\nu}A^*_\nu)\, d^3x \qquad 1(117)$$

is therefore conserved in time. As in the Klein–Gordon case, the charge can take on both positive and negative values.

Lorentz Invariance The field equations 1(113) will remain invariant under the infinitesimal Lorentz transformation

$$x'_\mu = x_\mu + \omega_{\mu\nu} x_\nu$$

if A_μ transforms as a four-vector

$$A'_\mu(x') = A_\mu(x) + \omega_{\mu\nu}A_\nu(x)$$

We can cast this transformation law into a form analogous to 1(75) by regarding A_ν as a column-vector and setting

$$\omega_{\mu\nu} = \frac{i}{2}\omega_{\rho\sigma}(S_{\rho\sigma})_{\mu\nu}$$

where the $S_{\rho\sigma}$ are the six 4×4 matrices

$$(S_{\rho\sigma})_{\mu\nu} = -i(\delta_{\rho\mu}\delta_{\sigma\nu} - \delta_{\rho\nu}\delta_{\sigma\mu}) \qquad 1(118)$$

Then the transformation law for $A(x)$ takes the form

$$A'(x') - A(x) = \frac{i}{2}\omega_{\rho\sigma}S_{\rho\sigma}A(x) \qquad 1(119)$$

$S_{\rho\sigma}$ is the spin-1 spin tensor. Under an ordinary space rotation the vector field **A** transforms according to

$$\mathbf{A}'(\mathbf{x}', t) - \mathbf{A}(\mathbf{x}, t) = \frac{i}{2}\omega_{ij}S_{ij}\mathbf{A}(\mathbf{x}, t)$$

where the S_{ij} are the three spin matrices

$$S_{12} = \begin{pmatrix} 0 & -i & 0 \\ i & 0 & 0 \\ 0 & 0 & 0 \end{pmatrix} \quad S_{31} = \begin{pmatrix} 0 & 0 & +i \\ 0 & 0 & 0 \\ -i & 0 & 0 \end{pmatrix} \quad S_{23} = \begin{pmatrix} 0 & 0 & 0 \\ 0 & 0 & -i \\ 0 & i & 0 \end{pmatrix} \qquad 1(120a)$$

A more compact notation for 1(120a) is

$$(S_k)_{ij} = -i\varepsilon_{kij} \qquad 1(120b)$$

where $S_1 = S_{23}$, $S_2 = S_{31}$, $S_3 = S_{12}$ and where ε_{kij} is the completely antisymmetric Levi–Civita symbol.

Hamiltonian The supplementary condition 1(112) indicates that the four components of A_ν are not all independent dynamical variables. To isolate the latter and cast the vector theory into Hamiltonian form, it is convenient to introduce a three-dimensional notation, setting $A_\nu = (\mathbf{A}, iA_0)$, $A^*_\nu = (\mathbf{A}^*, iA_0^*)$, and

$$F_{\lambda\nu} = \begin{cases} F_{ij} = \varepsilon_{ijk}B_k \\ F_{k4} = -iE_k \end{cases} \qquad F^*_{\lambda\nu} = \begin{cases} F^*_{ij} = \varepsilon_{ijk}B_k^* \\ F^*_{k4} = -iE_k^* \end{cases} \qquad 1(121)$$

In this notation the equations of motion 1(110a) and 1(110b) may be written in the form

$$\mathbf{B} = \mathbf{curl\ A} \tag{1(122a)}$$

$$\mathbf{E} = -\ \mathbf{grad}\ A_0 - \partial_t \mathbf{A} \tag{1(122b)}$$

$$\mathbf{curl\ B} - \partial_t \mathbf{E} = -\mu^2 \mathbf{A} \tag{1(122c)}$$

$$\text{div}\ \mathbf{E} = -\mu^2 A_0 \tag{1(122d)}$$

Examining this system, we see that 1(122b) and 1(122c) are true equations of motion involving the time derivatives $\partial_t \mathbf{A}$ and $\partial_t \mathbf{E}$. Equations 1(122a) and 1(122d) are conditions of constraint; we may regard them as equations *defining* **B** and A_0 in terms of the dynamical variables **A** and **E**. Eliminating **B** and A_0 we get the following system of coupled equations of motion for **A** and **E**:

$$\partial_t \mathbf{A} = -\mathbf{E} + \frac{1}{\mu^2}\mathbf{grad}(\text{div}\ \mathbf{E}) \tag{1(123a)}$$

$$\partial_t \mathbf{E} = \mu^2 \mathbf{A} + \mathbf{curl}(\mathbf{curl\ A}) \tag{1(123b)}$$

or, using the general vector relation **curl curl** = **grad** div $-\boldsymbol{\nabla}^2$

$$i\partial_t \mathbf{A} = -i\mathbf{E} + \frac{i}{\mu^2}\mathbf{grad}(\text{div}\ \mathbf{E})$$

$$i\partial_t \mathbf{E} = -i\boldsymbol{\nabla}^2 \mathbf{A} + i\mu^2 \mathbf{A} + i\ \mathbf{grad}(\text{div}\ \mathbf{A})$$

The above equations are already in Hamiltonian form. To write them in a compact form analogous to 1(20), we introduce the linear combinations

$$\boldsymbol{\Psi}_1 = -\frac{1}{\sqrt{2}}\left(\mathbf{A} - \frac{i}{\mu}\mathbf{E}\right)$$

$$\boldsymbol{\Psi}_2 = -\frac{1}{\sqrt{2}}\left(\mathbf{A} + \frac{i}{\mu}\mathbf{E}\right)$$

which satisfy the coupled equations

$$i\partial_t \boldsymbol{\Psi}_1 = -\frac{\boldsymbol{\nabla}^2}{2\mu}(\boldsymbol{\Psi}_1 + \boldsymbol{\Psi}_2) + \mu\boldsymbol{\Psi}_1 + \frac{1}{\mu}\mathbf{grad}\ \text{div}\ \boldsymbol{\Psi}_2 \tag{1(124a)}$$

$$i\partial_t \boldsymbol{\Psi}_2 = \frac{\boldsymbol{\nabla}^2}{2\mu}(\boldsymbol{\Psi}_1 + \boldsymbol{\Psi}_2) - \mu\boldsymbol{\Psi}_2 - \frac{1}{\mu}\mathbf{grad}\ \text{div}\ \boldsymbol{\Psi}_1 \tag{1(124b)}$$

The verification of 1(124a) and 1(124b) is left as an exercise. Moreover, if $\mathbf{a}$ is a 3-component column vector, one can show that

$$\mathbf{grad}\,\mathrm{div}\,\mathbf{a} = [\nabla^2-(\mathbf{S}\,.\,\nabla)^2]\mathbf{a} \qquad 1(125)$$

where $\mathbf{S} = (S_{23}, S_{31}, S_{12})$. (See Problem 9.) We may therefore write 1(124a) and 1(124b) in the compact form

$$i\partial_t\mathbf{\Psi} = \mathbf{H\Psi} \qquad 1(126a)$$

where $\mathbf{\Psi}$ is the six-component object

$$\mathbf{\Psi} = \begin{pmatrix}\mathbf{\Psi}_1\\ \mathbf{\Psi}_2\end{pmatrix} \qquad 1(126b)$$

and the Hamiltonian $\mathbf{H}$ is given by

$$\mathbf{H} = (\tau_3+i\tau_2)\frac{\mathbf{k}^2}{2\mu}+\mu\tau_3-i\tau_2\frac{\mathbf{k}^2-(\mathbf{S}\,.\,\mathbf{k})^2}{\mu} \qquad 1(126c)$$

in an obvious notation, the τ_i being the usual 2×2 Pauli matrices.

A short calculation shows that the total ‘ charge ’ 1(117) can be expressed in the same notation as

$$-i\int_V j_4(x)\,d^3x = \mu\int_V \mathbf{\Psi}^*\tau_3\,\mathbf{\Psi}d^3x \qquad 1(127)$$

in analogy with 1(21). As in the Klein–Gordon theory, particular interest is attached to the expectation value (see Problem 10),

$$\int_V \mathbf{\Psi}^*\tau_3\mathbf{H}\,\mathbf{\Psi}d^3x$$

$$=\frac{1}{\mu}\int_V\left(\mathbf{E}^*\,.\,\mathbf{E}+\mu^2\mathbf{A}^*\,.\,\mathbf{A}+\mathbf{A}^*\,.\,\mathbf{curl}\,\mathbf{curl}\,\mathbf{A}-\frac{1}{\mu^2}\mathbf{E}^*\,.\,\mathbf{grad}\,\mathrm{div}\,\mathbf{E}\right)d^3x \qquad 1(128)$$

which is *positive-definite* and will be identified in the following chapter as the field theoretic Hamiltonian of the vector field, to within a factor μ^{-1}.

Plane Wave Solutions Plane wave solutions to the vector field equations 1(113a) and 1(113b) are of the form

$$\varepsilon_{\mathbf{k}}\,e^{i\mathbf{k}.\mathbf{x}\mp ik_0t} \qquad 1(129)$$

with

$$k_0 = \omega_{\mathbf{k}} = (\mathbf{k}^2+\mu^2)^{1/2} > 0 \qquad 1(130)$$

and where the four-vector $\varepsilon_{\mathbf{k}}$ satisfies the constraint

$$k \,.\, \varepsilon_{\mathbf{k}} = 0 \tag{1(131)}$$

Both positive and negative energy solutions are included in 1(129).

For each $\mathbf{k}$, we can construct a set of three linearly independent four-vectors satisfying 1(131) by referring to 1(60). If $\boldsymbol{\varepsilon}_{\mathbf{k}\lambda}$ ($\lambda = 1, 2, 3$) is any triad of three-vectors satisfying the orthonormality relations

$$\boldsymbol{\varepsilon}_{\mathbf{k}\lambda} \cdot \boldsymbol{\varepsilon}_{\mathbf{k}\lambda'} = \delta_{\lambda\lambda'} \tag{1(132)}$$

then the three four-vectors

$$\varepsilon^{\alpha}_{\mathbf{k}\lambda} = \begin{cases} \boldsymbol{\varepsilon}_{\mathbf{k}\lambda} + \dfrac{\mathbf{k}(\mathbf{k} \,.\, \boldsymbol{\varepsilon}_{\mathbf{k}\lambda})}{\mu(\omega_{\mathbf{k}} + \mu)} & (\alpha = 1, 2, 3) \\ i\dfrac{\mathbf{k} \,.\, \boldsymbol{\varepsilon}_{\mathbf{k}\lambda}}{\mu} & (\alpha = 4) \end{cases} \tag{1(133)}$$

satisfy both 1(131) and the orthonormality relations

$$\varepsilon_{\mathbf{k}\lambda} \cdot \varepsilon_{\mathbf{k}\lambda'} = \delta_{\lambda\lambda'} \tag{1(134)}$$

The verification of 1(134) is straightforward and is left to the reader. A complete set of solutions to 1(113a) and 1(113b) is therefore given by

$$\frac{1}{\sqrt{V}}\frac{1}{\sqrt{2\omega_{\mathbf{k}}}}\varepsilon_{\mathbf{k}\lambda}\, e^{ik.x} \qquad (\lambda = 1, 2, 3) \tag{1(135a)}$$

and

$$\frac{1}{\sqrt{V}}\frac{1}{\sqrt{2\omega_{\mathbf{k}}}}\varepsilon_{\mathbf{k}\lambda}\, e^{-ik.x} \qquad (\lambda = 1, 2, 3) \tag{1(135b)}$$

A convenient choice for the triad of orthogonal unit vectors $\boldsymbol{\varepsilon}_{\mathbf{k}\lambda}$ is to take $\boldsymbol{\varepsilon}_{\mathbf{k}3}$ pointing along $\mathbf{k}$ with $\boldsymbol{\varepsilon}_{\mathbf{k}1}$ and $\boldsymbol{\varepsilon}_{\mathbf{k}2}$ orthogonal both to $\boldsymbol{\varepsilon}_{\mathbf{k}3}$ and to each other, i.e.

$$\boldsymbol{\varepsilon}_{\mathbf{k}3} = \frac{\mathbf{k}}{|\mathbf{k}|}$$

$$\boldsymbol{\varepsilon}_{\mathbf{k}1} \cdot \mathbf{k} = \boldsymbol{\varepsilon}_{\mathbf{k}2} \,.\, \mathbf{k} = \boldsymbol{\varepsilon}_{\mathbf{k}1} \cdot \boldsymbol{\varepsilon}_{\mathbf{k}2} = 0 \tag{1(136a)}$$

with the completeness relation

$$\sum_{\lambda=1}^{2} \varepsilon^{i}_{\mathbf{k}\lambda}\varepsilon^{j}_{\mathbf{k}\lambda} + \frac{k^i k^j}{\mathbf{k}^2} = \delta^{ij} \tag{1(136b)}$$

The polarization vectors $\boldsymbol{\varepsilon}_{\mathbf{k}1}$ and $\boldsymbol{\varepsilon}_{\mathbf{k}2}$ represent states of *transverse* polarization, while $\boldsymbol{\varepsilon}_{\mathbf{k}3}$ represents *longitudinal* polarization. Note that

$$\boldsymbol{\varepsilon}_{\mathbf{k}3} = -\boldsymbol{\varepsilon}_{-\mathbf{k}3} \tag{1(137a)}$$

to which we add, by convention

$$\begin{aligned} \boldsymbol{\varepsilon}_{\mathbf{k}1} &= -\boldsymbol{\varepsilon}_{-\mathbf{k}1} \\ \boldsymbol{\varepsilon}_{\mathbf{k}2} &= -\boldsymbol{\varepsilon}_{-\mathbf{k}2} \end{aligned} \tag{1(137b)}$$

The transverse and longitudinal polarization vectors are related to the eigenvectors of the helicity operator $\mathbf{S}\,.\,\mathbf{k}/|\mathbf{k}|$. This is most easily seen by choosing the reference system such that $\mathbf{k}$ points along the positive z-axis. The helicity operator is then

$$S_{12} = \begin{pmatrix} 0 & -i & 0 \\ i & 0 & 0 \\ 0 & 0 & 0 \end{pmatrix}$$

and we see that the longitudinal vector $\boldsymbol{\varepsilon}_{\mathbf{k}3}$ is an eigenvector of S_{12} with eigenvalue $s_z = 0$. In the same reference system we could also choose the transverse states to be the eigenvectors

$$\boldsymbol{\varepsilon}_{\mathbf{k}1} = \frac{1}{\sqrt{2}}\begin{pmatrix} 1 \\ i \\ 0 \end{pmatrix} \qquad \boldsymbol{\varepsilon}_{\mathbf{k}2} = \frac{1}{\sqrt{2}}\begin{pmatrix} 1 \\ -i \\ 0 \end{pmatrix} \tag{1(138)}$$

with eigenvalues $s_z = +1$ and $s_z = -1$ respectively, but the convention 1(137b) would then have to be modified, since the spin up and spin down states are interchanged under $\mathbf{k} \to -\mathbf{k}$.

Corresponding to the choice 1(136a) we have, for the polarization four-vectors 1(133)

$$\varepsilon_{\mathbf{k}1} = \begin{cases} \boldsymbol{\varepsilon}_{\mathbf{k}1} \\ 0 \end{cases} \qquad \varepsilon_{\mathbf{k}2} = \begin{cases} \boldsymbol{\varepsilon}_{\mathbf{k}2} \\ 0 \end{cases} \qquad \varepsilon_{\mathbf{k}3} = \begin{cases} \dfrac{\omega_{\mathbf{k}}}{\mu}\dfrac{\mathbf{k}}{|\mathbf{k}|} \\ i\dfrac{|\mathbf{k}|}{\mu} \end{cases} \tag{1(139)}$$

The three four-vectors $\varepsilon_{\mathbf{k}\lambda}$ do not of course form a complete set. To evaluate the spin sum

$$\Lambda^{\alpha\beta}(k) = \sum_{\lambda=1}^{3} \varepsilon_{\mathbf{k}\lambda}^{\alpha}\varepsilon_{\mathbf{k}\lambda}^{\beta} \tag{1(140)}$$

we note that $\varepsilon_{\mathbf{k}1}$, $\varepsilon_{\mathbf{k}2}$, $\varepsilon_{\mathbf{k}3}$, and ik/μ form a quartet of orthonormal four-vectors. From the completeness relation for this quartet we deduce that

$$\sum_{\lambda=1}^{3} \varepsilon_{\mathbf{k}\lambda}^{\alpha}\varepsilon_{\mathbf{k}\lambda}^{\beta} = \delta^{\alpha\beta} + \frac{k^{\alpha}k^{\beta}}{\mu^{2}} \qquad 1(141)$$

For future reference we write down the plane wave solutions for the three-vectors $\mathbf{A}(x)$ and $\mathbf{E}(x)$. With the choice 1(136a) for the $\boldsymbol{\varepsilon}_{\mathbf{k}\lambda}$, the wave functions for $\mathbf{A}$ follow by inspection of 1(135a), 1(135b) and 1(139). A complete set is

$$\begin{array}{ccc} (\lambda = 1, 2) & & (\lambda = 3) \\ \dfrac{1}{\sqrt{V}}\dfrac{1}{\sqrt{2\omega_{\mathbf{k}}}}\boldsymbol{\varepsilon}_{\mathbf{k}\lambda}\,\mathrm{e}^{\pm ik.x} & \text{and} & \dfrac{1}{\sqrt{V}}\dfrac{1}{\sqrt{2\omega_{\mathbf{k}}}}\dfrac{\omega_{\mathbf{k}}}{\mu}\dfrac{\mathbf{k}}{|\mathbf{k}|}\mathrm{e}^{\pm ik.x} \end{array} \qquad 1(142)$$

Correspondingly, we find for $\mathbf{E}(x)$, using 1(122b) and 1(139)

$$\begin{array}{ccc} (\lambda = 1, 2) & & (\lambda = 3) \\ \pm\dfrac{i}{\sqrt{V}}\sqrt{\dfrac{\omega_{\mathbf{k}}}{2}}\boldsymbol{\varepsilon}_{\mathbf{k}\lambda}\,\mathrm{e}^{\pm ik.x} & \text{and} & \pm\dfrac{i}{\sqrt{V}}\sqrt{\dfrac{\omega_{\mathbf{k}}}{2}}\dfrac{\mu}{\omega_{\mathbf{k}}}\dfrac{\mathbf{k}}{|\mathbf{k}|}\mathrm{e}^{\pm ik.x} \end{array} \qquad 1(143)$$

Observe that the normalization factors are such as to ensure that

$$\begin{aligned} -i\int j_4(x)\,d^3x &= i\int (\mathbf{E}^*.\mathbf{A} - \mathbf{A}^*.\mathbf{E})\,d^3x \\ &= +1 \end{aligned} \qquad 1(144)$$

for positive energy solutions.

1-6 The Maxwell field

Maxwell's Equations The vector theory for the case $\mu = 0$ is of particular significance, since it describes the electromagnetic field. The equations of motion 1(110a), 1(110b) and 1(111) become, in this case*

$$F_{\lambda\nu} = \partial_\lambda A_\nu - \partial_\nu A_\lambda \qquad 1(145a)$$

$$\partial_\lambda F_{\lambda\nu} = 0 \qquad 1(145b)$$

and

$$\Box A_\nu - \partial_\nu(\partial_\lambda A_\lambda) = 0 \qquad 1(146)$$

* Note that when $\mu = 0$, one can no longer derive the set 1(145a), 1(145b) directly from the Bargmann–Wigner equations, but one *can* derive the Maxwell set 1(149a), 1(149b) (see Problem 11).

respectively. Note that the extra constraint 1(112) can no longer be derived when $\mu = 0$; it can however be *imposed* by selecting a special *gauge* (see below).

In terms of the three-dimensional notation introduced in 1(121), the field equations take the form

$$\mathbf{B} = \mathbf{curl\ A} \tag{1(147a)}$$

$$\mathbf{E} = -\mathbf{grad}\ A_0 - \partial_t \mathbf{A} \tag{1(147b)}$$

$$\mathbf{curl\ B} - \partial_t \mathbf{E} = 0 \tag{1(147c)}$$

$$\text{div}\ \mathbf{E} = 0 \tag{1(147d)}$$

or, equivalently

$$\text{div}\ \mathbf{B} = 0 \tag{1(148a)}$$

$$\mathbf{curl\ E} + \partial_t \mathbf{B} = 0 \tag{1(148b)}$$

$$\mathbf{curl\ B} - \partial_t \mathbf{E} = 0 \tag{1(148c)}$$

$$\text{div}\ \mathbf{E} = 0 \tag{1(148d)}$$

where the vector potential no longer appears. Equations 1(148a), 1(148b), 1(148c), 1(148d) are the equations originally written down by Maxwell; they may be written in the covariant notation

$$\partial_\lambda \hat{F}_{\lambda\nu} = 0 \tag{1(149a)}$$

$$\partial_\lambda F_{\lambda\nu} = 0 \tag{1(149b)}$$

where $\hat{F}_{\lambda\nu}$ is the dual tensor

$$\hat{F}_{\lambda\nu} = \tfrac{1}{2}\varepsilon_{\lambda\nu\rho\sigma} F_{\rho\sigma} \tag{1(150)}$$

Gauges The vector field acquires an added degree of complexity in the massless case, due to the phenomenon of *gauge invariance*. Indeed the field equations 1(145a) and 1(145b) are invariant under the substitution

$$A'_\mu(x) = A_\mu(x) - \partial_\mu \phi(x) \tag{1(151)}$$

where $\phi(x)$ is an arbitrary Lorentz scalar. The transformations 1(151) are known as gauge transformations with different choices of A_μ representing different gauges. Of particular importance are the *Lorentz* gauge characterized by the Lorentz condition

$$\partial_\mu A_\mu = 0 \tag{1(152)}$$

and the *radiation* gauge characterized by

$$\operatorname{div} \mathbf{A} = 0 \qquad 1(153a)$$

$$A_0 = 0 \qquad 1(153b)$$

To obtain the Lorentz gauge we note that if a given A_μ does not satisfy the Lorentz condition 1(152), then the regauged potential $A'_\mu = A_\mu - \partial_\mu \phi$ with $\Box \phi = \partial_\mu A_\mu$ will satisfy it. There are, of course, many Lorentz gauges obtained by regauging A'_μ by means of ϕ's satisfying $\Box \phi = 0$. In any Lorentz gauge the wave equation 1(146) for A_μ reduces simply to

$$\Box A_\mu = 0 \qquad 1(154)$$

Though manifestly covariant, the Lorentz gauge is rather uneconomical, in that it fails to eliminate all but the essential degrees of freedom of the electromagnetic field. For the massive vector field we have seen the essential degrees of freedom are given by $\mathbf{A}(x)$ and $\mathbf{E}(x)$, the three degrees of freedom of the vector $\mathbf{A}$ being connected to the three possible polarization states of a spin 1 particle. Since, as indicated in Section 1-4, the photon has only *two* possible polarization states, we should expect that, for $\mu = 0$, not all components of $\mathbf{A}$ and $\mathbf{E}$ represent independent dynamical degrees of freedom. Now an arbitrary vector field $\mathbf{V}(\mathbf{x})$ may be decomposed into transverse and longitudinal parts according to

$$\mathbf{V} = \mathbf{V}^t + \mathbf{V}^l \qquad 1(155)$$

with

$$\operatorname{div} \mathbf{V}^t = 0 \qquad 1(156a)$$

$$\mathbf{curl}\, \mathbf{V}^l = 0 \qquad 1(156b)$$

Explicitly

$$V_i^t = \left(\delta_{ij} - \frac{\partial_i \partial_j}{\nabla^2}\right) V_j \qquad 1(157a)$$

$$V_i^l = \frac{\partial_i \partial_j}{\nabla^2} V_j = \frac{\partial_i}{\nabla^2} \operatorname{div} \mathbf{V} \qquad 1(157b)$$

Here the symbol ∇^{-2} represents the operation

$$\frac{1}{\nabla^2} \mathbf{V}(\mathbf{x}) = \int D(\mathbf{x} - \mathbf{x}') \mathbf{V}(\mathbf{x}')\, d^3 x' \qquad 1(158)$$

with

$$\nabla^2 D(\mathbf{x}) = \delta^{(3)}(\mathbf{x}) \qquad 1(159a)$$

or explicitly

$$D(\mathbf{x}) = -\frac{1}{4\pi|\mathbf{x}|} \qquad 1(159)$$

Let us examine the longitudinal and transverse parts of **E** and **A**. By the Maxwell equation 1(147d) we see that the longitudinal part of **E** vanishes identically, that is, we have

$$\mathbf{E} = \mathbf{E}^t \qquad 1(160)$$

Moreover, the longitudinal part of **A** may be gauged away by selecting the radiation gauge: if **A** does not satisfy the condition div **A** = 0, then the regauged potential $\mathbf{A}' = \mathbf{A} - \nabla\phi$ with $\nabla^2\phi = \text{div}\,\mathbf{A}$ will satisfy it. When div **A** = 0, the equations of motion 1(146) for A_ν take the form

$$\Box\mathbf{A} - \mathbf{grad}\,\partial_t A_0 = 0 \qquad 1(161a)$$

$$\nabla^2 A_0 = 0 \qquad 1(161b)$$

The second equation yields 1(153b). Thus in the radiation gauge $\mathbf{A}^l$ and A_0 are completely eliminated from the picture and only the two transverse degrees of freedom represented by $\mathbf{A}^t$ remain. The equation of motion for the transverse potential is simply

$$\Box\mathbf{A}^t = 0 \qquad 1(162)$$

from 1(161a). The electric field $\mathbf{E}(x) = \mathbf{E}^t(x)$ is given in terms of $\mathbf{A}^t(x)$ by 1(147b), i.e.

$$\mathbf{E}^t(x) = -\dot{\mathbf{A}}^t(x) \qquad 1(163)$$

Plane Wave Solutions The physical nature of the electromagnetic field is best exhibited in the radiation gauge. In this gauge, a complete set of plane waves for the vector potential $\mathbf{A}^t$ is given by

$$\frac{1}{\sqrt{V}}\frac{1}{\sqrt{2|\mathbf{k}|}}\boldsymbol{\varepsilon}_{\mathbf{k}\lambda}\, e^{ik.x} \qquad (\lambda = 1, 2) \qquad 1(164a)$$

and

$$\frac{1}{\sqrt{V}}\frac{1}{\sqrt{2|\mathbf{k}|}}\boldsymbol{\varepsilon}_{\mathbf{k}\lambda}\, e^{-ik.x} \qquad (\lambda = 1, 2) \qquad 1(164b)$$

with $k_0 = |\mathbf{k}|$ and

$$\mathbf{k}\,.\,\boldsymbol{\varepsilon}_{\mathbf{k}\lambda} = 0 \qquad 1(165)$$

The positive energy solutions are normalized according to 1(144). The transversality condition 1(165) is required in order to satisfy the constraint div $\mathbf{A}(x) = 0$, or $\mathbf{A}(x) = \mathbf{A}^t(x)$. Thus, in the radiation gauge, the longitudinal states $\boldsymbol{\varepsilon}_{\mathbf{k}3}$ associated with the $s = 0$ eigenvalue of helicity are completely suppressed, in line with the remarks following 1(154). By virtue of 1(136b), the spin sum $\Sigma_{\lambda=1}^{2}\varepsilon_{\mathbf{k}\lambda}^{i}\varepsilon_{\mathbf{k}\lambda}^{j}$ is given by

$$\sum_{\lambda=1}^{2} \varepsilon_{\mathbf{k}\lambda}^{i}\varepsilon_{\mathbf{k}\lambda}^{j} = \delta^{ij} - \frac{k^i k^j}{\mathbf{k}^2} \qquad 1(166)$$

The radiation gauge is 3-dimensional in structure but it is formally possible to write the spin sum 1(166) in 4-dimensional notation by extending $\boldsymbol{\varepsilon}_{\mathbf{k}1}$ and $\boldsymbol{\varepsilon}_{\mathbf{k}2}$ into 4-vectors

$$e_{\mathbf{k}1} = (\boldsymbol{\varepsilon}_{\mathbf{k}1}, 0) \qquad 1(167)$$

$$e_{\mathbf{k}2} = (\boldsymbol{\varepsilon}_{\mathbf{k}2}, 0) \qquad 1(168)$$

To get a completeness relation, we must form a quartet of orthonormal 4-vectors. We therefore add the longitudinal vector $e_{\mathbf{k}3} = \left(\frac{\mathbf{k}}{|\mathbf{k}|}, 0\right)$ or, in an arbitrary frame

$$e_{\mathbf{k}3} = -\frac{k + \eta(k\,.\,\eta)}{k\,.\,\eta} \qquad 1(169)$$

where $k = (\mathbf{k}, i|\mathbf{k}|)$ and where we have introduced a timelike unit vector η which reduces to

$$\eta = (0, 0, 0, i)$$

in the Lorentz frame in which $e_{\mathbf{k}1}$ and $e_{\mathbf{k}2}$ have the form 1(167) and 1(168). To complete the quartet, we add

$$e_{\mathbf{k}4} = i\eta \qquad 1(170)$$

By construction $e_{\mathbf{k}3}$ and $e_{\mathbf{k}4}$ are orthogonal to each other and to $e_{\mathbf{k}1}$ and $e_{\mathbf{k}2}$. Moreover, all four vectors $e_{\mathbf{k}\lambda}$ ($\lambda = 1 \ldots 4$) are normalized to 1. Thus, we have

$$e_{\mathbf{k}\lambda}\,.\,e_{\mathbf{k}\lambda'} = \delta_{\lambda\lambda'} \qquad (\lambda, \lambda' = 1 \ldots 4) \qquad 1(171)$$

and the completeness relation

$$\sum_{\lambda=1}^{4} e_{\mathbf{k}\lambda}^{\alpha} e_{\mathbf{k}\lambda}^{\beta} = \delta^{\alpha\beta} \qquad 1(172)$$

From 1(172), we derive the equality

$$\sum_{\lambda=1}^{2} e^{\alpha}_{\mathbf{k}\lambda} e^{\beta}_{\mathbf{k}\lambda} = \delta^{\alpha\beta} - \frac{k^{\alpha} k^{\beta}}{(k \cdot \eta)^2} - \frac{\eta^{\alpha} k^{\beta} + k^{\alpha} \eta^{\beta}}{(k \cdot \eta)} \qquad 1(173)$$

which represents the 4-dimensional form of 1(166). In contrast to the corresponding expression 1(141) for the massive vector case, 1(173) is *not* Lorentz covariant, owing to the explicit dependence of the right-hand side on the timelike vector η. The latter depends on the frame to which the radiation gauge is referred.

Since the real dynamical degrees of freedom of the electromagnetic field are those of the transverse radiation field, additional, unphysical, degrees of freedom are necessarily introduced if one attempts to maintain manifest Lorentz covariance. Thus, in the Lorentz gauge, a complete set of solutions to 1(154), subject to the subsidiary Lorentz condition 1(152), is given by

$$\frac{1}{\sqrt{V}} \frac{1}{\sqrt{2|\mathbf{k}|}} \varepsilon_{\mathbf{k}\lambda} \, \mathrm{e}^{\pm ik.x} \qquad (\lambda = 1, 2, 3) \qquad 1(174)$$

where the polarization vectors $\varepsilon_{\mathbf{k}\lambda}$ satisfy the covariant constraint

$$k \cdot \varepsilon_{\mathbf{k}\lambda} = 0 \qquad 1(175)$$

Two polarization vectors, say $\varepsilon_{\mathbf{k}1}$ and $\varepsilon_{\mathbf{k}2}$, can be identified with 1(167) and 1(168), but it is impossible to construct a third vector *of finite norm* satisfying 1(175)*. The only candidate is

$$\varepsilon^{\mu}_{\mathbf{k}\lambda} \sim k^{\mu} \qquad 1(176)$$

which satisfies

$$k \cdot \varepsilon_{\mathbf{k}3} = 0 \qquad 1(177a)$$

and

$$\varepsilon_{\mathbf{k}3} \cdot \varepsilon_{\mathbf{k}3} = 0 \qquad 1(177b)$$

since $k^2 = 0$. The corresponding solutions

$$\frac{1}{\sqrt{V}} \frac{1}{\sqrt{2|\mathbf{k}|}} \varepsilon_{\mathbf{k}3} \, \mathrm{e}^{\pm ik.x} \qquad 1(178)$$

have zero norm in Hilbert space and are therefore unobservable. This is to be expected since the states corresponding to 1(176) can be gauged

* Note that when $\mu = 0$ the set of vectors 1(133) can no longer be used as solutions to 1(175).

away by a transformation of the type 1(151) or equivalently $\varepsilon_{\mathbf{k}}^{\mu'} = \varepsilon_{\mathbf{k}}^{\mu} - ak^{\mu}$ where a is a suitable constant.

1-7 Rarita–Schwinger field

Completely Symmetric Third Rank Spinor For $s = \frac{3}{2}$, the Bargmann–Wigner equations reduce to the following set of three equations for the symmetric rank 3 spinor $\psi_{\alpha\beta\gamma}$

$$(\gamma \,.\, \partial + m)_{\alpha\alpha'}\psi_{\alpha'\beta\gamma}(x) = 0 \tag{1(179a)}$$

$$(\gamma \,.\, \partial + m)_{\beta\beta'}\psi_{\alpha\beta'\gamma}(x) = 0 \tag{1(179b)}$$

$$(\gamma \,.\, \partial + m)_{\gamma\gamma'}\psi_{\alpha\beta\gamma'}(x) = 0 \tag{1(179c)}$$

We seek to write $\psi_{\alpha\beta\gamma}(x)$ in a form analogous to 1(109) (Salam, 1965). Taking into account the symmetry requirement in the first two indices, we write

$$\psi_{\alpha\beta\gamma}(x) = (\gamma_\mu C)_{\alpha\beta}\psi^{\mu}_{\gamma}(x) + \tfrac{1}{2}(\Sigma_{\mu\nu}C)_{\alpha\beta}\psi^{\mu\nu}_{\gamma}(x) \tag{1(180)}$$

where ψ^{μ}_{γ} is a vector-spinor, i.e. transforms as the product of a four-vector and Dirac spinor. $\psi^{\mu\nu}_{\gamma}$ transforms as the product of an antisymmetric second-rank tensor and spinor. Total symmetry in all three indices will be ensured by requiring that the contractions of $\psi_{\alpha\beta\gamma}$ with the three independent antisymmetric second rank spinors $C^{-1}_{\beta\gamma}$, $(iC^{-1}\gamma_5)_{\beta\gamma}$ and $(iC^{-1}\gamma_5\gamma_\lambda)_{\beta\gamma}$ vanish. This yields the three conditions

$$\gamma_\mu\psi_\mu + \tfrac{1}{2}\Sigma_{\mu\nu}\psi_{\mu\nu} = 0 \tag{1(181a)}$$

$$\gamma_\mu\gamma_5\psi_\mu + \tfrac{1}{2}\Sigma_{\mu\nu}\gamma_5\psi_{\mu\nu} = 0 \tag{1(181b)}$$

$$\gamma_\mu\gamma_5\gamma_\lambda\psi_\mu + \tfrac{1}{2}\Sigma_{\mu\nu}\gamma_5\gamma_\lambda\psi_{\mu\nu} = 0 \tag{1(181c)}$$

where each term carries a single suppressed spinor index; for example $\gamma_\mu\psi_\mu \equiv (\gamma_\mu)_{\alpha\gamma}\psi^{\mu}_{\gamma}$.

We can replace the first two conditions by the equivalent set

$$\gamma_\mu\psi_\mu = 0 \tag{1(182a)}$$

$$\Sigma_{\mu\nu}\psi_{\mu\nu} = 0 \tag{1(182b)}$$

by multiplying 1(181c) by γ_5 and taking the sum and difference with 1(181b). Using 1(182a), the third condition can be transformed to

$$4\psi_\lambda - \Sigma_{\mu\nu}\gamma_\lambda\psi_{\mu\nu} = 0$$

which can be further simplified by using

$$[\Sigma_{\mu\nu}, \gamma_\lambda] = 2i\delta_{\mu\lambda}\gamma_\nu - 2i\delta_{\nu\lambda}\gamma_\mu$$

and the condition 1(182b). We obtain

$$\psi_\lambda + i\gamma_\mu \psi_{\mu\lambda} = 0 \tag{1(182c)}$$

We now note that 1(182b) follows as a consequence of 1(182a) and 1(182c), so that there are only two independent conditions, namely

$$\gamma_\mu \psi_\mu = 0 \tag{1(183a)}$$

$$\gamma_\mu \psi_{\mu\nu} - i\psi_\nu = 0 \tag{1(183b)}$$

for 1(180) to represent a fully symmetric third rank spinor. Together, 1(183a) and 1(183b) represent a total of $4+16 = 20$ constraints on the $16+24 = 40$ components of ψ_μ and $\psi_{\lambda\nu}$, bringing the number of independent components down to 20; this is the correct number of components for a completely symmetric third rank tensor in a four-dimensional vector space*.

Rarita–Schwinger Equations We now apply the Bargmann–Wigner equations to the expression 1(180). The first equation 1(179a) reads

$$(\gamma . \partial + m)_{\alpha\alpha'}(\gamma_\mu C)_{\alpha'\beta}\psi^\mu_\gamma(x) + \tfrac{1}{2}(\gamma . \partial + m)_{\alpha\alpha'}(\Sigma_{\mu\nu}C)_{\alpha'\beta}\psi^{\mu\nu}_\gamma(x) = 0$$

Let us contract this equation with $(C^{-1}\Sigma_{\lambda\rho})_{\beta\alpha}$. To evaluate the traces which appear, we apply the formulae†

$$Tr\gamma_\nu\gamma_\mu\Sigma_{\lambda\rho} = 4i(\delta_{\nu\lambda}\delta_{\mu\rho} - \delta_{\mu\lambda}\delta_{\nu\rho})$$

$$Tr\Sigma_{\mu\nu}\Sigma_{\lambda\rho} = 4(\delta_{\mu\lambda}\delta_{\nu\rho} - \delta_{\nu\lambda}\delta_{\mu\rho})$$

to obtain the result

$$im\psi_{\lambda\rho} = \partial_\lambda \psi_\rho - \partial_\rho \psi_\lambda \tag{1(184)}$$

which determines $\psi_{\lambda\rho}$ in terms of ψ_μ. For 1(184) to be compatible with the symmetry conditions 1(183a) and 1(183b) it follows that

$$(\gamma . \partial + m)\psi_\lambda(x) = 0 \tag{1(185)}$$

Thus each component of ψ_λ must satisfy the Dirac equation. The same result may be derived directly from the Bargmann–Wigner equation 1(179c) which, when applied to 1(180) reads

$$(\gamma_\mu C)_{\alpha\beta}[(\gamma . \partial + m)\psi_\mu]_\gamma + \tfrac{1}{2}(\Sigma_{\mu\nu}C)_{\alpha\beta}[(\gamma . \partial + m)\psi_{\mu\nu}]_\gamma = 0$$

* The general formula is $\binom{n+r-1}{r}$ where n is the dimensionality of the vector space and r is the rank of the completely symmetric tensor.

† See the theorems on traces of γ-matrices proved in Section 6-4.

Equation 1(185) then follows upon contracting the above equation with $(C^{-1}\gamma_\lambda)_{\beta\alpha}$.

Hence, the components of the tensor–spinor $\psi_{\mu\nu}$ can be regarded as derived variables, defined in terms of ψ_λ by 1(184). The spin $\frac{3}{2}$ field may be described entirely in terms of the vector–spinor ψ_λ satisfying 1(185) and 1(183a); there are no further independent equations.

The two equations

$$(\gamma \,.\, \partial + m)\psi_\lambda(x) = 0 \tag{1(186a)}$$

$$\gamma_\lambda \psi_\lambda(x) = 0 \tag{1(186b)}$$

are known as the Rarita–Schwinger equations (Rarita, 1941). They yield the further constraint

$$\partial_\lambda \psi_\lambda(x) = 0 \tag{1(187)}$$

since, from 1(186a) and 1(186b) we have

$$0 = \gamma_\lambda \gamma_\nu \partial_\nu \psi_\lambda = (2\delta_{\lambda\nu} - \gamma_\nu \gamma_\lambda)\partial_\nu \psi_\lambda = 2\partial_\lambda \psi_\lambda$$

Moreover, one can combine the two equations 1(186a) and 1(186b) into the single equation of motion

$$-[(\gamma \,.\, \partial + m)\delta_{\mu\lambda} - \tfrac{1}{3}(\gamma_\mu \partial_\lambda + \gamma_\lambda \partial_\mu) + \tfrac{1}{3}\gamma_\mu(\gamma \,.\, \partial - m)\gamma_\lambda]\psi_\lambda(x) = 0 \tag{1(188)}$$

Indeed, multiplying 1(188) on the left by γ_μ and ∂_μ we find

$$\tfrac{2}{3}\partial_\lambda \psi_\lambda - \tfrac{1}{3}m\gamma_\lambda \psi_\lambda = 0 \tag{1(189a)}$$

and

$$m\partial_\lambda \psi_\lambda + \gamma \,.\, \partial(\tfrac{2}{3}\partial_\lambda \psi_\lambda - \tfrac{1}{3}m\gamma_\lambda \psi_\lambda) = 0 \tag{1(189b)}$$

respectively. From 1(189a) and 1(189b) we deduce that $\partial_\lambda \psi_\lambda = \gamma_\lambda \psi_\lambda = 0$ and from 1(188) we recover 1(186a).

Lorentz Invariance The invariance of the Rarita–Schwinger equation 1(188) under a Lorentz transformation

$$x'_\mu = x_\mu + \omega_{\mu\nu} x_\nu$$

requires $\psi_\mu(x)$ to undergo the vector–spinor transformation

$$\psi'_\mu(x') = \psi_\mu(x) + \omega_{\mu\nu}\psi_\nu(x) + \frac{i}{4}\omega_{\rho\sigma}\Sigma_{\rho\sigma}\psi_\mu(x)$$

or, in analogy with 1(75) and 1(119)

$$\psi'(x') = \psi(x) + \frac{i}{2}\omega_{\rho\sigma}(S_{\rho\sigma} + \tfrac{1}{2}\Sigma_{\rho\sigma})\psi(x)$$

where $S_{\rho\sigma}$ and $\Sigma_{\rho\sigma}$ operate on the vector and spinor indices of ψ respectively. $S_{\rho\sigma}+\frac{1}{2}\Sigma_{\rho\sigma}$ is the spin tensor of the Rarita–Schwinger theory.

Conserved Current Setting

$$\psi_\lambda = (\psi_1, \psi_2, \psi_3, i\psi_0) \tag{1(190a)}$$

we define the hermitian conjugate of ψ_λ with respect to the spinor indices, by

$$\psi^\dagger{}_\lambda = (\psi_1{}^\dagger, \psi_2{}^\dagger, \psi_3{}^\dagger, i\psi_0{}^\dagger) \tag{1(190b)}$$

As in 1(114) we have taken care *not* to change the sign of the imaginary unit in the metric. The adjoint field is defined in analogy with 1(31) by

$$\bar{\psi}_\lambda = \psi^\dagger{}_\lambda \gamma_4 \tag{1(191)}$$

With this convention, the adjoint field satisfies the equations

$$\bar{\psi}_\lambda(x)(\gamma \,.\, \overleftarrow{\partial} - m) = 0 \tag{1(192a)}$$

$$\bar{\psi}_\lambda(x)\gamma_\lambda = 0 \tag{1(192b)}$$

or, combining 1(192a) and 1(192b)

$$-\bar{\psi}_\lambda(x)[(\gamma \,.\, \overleftarrow{\partial} - m)\delta_{\lambda\mu} - \tfrac{1}{3}(\gamma_\lambda \overleftarrow{\partial}_\mu + \gamma_\mu \overleftarrow{\partial}_\lambda) + \tfrac{1}{3}\gamma_\lambda(\gamma \,.\, \overleftarrow{\partial} + m)\gamma_\mu] = 0 \tag{1(193)}$$

From 1(188) and 1(193) one can derive the current conservation law

$$\partial_\mu j_\mu(x) = 0 \tag{1(194a)}$$

$$\begin{aligned} j_\mu &= i(\bar{\psi}_\lambda \gamma_\mu \psi_\lambda - \tfrac{1}{3}\bar{\psi}_\lambda \gamma_\lambda \psi_\mu - \tfrac{1}{3}\bar{\psi}_\mu \gamma_\lambda \psi_\lambda + \tfrac{1}{3}\bar{\psi}_\lambda \gamma_\lambda \gamma_\mu \gamma_\nu \psi_\nu) \\ &= i\bar{\psi}_\lambda \gamma_\mu \psi_\lambda \end{aligned} \tag{1(194b)}$$

by application of the standard procedure. The total charge is

$$\begin{aligned} -i \int j_4(x)\, d^3x &= \int \psi^\dagger{}_\lambda(x)\psi_\lambda(x)\, d^3x \\ &= \int (\psi_1{}^\dagger \psi_1 + \psi_2{}^\dagger \psi_2 + \psi_3{}^\dagger \psi_3 - \psi_0{}^\dagger \psi_0)\, d^3x \end{aligned} \tag{1(195)}$$

Although the charge density is not a positive-definite form, the charge is necessarily positive, since ψ_0 vanishes in the rest system, as we shall see presently.

Plane Wave Solutions We seek solutions to the Rarita–Schwinger equations of the form

$$\psi^\lambda(x) \sim w^\lambda(\mathbf{k},\quad E)\, e^{i\mathbf{k}.\mathbf{x} - iEt}$$

Solutions exist for both positive and negative energies. For positive energies $E = E_{\mathbf{k}}$ we have, setting $w^{\lambda}(\mathbf{k}, E_{\mathbf{k}}) = u^{\lambda}_{\mathbf{k}}$,

$$(\gamma \,.\, k - im)u^{\lambda}_{\mathbf{k}} = 0 \qquad 1(196a)$$

$$\gamma^{\lambda}u^{\lambda}_{\mathbf{k}} = 0 \qquad 1(196b)$$

Let us examine these equations in the rest frame. For $\mathbf{k} = 0$, 1(196a) reduces to

$$\gamma_4 u^{\lambda} = u^{\lambda} \qquad (\lambda = 1 \ldots 4) \qquad 1(197a)$$

indicating that only the top two components of u^{λ} survive as in 1(50). The subsidiary condition 1(196b) is then

$$u^4 + \gamma^i u^i = 0 \qquad 1(197b)$$

On the other hand, multiplying 1(196b) on the left by γ^4 and using 1(197a), we get

$$u^4 - \gamma^i u^i = 0$$

so that in the rest frame we get $u^0 = 0$ (confirming our statement following 1(195)*) and $\gamma^i u^i = 0$ or

$$\boldsymbol{\sigma} \,.\, \mathbf{u} = 0 \qquad 1(198)$$

Thus, in the rest frame, we need only consider the vector–spinor $\mathbf{u}$. The condition 1(198) provides two constraints on the six non-zero components of $\mathbf{u}$, leaving $4 = 2 \times \frac{3}{2} + 1$ independent components as desired. A set of four orthogonal solutions can be built up from the eigenstates

$$\boldsymbol{\varepsilon}_1 = \frac{1}{\sqrt{2}}\begin{pmatrix} 1 \\ i \\ 0 \end{pmatrix} \qquad \boldsymbol{\varepsilon}_2 = \frac{1}{\sqrt{2}}\begin{pmatrix} 1 \\ -i \\ 0 \end{pmatrix} \qquad \boldsymbol{\varepsilon}_3 = \begin{pmatrix} 0 \\ 0 \\ 1 \end{pmatrix}$$

of S_z with eigenvalues $+1$, -1 and 0 respectively. From these we construct the four vector-spinors

$$\mathbf{u}_1 = \boldsymbol{\varepsilon}_1 u_+ \qquad 1(200a)$$

$$\mathbf{u}_2 = \frac{1}{\sqrt{3}}\boldsymbol{\varepsilon}_1 u_- - \sqrt{\tfrac{2}{3}}\boldsymbol{\varepsilon}_3 u_+ \qquad 1(200b)$$

$$\mathbf{u}_3 = \frac{1}{\sqrt{3}}\boldsymbol{\varepsilon}_2 u_+ + \sqrt{\tfrac{2}{3}}\boldsymbol{\varepsilon}_3 u_- \qquad 1(200c)$$

$$\mathbf{u}_4 = \boldsymbol{\varepsilon}_2 u_- \qquad 1(200d)$$

* A similar argument shows that $\psi^0 = 0$ also for negative energy solutions in the rest frame.

which are eigenstates of $S_z+\frac{1}{2}\sigma_z$ with eigenvalues $+\frac{3}{2}$, $\frac{1}{2}$, $-\frac{1}{2}$ and $-\frac{3}{2}$ respectively. The solutions 1(200) are orthogonal, normalized, and satisfy the subsidiary condition 1(198). For $\mathbf{u}_1$ and $\mathbf{u}_2$, for example, the subsidiary condition yields the relations

$$(\sigma_1+i\sigma_2)u_+ = 0$$

and

$$\tfrac{1}{2}(\sigma_1+i\sigma_2)u_- - \sigma_3 u_+ = 0$$

which are readily seen to be satisfied. The coefficients in 1(200) are just the Clebsch–Gordon coefficients for coupling spin 1 and $\frac{1}{2}$ to yield spin $\frac{3}{2}$. This is in fact the meaning of the subsidiary condition $\boldsymbol{\sigma}\,.\,\mathbf{u} = 0$; it ensures that $\mathbf{u}$ is an eigenstate of the total spin operator $(\mathbf{S}+\frac{1}{2}\boldsymbol{\sigma})^2$ with eigenvalue $\frac{3}{2}(\frac{3}{2}+1) = \frac{15}{4}$ (see Problem 13). In general, the subsidiary condition 1(196b) serves to project out that part of the vector–spinor u^λ which refers to spin $\frac{3}{2}$*.

The same principle can be applied to construct solutions for finite $\mathbf{k}$ but, in this case we must use polarization *four*-vectors which satisfy $k\,.\,\varepsilon_{\mathbf{k}} = 0$, as in 1(139) for example. If the z-axis points in the direction of $\mathbf{k}$, the four independent solutions for positive energy are easily seen to be

$$\begin{aligned}
\mathbf{u}_{\mathbf{k}1} &= \boldsymbol{\varepsilon}_1 u_{\mathbf{k}1} & u^4_{\mathbf{k}1} &= 0\\
\mathbf{u}_{\mathbf{k}2} &= \frac{1}{\sqrt{3}}\boldsymbol{\varepsilon}_1 u_{\mathbf{k}2} - \sqrt{\frac{2}{3}\frac{E_{\mathbf{k}}}{m}}\,\boldsymbol{\varepsilon}_3 u_{\mathbf{k}1} & u^4_{\mathbf{k}2} &= -i\sqrt{\frac{2}{3}\frac{|\mathbf{k}|}{m}}\,u_{\mathbf{k}1}\\
\mathbf{u}_{\mathbf{k}3} &= \frac{1}{\sqrt{3}}\boldsymbol{\varepsilon}_2 u_{\mathbf{k}1} + \sqrt{\frac{2}{3}\frac{E_{\mathbf{k}}}{m}}\,\boldsymbol{\varepsilon}_3 u_{\mathbf{k}2} & u^4_{\mathbf{k}3} &= i\sqrt{\frac{2}{3}\frac{|\mathbf{k}|}{m}}\,u_{\mathbf{k}2}\\
\mathbf{u}_{\mathbf{k}4} &= \boldsymbol{\varepsilon}_2 u_{\mathbf{k}2} & u^4_{\mathbf{k}4} &= 0
\end{aligned} \qquad 1(201)$$

where $\boldsymbol{\varepsilon}_1$, $\boldsymbol{\varepsilon}_2$ and $\boldsymbol{\varepsilon}_3$ are given by 1(199) and where $u_{\mathbf{k}1}$ and $u_{\mathbf{k}2}$ are the Dirac spinors 1(51a) for $\mathbf{k} = (0, 0, |\mathbf{k}|)$ (Kusaka, 1941). The same result holds for arbitrary $\mathbf{k}$ if $u_{\mathbf{k}1}$, and $u_{\mathbf{k}2}$ are replaced by the corresponding eigenstates of $\boldsymbol{\Sigma}\,.\,\mathbf{k}/|\mathbf{k}|$ (as in the neutrino case), and the ε_i $(i = 1, 2, 3)$ by the corresponding eigenstates of $\mathbf{S}\,.\,\mathbf{k}/|\mathbf{k}|$. In this way a complete set of solutions

$$\frac{1}{\sqrt{V}}\sqrt{\frac{m}{E_{\mathbf{k}}}}\,u^\lambda_{\mathbf{k}\sigma}\,e^{ik.x} \qquad \sigma = 1\ldots 4 \qquad 1(202a)$$

* The isolation of the spin $\frac{3}{2}$ part of ψ_μ from its spin $\frac{1}{2}$ part is the chief difficulty in setting up the Hamiltonian formulation of the Rarita–Schwinger theory (see Moldauer, 1956).

and

$$\frac{1}{\sqrt{V}}\sqrt{\frac{m}{E_{\mathbf{k}}}}v^{\lambda}_{\mathbf{k}\sigma}\,e^{-ik.x} \qquad \sigma = 1\ldots 4 \qquad 1(202b)$$

can be constructed, where the vector–spinors $u^{\lambda}_{\mathbf{k}\sigma}$ and $v^{\lambda}_{\mathbf{k}\sigma}$ satisfy the orthogonality and normalization conditions

$$\Sigma_{\lambda}\bar{u}^{\lambda}_{\mathbf{k}\sigma}u^{\lambda}_{\mathbf{k}\sigma'} = \delta_{\sigma\sigma'} \qquad 1(202a)$$

$$\Sigma_{\lambda}\bar{v}^{\lambda}_{\mathbf{k}\sigma}v^{\lambda}_{\mathbf{k}\sigma'} = -\delta_{\sigma\sigma'} \qquad 1(202b)$$

$$\Sigma_{\lambda}\bar{u}^{\lambda}_{\mathbf{k}\sigma}v^{\lambda}_{\mathbf{k}\sigma'} = \Sigma_{\lambda}\bar{v}^{\lambda}_{\mathbf{k}\sigma}u^{\lambda}_{\mathbf{k}\sigma'} = 0 \qquad 1(202c)$$

$$\Sigma_{\lambda}\bar{u}^{\lambda}_{\mathbf{k}\sigma}\gamma_4 v^{\lambda}_{-\mathbf{k}\sigma'} = \Sigma_{\lambda}\bar{v}^{\lambda}_{\mathbf{k}\sigma}\gamma_4 u^{\lambda}_{-\mathbf{k}\sigma'} = 0 \qquad 1(202d)$$

To evaluate the spin sums

$$P^{\mu\nu}_{+}(\mathbf{k}) = \sum_{\sigma} u^{\mu}_{\mathbf{k}\sigma}\bar{u}^{\nu}_{\mathbf{k}\sigma} \quad \text{and} \quad P^{\mu\nu}_{-}(\mathbf{k}) = -\sum_{\sigma} v^{\mu}_{\mathbf{k}\sigma}\bar{v}^{\nu}_{\mathbf{k}\sigma}$$

we first note that on grounds of Lorentz invariance they must be linear combinations of the following ten tensors

$$\delta_{\mu\nu} \qquad \gamma_{\mu}\gamma_{\nu} \qquad \frac{1}{m}\gamma_{\mu}k_{\nu} \qquad \frac{1}{m}\gamma_{\nu}k_{\mu} \qquad \frac{1}{m^2}k_{\mu}k_{\nu}$$

$$\frac{\gamma\,.\,k}{m}\delta_{\mu\nu} \quad \frac{\gamma\,.\,k}{m}\gamma_{\mu}\gamma_{\nu} \quad \frac{\gamma\,.\,k}{m}\gamma_{\mu}k_{\nu} \quad \frac{\gamma\,.\,k}{m}\gamma_{\nu}k_{\mu} \quad \frac{\gamma\,.\,k}{m^3}k_{\mu}k_{\nu} \qquad 1(203)$$

with coefficients which depend only on $k^2 = -m^2$. No other terms with the correct tensor properties can be constructed. Moreover, $P^{\mu\nu}_{\pm}$ must satisfy the conditions

$$k_{\mu}P^{\mu\nu}_{\pm} = P^{\mu\nu}_{\pm}k_{\nu} = 0 \qquad 1(204a)$$

$$\gamma_{\mu}P^{\mu\nu}_{\pm} = P^{\mu\nu}_{\pm}\gamma_{\nu} = 0 \qquad 1(204b)$$

$$P^{\mu\lambda}_{\pm}P^{\lambda\nu}_{\pm} = P^{\mu\nu}_{\pm} \qquad 1(204c)$$

$$P^{\mu\lambda}_{\pm}P^{\lambda\nu}_{\mp} = 0 \qquad 1(204d)$$

In the rest frame $k_{\mu} = (0, 0, 0, im)$, 1(204a) yields

$$mP^{4\nu}_{\pm} = mP^{\mu 4}_{\pm} = 0$$

so that $P^{\mu\nu}_{\pm}$ reduces to a three-dimensional tensor $P^{ij}_{\pm}$. By 1(203), $P^{ij}_{\pm}$ must be of the general form

$$P^{ij}_{\pm} = a\delta_{ij} + b\gamma_i\gamma_j + c\delta_{ij}\gamma_4 k_4 + d\gamma_i\gamma_j\gamma_4 k_4$$

with $k_4 = im$. The determination of the coefficients is straightforward and one finds

$$P^{ij}_{\pm} = \pm \frac{\gamma_4 k_4 \pm im}{2im} (\delta^{ij} - \tfrac{1}{3}\gamma^i \gamma^j) \qquad 1(205)$$

We can now generalize 1(205) to an arbitrary frame by noting that in the rest system

$$(\boldsymbol{\gamma}, 0) = \left(\delta_{\mu\nu} - \frac{k_\mu k_\nu}{k^2}\right)\gamma_\nu$$

$$\begin{pmatrix} 1 & & & \\ & 1 & & \\ & & 1 & \\ & & & 0 \end{pmatrix} = \left(\delta_{\mu\nu} - \frac{k_\mu k_\nu}{k^2}\right)$$

and

$$\gamma_4 k_4 = \gamma \, . \, k$$

The covariant form of the projection operators is therefore found to be

$$P^{\mu\nu}_{+}(\mathbf{k}) = \frac{\gamma \, . \, k + im}{2im}\left[\delta^{\mu\nu} - \frac{k^\mu k^\nu}{k^2} - \frac{1}{3}\left(\delta^{\mu\rho} - \frac{k^\mu k^\rho}{k^2}\right)\left(\delta^{\nu\sigma} - \frac{k^\nu k^\sigma}{k^2}\right)\gamma^\rho \gamma^\sigma\right]$$

$$= \frac{\gamma \, . \, k + im}{2im}\left[\delta^{\mu\nu} - \tfrac{1}{3}\gamma^\mu \gamma^\nu + \frac{i}{3m}(\gamma^\mu k^\nu - \gamma^\nu k^\mu) + \frac{2}{3m^2} k^\mu k^\nu\right] \qquad 1(206a)$$

and

$$P^{\mu\nu}_{-}(\mathbf{k}) = -\frac{\gamma \, . \, k - im}{2im}\left[\delta^{\mu\nu} - \tfrac{1}{3}\gamma^\mu \gamma^\nu - \frac{i}{3m}(\gamma^\mu k^\nu - \gamma^\nu k^\mu) + \frac{2}{3m^2} k^\mu k^\nu\right] \qquad 1(206b)$$

PROBLEMS

1. Check Equation 1(23).
2. Verify the normalization and orthogonality relations 1(54a), 1(54b) and 1(56a), 1(56b).
3. Establish the relations 1(58a), 1(58b), using only 1(52a), 1(52b) and 1(55a), 1(55b).
4. Exhibit the spin matrices $\Sigma_{\mu\nu}$ explicitly, using the representation 1(28). Setting $\psi = \begin{pmatrix} \varphi \\ \chi \end{pmatrix}$, express the transformation law 1(75) in terms of the two-component objects φ and χ. How does the result differ from ordinary three-dimensional spin transformations on φ and χ?

5. Verify explicitly that, under a Lorentz transformation, the bilinear quantities $\bar{\psi}\psi$, $\bar{\psi}\gamma_5\psi$, $\bar{\psi}\gamma_\mu\psi$, $\bar{\psi}\gamma_\mu\gamma_5\psi$ and $\bar{\psi}\Sigma_{\mu\nu}\psi$ transform as scalar, pseudoscalar, vector, pseudovector and second rank tensor respectively.
6. Construct a complete set of solutions to the *two-component* theory of the neutrino, using the representation 1(95a) for the γ-matrices.
7. Show that, in analogy with 1(75), the spin tensor 1(101) is the generator of infinitesimal Lorentz transformations on the Bargmann–Wigner multispinor $\psi_{\alpha\beta\gamma\ldots}(x)$.
8. Derive 1(124a) and 1(124b).
9. Prove the identity 1(125). [Hint: use Equation 1(120b).]
10. Check the equalities 1(127) and 1(128).
11. Derive the Maxwell equations 1(149a) and 1(149b) from the Bargmann–Wigner equations for zero rest mass.
12. Derive the current conservation law 1(194a), 1(194b).
13. If u_1, u_2 and u_3 are two-component spinors, show that the spinor–vector

$$\mathbf{u} = \begin{pmatrix} u_1 \\ u_2 \\ u_3 \end{pmatrix}$$

is an eigenstate of $(\mathbf{S}+\frac{1}{2}\boldsymbol{\sigma})^2$ with eigenvalue $\frac{15}{4}$ if it satisfies $\boldsymbol{\sigma}\,.\,\mathbf{u} = 0$. [Hint: use the relation $\varepsilon_{kij}\varepsilon_{krs} = \delta_{ir}\delta_{js}-\delta_{is}\delta_{jr}$.]
14. Derive the result 1(205).

REFERENCES

Bargmann, V. (1948) (with E. P. Wigner) *Proc. Nat. Acad. Sci.* (USA) **34**, 211
Dirac, P. A. M. (1928) *Proc. Roy. Soc.* **A 117**, 610
Dirac, P. A. M. (1936) *Proc. Roy. Soc.* **A 155**, 447
Feshbach, H. (1958) (with F. Villars) *Rev. Mod. Phys.* **30**, 24
Fierz, M. (1939a) *Helv. Phys. Acta* **12**, 3
Fierz, M. (1939b) (with W. Pauli) *Proc. Roy. Soc.* **A 173**, 211
Gordon, W. (1926) *Z., Phys.* **40**, 117, 121
Jacob, M. (1959) (with G. C. Wick) *Ann. Phys.* **7**, 404
Klein, O. (1926) *Z. Phys.* **37**, 895
Kusaka, S. (1941) *Phys. Rev.* **60**, 61
Lee, T. D. (1957) (with C. N. Yang) *Phys. Rev.* **105**, 1671
Lubanski, J. K. (1942) *Physica* **9**, 310
Moldauer, P. A. (1956) (with K. M. Case) *Phys. Rev.* **102**, 279
Pauli, W. (1934) (with V. Weisskopf) *Helv. Phys. Acta* **7**, 709
Rarita, W. (1941) (with J. Schwinger) *Phys. Rev.* **60**, 61
Salam, A. (1965) (with R. Delbourgo and J. Strathdee) *Proc. Roy. Soc.* **A 284**, 146

2

Lagrangian Field Theory

2-1 Hamilton's action principle

Canonical Coordinates for a Field To lay the basis for field quantization, we now reformulate classical field theories in the canonical language of Lagrangian and Hamiltonian mechanics. The essence of the method is to associate to a given field a Lagrangian, such that the equations of motion are recovered from Hamilton's variational principle.

We illustrate this in detail for the case of a real spin zero field

$$\phi(\mathbf{x}, t) = \phi^*(\mathbf{x}, t) \qquad 2(1)$$

satisfying

$$(\Box - \mu^2)\phi(\mathbf{x}, t) = 0 \qquad 2(2)$$

Let us regard the value of $\phi(\mathbf{x}, t)$ at each point $\mathbf{x}$ as an independent canonical coordinate. To deal with the continuously infinite number of canonical coordinates introduced in this picture, we begin by dividing 3-space into tiny cells of volume δV_i. The average value of $\phi(\mathbf{x}, t)$ in the ith cell will be denoted by $\phi_i(t)$ and the field Lagrangian which we seek will be a function

$$L(t) = L(\phi_i(t), \dot{\phi}_i(t))$$

of the canonical coordinates ϕ_i and their time derivatives $\dot{\phi}_i$. The key requirement is that the equation of motion 2(2) follow from Hamilton's variational principle

$$\delta W_{21} = \delta \int_{t_1}^{t_2} L(t)\, dt = 0 \qquad 2(3)$$

for variations $\delta\phi_i$ which vanish at t_1 and t_2 but are otherwise arbitrary.

The integral

$$W_{21} = \int_{t_1}^{t_2} L(t)\, dt$$

is known as the *action* integral. The variational principle 2(3) is referred to as the *principle of stationary action*; it states that the physical ' path ' followed by each coordinate ϕ_i from $\phi_i(t_1)$ to $\phi_i(t_2)$ is such that the action W_{21} remains unchanged under small variations $\delta\phi_i(t)$ with $\delta\phi_i(t_1) = \delta\phi_i(t_2) = 0$. From 2(3) one derives in the usual way the Euler–Lagrange equations

$$\partial_t \frac{\partial L(t)}{\partial \dot{\phi}_i} = \frac{\partial L(t)}{\partial \phi_i} \qquad 2(4)$$

by carrying out the variational derivatives and integrating by parts.* Thus $L(t)$ must be such that we recover 2(2) from the Euler–Lagrange equations in the continuum limit.

We stress that it is the values of the *field* which are regarded as the canonical coordinates. The space-time coordinates appear simply as parameters.

To pass from discrete to continuum notation, we introduce the concept of a functional and of functional differentiation. In the continuum limit the function of infinitely many variables

$$L(\phi_i(t), \dot{\phi}_i(t)) \qquad 2(5a)$$

becomes a *functional*

$$L[\phi(\mathbf{x}, t), \dot{\phi}(\mathbf{x}, t)] \qquad 2(5b)$$

of the *functions* $\phi(\mathbf{x}, t)$ and $\dot{\phi}(\mathbf{x}, t)$. Functional dependence will be denoted by square brackets. The characteristic feature of a functional $F[\varphi(\mathbf{x})]$ is that it depends on the value of a function $\varphi(\mathbf{x})$ not at any particular point $\mathbf{x}$ but rather on its values over a whole range. For example, 2(5b) depends on the values of the two functions ϕ and $\dot{\phi}$ over the whole of three-dimensional space. Let us now subject the values of $\varphi(\mathbf{x})$ to independent infinitesimal variations $\delta\varphi(\mathbf{x})$ at each point $\mathbf{x}$. The corresponding variation in the functional $F[\varphi]$ must be of the form

$$\delta F[\varphi] = F[\varphi + \delta\varphi] - F[\varphi] = \int \frac{\delta F[\varphi]}{\delta\varphi(\mathbf{x})} \delta\varphi(\mathbf{x})\, d^3x \qquad 2(6)$$

* See for example H. Goldstein, *Classical Mechanics*, Addison-Wesley, 1951. If higher order time derivatives of ϕ_i appear in L, 2(4) no longer applies, but we shall not need to consider this possibility.

where, by definition, $\delta F[\varphi]/\delta\varphi(\mathbf{x})$ is the *functional derivative* of $F[\varphi]$ with respect to the value of φ at the point $\mathbf{x}$. From the definition, one easily verifies that the functional derivative satisfies the usual properties associated with differentiation. If a, b and c are arbitrary functions, independent of $\varphi(\mathbf{x})$, one has

$$\frac{\delta}{\delta\varphi(\mathbf{x})} a = 0 \qquad 2(7a)$$

$$\frac{\delta}{\delta\varphi(\mathbf{x})}(aF[\varphi]+bG[\varphi]) = a\frac{\delta F[\varphi]}{\delta\varphi(\mathbf{x})}+b\frac{\delta G[\varphi]}{\delta\varphi(\mathbf{x})} \qquad 2(7b)$$

$$\frac{\delta}{\delta\varphi(\mathbf{x})}(F[\varphi]G[\varphi]) = \frac{\delta F[\varphi]}{\delta\varphi(\mathbf{x})}G[\varphi]+F[\varphi]\frac{\delta G[\varphi]}{\delta\varphi(\mathbf{x})} \qquad 2(7c)$$

and the rule for differentiating a functional of a functional

$$\frac{\delta}{\delta\varphi(\mathbf{x})}F[\eta[\varphi]] = \int \frac{\delta F}{\delta\eta(\mathbf{y})}\frac{\delta\eta(\mathbf{y})}{\delta\varphi(\mathbf{x})}\, d^3y \qquad 2(7d)$$

Extending 2(6) to a functional of two functions, we have, for $L(t)$

$$\delta L[\phi,\dot{\phi}] = \int \left(\frac{\delta L}{\delta\phi(\mathbf{x})}\delta\phi(\mathbf{x})+\frac{\delta L}{\delta\dot{\phi}(\mathbf{x})}\delta\dot{\phi}(\mathbf{x})\right) d^3x \qquad 2(8a)$$

On the other hand, in discrete notation

$$\begin{aligned}\delta L(\phi_i,\dot{\phi}_i) &= \sum_i \left(\frac{\partial L}{\partial\phi_i}\delta\phi_i+\frac{\partial L}{\partial\dot{\phi}_i}\delta\dot{\phi}_i\right)\\ &= \sum_i \left(\frac{1}{\delta V_i}\frac{\partial L}{\partial\phi_i}\delta\phi_i+\frac{1}{\delta V_i}\frac{\partial L}{\partial\dot{\phi}_i}\delta\dot{\phi}_i\right)\delta V_i \qquad 2(8b)\end{aligned}$$

Identifying 2(8a) with 2(8b) in the continuum limit we have, since variations at distinct points are independent of one another,

$$\frac{\delta L(t)}{\delta\phi(\mathbf{x},t)} = \lim_{\delta V_i\to 0}\frac{1}{\delta V_i}\frac{\partial L(t)}{\partial\phi_i(t)} \qquad 2(9a)$$

$$\frac{\delta L(t)}{\delta\dot{\phi}(\mathbf{x},t)} = \lim_{\delta V_i\to 0}\frac{1}{\delta V_i}\frac{\partial L(t)}{\partial\dot{\phi}_i(t)} \qquad 2(9b)$$

where $\mathbf{x}$ is in the ith cell. Thus the functional derivative $\delta L(t)/\delta\phi(\mathbf{x},t)$ is essentially proportional to the derivative of L with respect to the value of ϕ at the point $\mathbf{x}$. In continuum notation, the Euler–Lagrange equations

2(4) take the form

$$\partial_t \frac{\delta L}{\delta\dot{\phi}(\mathbf{x}, t)} = \frac{\delta L}{\delta\phi(\mathbf{x}, t)} \tag{2(10)}$$

Lagrangian Density The task of determining a suitable Lagrangian for the Klein–Gordon field is simplified by writing $L(t)$ as a sum of contributions from all points of 3-space:

$$L(t) = \int_V \mathscr{L}(\phi(\mathbf{x}, t), \dot{\phi}(\mathbf{x}, t), \nabla\phi(\mathbf{x}, t))\, d^3x \tag{2(11)}$$

where $\mathscr{L}(x)$, the *Lagrangian density*, is an ordinary function of the fields $\phi(\mathbf{x}, t)$ and their space and time derivatives, the latter being assumed to satisfy periodic boundary conditions at the surface of the normalization volume V. Note the appearance in $\mathscr{L}(x)$ of the space derivatives $\nabla\phi$ as well as ϕ. This follows from the requirement that $\mathscr{L}(x)$ be a Lorentz scalar which, in turn, is necessary to guarantee the Lorentz invariance of Hamilton's principle 2(3). We shall discuss Lorentz invariance in detail in Section 2-3.

Let us express the functional derivatives $\delta L/\delta\phi$ and $\delta L/\delta\dot{\phi}$ in terms of $\mathscr{L}(x)$. We have

$$\delta L = \int_V \left[\frac{\partial\mathscr{L}}{\delta\phi(\mathbf{x}, t)}\delta\phi(\mathbf{x}, t) + \frac{\partial\mathscr{L}}{\partial(\nabla\phi(\mathbf{x}, t))}\delta\nabla\phi(\mathbf{x}, t) + \frac{\partial\mathscr{L}}{\partial(\dot{\phi}(\mathbf{x}, t))}\delta\dot{\phi}(\mathbf{x}, t)\right] d^3x$$

$$= \int_V \left[\left(\frac{\partial\mathscr{L}}{\partial\phi(\mathbf{x}, t)} - \nabla\frac{\partial\mathscr{L}}{\partial(\nabla\phi(\mathbf{x}, t))}\right)\delta\phi(\mathbf{x}, t) + \frac{\partial\mathscr{L}}{\partial\dot{\phi}(x, t)}\delta\dot{\phi}(\mathbf{x}, t)\right] d^3x$$

where we have integrated by parts and dropped the surface term by virtue of the periodic boundary conditions. Comparing the above result with 2(8a) and taking into account that variations at distinct points are independent, we deduce that

$$\frac{\delta L(t)}{\delta\phi(\mathbf{x}, t)} = \frac{\partial\mathscr{L}(\mathbf{x}, t)}{\partial\phi(\mathbf{x}, t)} - \nabla\frac{\partial\mathscr{L}}{\partial(\nabla\phi(\mathbf{x}, t))} \tag{2(12a)}$$

$$\frac{\delta L(t)}{\delta\dot{\phi}(\mathbf{x}, t)} = \frac{\partial\mathscr{L}(\mathbf{x}, t)}{\partial\dot{\phi}(\mathbf{x}, t)} \tag{2(12b)}$$

Hence, in terms of $\mathscr{L}(x)$, the Euler–Lagrange equation of motion 2(10) becomes

$$\partial_t \frac{\partial \mathscr{L}(x)}{\partial(\partial_t \phi(x))} = \frac{\partial \mathscr{L}(x)}{\partial \phi(x)} - \nabla \frac{\partial \mathscr{L}(x)}{\partial(\nabla \phi(x))}$$

or

$$\partial_\mu \frac{\partial \mathscr{L}(x)}{\partial(\partial_\mu \phi(x))} = \frac{\partial \mathscr{L}(x)}{\partial \phi(x)} \tag{2(13)}$$

This last form can also be derived directly from Hamilton's principle, if the latter is written in the form

$$\delta \int_{t_1}^{t_2} dt \int_V d^3x \mathscr{L}(\phi(x), \partial_\mu \phi(x)) = 0 \tag{2(14)}$$

for variations $\delta\phi(\mathbf{x}, t)$ which vanish at t_1 and t_2 and satisfy the periodicity conditions on the boundary of V. Comparing 2(2) with 2(13) we see that a suitable Lagrangian density for the Klein–Gordon field is given by

$$\mathscr{L}_{\text{KG}}(x) = -\tfrac{1}{2}\partial_\mu \phi(x) \partial_\mu \phi(x) - \frac{\mu^2}{2}\phi(x)\phi(x) \tag{2(15)}$$

Case of more than one Independent Field To discuss fields of spin greater than 0, the above formalism must be extended to Lagrangian densities which are functions of several independent fields, say $\mathscr{L}(\phi_A, \partial_\mu \phi_A)$ with $A = 1, \ldots N$. Since the N fields are assumed to be independent, each field may be varied separately and Hamilton's principle yields a system of N Euler–Lagrange equations of the form

$$\partial_\mu \frac{\partial \mathscr{L}}{\partial(\partial_\mu \phi_A)} = \frac{\partial \mathscr{L}}{\partial \phi_A} \qquad (A = 1, \ldots N) \tag{2(16)}$$

We can also include complex fields $\phi_A \neq \phi_A^*$ in this formalism. Decomposing ϕ_A into real and imaginary parts

$$\begin{aligned} \phi_{A1} &= \frac{1}{\sqrt{2}}(\phi_A + \phi_A^*) \\ -i\phi_{A2} &= \frac{1}{\sqrt{2}}(\phi_A - \phi_A^*) \end{aligned} \tag{2(17)}$$

we can vary the real fields ϕ_{A1} and ϕ_{A2} independently to get the Euler–

Lagrange equations

$$\partial_\mu \frac{\partial \mathscr{L}}{\partial(\partial_\mu \phi_{Ai})} = \frac{\partial \mathscr{L}}{\partial \phi_{Ai}} \qquad (i = 1, 2) \tag{2(18a)}$$

Alternatively, we can vary ϕ_A and ϕ_A^* independently* and get

$$\partial_\mu \frac{\partial \mathscr{L}}{\partial(\partial_\mu \phi_A)} = \frac{\partial \mathscr{L}}{\partial \phi_A} \qquad \partial_\mu \frac{\partial \mathscr{L}}{\partial(\partial_\mu \phi^*_A)} = \frac{\partial \mathscr{L}}{\partial \phi_A^*} \tag{2(18b)}$$

The forms 2(18a) and 2(18b) are equivalent and may be derived from one another by using

$$\frac{\partial \mathscr{L}}{\partial \phi_A} = \frac{1}{\sqrt{2}}\left(\frac{\partial \mathscr{L}}{\partial \phi_{A1}} + i\frac{\partial \mathscr{L}}{\partial \phi_{A2}}\right)$$

$$\frac{\partial \mathscr{L}}{\partial \phi^*_A} = \frac{1}{\sqrt{2}}\left(\frac{\partial \mathscr{L}}{\partial \phi_{A1}} - i\frac{\partial \mathscr{L}}{\partial \phi_{A2}}\right)$$

and the corresponding expressions for $\partial\mathscr{L}/\partial(\partial_\mu\phi_A)$ and $\partial\mathscr{L}/\partial(\partial_\mu\phi_A^*)$.

Without further ado we now write down the Lagrangian densities corresponding to each of the fields studied in Chapter 1. The Lagrangian density for a real Klein–Gordon field has already been given. For a **complex Klein–Gordon field**, we can take $\mathscr{L}$ to be the sum of two terms of the form 2(15), i.e.

$$\mathscr{L}_{\mathrm{KG}} = -\tfrac{1}{2}\sum_{i=1}^{2} (\partial_\mu\phi_i\partial_\mu\phi_i + \mu^2\phi_i\phi_i) \tag{2(19a)}$$

or, alternatively, in terms of ϕ and ϕ^*

$$\mathscr{L}_{\mathrm{KG}} = -\partial_\mu\phi^*\partial_\mu\phi - \mu^2\phi^*\phi \tag{2(19b)}$$

A suitable Lagrangian density for the **Dirac field** is

$$\mathscr{L}_{\mathrm{Dirac}} = -\bar{\psi}(\gamma\,.\,\partial + m)\psi \tag{2(20)}$$

in which each component of ψ and of $\bar{\psi} = \psi^\dagger\gamma_4$ is to be varied independently, to yield the equations of motion 1(24) and 1(32). Note that $\mathscr{L}$ vanishes as a consequence of the equation of motion, but this simply means that the stationary value of the action integral is reached for $\mathscr{L} = 0$.

The Lagrangian formulation of the Bargmann–Wigner equations for arbitrary spin is extremely involved, [Kamefuchi (1966)]. For this reason

* Actually, ϕ_A and ϕ_A^* cannot really be varied independently, as they are related by complex conjugation. A more correct statement is that everything goes through *as if* ϕ_A and ϕ_A^* could be varied independently.

one always works with the tensor formalism, as given in Sections 1-5, 1-6, and 1-7, for spin 1 and $\frac{3}{2}$. For **real** and **complex vector fields** the Lagrangian densities are respectively

$$\mathscr{L}_V = -\tfrac{1}{4}F_{\mu\nu}F_{\mu\nu} - \tfrac{1}{2}\mu^2 A_\mu A_\mu$$
$$= -\tfrac{1}{4}(\partial_\mu A_\nu - \partial_\nu A_\mu)(\partial_\mu A_\nu - \partial_\nu A_\mu) - \tfrac{1}{2}\mu^2 A_\mu A_\mu \qquad 2(21)$$

and

$$\mathscr{L}_V = -\tfrac{1}{2}F^*_{\ \mu\nu}F_{\mu\nu} - \tfrac{1}{2}\mu^2 A^*_{\ \mu} A_\mu$$
$$= -\tfrac{1}{2}(\partial_\mu A^*_{\ \nu} - \partial_\nu A^*_{\ \mu})(\partial_\mu A_\nu - \partial_\nu A_\mu) - \tfrac{1}{2}\mu^2 A^*_{\ \mu} A_\mu \qquad 2(22)$$

For example, varying A_ν in 2(21) we get the Euler–Lagrange equations

$$0 = \frac{\partial \mathscr{L}}{\partial A_\nu} - \partial_\lambda \frac{\partial \mathscr{L}}{\partial(\partial_\lambda A_\nu)} = -\mu^2 A_\nu + \partial_\lambda(\partial_\lambda A_\nu - \partial_\nu A_\lambda)$$

in agreement with 1(111)*. In three-dimensional notation 2(21) and 2(22) take the form

$$\mathscr{L}_V = \tfrac{1}{2}(\mathbf{E}^2 - \mathbf{B}^2) - \tfrac{1}{2}\mu^2(\mathbf{A}^2 - A_0^2) \qquad 2(23a)$$

and

$$\mathscr{L}_V = (\mathbf{E}^*\mathbf{E} - \mathbf{B}^*\mathbf{B}) - \mu^2(\mathbf{A}^*\mathbf{A} - A_0^{\ *}A_0) \qquad 2(23b)$$

respectively. To get the Lagrangian density for the **Maxwell field**, we set $\mu = 0$ in the Lagrangian 2(21) for the real vector field.

$$\mathscr{L}_{em} = -\tfrac{1}{4}F_{\mu\nu}F_{\mu\nu} = -\tfrac{1}{4}(\partial_\mu A_\nu - \partial_\nu A_\mu)(\partial_\mu A_\nu - \partial_\nu A_\mu)$$

If we select the radiation gauge in which $\mathbf{A} = \mathbf{A}^t$, we can write $\mathscr{L}_{em}$ in three-dimensional notation as

$$\mathscr{L}_{em} = \tfrac{1}{2}(\mathbf{E}^2 - \mathbf{B}^2) = \tfrac{1}{2}[(\dot{\mathbf{A}}^t)^2 - (\mathbf{curl}\ \mathbf{A}^t)^2] \qquad 2(24)$$

Since in the radiation gauge the three components of $\mathbf{A}$ are not independent but are subject to the constraint div $\mathbf{A} = 0$, only $\mathbf{A}^t$ is to be varied in deriving the Euler–Lagrange equations. An alternative form for 2(24)

* An alternative approach is to treat the field strengths $F_{\mu\nu}$ on an equal footing with the potentials. A suitable Lagrangian is, in this case,

$$\mathscr{L} = -\tfrac{1}{2}F_{\mu\nu}(\partial_\mu A_\nu - \partial_\nu A_\mu) + \tfrac{1}{4}F_{\mu\nu}F_{\mu\nu} - \tfrac{1}{2}\mu^2 A_\mu A_\mu$$

in which both $F_{\mu\nu}$ and A_ν are to be varied independently to yield the system 1(110a) and 1(110b).

follows upon noting that

$$\begin{aligned}(\textbf{curl A}^t)^2 &= \varepsilon_{kji}\varepsilon_{klm}\nabla_j A_i^t \nabla_l A_m^t \\ &= (\delta_{jl}\delta_{im} - \delta_{jm}\delta_{il})\nabla_j A_i^t \nabla_l A_m^t \\ &= \boldsymbol{\nabla} A_i^t \boldsymbol{\nabla} A_i^t\end{aligned}$$

since div $\mathbf{A}^t = 0$. Hence

$$\mathscr{L}_{em} = -\tfrac{1}{2}\sum_{i=1}^{3} (\boldsymbol{\nabla} A_i^t \boldsymbol{\nabla} A_i^t - \dot{A}_i^t \dot{A}_i^t) \qquad 2(25)$$

In this form $\mathscr{L}_{em}$ appears as the direct generalization of 2(15) to the two independent degrees of freedom of the transverse vector potential.

For the **Rarita–Schwinger field** we take

$$\begin{aligned}\mathscr{L}_{\text{RS}} = -\{&\bar{\psi}_\mu(\gamma\,.\,\partial + m)\psi_\mu - \tfrac{1}{3}\bar{\psi}_\mu(\gamma_\mu\partial_\nu + \gamma_\nu\partial_\mu)\psi_\nu \\ &+ \tfrac{1}{3}\bar{\psi}_\mu\gamma_\mu(\gamma\,.\,\partial - m)\gamma_\nu\psi_\nu\}\end{aligned} \qquad 2(26)$$

which yields 1(188) and 1(193) upon varying ψ_μ and $\bar{\psi}_\mu$ independently. As in the Dirac case, $\mathscr{L}_{\text{RS}}$ vanishes as a consequence of the equations of motion. Notice that, unlike the gauge condition for the Maxwell field, the subsidiary condition 1(186b) is in this case a consequence of the equation of motion. Each component of ψ_μ may therefore be varied independently in deriving the Euler–Lagrange equations from Hamilton's principle.

Finally, we note that the non-relativistic Schrödinger equation

$$-\frac{1}{2m}\boldsymbol{\nabla}^2\psi + V\psi = i\dot{\psi} \qquad 2(27)$$

for a particle of mass m in a potential $V(x)$ may be regarded as a field equation and derived from the Lagrangian density

$$\mathscr{L}_S = i\psi^*\dot{\psi} - \frac{1}{2m}\boldsymbol{\nabla}\psi^*\boldsymbol{\nabla}\psi - V\psi^*\psi \qquad 2(28)$$

Fourier Analysis We have seen that a field may be viewed as a mechanical system with an infinite number of degrees of freedom by regarding the value of the field at each point in 3-space as an independent dynamical variable. An alternative mechanical representation of the field is obtained by decomposing it into normal modes. Assuming the Lagrangian density $\mathscr{L}(\phi_A, \partial_\mu\phi_A)$ to be a function of N fields ϕ_A ($A = 1, \ldots N$), let us expand

ϕ_A in terms of a complete orthonormal set of functions $\varphi_n(x)$:

$$\phi^A(\mathbf{x}, t) = \sum_{n=0}^{\infty} q_n^A(t)\varphi_n(\mathbf{x}) \qquad (A = 1, \ldots N) \tag{2(29)}$$

with

$$\int_V \varphi_m^*(\mathbf{x})\varphi_n(\mathbf{x})\, d^3x = \delta_{mn} \tag{2(30a)}$$

$$\sum_{n=0}^{\infty} \varphi_n^*(\mathbf{x})\varphi_n(\mathbf{x}') = \delta^{(3)}(\mathbf{x}-\mathbf{x}') \tag{2(30b)}$$

where V is the normalization volume. We now show that the coefficients $q_n^A(t)$ $(A = 1 \ldots N)$ form an infinite set of canonical coordinates for the field.

The Lagrangian

$$L = \int_V \mathscr{L}(\phi^A, \boldsymbol{\nabla}\phi^A, \dot{\phi}^A)\, d^3x$$

is evidently a function, in the ordinary sense, of the q_n^A and $\dot{q}_n^A$. The derivatives of L with respect to these coefficients are given by

$$\frac{\partial L}{\partial q_n^A} = \int_V d^3x \left(\frac{\partial \mathscr{L}}{\partial \phi^A}\varphi_n(\mathbf{x}) + \frac{\partial \mathscr{L}}{\partial(\boldsymbol{\nabla}\phi^A)}\boldsymbol{\nabla}\varphi_n(\mathbf{x}) \right) \tag{2(31a)}$$

and

$$\frac{\partial L}{\partial \dot{q}_n^A} = \int_V d^3x \frac{\partial \mathscr{L}}{\partial \dot{\phi}^A}\varphi_n(\mathbf{x}) \tag{2(31b)}$$

respectively, where we have used the expansion 2(29) and the corresponding expansion for $\dot{\phi}^A$. Performing an integration by parts and assuming periodic boundary conditions at the surface of V we can write 2(31a) in the form

$$\frac{\partial L}{\partial q_n^A} = \int d^3x \left(\frac{\partial \mathscr{L}}{\partial \phi^A} - \boldsymbol{\nabla}\frac{\partial \mathscr{L}}{\partial(\boldsymbol{\nabla}\phi^A)} \right)\varphi_n(\mathbf{x}) \tag{2(32)}$$

which, when combined with 2(31b), yields the Euler–Lagrange equations

$$\frac{\partial L}{\partial q_n^A} = \partial_t \frac{\partial L}{\partial \dot{q}_n^A} \tag{2(33)}$$

by virtue of 2(13) and the completeness of the $\varphi_n(\mathbf{x})$. From 2(33) we infer that the $q_n^A(t)$ form an infinite set of canonical coordinates for the field.

The most convenient choice for the $\varphi_n(\mathbf{x})$ is the set of all exponentials of the form

$$\frac{1}{\sqrt{V}} e^{i\mathbf{k}.\mathbf{x}} \tag{2(34)}$$

Then the expansion 2(29) takes the form

$$\phi^A(\mathbf{x}, t) = \frac{1}{\sqrt{V}} \sum_{\mathbf{k}} q_{\mathbf{k}}^A(t)\, e^{i\mathbf{k}.\mathbf{x}} \qquad (A = 1, \dots N) \tag{2(35)}$$

We note that the Fourier decomposition 2(35) can always be identified with the expansion of $\phi^A(x)$ in terms of a complete set of plane wave solutions to the equations of motion 2(16). For example, a real scalar field $\phi(\mathbf{x}, t)$ may be expanded in terms of the complete set of plane wave solutions 1(2a) and 1(2b) according to

$$\phi(\mathbf{x}, t) = \frac{1}{\sqrt{V}} \sum_{\mathbf{k}} \frac{1}{\sqrt{2\omega_{\mathbf{k}}}} (a_{\mathbf{k}}\, e^{i\mathbf{k}.\mathbf{x} - i\omega_{\mathbf{k}} t} + a_{\mathbf{k}}^*\, e^{-i\mathbf{k}.\mathbf{x} + i\omega_{\mathbf{k}} t}) \tag{2(36)}$$

where the reality of ϕ requires the coefficients of $e^{i\mathbf{k}.\mathbf{x}}$ and $e^{-i\mathbf{k}.\mathbf{x}}$ to be complex conjugates of one another. Identifying 2(36) with the Fourier decomposition

$$\phi(\mathbf{x}, t) = \frac{1}{\sqrt{V}} \sum_{\mathbf{k}} q_{\mathbf{k}}(t)\, e^{i\mathbf{k}.\mathbf{x}} \tag{2(37)}$$

we get, since $\omega_{\mathbf{k}} = \omega_{-\mathbf{k}}$

$$q_{\mathbf{k}}(t) = \frac{1}{\sqrt{2\omega_{\mathbf{k}}}} (a_{\mathbf{k}}\, e^{-i\omega_{\mathbf{k}} t} + a_{-\mathbf{k}}^*\, e^{i\omega_{\mathbf{k}} t}) \tag{2(38)}$$

Evidently many other sets of functions $\varphi_n(\mathbf{x})$ satisfying 2(30a) and 2(30b) are available for the expansion of $\phi^A(\mathbf{x}, t)$. For example, one can take the *real* functions

$$\frac{\sqrt{2}}{\sqrt{V}} \cos \mathbf{k}.\mathbf{x} \qquad \text{and} \qquad \frac{\sqrt{2}}{\sqrt{V}} \sin \mathbf{k}.\mathbf{x}$$

with the components of $\mathbf{k}$ restricted to the range $0 \leqslant k_z \leqslant \infty$ and $-\infty \leqslant k_x, k_y \leqslant +\infty$. However, the normal modes associated with the imaginary exponentials 2(34) have the considerable advantage that they carry a definite value of momentum.

2-2 Hamiltonian formalism

Conjugate Fields Let us pursue the canonical formalism of mechanics and pass from the Euler–Lagrange equations 2(10) to equations of motion of the Hamiltonian type. We shall illustrate the procedure in detail for the case of a real Klein–Gordon field. The generalization to more complicated cases is not necessarily straightforward and will be considered later.

To each canonical coordinate ϕ_i, representing the value of $\phi(\mathbf{x}, t)$ in the ith cell, we associate a conjugate momentum

$$p_i(t) = \frac{\partial L(t)}{\partial \dot{\phi}_i(t)} \tag{2(39)}$$

and define the Hamiltonian of the field according to the standard prescription

$$H = \sum_i p_i \dot{\phi}_i - L \tag{2(40)}$$

The Euler–Lagrange equations 2(4) may then be replaced by the Hamiltonian equations of motion*

$$\frac{\partial H}{\partial \phi_i} = -\dot{p}_i \qquad \frac{\partial H}{\partial p_i} = \dot{\phi}_i \tag{2(41)}$$

where H is a function of the ϕ_i and p_i, $\dot{\phi}_i$ having been eliminated by solving 2(39) for $\dot{\phi}_i$.

Going over to continuum notation, we have, combining 2(39) with 2(11),

$$p_i(t) = \frac{\partial \mathscr{L}(\mathbf{x}, t)}{\partial \dot{\phi}(\mathbf{x}, t)} \delta V_i = \pi(\mathbf{x}, t)\, \delta V_i \tag{2(42)}$$

in the continuum limit $\delta V_i \to d^3x$ for $\mathbf{x}$ in the ith cell.

The quantity

$$\pi(x) = \frac{\partial \mathscr{L}(x)}{\partial \dot{\phi}(x)} \tag{2(43)}$$

introduced in 2(42) is known as the *conjugate field*. Using 2(42) we can write the Hamiltonian 2(40) as

$$H(t) = \int_V (\pi(x)\dot{\phi}(x) - \mathscr{L}(x))\, d^3x \tag{2(44)}$$

* See for example H. Goldstein, *Classical Mechanics*, Addison-Wesley, 1951.

in the continuum limit. The integrand

$$\mathscr{H}(x) = \pi(x)\dot{\phi}(x) - \mathscr{L}(x) \tag{2(45)}$$

is known as the Hamiltonian density. For the real Klein–Gordon field we find, using 2(15)

$$\pi(x) = \dot{\phi}(x) \tag{2(46)}$$

and

$$H = \frac{1}{2}\int_V (\pi^2 + (\nabla\phi)^2 + \mu^2\phi^2)\, d^3x \tag{2(47)}$$

From the definition of the functional derivative we have, in analogy with 2(9a) and 2(9b)

$$\frac{\delta H(t)}{\delta\phi(\mathbf{x}, t)} = \lim_{\delta V_i \to 0} \frac{1}{\delta V_i} \frac{\partial H(t)}{\partial \phi_i(t)} \tag{2(48a)}$$

$$\frac{\delta H(t)}{\delta\pi(\mathbf{x}, t)} = \lim_{\delta V_i \to 0} \frac{1}{\delta V_i} \frac{\partial H(t)}{\partial \pi_i(t)} \tag{2(48b)}$$

for $\mathbf{x}$ in the ith cell. This allows us to write the Hamiltonian equations of motion in continuum notation. Combining 2(41), 2(42) and 2(48a), 2(48b) we find

$$\frac{\delta H(t)}{\delta\phi(x)} = -\dot{\pi}(x) \qquad \frac{\delta H(t)}{\delta\pi(x)} = \dot{\phi}(x) \tag{2(49)}$$

Fourier Decomposition Let us expand $\pi(\mathbf{x}, t)$ in terms of the complete set of functions $\varphi_n(\mathbf{x}, t)$ satisfying 2(30a) and 2(30b). If the expansion is written in the form

$$\pi(\mathbf{x}, t) = \sum_n p_n(t)\varphi_n^*(\mathbf{x}) \tag{2(50)}$$

we can identify the $p_n(t)$ as the momenta canonically conjugate to the coordinates $q_n(t)$ featured in 2(37). Indeed, inverting the above expansion with the aid of 2(30a) and using 2(43), we get

$$\begin{aligned} p_n(t) &= \int d^3x \pi(\mathbf{x}, t)\varphi_n(\mathbf{x}) \\ &= \int d^3x \frac{\partial \mathscr{L}}{\partial \dot{\phi}(x)}\varphi_n(\mathbf{x}) \\ &= \frac{\partial L}{\partial \dot{q}_n} \end{aligned}$$

by 2(31b). If the $\varphi_n(\mathbf{x})$ are identified with the set of exponentials 2(34), the expansion of $\pi(\mathbf{x}, t)$ takes the form

$$\pi(\mathbf{x}, t) = \frac{1}{\sqrt{V}} \sum_{\mathbf{k}} p_{\mathbf{k}}(t)\, e^{-i\mathbf{k}.\mathbf{x}} \tag{2(51)}$$

Poisson Brackets Let us consider a functional $F[\phi, \pi]$ which has no explicit t-dependence, i.e. which depends on t only through ϕ and π. Then

$$\begin{aligned} \dot{F}(t) &= \int_V d^3x \left(\frac{\delta F(t)}{\delta\phi(x)} \dot{\phi}(x) + \frac{\delta F(t)}{\delta\pi(x)} \dot{\pi}(x) \right) \\ &= \int_V d^3x \left(\frac{\delta F(t)}{\delta\phi(x)} \frac{\delta H(t)}{\delta\pi(x)} - \frac{\delta F(t)}{\delta\pi(x)} \frac{\delta H(t)}{\delta\phi(x)} \right) \end{aligned} \tag{2(52)}$$

We now define a functional *Poisson bracket*

$$[F, G]_{\mathrm{PB}} = \int_V \left(\frac{\delta F}{\delta\phi(x)} \frac{\delta G}{\delta\pi(x)} - \frac{\delta F}{\delta\pi(x)} \frac{\delta G}{\delta\phi(x)} \right) d^3x \tag{2(53)}$$

where F and G are arbitrary functionals of ϕ and π. To exhibit the relation of 2(53) to the ordinary Poisson bracket introduced in classical mechanics we revert to the discrete notation of Section 2-1. Then

$$\begin{aligned} [F, G]_{\mathrm{PB}} &= \sum_i \left(\frac{1}{\delta V_i} \frac{\partial F}{\partial\phi_i} \frac{1}{\delta V_i} \frac{\partial G}{\partial\pi_i} - \frac{1}{\delta V_i} \frac{\partial F}{\partial\pi_i} \frac{1}{\delta V_i} \frac{\partial G}{\partial\phi_i} \right) \delta V_i \\ &= \sum_i \left(\frac{\partial F}{\partial\phi_i} \frac{\partial G}{\partial p_i} - \frac{\partial F}{\partial p_i} \frac{\partial G}{\partial\phi_i} \right) \end{aligned}$$

where we have applied 2(48a) and 2(48b) to F and G and used 2(42). The right-hand side is just the usual Poisson bracket for a canonical system with an infinite number of degrees of freedom.

With the aid of the definition 2(53) we can write 2(52) as

$$\dot{F} = [F, H]_{\mathrm{PB}} \tag{2(54)}$$

If F depends *explicitly* on time, then we have, instead

$$\dot{F} = [F, H]_{\mathrm{PB}} + \frac{\partial F}{\partial t} \tag{2(55)}$$

We now remark that

$$\frac{\delta\phi(\mathbf{x}, t)}{\delta\phi(\mathbf{x}', t)} = \delta^{(3)}(\mathbf{x} - \mathbf{x}') = \frac{\delta\pi(\mathbf{x}, t)}{\delta\pi(\mathbf{x}', t)} \tag{2(56)}$$

Equation 2(56) follows upon writing

$$\phi(\mathbf{x}, t) = \int \delta^{(3)}(\mathbf{x}-\mathbf{x}')\phi(\mathbf{x}', t)\, d^3x$$

$$\pi(\mathbf{x}, t) = \int \delta^{(3)}(\mathbf{x}-\mathbf{x}')\pi(\mathbf{x}', t)\, d^3x$$

and applying the definition of the functional derivative. From 2(56) it follows that

$$[\phi(\mathbf{x}, t), H(t)]_{\mathrm{PB}} = \int_V d^3x' \frac{\delta\phi(\mathbf{x}, t)}{\delta\phi(\mathbf{x}', t)} \frac{\delta H(t)}{\delta\pi(\mathbf{x}', t)} = \frac{\delta H(t)}{\delta\pi(\mathbf{x}, t)} \qquad 2(57a)$$

and similarly

$$[\pi(\mathbf{x}, t), H(t)]_{\mathrm{PB}} = -\frac{\delta H(t)}{\delta\phi(\mathbf{x}, t)} \qquad 2(57b)$$

so that the Hamiltonian equations of motion 2(49) may be written as

$$\dot{\phi}(\mathbf{x}, t) = [\phi(\mathbf{x}, t), H(t)]_{\mathrm{PB}} \qquad 2(58a)$$

$$\dot{\pi}(\mathbf{x}, t) = [\pi(\mathbf{x}, t), H(t)]_{\mathrm{PB}} \qquad 2(58b)$$

These equations also follow as a particular case of 2(54).

From 2(53) and 2(56) we also derive the important relations

$$[\phi(\mathbf{x}, t), \pi(\mathbf{x}', t)]_{\mathrm{PB}} = \delta^{(3)}(\mathbf{x}-\mathbf{x}') \qquad 2(59)$$

$$[\phi(\mathbf{x}, t), \phi(\mathbf{x}', t)]_{\mathrm{PB}} = [\pi(\mathbf{x}, t), \pi(\mathbf{x}', t)]_{\mathrm{PB}} = 0 \qquad 2(60)$$

which can be used as a springboard for quantization, as we shall see in Chapter 3.

Extension to Several Independent Fields In principle the extension of the Hamiltonian formalism to systems consisting of more than one independent field is straightforward. To each independent field there corresponds a conjugate field

$$\pi_A(x) = \frac{\partial \mathscr{L}(x)}{\partial \dot{\phi}_A(x)} \qquad 2(61)$$

and the definition of the Hamiltonian density is given by

$$\mathscr{H}(x) = \sum_{A=1}^{N} \pi_A(x)\dot{\phi}_A(x) - \mathscr{L}(x) \qquad 2(62)$$

The Hamiltonian equations of motion 2(49) and 2(58a), 2(58b) generalize to

$$\frac{\delta H}{\delta \phi_A(x)} = -\dot{\pi}_A(x) \qquad \frac{\delta H}{\delta \pi_A(x)} = \dot{\phi}_A(x) \tag{2(63)}$$

and

$$\dot{\phi}_A(\mathbf{x}, t) = [\phi_A(\mathbf{x}, t), H(t)]_{\mathrm{PB}} \tag{2(64a)}$$

$$\dot{\pi}_A(\mathbf{x}, t) = [\pi_A(\mathbf{x}, t), H(t)]_{\mathrm{PB}} \tag{2(64b)}$$

respectively, where the Poisson bracket of two functions F and G is now defined by

$$[F, G]_{\mathrm{PB}} = \sum_{A=1}^{N} \int \left(\frac{\delta F}{\delta \phi_A(x)} \frac{\delta G}{\delta \pi_A(x)} - \frac{\delta F}{\delta \pi_A(x)} \frac{\delta G}{\delta \phi_A(x)} \right) d^3x \tag{2(65)}$$

Finally, the relations 2(59a) and 2(59b) can be expected to generalize to

$$[\phi_A(\mathbf{x}, t), \pi_{A'}(\mathbf{x}', t)]_{\mathrm{PB}} = \delta_{AA'}\delta^{(3)}(\mathbf{x}-\mathbf{x}') \tag{2(66a)}$$

$$[\phi_A(\mathbf{x}, t), \phi_{A'}(\mathbf{x}', t)]_{\mathrm{PB}} = [\pi_A(\mathbf{x}, t), \pi_{A'}(\mathbf{x}', t)] = 0 \tag{2(66b)}$$

The difficulty is that the above formalism is valid only if the ϕ_A represent truly independent field variables whereas, in practice, the components of fields for spin greater than zero are generally subject to one or more subsidiary conditions. For this reason, each case must be accorded separate treatment.

Complex Scalar Field For a complex Klein–Gordon field, the application of the canonical formalism is straightforward, since ϕ and ϕ^*, or alternatively ϕ_1 and ϕ_2 may be regarded as independent field variables. Defining the conjugate fields

$$\pi = \frac{\partial \mathcal{L}}{\partial \dot{\phi}} \qquad \pi^* = \frac{\partial \mathcal{L}}{\partial \dot{\phi}^*} \tag{2(67a)}$$

and

$$\pi_1 = \frac{\partial \mathcal{L}}{\partial \dot{\phi}_1} \qquad \pi_2 = \frac{\partial \mathcal{L}}{\partial \dot{\phi}_2} \tag{2(67b)}$$

we have the relations

$$\begin{aligned} \pi_1 &= \frac{1}{\sqrt{2}}(\pi + \pi^*) \\ i\pi_2 &= \frac{1}{\sqrt{2}}(\pi - \pi^*) \end{aligned} \tag{2(68)}$$

The Hamiltonian density can be defined either by

$$\mathscr{H} = \pi\dot{\phi} + \pi^*\dot{\phi}^* - \mathscr{L}$$

or equivalently

$$\mathscr{H} = \pi_1\dot{\phi}_1 + \pi_2\dot{\phi}_2 - \mathscr{L}$$

Using 2(19a) and 2(19b) we find

$$\begin{cases} \pi = \dot{\phi}^* \\ \pi^* = \dot{\phi} \end{cases} \quad \text{and} \quad \begin{cases} \pi_1 = \dot{\phi}_1 \\ \pi_2 = \dot{\phi}_2 \end{cases} \qquad 2(69)$$

and

$$\mathscr{H}(x) = \pi^*\pi + (\nabla\phi^*).(\nabla\phi) + \mu^2\phi^*\phi \qquad 2(70a)$$

or equivalently

$$\mathscr{H}(x) = \sum_{i=1}^{2} [\pi_i^2 + (\nabla\phi_i)^2 + \mu^2\phi_i^2] \qquad 2(70b)$$

The total Hamiltonian H is therefore given by

$$H = \int_V d^3x(\dot{\phi}^*\dot{\phi} - \phi^*\nabla^2\phi + \mu^2\phi^*\phi) \qquad 2(71)$$

where an integration by parts has been performed. To within a factor μ^{-1}, 2(71) is identical to the expectation value 1(23).* Observe that the field Hamiltonian (which we shall later identify with the field energy), is *positive-definite*.

We also note that the equal-time Poisson brackets for the canonically conjugate pairs (ϕ, π) and (ϕ^*, π^*) are given by

$$[\phi(\mathbf{x}, t), \pi(\mathbf{x}', t)]_{PB} = \delta^{(3)}(\mathbf{x} - \mathbf{x}') \qquad 2(72a)$$

$$[\phi^*(\mathbf{x}, t), \pi^*(\mathbf{x}', t)]_{PB} = \delta^{(3)}(\mathbf{x} - \mathbf{x}') \qquad 2(72b)$$

All other equal-time Poisson brackets of two field variables vanish.

Dirac Field Difficulties arise as soon as we leave the realm of spin zero fields. If we apply the prescription 2(61) to the Dirac Lagrangian 2(20), treating ψ and $\psi^\dagger$ as independent canonical coordinates we find

$$\pi_\psi = \frac{\partial\mathscr{L}}{\partial\dot{\psi}} = i\psi^\dagger \qquad 2(73a)$$

* The factor μ^{-1} could be eliminated by redefining the wave functions Ψ_1 and Ψ_2 in 1(18a) and 1(18b) to include an additional factor $\mu^{1/2}$.

and

$$\pi_{\psi^\dagger} = \frac{\partial \mathscr{L}}{\partial \dot{\psi}^\dagger} = 0 \tag{2(73b)}$$

It follows that we cannot regard ψ and $\psi^\dagger$ as independent canonical coordinates. In particular 2(73b) would imply that

$$[\psi^\dagger, \pi_{\psi^\dagger}] = 0$$

in contradiction with 2(66a).

We are therefore forced to conclude that the only independent pairs of canonically conjugate fields are the four components of ψ and of $\pi = i\psi^\dagger$. This type of situation arises whenever the Lagrangian density is of first order in $\partial_t \psi$. Defining the Hamiltonian density by

$$\mathscr{H} = \pi\dot{\psi} - \mathscr{L} = i\psi^\dagger\dot{\psi} - \mathscr{L} \tag{2(74)}$$

and using 2(20) we find

$$\begin{aligned} \mathscr{H} &= \psi^\dagger(\gamma_4 \boldsymbol{\gamma} \cdot \boldsymbol{\nabla} + m\gamma_4)\psi \\ &= \psi^\dagger \mathbf{H} \psi \end{aligned} \tag{2(75)}$$

where $\mathbf{H}$ is the Hamiltonian 1(36) of the one-particle theory. Hence the field Hamiltonian H is given by the expectation value*

$$\begin{aligned} H &= \int \psi^\dagger \mathbf{H} \psi \, d^3x \\ &= \int \psi^\dagger i\partial_t \psi \, d^3x \end{aligned} \tag{2(76)}$$

of the one-particle Hamiltonian, a result which is similar to the one derived earlier for the Klein–Gordon Hamiltonian. Note, however, that in contrast to 2(71), the Hamiltonian 2(76) can take on both positive and negative values.

The remainder of the canonical formalism can be carried through as in the general case. In particular, the equal-time Poisson brackets of the fields are given in terms of ψ and $\psi^\dagger$ by

$$\begin{aligned} [\psi_\alpha(\mathbf{x}, t), \psi_\beta^\dagger(\mathbf{x}', t)]_{\text{PB}} &= -i\delta_{\alpha\beta}\delta^{(3)}(\mathbf{x}-\mathbf{x}') \\ [\psi_\alpha(\mathbf{x}, t), \psi_\beta(\mathbf{x}', t)]_{\text{PB}} &= [\psi_\alpha^\dagger(\mathbf{x}, t), \psi_\beta^\dagger(\mathbf{x}', t]_{\text{PB}} = 0 \end{aligned} \tag{2(77)}$$

* The expression 2(76) also follows trivially from 2(74) upon noting that the Dirac Lagrangian vanishes as a consequence of the equations of motion. Note that the Dirac Hamiltonian density is real, even though $\mathscr{L}$, given by 2(20), is not.

where the spinor indices α and β have been exhibited explicitly and

$$[F, G]_{PB} = \sum_{\alpha=1}^{4} \int \left(\frac{\delta F}{\delta \psi_\alpha(x)} \frac{\delta G}{\delta \pi_\alpha(x)} - \frac{\delta F}{\delta \pi_\alpha(x)} \frac{\delta G}{\delta \psi_\alpha(x)} \right) d^3x$$

Real and Complex Vector Fields For a real vector field the Lagrangian density is given by 2(21). Defining the conjugate fields

$$\pi_\mu = \frac{\partial \mathscr{L}}{\partial \dot{A}_\mu}$$

we find

$$\pi_i = -iF_{i4} \tag{2(78a)}$$

but

$$\pi_4 = 0 \tag{2(78b)}$$

Again we must conclude that only the vector potential $\mathbf{A}(x)$ and its conjugate field $\boldsymbol{\pi}(x) = -\mathbf{E}(x)$ represent independent canonical variables. This is not too surprising in view of the remarks following 1(122). The Hamiltonian density is defined by

$$\mathscr{H} = -\mathbf{E} \,.\, \dot{\mathbf{A}} - \mathscr{L}$$

Using 2(23a) and 1(122b) we find

$$\begin{aligned}\mathscr{H} &= \mathbf{E}^2 + \mathbf{E} \,.\, \nabla A_0 - \tfrac{1}{2}(\mathbf{E}^2 - \mathbf{B}^2) + \tfrac{1}{2}\mu^2(\mathbf{A}^2 - A_0^2) \\ &= \tfrac{1}{2}(\mathbf{E}^2 + \mu^2 \mathbf{A}^2 + \mathbf{B}^2) + \mathbf{E} \,.\, \nabla A_0 - \frac{\mu^2}{2} A_0^2\end{aligned} \tag{2(79)}$$

Enlisting the aid of 1(122d) we have

$$\begin{aligned}\mathbf{E} \,.\, \nabla A_0 &= \nabla \,.\, (\mathbf{E} A_0) - A_0(\nabla \,.\, \mathbf{E}) \\ &= \nabla \,.\, (\mathbf{E} A_0) + \mu^2 A_0^2\end{aligned} \tag{2(80)}$$

so that the field Hamiltonian H may be written as

$$H = \int_V \mathscr{H} \, d^3x = \tfrac{1}{2} \int_V (\mathbf{E}^2 + \mu^2 \mathbf{A}^2 + \mathbf{B}^2 + \mu^2 A_0^2) \, d^3x$$

where the surface term arising from the space divergence in 2(80) has been dropped. Using 1(122a) and 1(122d) we obtain the final result

$$H = \tfrac{1}{2} \int \left[\mathbf{E}^2 + \mu^2 \mathbf{A}^2 + (\mathbf{curl}\ \mathbf{A})^2 + \frac{1}{\mu^2} (\text{div}\ \mathbf{E})^2 \right] d^3x \tag{2(81)}$$

The Hamiltonian equations of motion for the canonically conjugate pair **E** and $-$**A** read

$$\dot{\mathbf{A}}(\mathbf{x}, t) = [\mathbf{A}(\mathbf{x}, t), H(t)]_{\mathrm{PB}} = -\frac{\delta H(t)}{\delta \mathbf{E}(\mathbf{x}, t)} \qquad 2(82a)$$

$$\dot{\mathbf{E}}(\mathbf{x}, t) = [\mathbf{E}(\mathbf{x}, t), H(t)]_{\mathrm{PB}} = \frac{\delta H(t)}{\delta \mathbf{A}(\mathbf{x}, t)} \qquad 2(82b)$$

When the functional derivatives are evaluated the result is just the system of coupled equations 1(123a) and 1(123b), as the reader can easily verify. The equal-time Poisson brackets of **A** and **E** are given by

$$[A_i(\mathbf{x}, t), -E_j(\mathbf{x}', t)]_{\mathrm{PB}} = \delta_{ij}\delta^{(3)}(\mathbf{x}-\mathbf{x}') \qquad 2(83a)$$

$$[A_i(\mathbf{x}, t), A_j(\mathbf{x}', t)]_{\mathrm{PB}} = [E_i(\mathbf{x}, t), E_j(\mathbf{x}', t)]_{\mathrm{PB}} = 0 \qquad 2(83b)$$

in accordance with 2(66a) and 2(66b), the Poisson bracket $[F, G]_{\mathrm{PB}}$ being defined by

$$[F, G]_{\mathrm{PB}} = -\sum_{i=1}^{3} \int \left(\frac{\delta F}{\delta A_i(x)} \frac{\delta G}{\delta E_i(x)} - \frac{\delta F}{\delta E_i(x)} \frac{\delta G}{\delta A_i(x)} \right) d^3x$$

The generalization to a *complex* vector field described by the Lagrangian density 2(23b) is straightforward. There are six independent canonical coordinates, namely **A** and $\mathbf{A}^*$ together with their conjugate fields

$$\boldsymbol{\pi} = \frac{\partial \mathscr{L}}{\partial \dot{\mathbf{A}}} = -\mathbf{E}^* \qquad 2(84a)$$

$$\boldsymbol{\pi}^* = \frac{\partial \mathscr{L}}{\partial \dot{\mathbf{A}}^*} = -\mathbf{E} \qquad 2(84b)$$

The Hamiltonian density is defined by

$$\mathscr{H} = -\mathbf{E}^* . \dot{\mathbf{A}} - \mathbf{E} . \dot{\mathbf{A}}^* - \mathscr{L}$$

and one finds

$$H = \int_V \mathscr{H}\, d^3x$$

$$= \int_V \left[\mathbf{E}^* . \mathbf{E} + \mu^2 \mathbf{A}^* . \mathbf{A} + (\mathbf{curl}\, \mathbf{A}^*) . (\mathbf{curl}\, \mathbf{A}) + \frac{1}{\mu^2} (\mathrm{div}\, \mathbf{E}^*)(\mathrm{div}\, \mathbf{E}) \right] d^3x \qquad 2(85)$$

by following the same steps as in deriving 2(81). By performing an integration by parts on the last two terms, we can write the field Hamiltonian in the form

$$H = \int_V \left(\mathbf{E}^* . \mathbf{E} + \mu^2 \mathbf{A}^* . \mathbf{A} + \mathbf{A}^* . \mathbf{curl\ curl\ A} - \frac{1}{\mu^2} \mathbf{E}^* . \mathbf{grad}\ \mathrm{div}\ \mathbf{E} \right) d^3x$$

which is identical, to within a factor μ^{-1}, to the expectation value 1(128) of the Hamiltonian for the one-particle theory. Note that H is positive-definite.

The remainder of the canonical formalism presents no new features. In particular, the equal-time Poisson brackets for the fields and their conjugate momenta are given by

$$[A_i(\mathbf{x}, t), -E_j^*(\mathbf{x}', t)]_{\mathrm{PB}} = \delta_{ij}\delta^{(3)}(\mathbf{x}-\mathbf{x}') \qquad 2(86a)$$

$$[A_i^*(\mathbf{x}, t), -E_j(\mathbf{x}', t)]_{\mathrm{PB}} = \delta_{ij}\delta^{(3)}(\mathbf{x}-\mathbf{x}') \qquad 2(86b)$$

All other equal-time Poisson brackets of two fields vanish.

Maxwell Field As in the massive vector case, we see from the Maxwell field Lagrangian 2(24) that the momentum canonically conjugate to A_4 vanishes identically. Since the electromagnetic field is real, the independent canonical fields are reduced to the three components of the vector potential $\mathbf{A}(x)$ and the conjugate field $-\mathbf{E}(x)$.

A further reduction of the number of independent variables is called for in the electromagnetic case, since
(a) the Maxwell equation $\mathrm{div}\,\mathbf{E} = 0$ ensures that the longitudinal part of $\mathbf{E}(x)$ vanishes identically, and
(b) we are at liberty to choose the radiation gauge in which the longitudinal part of $\mathbf{A}(x)$ also vanishes identically. Thus, in the radiation gauge, we have

$$\mathbf{A}(x) = \mathbf{A}^t(x) \qquad 2(87a)$$

$$\mathbf{E}(x) = \mathbf{E}^t(x) \qquad 2(87b)$$

and the independent canonical fields are simply $\mathbf{A}^t$ and $\boldsymbol{\pi}^t = -\mathbf{E}^t$. This means, in particular, that 2(83a) cannot hold for the Maxwell field, since it is in contradiction with both 2(87a) and 2(87b). For example, it follows from 2(83a) that

$$[\mathrm{div}\,\mathbf{A}(\mathbf{x}, t), -E_j(\mathbf{x}', t)]_{\mathrm{PB}} = \partial_i\delta_{ij}\delta^{(3)}(\mathbf{x}-\mathbf{x}')$$
$$\neq 0$$

in contradiction with the constraint $\mathrm{div}\,\mathbf{A} = 0$. The correct form for the equal-time Poisson bracket is in fact

$$[A_i(\mathbf{x}, t), -E_j(\mathbf{x}', t)]_{\text{PB}} = \tau_{ij}(\mathbf{x})\delta^{(3)}(\mathbf{x}-\mathbf{x}') \qquad 2(88)$$

where $\tau_{ij}(\mathbf{x})$ is a projection operator satisfying

$$\partial_i \tau_{ij}(\mathbf{x}) = \partial_j \tau_{ij}(\mathbf{x}) = 0$$

However, 2(88) is best derived by working in terms of the canonical coordinates for the normal modes. This will be done for the quantum case in Section 4-2.

The Hamiltonian density for the radiation gauge is defined by

$$\mathscr{H} = -\mathbf{E}^t . \dot{\mathbf{A}}^t - \mathscr{L} \qquad 2(89)$$

Using 2(25) and the fact that $\mathbf{E}^t = -\dot{\mathbf{A}}^t$ we get*

$$\mathscr{H} = \tfrac{1}{2} \sum_{i=1}^{3} (E_i^t E_i^t + \nabla A_i^t . \nabla A_i^t) \qquad 2(90)$$

and

$$H = \tfrac{1}{2} \sum_{i=1}^{3} \int_V (E_i^t E_i^t + \nabla A_i^t . \nabla A_i^t)\, d^3x \qquad 2(91)$$

Rarita–Schwinger Field Defining the momenta canonically conjugate to ψ_μ by

$$\pi_\mu = \frac{\partial \mathscr{L}}{\partial \dot{\psi}_\mu}$$

and using 2(26) we find

$$\pi_\mu = i\bar{\psi}_\mu \gamma_4 - \frac{i}{3}\bar{\psi}_\lambda \gamma_\lambda \delta_{\mu 4} - \frac{i}{3}\bar{\psi}_4 \gamma_\mu + \frac{i}{3}\bar{\psi}_\lambda \gamma_\lambda \gamma_4 \gamma_\mu$$

$$= i(\bar{\psi}_\mu \gamma_4 - \tfrac{1}{3}\bar{\psi}_4 \gamma_\mu) \qquad 2(92)$$

where we have used the subsidiary condition 1(186b), i.e.

$$\gamma_\mu \psi_\mu = \bar{\psi}_\mu \gamma_\mu = 0$$

Owing to the subsidiary condition, the Hamiltonian formalism becomes rather unrewarding in the Rarita–Schwinger case and we shall not pursue it further. The principle difficulty is the isolation of the independent

* By using 2(24) we can also write the Hamiltonian density in the form $\mathscr{H} = \frac{1}{2}(\mathbf{E}^2 + \mathbf{B}^2)$. This is just the familiar expression for the energy density of the electromagnetic field in Heaviside–Lorentz rationalized units.

dynamical fields, which involves the decomposition of the vector spinor ψ_μ into its irreducible spin $\frac{1}{2}$ and $\frac{3}{2}$ parts, and the isolation of the latter. The problems are essentially the same as those encountered in setting up the Hamiltonian form of the one-particle theory. [Moldauer (1956).]

Although the ψ_μ are not independent, the energy density of the field is still given by

$$\begin{aligned}\mathscr{H} &= \pi_\mu \dot{\psi}_\mu - \mathscr{L} \\ &= i\bar{\psi}_\mu \gamma_4 \dot{\psi}_\mu - \frac{i}{3}\bar{\psi}_4 \gamma_\mu \dot{\psi}_\mu - \mathscr{L}\end{aligned} \qquad 2(93)$$

(See the remarks following 2(117a) and 2(117b).) Since $\gamma_\mu \psi_\mu = 0$ and $\mathscr{L}$ vanishes by virtue of the equations of motion, we get, as in 2(76)

$$H = \int_V \mathscr{H}(x)\, d^3x = \int_V \psi^\dagger{}_\mu i\partial_t \psi_\mu \, d^3x \qquad 2(94)$$

Schrödinger Field The canonical treatment of the non-relativistic Schrödinger field is rather similar to that of the Dirac field. Using 2(28) we see that the momentum π canonically conjugate to ψ is given by

$$\pi = \frac{\partial \mathscr{L}}{\partial \dot{\psi}} = i\psi^* \qquad 2(95)$$

while the momentum conjugate to ψ^* vanishes identically. There is therefore only one independent pair of canonically conjugate fields, i.e. ψ and $\pi = i\psi^*$. The Hamiltonian is defined by

$$\begin{aligned}\mathscr{H} &= \pi\dot{\psi} - \mathscr{L} \\ &= \frac{1}{2m}\nabla\psi^* . \nabla\psi + V\psi^*\psi\end{aligned} \qquad 2(96)$$

The total integrated Hamiltonian is therefore given by

$$H = \int_V d^3x \mathscr{H}(x) = \int_V d^3x \psi^*(x)\left[-\frac{1}{2m}\nabla^2 + V(x)\right]\psi(x) \qquad 2(97)$$

after integration by parts. The quantity in square brackets is just the well-known Hamiltonian of the one-particle theory.

The equal-time Poisson brackets for ψ and ψ^* are given by

$$\begin{aligned}[\psi(\mathbf{x}, t), \psi^*(\mathbf{x}', t)]_{\mathrm{PB}} &= -i\delta^{(3)}(\mathbf{x}-\mathbf{x}') \\ [\psi(\mathbf{x}, t), \psi(\mathbf{x}', t)]_{\mathrm{PB}} &= [\psi^*(\mathbf{x}, t), \psi^*(\mathbf{x}', t)]_{\mathrm{PB}} = 0\end{aligned} \qquad 2(98)$$

and the remainder of the canonical formalism goes through in a straightforward way.

2-3 Symmetries and conservation laws

Noether's Theorem In this section we explore the connection between symmetry transformations and conservation laws for classical fields. This connection, known as Noether's theorem, is a basic property of particle and field mechanics. We shall show that continuous symmetry transformations, which leave the Lagrangian density invariant, generate conservation laws which serve to identify constants of the motion.

Let us assume that the Lagrangian density $\mathscr{L}(\phi_A, \partial_\mu \phi_A)$ $(A = 1, \ldots N)$ is invariant under the combined effect of the infinitesimal change of coordinates

$$x'_\mu = x_\mu + \delta x_\mu \qquad 2(99a)$$

and an associated linear field transformation

$$\phi'_A(x') = \phi_A(x) + \sum_{B=1}^{N} S_{AB}\phi_B(x) \qquad 2(99b)$$

The invariance of the Lagrangian density is to be understood as the statement that

$$\mathscr{L}(x) = \mathscr{L}'(x') \qquad 2(100)$$

where $\mathscr{L}'(x')$ stands for $\mathscr{L}(\phi'_A(x'), \partial'_\mu \phi'_A(x'))$. Specific examples will be considered later. For the present we simply assume the existence of a symmetry transformation of the type 2(99a) and 2(99b). We shall make the further assumption that the Jacobian of the coordinate change 2(99a) is equal to 1, i.e.

$$\left|\frac{\partial x'_\mu}{\partial x_\nu}\right| = 1 \qquad 2(101)$$

From these assumptions, it follows that the action integral

$$W_{21} = \int_{t_1}^{t_2} L(t)\, dt$$

remains invariant under 2(99a) and 2(99b) since

$$\int_{\Omega'} \mathscr{L}'(x')\, d^4x' = \int_{\Omega} \mathscr{L}'(x') \left|\frac{\partial x'}{\partial x}\right| d^4x = \int_{\Omega} \mathscr{L}(x)\, d^4x \qquad 2(102)$$

where $d^4x = d^3x\, dt$ and where Ω is the space-time region enclosed in the 4-volume $V \times (t_2 - t_1)$. Ω' denotes the transformed region. We now manipulate the invariance statement 2(102) to derive conservation laws.

Setting

$$\delta W_{21} = \int_{\Omega'} \mathscr{L}'(x')\, d^4x' - \int_{\Omega} \mathscr{L}(x)\, d^4x$$

$$= \int_{\Omega'} \mathscr{L}'(x)\, d^4x - \int_{\Omega} \mathscr{L}(x)\, d^4x$$

we add and subtract the integral $-\int_\Omega \mathscr{L}'(x)\, d^4x$ to get

$$\delta W_{21} = \left(\int_{\Omega'} - \int_{\Omega}\right)\mathscr{L}'(x) + \int_{\Omega} (\mathscr{L}'(x) - \mathscr{L}(x))\, d^4x$$

$$= \int_{\Sigma} \mathscr{L}(x)\delta x_\mu\, d\sigma_\mu + \int_{\Omega} \delta\mathscr{L}(x)\, d^4x \qquad 2(103)$$

where Σ is the surface of V and where $d\sigma_\mu$ is the normal to the surface

$$d\sigma_\mu = \left(dx_2\, dx_3\, dt,\ dx_1\, dx_3\, dt,\ dx_1\, dx_2\, dt,\ \frac{1}{i}\, dx_1\, dx_2\, dx_3\right) \qquad 2(104)$$

whose length is equal to the area of an infinitesimal surface element. In writing 2(103) we have approximated $\mathscr{L}'(x)$ by $\mathscr{L}(x)$ since it multiplies the first order infinitesimal δx_μ. The variation $\delta\mathscr{L}(x) = \mathscr{L}'(x) - \mathscr{L}(x)$ represents the change in $\mathscr{L}$ due to the variations

$$\delta\phi_A(x) = \phi'_A(x) - \phi_A(x) \qquad 2(105)$$

and can be expressed in the form

$$\delta\mathscr{L} = \sum_A \left[\frac{\partial\mathscr{L}}{\partial\phi_A}\delta\phi_A + \frac{\partial\mathscr{L}}{\partial(\partial_\mu\phi_A)}\delta\partial_\mu\phi_A\right]$$

$$= \partial_\mu \sum_A \left(\frac{\partial\mathscr{L}}{\partial(\partial_\mu\phi_A)}\delta\phi_A\right) \qquad 2(106)$$

where we have used the Euler–Lagrange equations 2(16) and the fact that $\delta\partial_\mu\phi_A = \partial_\mu\delta\phi_A$. The latter property holds for all variations which we shall consider.

Inserting 2(106) into 2(103) we can derive the conservation laws by either of two methods. If we convert the second term of 2(103) to a surface integral by means of Gauss's theorem

$$\int_{\Omega} \partial_\mu f_\mu\, d^4x = \int_{\Sigma} f_\mu\, d\sigma_\mu \qquad 2(107)$$

we get

$$\delta W_{21} = \int_{\Sigma} \left(\mathscr{L}(x)\delta x_\mu + \sum_A \frac{\partial \mathscr{L}}{\partial(\partial_\mu \phi_A)} \delta\phi_A \right) d\sigma_\mu \qquad 2(108)$$

Inspection of this integral reveals that only the $\mu = 4$ term yields a non-vanishing result, representing the contributions of the surface planes $t = t_1$ and $t = t_2$. The remainder of the surface integral can be seen to vanish by virtue of the periodic boundary conditions on the surface of the 3-volume V. Thus

$$\delta W_{21} = \left[\int_{t_2} - \int_{t_1} \right] d^3x \left(\mathscr{L}(x)\delta t + \sum_A \frac{\partial \mathscr{L}}{\partial \dot{\phi}_A} \delta\phi_A \right) \qquad 2(109)$$

Equation 2(109) is the desired result. It states that when $\delta W_{21} = 0$ the quantity

$$G(t) = \int_V d^3x \left(\mathscr{L}(x)\delta t + \sum_A \frac{\partial \mathscr{L}}{\partial \dot{\phi}_A} \delta\phi_A \right) \qquad 2(110)$$

is conserved in time, i.e.

$$G(t_1) = G(t_2) \qquad 2(111)$$

Alternatively, we can apply Gauss's theorem in reverse and convert the first term in 2(103) to a volume integral. In this case we obtain

$$\delta W_{21} = \int_{\Omega} \partial_\mu \left(\mathscr{L}(x)\delta x_\mu + \sum_A \frac{\partial \mathscr{L}}{\partial(\partial_\mu \phi_A)} \delta\phi_A \right) d^4x \qquad 2(112)$$

which, for $\delta W_{21} = 0$, yields the differential conservation law

$$\partial_\mu \left(\mathscr{L}(x)\delta x_\mu + \sum_A \frac{\partial \mathscr{L}}{\partial(\partial_\mu \phi_A)} \delta\phi_A \right) = 0 \qquad 2(113)$$

since the infinitesimal parameters which characterize the transformations 2(99a) and 2(99b) can be varied arbitrarily within their range of definition (see the examples given below). From 2(113) we recover the conservation law for $G(t)$ in the usual fashion.

Space–Time Displacements With the exception of 2(28), each of the Lagrangian densities considered in Section 2-1 has no explicit dependence on the space–time coordinates and is therefore invariant under the infinitesimal space–time translations

$$x'_\mu = x_\mu + \varepsilon_\mu \qquad 2(114a)$$

and

$$\phi'_A(x') = \phi_A(x) \qquad (A = 1, \dots N) \tag{2(114b)}$$

where the ε_μ are constants. To apply 2(110) and 2(111) we need the explicit form of $\delta\phi_A(x)$. This may be inferred from 2(114b) by writing

$$0 = \phi'_A(x') - \phi_A(x) = \phi'_A(x') - \phi_A(x') + \phi_A(x') - \phi_A(x)$$
$$= \delta\phi_A(x') + \delta x_\mu \partial_\mu \phi_A(x)$$

or to terms of first order in δx_μ

$$\delta\phi_A(x) = -\delta x_\mu \partial_\mu \phi_A(x)$$
$$= -\varepsilon_\mu \partial_\mu \phi_A(x) \tag{2(115)}$$

Applying 2(110) we conclude that

$$G(t) = \int_V d^3x \left(-i\varepsilon_4 \mathscr{L}(x) - \sum_A \frac{\partial \mathscr{L}}{\partial \dot{\phi}_A} \varepsilon_\mu \partial_\mu \phi_A \right)$$
$$= \varepsilon_\mu \int d^3x \sum_A (-\pi_A \partial_\mu \phi_A - i\delta_{\mu 4} \mathscr{L}(x))$$

Since the ε_μ are arbitrary, each component of the four-vector

$$P_\mu = \int \mathscr{P}_\mu(x)\, d^3x$$
$$\mathscr{P}_\mu = -\sum_A \pi_A \partial_\mu \phi_A - i\delta_{\mu 4} \mathscr{L}(x) \tag{2(116)}$$

is separately conserved. The spatial part

$$\mathbf{P} = -\int \sum_A \pi_A \nabla \phi_A\, d^3x \tag{2(117a)}$$

is identified as the momentum of the field while

$$P_0 = -iP_4 = \int d^3x \sum_A (\pi_A \dot{\phi}_A - \mathscr{L}) \tag{2(117b)}$$

is identified as the energy. Comparison of 2(117b) and 2(62) reveals that the energy of the field is always equal to the Hamiltonian H as given by the canonical prescription 2(62). This applies even when the ϕ_A are not all independent field variables, as for example in the Rarita-Schwinger case.

The differential form of the energy-momentum conservation law may be obtained from 2(113). We get

$$\partial_\mu \mathcal{T}_{\mu\nu} = 0 \tag{2(118)}$$

where $\mathcal{T}_{\mu\nu}$, the *energy-momentum tensor*, is given by

$$\mathcal{T}_{\mu\nu} = -\sum_A \frac{\partial \mathcal{L}}{\partial(\partial_\mu \phi_A)} \partial_\nu \phi_A + \mathcal{L}\delta_{\mu\nu} \tag{2(119)}$$

From 2(117) we recover the conservation of the energy-momentum

$$P_\mu = \int \mathcal{P}_\mu(x)\, d^3x = -i \int \mathcal{T}_{4\mu}(x)\, d^3x$$

From the physical point of view the invariance of the Lagrangian density under space–time displacements means that the field system does not exchange energy and momentum with applied external sources. This clearly fails to apply to the Lagrangian 2(28) for the Schrödinger field in the presence of an arbitrary external potential $V(x)$. If $V(x)$ is time-independent, however, energy (though not momentum) conservation will be preserved.

Space–Time Rotations The reader can easily verify that each Lagrangian density considered in Section 2-1, (with the exception of the non-relativistic Schrödinger Lagrangian), remains invariant under an infinitesimal space–time rotation or Lorentz transformation,

$$x'_\mu = x_\mu + \omega_{\mu\nu} x_\nu \tag{2(120a)}$$

if, at the same time, the fields $\phi_A(x)$ are subjected to the corresponding spin transformation

$$\phi'_A(x') = \phi_A(x) + \frac{i}{2} \sum_B \omega_{\rho\sigma} (s_{\rho\sigma})_{AB} \phi_B(x) \tag{2(120b)}$$

where summation over all values of ρ and σ is understood and where $s_{\rho\sigma}$ is the spin tensor of the theory, i.e.

$$s_{\rho\sigma} = \tfrac{1}{2}\Sigma_{\rho\sigma} \qquad \text{(spin } \tfrac{1}{2})$$

$$s_{\rho\sigma} = S_{\rho\sigma} \qquad \text{(spin 1)}$$

$$s_{\rho\sigma} = S_{\rho\sigma} + \tfrac{1}{2}\Sigma_{\rho\sigma} \qquad \text{(spin } \tfrac{3}{2})$$

For spinless fields of course the second term in 2(120b) is absent. In each case, the set of transformations 2(120a) and 2(120b) is identical with the set which leaves the field equations invariant.

The form of $\delta\phi_A(x)$ can be inferred from 2(120b) by following the same steps as in the derivation of 2(115), with the result

$$\delta\phi_A(x) = -\delta x_\mu \partial_\mu \phi_A(x) + \frac{i}{2}\sum_B \omega_{\rho\sigma}(s_{\rho\sigma})_{AB}\phi_B(x)$$

$$= -\omega_{\mu\nu}x_\nu \partial_\mu \phi_A(x) + \frac{i}{2}\sum_B \omega_{\rho\sigma}(s_{\rho\sigma})_{AB}\phi_B(x) \qquad 2(121)$$

Note that 2(121) is identical to the infinitesimal form of the 'active' Lorentz transformation

$$\phi'_A(x) = \sum_B L_{AB}\phi_B(a^{-1}x)$$

as given for, say, the Dirac theory by 1(77). Applying Noether's theorem we conclude that

$$G(t) = \int_V d^3x \left(-i\omega_{4\nu}x_\nu \mathscr{L} - \omega_{\mu\nu}\sum_A \frac{\partial \mathscr{L}}{\partial \dot{\phi}_A} x_\nu \partial_\mu \phi_A + \frac{i}{2}\omega_{\mu\nu}\sum_{AB}\frac{\partial \mathscr{L}}{\partial \dot{\phi}_A}(s_{\mu\nu})_{AB}\phi_B \right)$$

$$= \int_V d^3x \left[\omega_{\mu\nu}x_\nu \left(-i\mathscr{L}\delta_{4\mu} - \sum_A \pi_A \partial_\mu \phi_A \right) + \frac{i}{2}\omega_{\mu\nu}\sum_{AB}\pi_A(s_{\mu\nu})_{AB}\phi_B \right]$$

$$= \int_V d^3x \left[\omega_{\mu\nu}x_\nu \mathscr{P}_\mu + \frac{i}{2}\omega_{\mu\nu}\sum_{AB}\pi_A(s_{\mu\nu})_{AB}\phi_B \right]$$

is conserved in time. Since the $\omega_{\mu\nu}$ are independent parameters satisfying the antisymmetry condition

$$\omega_{\mu\nu} = -\omega_{\nu\mu} \qquad 2(122)$$

we deduce that the six components of the antisymmetric tensor

$$M_{\mu\nu} = \int d^3x \left[x_\mu \mathscr{P}_\nu - x_\nu \mathscr{P}_\mu - i\sum_{AB}\pi_A(s_{\mu\nu})_{AB}\phi_B\right] \qquad 2(123)$$

are separately conserved. $M_{\mu\nu}$ is known as the total angular momentum tensor of the field and is exhibited in 2(123) as the sum of an orbital part and a spin term. For the Dirac field, for example, we have, using 2(117a) and 2(73a)

$$M_{ij} = \int d^3x\psi^\dagger[-i(x_i\partial_j - x_j\partial_i) + \tfrac{1}{2}\Sigma_{ij}]\psi \qquad 2(124)$$

$$(i, j = 1, 2, 3)$$

for the three spatial components. The quantity in square brackets is just the total angular momentum operator of the one-particle Dirac theory*.

For completeness we give the differential conservation law for angular momentum which follows from 2(113). We have

$$\partial_\lambda \mathscr{M}_{\mu\nu\lambda}(x) = 0 \qquad 2(125)$$

where the angular momentum tensor density $\mathscr{M}_{\mu\nu\lambda}(x)$ is given by

$$\mathscr{M}_{\mu\nu\lambda} = x_\mu \mathscr{T}_{\lambda\nu} - x_\nu \mathscr{T}_{\lambda\mu} - i\sum_{AB} \frac{\partial \mathscr{L}}{\partial(\partial_\lambda \phi_A)}(s_{\mu\nu})_{AB}\phi_B \qquad 2(126)$$

From 2(125) we recover the result that the angular momentum tensor

$$M_{\mu\nu} = -i\int \mathscr{M}_{\mu\nu4}(x)\, d^3x \qquad 2(127)$$

is conserved in time.

Internal Symmetries Additional conservation laws are obtained if the Lagrangian possesses so-called ‘internal’ symmetries, that is, symmetries which are not linked to the Lorentz group. The simplest symmetry transformation of this type is the phase transformation

$$\phi_A \to e^{i\alpha}\phi_A \qquad \phi_A^* \to e^{-i\alpha}\phi_A^* \qquad 2(128)$$

for complex fields $\phi_A \neq \phi_A^*$. Examples of Lagrangian densities which are invariant under the phase transformation are given by 2(19b), 2(20), 2(22), 2(26) and 2(28). In its infinitesimal form, 2(128) may be written as

$$\phi_A \to \phi_A + i\alpha\phi_A \qquad \phi_A^* \to \phi_A^* - i\alpha\phi_A^* \qquad 2(129)$$

and we have

$$0 = \frac{\partial \mathscr{L}}{\partial \alpha} = \sum_A \left(\frac{\partial \mathscr{L}}{\partial \phi_A} i\phi_A + \frac{\partial \mathscr{L}}{\partial(\partial_\mu \phi_A)} i\partial_\mu \phi_A - \frac{\partial \mathscr{L}}{\partial \phi_A^*} i\phi_A^* - \frac{\partial \mathscr{L}}{\partial(\partial_\mu \phi_A^*)} i\partial_\mu \phi_A^* \right)$$

* The physical significance of the conservation law for the ‘time’ components of 2(123) may be grasped by considering the corresponding quantity $i(\mathbf{x}p_0 - x_0\mathbf{p})$ for point particles in classical mechanics. For a single point particle, the conservation of this quantity adds nothing new, but for a collection of particles the conservation law states that the relativistic centre of inertia vector $\sum_n \mathbf{x}^{(n)}p_0^{(n)}/\sum_n p_0^{(n)}$ moves with the uniform velocity $\sum_n \mathbf{p}^{(n)}/\sum_n p_0^{(n)}$.

Using the Euler–Lagrange equations we get

$$0 = \partial_\mu \sum_A \left(\frac{\partial \mathscr{L}}{\partial(\partial_\mu \phi_A)} \phi_A - \frac{\partial \mathscr{L}}{\partial(\partial_\mu \phi_A{}^*)} \phi_A{}^* \right)$$

so that the current

$$j_\mu = -i \sum_A \left(\frac{\partial \mathscr{L}}{\partial(\partial_\mu \phi_A)} \phi_A - \frac{\partial \mathscr{L}}{\partial(\partial_\mu \phi_A{}^*)} \phi_A{}^* \right) \qquad 2(130)$$

is conserved. Computing 2(130) for the Lagrangian densities 2(19b), 2(20), 2(22), 2(26) and 2(28), we recover the conserved currents

$$j_\mu^{\mathrm{KG}} = i((\partial_\mu \phi^*)\phi - (\partial_\mu \phi) \,.\, \phi^*) \qquad 2(131\mathrm{a})$$

$$j_\mu^{\mathrm{Dirac}} = i\bar{\psi}\gamma_\mu \psi \qquad 2(131\mathrm{b})$$

$$j_\mu^V = i(F^*{}_{\mu\nu} A_\nu - F_{\mu\nu} A^*{}_\nu) \qquad 2(131\mathrm{c})$$

$$j_\mu^{\mathrm{RS}} = i\bar{\psi}_\lambda \gamma_\mu \psi_\lambda \qquad 2(131\mathrm{d})$$

$$j_\mu^{\mathrm{S}} = (\mathbf{j}^{\mathrm{S}}, i\rho^{\mathrm{S}}) = \left(\mathrm{Re} \frac{1}{im} \psi^* \,\mathbf{grad}\, \psi, i\psi^* \psi \right) \qquad 2(131\mathrm{e})$$

for the Klein–Gordon, Dirac, Vector, Rarita–Schwinger, and Schrödinger theories. In each case the 'charge'

$$Q = -i \int_V j_4(x)\, d^3x = -i \sum_A \int_V (\pi_A \phi_A - \pi_A{}^* \phi_A{}^*)\, d^3x \qquad 2(132)$$

is conserved in time.

As a more general example of an internal symmetry, let us suppose that a given Lagrangian density $\mathscr{L}(\phi_A, \partial_\mu \phi)$ is invariant under an infinitesimal transformation which mixes the $\phi_A(x)$ at each point, i.e.

$$\phi_A(x) \rightarrow \phi_A(x) + \alpha \sum_B \lambda_{AB} \phi_B(x) \qquad 2(133)$$

where the λ_{AB} are a set of constants and α is an infinitesimal parameter. By following the same steps as before we get the conserved current

$$j_\mu = -\sum_{AB} \frac{\partial \mathscr{L}}{\partial(\partial_\mu \phi_A)} \lambda_{AB} \phi_B \qquad 2(134)$$

and conclude that

$$Q(\lambda) = -\int_V \pi_A \lambda_{AB} \phi_B \, d^3x \qquad 2(135)$$

is conserved in time. The simplest example of this type of symmetry is exhibited by the Lagrangian density 2(19a), which is invariant under an infinitesimal rotation of the two-component vector (ϕ_1, ϕ_2):

$$\begin{aligned} \phi_1'(x) &= \phi_1(x)+\alpha\phi_2(x) \\ \phi_2'(x) &= \phi_2(x)-\alpha\phi_1(x) \end{aligned} \qquad 2(136)$$

The corresponding current

$$j_\mu^{\text{KG}}(x) = +(\partial_\mu\phi_1)\phi_2-(\partial_\mu\phi_2)\phi_1 \qquad 2(137)$$

is identical to 2(131a), and the conserved 'charge' is given by

$$Q = +\int_V (\pi_2\phi_1-\pi_1\phi_2)\, d^3x \qquad 2(138a)$$

or equivalently

$$Q = -i\int_V (\pi\phi-\pi^*\phi^*)\, d^3x \qquad 2(138b)$$

Generators We have seen that to each continuous symmetry transformation of the fields there corresponds a conserved quantity. For space–time symmetries this quantity is given by

$$G(t) = \varepsilon_\mu P_\mu-\omega_{\mu\nu}M_{\mu\nu} \qquad 2(139)$$

for arbitrary space–time displacements and rotations characterized by the infinitesimal parameters ε_μ and $\omega_{\mu\nu}$. For internal symmetries on the other hand, the conserved quantity is some generalized 'charge' of the form 2(132) or 2(135). We shall refer to the conserved quantity corresponding to a given symmetry transformation as the *generator* of that transformation. The justification for this nomenclature is the fact that, in each case, the variations $\delta\phi_A$ are 'generated' by forming the equal-time Poisson bracket of ϕ_A with the generator of the transformation. In particular, for space–time symmetries,

$$\delta\phi_A(\mathbf{x}, t) = [\phi_A(\mathbf{x}, t), G(t)]_{\text{PB}} \qquad 2(140a)$$

$$\delta\pi_A(\mathbf{x}, t) = [\pi_A(\mathbf{x}, t), G(t)]_{\text{PB}} \qquad 2(140b)$$

where $\delta\phi_A$ is the 'active' variation 2(121) and $\delta\pi_A$ is the corresponding variation of π_A. We assume here that the ϕ_A form a set of N independent fields.

For time displacements, 2(140a) and 2(140b) is a direct consequence of the Hamiltonian equations of motion 2(64a) and 2(64b). We have

$$\delta\phi_A(\mathbf{x}, t) = -\varepsilon_0\partial_0\phi_A(\mathbf{x}, t) = [\phi_A(\mathbf{x}, t), -\varepsilon_0 P_0]_{\text{PB}}$$

and similarly for $\delta\pi_A(\mathbf{x}, t)$. For spatial displacements we note that

$$[\phi_A(\mathbf{x}, t), \mathbf{P}(t)]_{\text{PB}} = \frac{\delta\mathbf{P}(t)}{\delta\pi_A(\mathbf{x}, t)} = -\boldsymbol{\nabla}\phi_A(\mathbf{x}, t)$$

$$[\pi_A(\mathbf{x}, t), \mathbf{P}(t)]_{\text{PB}} = -\frac{\delta\mathbf{P}(t)}{\delta\phi_A(\mathbf{x}, t)} = -\boldsymbol{\nabla}\pi_A(\mathbf{x}, t)$$

where we have used 2(117a) and a straightforward generalization of 2(57a) and 2(57b). Thus 2(140) reads

$$\delta\phi_A(\mathbf{x}, t) = [\phi_A(\mathbf{x}, t), \boldsymbol{\varepsilon}\,.\,\mathbf{P}(t)]_{\text{PB}} = -\boldsymbol{\varepsilon}\,.\,\boldsymbol{\nabla}\phi_A(\mathbf{x}, t)$$

$$\delta\pi_A(\mathbf{x}, t) = [\pi_A(\mathbf{x}, t), \boldsymbol{\varepsilon}\,.\,\mathbf{P}(t)]_{\text{PB}} = -\boldsymbol{\varepsilon}\,.\,\boldsymbol{\nabla}\pi_A(\mathbf{x}, t)$$

confirming that $\boldsymbol{\varepsilon}\,.\,\mathbf{P}$ is the generator of spatial displacements in the direction $\boldsymbol{\varepsilon}$. The verification of 2(140a) and 2(140b) for space–time rotations is left as an exercise to the reader. [Problem 2.]

For internal symmetries we have, in place of 2(140a) and 2(140b)

$$\delta\phi_A(\mathbf{x}, t) = [\phi_A(\mathbf{x}, t), -\alpha Q]_{\text{PB}} \qquad 2(141\text{a})$$

$$\delta\pi_A(\mathbf{x}, t) = [\pi_A(\mathbf{x}, t), -\alpha Q]_{\text{PB}} \qquad 2(141\text{b})$$

where Q is the conserved charge as given by 2(132) or 2(135). For the latter case, for example, we have

$$[\phi_A(\mathbf{x}, t), -\alpha Q]_{\text{PB}} = \alpha\sum_B \lambda_{AB}\phi_B(\mathbf{x}, t) = \delta\phi_A(\mathbf{x}, t)$$

in agreement with 2(141a).

In this language, Noether's theorem may be rephrased as the statement that the generator of a symmetry transformation, in the sense 2(140) or 2(141), is a constant of the motion. Let us recover this theorem without recourse to the Lagrangian formulation. Under an arbitrary transformation

$$\delta\phi_A(\mathbf{x}, t) = [\phi_A(\mathbf{x}, t), G(t)]_{\text{PB}} = \frac{\delta G(t)}{\delta\pi_A(\mathbf{x}, t)}$$

$$\delta\pi_A(\mathbf{x}, t) = [\pi_A(\mathbf{x}, t), G(t)]_{\text{PB}} = -\frac{\delta G(t)}{\delta\phi_A(\mathbf{x}, t)}$$

the Hamiltonian functional transforms according to

$$\begin{aligned}\delta H &= \sum_A \int \left(\frac{\delta H}{\delta \phi_A(\mathbf{x}, t)} \delta \phi_A(\mathbf{x}, t) + \frac{\delta H}{\delta \pi_A(\mathbf{x}, t)} \delta \pi_A(\mathbf{x}, t) \right) d^3x \\ &= \sum_A \int \left(\frac{\delta H}{\delta \phi_A(\mathbf{x}, t)} \frac{\delta G(t)}{\delta \pi_A(\mathbf{x}, t)} - \frac{\delta H}{\delta \pi_A(\mathbf{x}, t)} \frac{\delta G(t)}{\delta \phi_A(\mathbf{x}, t)} \right) d^3x \\ &= [H, G(t)]_{\mathrm{PB}} \end{aligned} \qquad 2(142)$$

For space–time displacements, space rotations, and internal symmetries, the symmetry character of the transformation is expressed by the invariance condition

$$\delta H = 0 \qquad 2(143)$$

Thus by 2(54) and 2(142)

$$\dot{G}(t) = [G(t), H]_{\mathrm{PB}} = 0 \qquad 2(144)$$

For pure Lorentz transformations, on the other hand, H transforms as the fourth component of the momentum four-vector

$$\delta H = i\omega_{i4} P_i$$

if the theory is Lorentz-invariant. By 2(142) and 2(139) it follows that

$$[H, M_{i4}]_{\mathrm{PB}} = -iP_i \qquad 2(145)$$

and since M_{i4} as given by 2(123) carries an explicit time dependence, we can apply 2(55) to get

$$\begin{aligned}\dot{M}_{i4} &= [M_{i4}, H]_{\mathrm{PB}} + \frac{\partial M_{i4}}{\partial t} \\ &= iP_i - iP_i \\ &= 0\end{aligned} \qquad 2(146)$$

PROBLEMS

1. Evaluate the functional derivatives in 2(82a) and 2(82b) and show that the results agree with 1(123a) and 1(123b).
2. (a) Derive the transformation law of $\pi_A(x)$ for space rotations and pure Lorentz transformations.
   (b) Complete the proof of the relations 2(140a) and 2(140b).

REFERENCES

Kamefuchi, S. (1966) (with Y. Takahashi) *Nuovo Cimento* **44**, 1
Moldauer, P. A. (1956) (with K. M. Case) *Phys. Rev.* **102**, 279

3

Quantum Fields

3-1 Introduction

The formulation of relativistic quantum mechanics given in Chapter 1 suffers from two very serious defects. The first is the failure of the probability interpretation for particles of spin 0 and 1 (and for integer spin particles in general) owing to the appearance of states with negative probability. The second, more basic, difficulty is the appearance, in all cases, of negative energy states. Our task now is to construct a theory which is free from these defects.

The key to the correct procedure is provided by the Lagrangian formalism. In Chapter 2 we ignored the quantum, or particle, aspects of relativistic wave equations and concentrated simply on the field aspect. By treating the field as a mechanical system we identified certain basic constants of the motion as the energy, momentum, charge, etc. of the field. We now ask: how are these field quantities connected to the properties of single particle states? The formalism presented so far makes no provision for such a connection. In fact there seems to be a contradiction between the particle and field aspects since, for integer spins (though not for half integer spins), the *field* energy is positive-definite. The solution is provided by *field quantization*, whereby the field is reinterpreted as a *quantum* rather than classical mechanical system, with an infinite number of degrees of freedom.

We shall find that to ensure the positive-definiteness of the energy we must treat integer and half-integer spin fields on different footings. Integer spin fields can be quantized by a straightforward extension of the canonical quantization method used in setting up ordinary quantum mechanics, whereas for half-integer spin fields a modified approach is necessary.

In this chapter we shall quantize the Klein–Gordon, Dirac, and non-relativistic Schrödinger fields and discuss general features of quantum field theory, such as the quantum action principle, particle localizability and the connection between spin and statistics. The quantization of spin 1 and spin $\frac{3}{2}$ fields will be taken up in Chapter 4.

3-2 Quantization of the Klein–Gordon field

Real Klein–Gordon Field The simplest example of a relativistic field is the real scalar field. To quantize it we apply the prescription which is used to quantize non-relativistic mechanics, and make the replacement

$$[A, B]_{\text{PB}} \to \frac{1}{i}[A, B] \qquad 3(1)$$

where $[A, B]$ is the commutator $AB - BA$. Effecting this substitution in 2(59) and 2(60) we get the canonical *equal-time commutation relations*

$$[\phi(\mathbf{x}, t), \pi(\mathbf{x}', t)] = i\delta^{(3)}(\mathbf{x} - \mathbf{x}') \qquad 3(2a)$$

$$[\phi(\mathbf{x}, t), \phi(\mathbf{x}', t)] = [\pi(\mathbf{x}, t), \pi(\mathbf{x}', t)] = 0 \qquad 3(2b)$$

In terms of the discrete notation introduced in Section 2.1, these commutation relations are equivalent to the standard quantum mechanical commutation rules

$$[\phi_i, p_j] = i\delta_{ij} \qquad 3(3a)$$

$$[\phi_i, \phi_j] = [p_i, p_j] = 0 \qquad 3(3b)$$

for the infinite set of coordinates ϕ_i and canonically conjugate momenta p_i. To recover 3(2) from 3(3), we simply apply 2(42) and note that, in the continuum limit,

$$\frac{1}{\delta V_i}\delta_{ij} \to \delta^{(3)}(\mathbf{x} - \mathbf{x}')$$

where $\mathbf{x}$ and $\mathbf{x}'$ lie in the ith and jth cell respectively. Alternatively we can decompose the field into normal modes and work in terms of the coordinates $q^{(n)}$ and their conjugate momenta $p^{(n)}$. Then, the commutation rules 3(2a, b) are again equivalent to the standard relations

$$[q_n, p_{n'}] = i\delta_{nn'} \qquad 3(4a)$$

$$[q_n, q_{n'}] = [p_n, p_{n'}] = 0 \qquad 3(4b)$$

Indeed, applying 2(29) and 2(50) we get from 3(4a), for example

$$\begin{aligned}[\phi(\mathbf{x}, t), \pi(\mathbf{x}', t)] &= \sum_{nn'} [q_n(t), p_{n'}(t)]\varphi_n(\mathbf{x})\varphi_{n'}^*(\mathbf{x}') \\ &= i \sum_{n} \delta_{nn'}\varphi_n(\mathbf{x})\varphi_{n'}^*(\mathbf{x}') \\ &= i\delta^{(3)}(\mathbf{x}-\mathbf{x}')\end{aligned}$$

where we have used the completeness relation 2(30b). This checks 3(2a). One can also reverse the argument and establish 3(4a) starting from 3(2a) by using the orthonormality conditions 2(30a).

Since $\varphi(x)$ and $\pi(x)$ are real fields, classically, they become hermitian operators upon quantization. The Hamiltonian

$$H = \tfrac{1}{2} \int_V (\pi^2 + (\nabla\phi)^2 + \mu^2\phi^2)\, d^3x \tag{3(5a)}$$

and the field momentum, given classically by 2(117a),

$$\mathbf{P} = -\int_V \pi\nabla\phi\, d^3x \tag{3(5b)}$$

become hermitian operators, and the Hamiltonian equations of motion 2(58a) and 2(58b) become

$$\dot{\phi}(\mathbf{x}, t) = \frac{1}{i}[\phi(\mathbf{x}, t), H(t)] \tag{3(6a)}$$

$$\dot{\pi}(\mathbf{x}, t) = \frac{1}{i}[\pi(\mathbf{x}, t), H(t)] \tag{3(6b)}$$

upon effecting the replacement 3(1). Equations 3(6a) and 3(6b) are the equations of motion in the *Heisenberg picture* in which the time dependence is carried by the operators. As in ordinary quantum mechanics, we could also describe the evolution of the system in the *Schrödinger picture*, in which all the time dependence is thrown onto the state vectors, but for the present we adhere to the Heisenberg picture.

Let us evaluate the commutators on the right-hand side, using 3(2a) and 3(2b). We find

$$\dot{\phi}(\mathbf{x}, t) = \pi(\mathbf{x}, t) \tag{3(7a)}$$

$$\dot{\pi}(\mathbf{x}, t) = (\nabla^2 - \mu^2)\phi(\mathbf{x}, t) \tag{3(7b)}$$

The first equation reveals that the relation between π and ϕ is the same as in the classical case. Eliminating π we get

$$\ddot{\phi} = (\nabla^2 - \mu^2)\phi \qquad 3(8)$$

so that the quantized field operator still satisfies the Klein–Gordon equation. Thus we are assured that the Hamiltonian equations of motion with the commutation relations 3(2a) and 3(2b) are consistent with the Euler–Lagrange equation of motion

$$\frac{\partial \mathscr{L}}{\partial \phi} = \partial_\mu \frac{\partial \mathscr{L}}{\partial(\partial_\mu \phi)}$$

derived from

$$\mathscr{L}(x) = -\tfrac{1}{2}\partial_\mu \phi \partial_\mu \phi - \tfrac{1}{2}\mu^2 \phi^2$$

by means of Hamilton's principle*.

As the classical and quantum equations of motion are identical, we can expand $\phi(x)$ in terms of the complete set of plane-wave solutions 1(2a) and 1(2b). Instead of 2(36) we now have

$$\phi(\mathbf{x}, t) = \frac{1}{\sqrt{V}} \sum_{\mathbf{k}} \frac{1}{\sqrt{2\omega_{\mathbf{k}}}} (a_{\mathbf{k}}\, e^{i\mathbf{k}.\mathbf{x} - i\omega_{\mathbf{k}} t} + a_{\mathbf{k}}^\dagger\, e^{-i\mathbf{k}.\mathbf{x} + i\omega_{\mathbf{k}} t}) \qquad 3(9)$$

where the *operators* $a_{\mathbf{k}}$ and $a_{\mathbf{k}}^\dagger$ are taken to be hermitian adjoints of each other, to ensure the hermicity of $\phi(\mathbf{x}, t)$. The expansion 3(9) can be inverted to give $a_{\mathbf{k}}$ and $a_{\mathbf{k}}^\dagger$ in terms of ϕ:

$$a_{\mathbf{k}} = \frac{i}{\sqrt{V}} \frac{1}{\sqrt{2\omega_{\mathbf{k}}}} \int_V d^3x\, e^{-i\mathbf{k}.\mathbf{x} + i\omega_{\mathbf{k}} t} \overleftrightarrow{\frac{\partial}{\partial t}} \phi(\mathbf{x}, t) \qquad 3(10a)$$

$$a_{\mathbf{k}}^\dagger = \frac{i}{\sqrt{V}} \frac{1}{\sqrt{2\omega_{\mathbf{k}}}} \int_V d^3x \phi(\mathbf{x}, t) \overleftrightarrow{\frac{\partial}{\partial t}}\, e^{i\mathbf{k}.\mathbf{x} - i\omega_{\mathbf{k}} t} \qquad 3(10b)$$

where the symbol $\overleftrightarrow{\partial}_t$ is defined by

$$A\overleftrightarrow{\partial}_t B = A\partial_t B - (\partial_t A)B \qquad 3(11)$$

* $\phi(x)$ being an operator, we define

$$\frac{\partial \mathscr{L}}{\partial \phi} = \lim_{\varepsilon \to 0} \frac{\mathscr{L}(\phi + \varepsilon) - \mathscr{L}(\phi)}{\varepsilon}$$

where ε is a *c*-number. One must take care to maintain the order of operator factors when taking the variational derivative corresponding to 2(14). For example, the variation of ϕ^2 is given by $\phi\delta\phi + \delta\phi\phi$. However, closer inspection reveals that this complication may be ignored, since, by subjecting ϕ and π in 3(2a) and 3(2b) to independent variations, we deduce that

$$[\delta\phi(x), \pi(x')]_{t=t'} = [\phi(x), \delta\pi(x')]_{t=t'} = [\phi(x), \delta\phi(x')]_{t=t'} = [\pi(x), \delta\pi(x')]_{t=t'} = 0$$

The verification of 3(10a) and 3(10b) is left as an exercise [Problem 1]. The commutation relations for the $a_{\mathbf{k}}$ and $a_{\mathbf{k}}^\dagger$ now follow from 3(2a) and 3(2b) and 3(10a) and 3(10b). Setting

$$f_{\mathbf{k}}(\mathbf{x}, t) = \frac{1}{\sqrt{V}} \frac{1}{\sqrt{2\omega_{\mathbf{k}}}} e^{i\mathbf{k}.\mathbf{x} - i\omega_{\mathbf{k}}t} \qquad 3(12)$$

and using 1(11), we get

$$\begin{aligned}
[a_{\mathbf{k}}, a_{\mathbf{k}'}^\dagger] &= -\int d^3x\, d^3x' [f_{\mathbf{k}}^*(\mathbf{x}, t)\overleftrightarrow{\partial}_t \phi(\mathbf{x}, t), \phi(\mathbf{x}', t)\overleftrightarrow{\partial}_t f_{\mathbf{k}'}(\mathbf{x}', t)] \\
&= \int d^3x\, d^3x' f_{\mathbf{k}}^*(\mathbf{x}, t)\overleftrightarrow{\partial}_t f_{\mathbf{k}'}(\mathbf{x}', t)[\phi(\mathbf{x}, t), \pi(\mathbf{x}', t)] \\
&= \int d^3x f_{\mathbf{k}}^*(\mathbf{x}, t) i\overleftrightarrow{\partial}_t f_{\mathbf{k}'}(\mathbf{x}, t) \\
&= \delta_{\mathbf{k}\mathbf{k}'}
\end{aligned} \qquad 3(13a)$$

Similarly we find

$$[a_{\mathbf{k}}, a_{\mathbf{k}'}] = \int d^3x f_{\mathbf{k}}^*(\mathbf{x}, t) i\overleftrightarrow{\partial}_t f_{\mathbf{k}'}^*(\mathbf{x}, t) = 0 \qquad 3(13b)$$

and

$$[a_{\mathbf{k}}^\dagger, a_{\mathbf{k}'}^\dagger] = 0 \qquad 3(13c)$$

For reasons which will become clear presently, the $a_{\mathbf{k}}^\dagger$ and $a_{\mathbf{k}}$ are known as creation and destruction operators respectively. Let us express the Hamiltonian in terms of these operators. Inserting 3(9) and

$$\pi(\mathbf{x}, t) = \frac{-i}{\sqrt{V}} \sum_{\mathbf{k}} \sqrt{\frac{\omega_{\mathbf{k}}}{2}} (a_{\mathbf{k}} e^{i\mathbf{k}.\mathbf{x} - i\omega_{\mathbf{k}}t} - a_{\mathbf{k}}^\dagger e^{-i\mathbf{k}.\mathbf{x} + i\omega_{\mathbf{k}}t})$$

into 3(5a), and carrying out the integration, we get the simple result

$$H = \tfrac{1}{2} \sum_{\mathbf{k}} (a_{\mathbf{k}}^\dagger a_{\mathbf{k}} + a_{\mathbf{k}} a_{\mathbf{k}}^\dagger)\omega_{\mathbf{k}}$$

or, using 3(13a)

$$H = \sum_{\mathbf{k}} (a_{\mathbf{k}}^\dagger a_{\mathbf{k}} + \tfrac{1}{2})\omega_{\mathbf{k}}$$

where $\frac{1}{2}\sum_{\mathbf{k}} \omega_{\mathbf{k}}$—the so-called *zero point energy*—is an infinite constant. We can proceed in the same way to evaluate the momentum given by

3(5b), with the result

$$\mathbf{P} = \sum_{\mathbf{k}} (a_{\mathbf{k}}{}^{\dagger} a_{\mathbf{k}} + \tfrac{1}{2})\mathbf{k}$$

$$= \sum_{\mathbf{k}} a_{\mathbf{k}}{}^{\dagger} a_{\mathbf{k}} \mathbf{k} \qquad 3(14)$$

Note that the zero point momentum $\frac{1}{2}\sum_{\mathbf{k}} \mathbf{k}$ vanishes by cancellation of $\mathbf{k}$ with $-\mathbf{k}$. The embarrassing zero point energy term $E_0 = \frac{1}{2}\sum_{\mathbf{k}} \omega_{\mathbf{k}}$ is conventionally removed simply by redefining the Hamiltonian to be*

$$H = \sum_{\mathbf{k}} a_{\mathbf{k}}{}^{\dagger} a_{\mathbf{k}} \omega_{\mathbf{k}} \qquad 3(15)$$

Formally this redefinition is known as *normal ordering*. An operator product is in normal ordered form if all creation operators stand to the left of all destruction operators, as in 3(15). Indicating normal ordering by the double-dot notation : :, our redefinition of H is the replacement of 3(5a) by

$$H = \tfrac{1}{2} \int_V :(\pi^2 + (\nabla\phi)^2 + \mu^2\phi^2): d^3x \qquad 3(16)$$

Note that by 3(9), the destruction and creation parts of $\phi(x)$ are associated respectively with the positive and negative frequency (energy) plane wave solutions to the Klein–Gordon equation. Denoting these positive and negative frequency parts by $\phi^{(+)}$ and $\phi^{(-)}$ respectively we have

$$\phi(x) = \phi^{(+)}(x) + \phi^{(-)}(x)$$

and

$$:\phi(x)\phi(y): = \phi^{(+)}(x)\phi^{(+)}(y) + \phi^{(-)}(x)\phi^{(+)}(y)$$

$$+ \phi^{(-)}(x)\phi^{(-)}(y) + \phi^{(-)}(y)\phi^{(+)}(x) \qquad 3(17)$$

Particle Interpretation By means of 3(14) and 3(15), the momentum and energy of the field are exhibited as an infinite sum of terms featuring the operators

$$N_{\mathbf{k}} = a_{\mathbf{k}}{}^{\dagger} a_{\mathbf{k}} \qquad 3(18)$$

* One can show that a nonzero E_0 contradicts the requirement of relativistic invariance [see Problem 13]. Actually, in the physics of material media and, in particular, in the theory of superconductivity, the analog of the zero point energy plays a crucial role, but in relativistic quantum field theory its physical significance, if any, is still obscure. For further discussion of this term see E. A. Power, *Introductory Quantum Electrodynamics*, Longmans, 1964, pp. 31–35.

This crucial result provides the particle interpretation absent in classical Lagrangian field theory. We shall show presently that the $N_{\mathbf{k}}$ are simultaneously diagonalizable and have integer eigenvalues

$$n_{\mathbf{k}} = 0, 1, 2, \ldots$$

so that the field momentum and energy may be written as

$$\mathbf{P} = \sum_{\mathbf{k}} n_{\mathbf{k}} \mathbf{k}$$

$$H = \sum_{\mathbf{k}} n_{\mathbf{k}} \omega_{\mathbf{k}}$$

In this form $\mathbf{P}$ and H are just the momentum and energy of an assembly of spinless particles of mass μ. In this assembly are contained $n_{\mathbf{k}_1}$ particles of momentum $\mathbf{k}_1$, $n_{\mathbf{k}_2}$ particles of momentum $\mathbf{k}_2$, etc. The operators 3(18) are known as *particle number* operators.

The statement that the $N_{\mathbf{k}}$ form a commuting—and, hence, simultaneously diagonalizable—set of operators, follows immediately upon using 3(13)

$$\begin{aligned}[N_{\mathbf{k}}, N_{\mathbf{k}'}] &= a_{\mathbf{k}}{}^{\dagger}[a_{\mathbf{k}}, a_{\mathbf{k}'}{}^{\dagger}]a_{\mathbf{k}'} + a_{\mathbf{k}'}{}^{\dagger}[a_{\mathbf{k}}{}^{\dagger}, a_{\mathbf{k}'}]a_{\mathbf{k}} \\ &= (a_{\mathbf{k}}{}^{\dagger}a_{\mathbf{k}} - a_{\mathbf{k}}{}^{\dagger}a_{\mathbf{k}})\delta_{\mathbf{k}\mathbf{k}'} = 0\end{aligned}$$

The second statement—that the eigenvalues $n_{\mathbf{k}}$ range over the set of all non-negative integers—is actually a well known result if it is realized that

$$H_{\mathbf{k}} = \tfrac{1}{2}(a_{\mathbf{k}}{}^{\dagger}a_{\mathbf{k}} + a_{\mathbf{k}}a_{\mathbf{k}}{}^{\dagger})\omega_{\mathbf{k}}$$

is just the Hamiltonian of a quantum-mechanical harmonic oscillator.* We shall nevertheless rederive the result, since the proof serves to exhibit certain important properties of the $a_{\mathbf{k}}$ and $a_{\mathbf{k}}{}^{\dagger}$. Let us first show that $n_{\mathbf{k}}$ must be non-negative. If $|n_{\mathbf{k}}\rangle$ denotes an eigenstate with eigenvalue $n_{\mathbf{k}}$, i.e.

$$N_{\mathbf{k}}|n_{\mathbf{k}}\rangle = n_{\mathbf{k}}|n_{\mathbf{k}}\rangle \qquad 3(19)$$

then

$$\begin{aligned}\|a_{\mathbf{k}}|n_{\mathbf{k}}\rangle\| &= \langle n_{\mathbf{k}}|a_{\mathbf{k}}{}^{\dagger}a_{\mathbf{k}}|n_{\mathbf{k}}\rangle \\ &= n_{\mathbf{k}}\langle n_{\mathbf{k}}|n_{\mathbf{k}}\rangle \\ &= n_{\mathbf{k}}\||n_{\mathbf{k}}\rangle\|\end{aligned}$$

* The more familiar form $H_{\mathbf{k}} = \frac{1}{2}(P_{\mathbf{k}}^2 + \omega_{\mathbf{k}}^2 Q_{\mathbf{k}}^2)$ can be recovered by setting

$$P_{\mathbf{k}} = \left(\frac{\omega_{\mathbf{k}}}{2}\right)^{1/2}(a_{\mathbf{k}} + a_{\mathbf{k}}{}^{\dagger}) \qquad Q_{\mathbf{k}} = \frac{i}{(2\omega_{\mathbf{k}})^{1/2}}(a_{\mathbf{k}} - a_{\mathbf{k}}{}^{\dagger})$$

The $P_{\mathbf{k}}$ and $Q_{\mathbf{k}}$ are easily seen to satisfy canonical commutation rules of the type 3(4). The representation of a quantum field as an assembly of quantum oscillators is the basis of Planck's law for the radiation from a black body cavity.

where $\| |a\rangle \|$ denotes the norm squared of the state $|a\rangle$. Hence the field energy 3(5a) which was positive-definite classically, retains this desirable feature in quantum field theory. To construct the eigenstates and eigenvalues of $N_{\mathbf{k}}$ we enlist the aid of the commutation rules

$$[N_{\mathbf{k}}, a_{\mathbf{k}}^{\dagger}] = a_{\mathbf{k}}^{\dagger}$$
$$[N_{\mathbf{k}}, a_{\mathbf{k}}] = -a_{\mathbf{k}} \tag{3(20)}$$

to derive

$$N_{\mathbf{k}} a_{\mathbf{k}}^{\dagger}|n_{\mathbf{k}}\rangle = a_{\mathbf{k}}^{\dagger} N_{\mathbf{k}}|n_{\mathbf{k}}\rangle + a_{\mathbf{k}}^{\dagger}|n_{\mathbf{k}}\rangle$$
$$= (n_{\mathbf{k}}+1) a_{\mathbf{k}}^{\dagger}|n_{\mathbf{k}}\rangle \tag{3(21a)}$$
$$N_{\mathbf{k}} a_{\mathbf{k}}|n_{\mathbf{k}}\rangle = a_{\mathbf{k}} N_{\mathbf{k}}|n_{\mathbf{k}}\rangle - a_{\mathbf{k}}|n_{\mathbf{k}}\rangle$$
$$= (n_{\mathbf{k}}-1) a_{\mathbf{k}}|n_{\mathbf{k}}\rangle \tag{3(21b)}$$

for an arbitrary eigenstate $|n_{\mathbf{k}}\rangle$. Equations 3(21a) and 3(21b) state that $a_{\mathbf{k}}^{\dagger}|n_{\mathbf{k}}\rangle$ and $a_{\mathbf{k}}|n_{\mathbf{k}}\rangle$ are eigenstates with eigenvalues $n_{\mathbf{k}}+1$ and $n_{\mathbf{k}}-1$ respectively. For this reason $a_{\mathbf{k}}^{\dagger}$ and $a_{\mathbf{k}}$ are known as particle *creation* and *destruction* operators; they respectively create and destroy one quantum of momentum **k** and energy $\omega_{\mathbf{k}}$. Successive applications of $a_{\mathbf{k}}^{\dagger}$ and $a_{\mathbf{k}}$ generate a series of eigenstates with eigenvalues $n_{\mathbf{k}}+1, n_{\mathbf{k}}+2, \ldots, n_{\mathbf{k}}-1, n_{\mathbf{k}}-2, \ldots$. Since all eigenvalues must be positive, the sequence must terminate at the lower limit with the eigenstate $|n_{\mathbf{k}}^0\rangle$ for which $a_{\mathbf{k}}|n_{\mathbf{k}}^0\rangle = 0$. Hence,

$$N_{\mathbf{k}}|n_{\mathbf{k}}^0\rangle = 0$$

and we conclude that $n_{\mathbf{k}}^0 = 0$. This completes the proof of 3(19). The sequence of eigenstates of $N_{\mathbf{k}}$ is given by

$$|n_{\mathbf{k}}\rangle = \frac{1}{(n_{\mathbf{k}}!)^{1/2}} a_{\mathbf{k}}^{\dagger} \ldots a_{\mathbf{k}}^{\dagger}|0\rangle$$

with the factor $(n_{\mathbf{k}}!)^{-1/2}$ ensuring that $\langle n_{\mathbf{k}}|n_{\mathbf{k}}\rangle = 1$. Simultaneous eigenstates

$$|n_{\mathbf{k}_1}, n_{\mathbf{k}_2}, \ldots\rangle$$

of $N_{\mathbf{k}_1}, N_{\mathbf{k}_2} \ldots$ may be constructed by operating on the 'vacuum' state $|0\rangle = |0, 0, \ldots\rangle$ with the appropriate number of creation operators

$$|n_{\mathbf{k}_1}, n_{\mathbf{k}_2}, \ldots\rangle = \frac{1}{(n_{\mathbf{k}_1}!\, n_{\mathbf{k}_2}! \ldots)^{1/2}} (a_{\mathbf{k}_1}^{\dagger})^{n_{\mathbf{k}_1}} (a_{\mathbf{k}_2}^{\dagger})^{n_{\mathbf{k}_2}} \ldots |0\rangle \tag{3(22)}$$

These states form a non-denumerably infinite set and provide a basis for the Hilbert space in which the field operators are defined*.

The matrix elements of the $a_{\mathbf{k}}$ and $a_{\mathbf{k}}^{\dagger}$ in the basis spanned by the states 3(22) can be derived by observing that

$$a_{\mathbf{k}_i}|n_{\mathbf{k}_1}, n_{\mathbf{k}_2}, \ldots, n_{\mathbf{k}_i}, \ldots\rangle = c(n_{\mathbf{k}_i})|n_{\mathbf{k}_1}, n_{\mathbf{k}_2}, \ldots, n_{\mathbf{k}_i}-1, \ldots\rangle$$

where, since all states are normalized to 1,

$$\begin{aligned} |c(n_{\mathbf{k}_i})|^2 &= \langle n_{\mathbf{k}_1}, n_{\mathbf{k}_2}, \ldots, n_{\mathbf{k}_i}, \ldots|a_{\mathbf{k}_i}{}^{\dagger}a_{\mathbf{k}_i}|n_{\mathbf{k}_1}, n_{\mathbf{k}_2}, \ldots, n_{\mathbf{k}_i}, \ldots\rangle \\ &= n_{\mathbf{k}_i} \end{aligned}$$

Hence $c(n_{\mathbf{k}_i}) = \sqrt{n_{\mathbf{k}_i}}$ to within a phase factor, and we get

$$a_{\mathbf{k}_i}|n_{\mathbf{k}_1}, \ldots, n_{\mathbf{k}_i}, \ldots\rangle = \sqrt{n_{\mathbf{k}_i}}|n_{\mathbf{k}_1}, \ldots, n_{\mathbf{k}_i}-1, \ldots\rangle \qquad 3(23a)$$

and also

$$a_{\mathbf{k}_i}{}^{\dagger}|n_{\mathbf{k}_1}, \ldots n_{\mathbf{k}_i}, \ldots\rangle = \sqrt{n_{\mathbf{k}_i}+1}|n_{\mathbf{k}_1}, \ldots, n_{\mathbf{k}_i}+1, \ldots\rangle \qquad 3(23b)$$

which follows easily from 3(23a). The non-zero matrix elements of $a_{\mathbf{k}_i}$ and $a_{\mathbf{k}_i}{}^{\dagger}$ are therefore

$$\langle n'_{\mathbf{k}_1}, \ldots, n'_{\mathbf{k}_i}, \ldots|a_{\mathbf{k}_i}|n_{\mathbf{k}_1}, \ldots, n_{\mathbf{k}i}, \ldots\rangle = \sqrt{n_{\mathbf{k}_i}}\,\delta(n'_{\mathbf{k}_1}, n_{\mathbf{k}_1}) \ldots \delta(n'_{\mathbf{k}_i}, n_{\mathbf{k}_i}-1) \ldots \qquad 3(24a)$$

$$\langle n'_{\mathbf{k}_1}, \ldots, n'_{\mathbf{k}_i}, \ldots|a_{\mathbf{k}_i}{}^{\dagger}|n_{\mathbf{k}_1}, \ldots, n_{\mathbf{k}_i}, \ldots\rangle = \sqrt{n_{\mathbf{k}_i}+1}\,\delta(n'_{\mathbf{k}_1}, n_{\mathbf{k}_1}) \ldots \delta(n'_{\mathbf{k}_i}, n_{\mathbf{k}_i}+1) \ldots \qquad 3(24b)$$

Observe that although $\mathbf{P}$ and H are diagonal in the basis formed by the simultaneous eigenstates of the number operators 3(18), the field ϕ itself is *not*, in view of 3(9) and 3(24a) and 3(24b). In the vacuum state, in particular, $\phi(\mathbf{x}, t)$ must undergo quantum fluctuations and these lead to experimentally detectable effects, notably the Lamb shift.†

To conclude, we have found that the quantized field is equivalent to an assembly of particles, or *quanta*. These quanta are indistinguishable, since state vectors of the form 3(22) are characterized only by the *number* of quanta corresponding to each mode k. We can infer that these quanta obey *Bose–Einstein statistics* by observing that the state vectors 3(22) are left unaltered by a rearrangement of the various particle creation operators $a_{\mathbf{k}}{}^{\dagger}$. Indeed, from the commutation relations 3(13) we see that all $a_{\mathbf{k}}{}^{\dagger}$ commute with each other so that their ordering in 3(22) is immaterial.

* In contrast, the Hilbert space of ordinary quantum mechanics is spanned by a denumerably infinite basis. Hilbert spaces which are spanned by a non-denumerably infinite basis are called non-separable. One consequence of the infinite number of field oscillators is the existence of a denumerable infinity of irreducible representations of the commutation rules 3(13) other than the one constructed here. [See Gärding (1954), Wightman (1955) and Haag (1955).]

† See Welton (1948).

Normalization Conventions As a result of 3(13), one-particle states

$$|\mathbf{k}\rangle = a_{\mathbf{k}}^{\dagger}|0\rangle$$

are normalized according to

$$\begin{aligned}\langle \mathbf{k}'|\mathbf{k}\rangle &= \langle 0|a_{\mathbf{k}'}\ a_{\mathbf{k}}^{\dagger}|0\rangle \\ &= \langle 0|[a_{\mathbf{k}'}, a_{\mathbf{k}}^{\dagger}]|0\rangle \\ &= \delta_{\mathbf{k}\mathbf{k}'} \end{aligned} \qquad 3(25)$$

An alternative convention is the covariant prescription

$$\langle \mathbf{k}'|\mathbf{k}\rangle = 2k_0\delta_{\mathbf{k}\mathbf{k}'} \qquad 3(26)$$

but this requires that the creation and destruction operators be redefined so as to satisfy

$$[\bar{a}_{\mathbf{k}}, \bar{a}_{\mathbf{k}'}^{\dagger}] = 2k_0\delta_{\mathbf{k}\mathbf{k}'}$$

the relation between the two conventions being provided by

$$\bar{a}_{\mathbf{k}} = \sqrt{2k_0}\, a_{\mathbf{k}}$$

The latter convention has the advantage that the matrix element of $\phi(x)$ between the one-particle state and the vacuum is simply

$$\langle 0|\phi(x)|\mathbf{k}\rangle = \frac{1}{\sqrt{V}}\,\mathrm{e}^{ik.x} \qquad 3(27)$$

[see Problem 2] whereas in our convention

$$\langle 0|\phi(x)|\mathbf{k}\rangle = \frac{1}{\sqrt{V}}\frac{1}{\sqrt{2\omega_{\mathbf{k}}}}\,\mathrm{e}^{ik.x} \qquad 3(28)$$

Complex Klein–Gordon Field The extension of field quantization to complex scalar fields is straightforward. To quantize we replace the Poisson brackets in 2(72a) and 2(72b) by commutators in accordance with 3(1). This yields the equal-time commutation rules

$$[\phi(\mathbf{x}, t), \pi(\mathbf{x}', t)] = i\delta^{(3)}(\mathbf{x}-\mathbf{x}') \qquad 3(29a)$$

$$[\phi^{\dagger}(\mathbf{x}, t), \pi^{\dagger}(\mathbf{x}', t)] = i\delta^{(3)}(\mathbf{x}-\mathbf{x}') \qquad 3(29b)$$

with all other equal-time commutators vanishing. Equivalently, we can work with two independent hermitian field operators ϕ_1 and ϕ_2 given by

$$\phi_1 = \frac{1}{\sqrt{2}}(\phi+\phi^\dagger)$$
$$-i\phi_2 = \frac{1}{\sqrt{2}}(\phi-\phi^\dagger) \qquad 3(30)$$

The non-vanishing equal-time commutators are then

$$[\phi_i(\mathbf{x}, t), \pi_j(\mathbf{x}', t)] = i\delta_{ij}\delta^{(3)}(\mathbf{x}-\mathbf{x}') \qquad 3(31)$$

The latter notation is more convenient, as it allows us to generalize to the complex case all results derived for the real scalar field. In particular, the Hamiltonian, calculated with the aid of 2(70b), becomes a sum of two terms of the form 3(15), namely

$$H = \sum_{\mathbf{k}} (a_{\mathbf{k}1}, a_{\mathbf{k}2} + a_{\mathbf{k}1}{}^\dagger, a_{\mathbf{k}2}{}^\dagger)\omega_{\mathbf{k}} \qquad 3(32)$$

where $a_{\mathbf{k}i}{}^\dagger$ and $a_{\mathbf{k}i}$ are the creation and destruction operators for particles of type i ($i = 1, 2$) appearing in the Fourier decompositions

$$\phi_i(\mathbf{x}, t) = \frac{1}{\sqrt{V}}\sum_{\mathbf{k}}\frac{1}{\sqrt{2\omega_{\mathbf{k}}}}(a_{\mathbf{k}i}\, e^{i\mathbf{k}.\mathbf{x}-i\omega_{\mathbf{k}}t} + a_{\mathbf{k}i}{}^\dagger\, e^{-i\mathbf{k}.\mathbf{x}+i\omega_{\mathbf{k}}t}) \qquad 3(33)$$

$(i = 1, 2)$

The $a_{\mathbf{k}i}$ and $a_{\mathbf{k}i}{}^\dagger$ satisfy the commutation rules

$$[a_{\mathbf{k}i}, a_{\mathbf{k}'j}{}^\dagger] = \delta_{ij}\delta_{\mathbf{k}\mathbf{k}'}$$
$$[a_{\mathbf{k}i}, a_{\mathbf{k}j}] = [a_{\mathbf{k}i}{}^\dagger, a_{\mathbf{k}j}{}^\dagger] = 0 \qquad 3(34)$$

generalizing 3(13). By lifting the reality constraint on $\phi(\mathbf{x}, t)$ we simply double the number of degrees of freedom.

Physically, the additional degree of freedom is connected with the charge of the field, given classically by 2(138b) and quantum mechanically by

$$Q = -i\int (\pi\phi - \pi^\dagger\phi^\dagger)\, d^3x \qquad 3(35)$$

The expansions of ϕ and $\phi^\dagger$ obtained from 3(33) by inverting 3(30) are given by

$$\phi(x) = \frac{1}{\sqrt{V}}\sum_{\mathbf{k}}\frac{1}{\sqrt{2\omega_{\mathbf{k}}}}[a_{\mathbf{k}}\, e^{i\mathbf{k}.\mathbf{x}-i\omega_{\mathbf{k}}t} + b_{\mathbf{k}}{}^\dagger\, e^{-i\mathbf{k}.\mathbf{x}+i\omega_{\mathbf{k}}t}] \qquad 3(36a)$$

$$\phi^\dagger(x) = \frac{1}{\sqrt{V}}\sum_{\mathbf{k}}\frac{1}{\sqrt{2\omega_{\mathbf{k}}}}[b_{\mathbf{k}}\,\mathrm{e}^{i\mathbf{k}.\mathbf{x}-i\omega_{\mathbf{k}}t}+a_{\mathbf{k}}{}^\dagger\,\mathrm{e}^{-i\mathbf{k}.\mathbf{x}+i\omega_{\mathbf{k}}t}] \qquad 3(36b)$$

where we have defined

$$a_{\mathbf{k}} = \frac{1}{\sqrt{2}}(a_{\mathbf{k}1}-ia_{\mathbf{k}2})$$
$$b_{\mathbf{k}} = \frac{1}{\sqrt{2}}(a_{\mathbf{k}1}+ia_{\mathbf{k}2}) \qquad 3(37)$$

From 3(37), the $a_{\mathbf{k}}$, $b_{\mathbf{k}}$ and their hermitian adjoints satisfy

$$[a_{\mathbf{k}}, a_{\mathbf{k}'}{}^\dagger] = \delta_{\mathbf{k}\mathbf{k}'} \qquad [b_{\mathbf{k}}, b_{\mathbf{k}'}{}^\dagger] = \delta_{\mathbf{k}\mathbf{k}'} \qquad 3(38)$$

with all other commutators vanishing. The charge can now be computed with the aid of 3(36a) and 3(36b) and the identifications

$$\pi = \dot{\phi}^\dagger \qquad \pi^\dagger = \dot{\phi} \qquad 3(39)$$

which may be obtained in the same way as 3(7a). A straightforward calculation yields the diagonal form

$$Q = \sum_{\mathbf{k}}(a_{\mathbf{k}}{}^\dagger a_{\mathbf{k}} - b_{\mathbf{k}}{}^\dagger b_{\mathbf{k}}) \qquad 3(40)$$

On the other hand, the Hamiltonian 3(32), when expressed in terms of the $a_{\mathbf{k}}$ and $b_{\mathbf{k}}$, becomes

$$H = \sum_{\mathbf{k}}(a_{\mathbf{k}}{}^\dagger a_{\mathbf{k}} + b_{\mathbf{k}}{}^\dagger b_{\mathbf{k}}) \qquad 3(41)$$

Hence, the $a_{\mathbf{k}}{}^\dagger$ and $b_{\mathbf{k}}{}^\dagger$ can be interpreted as creation operators for positively and negatively charged particles respectively. We have confirmed that the total probability of the one-particle theory becomes the total charge of an assembly of particles and anti-particles in quantum field theory. The charge in question may be either electric charge or some other conserved quantum number, i.e. strangeness. Depending on the interpretation which is placed on Q, the complex scalar field can be used to represent π^+ and π^- particles, K^+ and K^- particles, or K^0 and $\overline{K^0}$ particles. The real scalar field, on the other hand, can be used to represent particles like the π^0, which carry no charge, electric or otherwise.

Conservation Laws and Generators As in the classical field case, the conservation laws can be derived either from the invariance of the action integral or from the transformation law of the Hamiltonian. In the former case, one uses the quantum analogue of 2(109), namely

$$\delta W_{21} = G(t_2) - G(t_1) \tag{3(42)}$$

where $G(t)$ is now a hermitian operator

$$G(t) = \int_V d^3x(\mathscr{L}(x)\delta t + \pi\delta\phi + \pi^\dagger\delta\phi^\dagger) \tag{3(43)}$$

and where we assume, for definiteness, that we are dealing with a charged scalar field, i.e.

$$\mathscr{L}(x) = -\partial_\mu\phi^\dagger\partial_\mu\phi - \mu^2\phi^\dagger\phi \tag{3(44)}$$

The conservation laws for energy, momentum and angular momentum flow from 3(42) exactly as in the classical case, conservation of

$$G(t) = \varepsilon_\mu P_\mu - \omega_{\mu\nu} M_{\mu\nu} \tag{3(45)}$$

being assured by the invariance of the action integral under infinitesimal space-time displacements and rotations.

The active variations $\delta\phi(x)$ and $\delta\pi(x)$ are generated by $G(t)$ according to

$$\delta\phi(\mathbf{x}, t) = \frac{1}{i}[\phi(\mathbf{x}, t), G(t)] \tag{3(46a)}$$

$$\delta\pi(\mathbf{x}, t) = \frac{1}{i}[\pi(\mathbf{x}, t), G(t)] \tag{3(46b)}$$

These quantum relations replace the classical equations 2(140a) and 2(140b). As a check we consider space translations $\mathbf{x}' = \mathbf{x} + \boldsymbol{\epsilon}$ for which the generator is the hermitian operator

$$G(t) = \boldsymbol{\epsilon} . \mathbf{P} = -\boldsymbol{\epsilon}\int(\pi\nabla\phi + \pi^\dagger\nabla\phi^\dagger)\, d^3x$$

Applying the equal-time commutation rules, we have

$$[\phi(\mathbf{x}, t), G(t)] = -\boldsymbol{\epsilon}\int[\phi(\mathbf{x}, t), \pi(\mathbf{x}', t)]\nabla\phi(\mathbf{x}', t)\, d^3x'$$

$$= -i\boldsymbol{\epsilon} . \nabla\phi(\mathbf{x}, t) = i\delta\phi(\mathbf{x}, t)$$

in accordance with 3(46a). The corresponding relations

$$\delta\phi^\dagger(\mathbf{x}, t) = \frac{1}{i}[\phi^\dagger(\mathbf{x}, t), G(t)]$$

$$\delta\pi^\dagger(\mathbf{x}, t) = \frac{1}{i}[\pi^\dagger(\mathbf{x}, t), G(t)]$$

for the canonically conjugate pair $\phi^\dagger$ and $\pi^\dagger$ are immediate consequences of 3(46a) and 3(46b) and the hermiticity of $G(t)$.

Turning to internal symmetries, we observe that the conserved charge

$$-\alpha Q = i\alpha \int (\pi\phi - \pi^\dagger\phi^\dagger)\, d^3x$$

is the generator of infinitesimal phase transformations

$$\delta\phi = i\alpha\phi = \frac{1}{i}[\phi, -\alpha Q] \tag{3(47a)}$$

$$\delta\pi = -i\alpha\pi = \frac{1}{i}[\pi, -\alpha Q] \tag{3(47b)}$$

Note that from 3(47a) and 3(47b) we recover the statement that $\phi(\phi^\dagger)$ destroys (creates) one unit of positive charge, since

$$Q\phi|n\rangle = \phi Q|n\rangle - \phi|n\rangle = (n-1)\phi|n\rangle \tag{3(48a)}$$

$$Q\phi^\dagger|n\rangle = \phi^\dagger Q|n\rangle + \phi^\dagger|n\rangle = (n+1)\phi^\dagger|n\rangle \tag{3(48b)}$$

where $|n\rangle$ denotes a state of charge $+n$.

The form of 3(46a), 3(46b) and 3(47a), 3(47b) has special significance in quantum theory for it implies that the variations of the fields may be expressed as infinitesimal unitary transformations in Hilbert space. Indeed, setting

$$\delta\phi = \phi' - \phi \qquad \delta\pi = \pi' - \pi$$

we have

$$\phi' = \phi - i[\phi, G] = U^\dagger\phi U \tag{3(49a)}$$

and

$$\pi' = U^\dagger\pi U \tag{3(49b)}$$

where U is the unitary operator

$$U = 1 - iG \tag{3(50)}$$

with G infinitesimal and hermitian. Since G is time-independent, so is U. Selecting a basis in Hilbert space we may regard the variations $\delta\phi$ as variations of the matrix elements of $\phi(x)$ in the specified basis, i.e.

$$\langle\alpha|\phi(x)|\beta\rangle \to \langle\alpha|U^\dagger\phi(x)U|\beta\rangle$$

Naturally, we can also view this as a transformation of the state vectors

$$|\alpha\rangle \to U|\alpha\rangle \qquad |\beta\rangle \to U|\beta\rangle$$

or in differential form

$$\begin{aligned}\delta|\alpha\rangle &= -iG|\alpha\rangle \\ \delta|\beta\rangle &= -iG|\beta\rangle\end{aligned} \qquad 3(51)$$

From the fact that the active variations of $\phi(x)$ and $\pi(x)$ under infinitesimal space-time displacements and rotations can be represented as unitary transformations, we can directly deduce the invariance of the equal-time commutation relations. Multiplying 3(29a), for example, by $U^\dagger$ on the left and U on the right, we have

$$\begin{aligned}U^\dagger[\phi(\mathbf{x}, t), \pi(\mathbf{x}', t)]U &= [U^\dagger\phi(\mathbf{x}, t)U, U^\dagger\pi(\mathbf{x}', t)U] \\ &= [\phi'(\mathbf{x}, t), \pi'(\mathbf{x}', t)] = i\delta^{(3)}(\mathbf{x}-\mathbf{x}')\end{aligned}$$

Similarly

$$U^\dagger[\phi(\mathbf{x}, t), \phi^\dagger(\mathbf{x}', t)]U = [\phi'(\mathbf{x}, t), \phi'^\dagger(\mathbf{x}', t)] = 0$$

Note that under infinitesimal space-time displacements and rotations

$$x'_\mu = a_{\mu\nu}x_\nu + \varepsilon_\mu$$

with $a_{\mu\nu} = \delta_{\mu\nu} + \omega_{\mu\nu}$, the active variation

$$\phi'(x) = \phi(x) - \omega_{\mu\nu}x_\nu\partial_\mu\phi(x) - \varepsilon_\mu\partial_\mu\phi(x) \qquad 3(52)$$

may be written as*

$$\phi'(x) = \phi(a^{-1}x - \varepsilon) \qquad 3(53)$$

The transformation 3(49a) may therefore be presented in the form

$$\phi(a^{-1}x - \varepsilon) = U(a, \varepsilon)^\dagger\phi(x)U(a, \varepsilon) \qquad 3(54a)$$

or equivalently, substituting $x \to ax + \varepsilon$

$$U(a, \varepsilon)\phi(x)U(a, \varepsilon)^\dagger = \phi(ax + \varepsilon) \qquad 3(54b)$$

This is the common notation for the variation of a scalar field under infinitesimal space-time rotations and displacements. It is also useful to have the explicit form of the transformation law for *finite* space-time translations $x'_\mu = x_\mu + b_\mu$:

$$\mathrm{e}^{-ib.P}\phi(x)\,\mathrm{e}^{ib.P} = \phi(x + b) \qquad 3(55)$$

* See Equation 1(15).

This may be derived by expanding $\phi(x+b)$ in a Taylor series

$$\phi(x+b) = \phi(x)+b_\mu\partial_\mu\phi(x)+\frac{1}{2!}b_\mu b_\nu\partial_\mu\partial_\nu\phi(x)+\ldots$$

$$= \phi(x)+i[\phi(x), b\,.\,P]+\frac{i^2}{2!}[[\phi(x), b\,.\,P], b\,.\,P]+\ldots$$

and using the identity

$$\mathrm{e}^{-i\lambda S}O\,\mathrm{e}^{i\lambda S} = O+i\lambda[O, S]+\frac{i^2\lambda^2}{2!}[[O, S], S]+\ldots \qquad 3(56)$$

valid for arbitrary operators O and S.

To recover the connection between symmetries and conservation laws in the language of generators, we borrow the corresponding argument for the classical case. We observe that the condition for symmetry under infinitesimal space-time displacements, spatial rotations and internal transformations is the invariance of the Hamiltonian

$$H[\phi, \pi]$$

under the change 3(49a) and 3(49b). As H is a polynomial functional of ϕ and π it will undergo the alteration

$$H \to U^\dagger HU = H-i[H, G]$$

so that the symmetry condition is $[H, G] = 0$, implying that G is a constant of the motion. On the other hand, for pure Lorentz transformations generated by $-\omega_{i4}M_{i4}$, the requirement for symmetry is that $(\mathbf{P}, iH)$ transform as a 4-vector, i.e. that

$$\delta H = \frac{1}{i}[H, -\omega_{i4}M_{i4}] = i\omega_{i4}P_i$$

$$\delta P_j = \frac{1}{i}[P_j, -\omega_{i4}M_{i4}] = i\omega_{j4}H$$

or

$$[H, M_{i4}] = P_i \qquad 3(57a)$$

$$[P_j, M_{i4}] = \delta_{ij}H \qquad 3(57b)$$

Equation 3(57a) can be used to show that $\dot{M}_{i4} = 0$ by the same steps as in 2(146).

Equations 3(57a) and 3(57b) can be verified by explicit calculation with the aid of the equal-time commutation relations. [See Problem 3.]

Hence we are assured that the quantized scalar theory, though based on 3-dimensional equal-time commutation rules, is properly Lorentz-invariant.

Discrete Symmetries We have dealt so far with continuous symmetry transformations. We now turn to discrete symmetries, i.e. symmetries which cannot be represented in infinitesimal form. These include space-reflection, time-reflection and charge conjugation.

We consider space-reflection first. The Klein–Gordon equation and the equal-time commutation relations are invariant under the discrete operations

$$\phi(\mathbf{x}, t) \rightarrow +\phi(-\mathbf{x}, t) \tag{3(58a)}$$

and

$$\phi(\mathbf{x}, t) \rightarrow -\phi(-\mathbf{x}, t) \tag{3(58b)}$$

The transformations 3(58a) and 3(58b) are known as the (active) space-reflection or *parity* transformations. The choice between the alternatives 3(58a) and 3(58b) fixes the so-called ' intrinsic parity ' of the field. In the first case, the field is scalar and, in the second, pseudoscalar. The intrinsic parity can only be fixed by appealing to experiment. Thus, if ϕ represents the π-meson or K-meson field, its intrinsic parity must be taken to be odd*.

We wish to represent 3(58a) and 3(58b) by unitary transformations in Hilbert space, as in 3(54b). We therefore seek unitary operators $\mathscr{P}_s$ and $\mathscr{P}_{ps}$ such that

$$\mathscr{P}_s\phi(\mathbf{x}, t)\mathscr{P}_s^{-1} = +\phi(-\mathbf{x}, t)$$

$$\mathscr{P}_{ps}\phi(\mathbf{x}, t)\mathscr{P}_{ps}^{-1} = -\phi(-\mathbf{x}, t)$$

It is easiest to treat the problem in momentum space and determine $\mathscr{P}$ in terms of creation and destruction operators [Federbush (1958)]. Using 3(36a) we get, for $\mathscr{P}_s$ for example

$$\mathscr{P}_s a_{\mathbf{k}} \mathscr{P}_s^{-1} = a_{-\mathbf{k}} \tag{3(59a)}$$

$$\mathscr{P}_s b_{\mathbf{k}} \mathscr{P}_s^{-1} = b_{-\mathbf{k}} \tag{3(59b)}$$

* Since parity determination experiments always involve an interaction among particles, the intrinsic parity of a particle is meaningful only when specified *relative to the other particles with which it interacts.* For example, the negative parity assignment for the charged π is always made under the assumption that the parity of the proton relative to that of the neutron is even. The former cannot be determined independently of the latter. (See for example W. S. C. Williams, *An Introduction to Elementary Particles*, Academic Press, 1961, Chapter 7).

Since we seek a unitary operator, let us set

$$\mathscr{P}_s = e^{i\lambda S} \qquad 3(60)$$

A certain amount of trial and error and the use of the identity 3(56) or equivalently

$$e^{i\lambda S} O\, e^{-i\lambda S} = O + i\lambda[S, O] - \frac{\lambda^2}{2!}[S, [S, O]] + \ldots \qquad 3(61)$$

suggests the ansatz

$$S = \sum_{\mathbf{k}} (a_{\mathbf{k}}{}^{\dagger} a_{-\mathbf{k}} + b_{\mathbf{k}}{}^{\dagger} b_{-\mathbf{k}}) \qquad 3(61)$$

for which

$$[S, a_{\mathbf{k}}] = -a_{-\mathbf{k}}$$

$$[S, [S, a_{\mathbf{k}}]] = a_{\mathbf{k}}$$

$$\vdots$$

and

$$e^{i\lambda S} a_{\mathbf{k}}\, e^{-i\lambda S} = a_{\mathbf{k}} - i\lambda a_{-\mathbf{k}} - \frac{\lambda^2}{2!} a_{\mathbf{k}} + \frac{i\lambda^3}{3!} a_{-\mathbf{k}} + \ldots$$

$$= a_{\mathbf{k}} \cos \lambda - i a_{-\mathbf{k}} \sin \lambda$$

with a similar result for $b_{\mathbf{k}}$. We can dispose of the term in $a_{\mathbf{k}}$ by setting $\lambda = \pi/2$. This yields

$$e^{i\frac{\pi}{2}S} a_{\mathbf{k}}\, e^{-i\frac{\pi}{2}S} = -i a_{-\mathbf{k}}$$

$$e^{i\frac{\pi}{2}S} b_{\mathbf{k}}\, e^{-i\frac{\pi}{2}S} = -i b_{-\mathbf{k}}$$

This is very close to 3(59a) and 3(59b). To get exactly 3(59a) and 3(59b) we must multiply our ansatz 3(60) by an operator

$$e^{i\lambda' S'}$$

such that

$$e^{i\lambda' S'} a_{\mathbf{k}}\, e^{-i\lambda' S'} = i a_{\mathbf{k}}$$

$$e^{i\lambda' S'} b_{\mathbf{k}}\, e^{-i\lambda' S'} = i b_{\mathbf{k}}$$

Our experience with 3(61) now indicates that we should take

$$S' = \sum_{\mathbf{k}} (a_{\mathbf{k}}{}^{\dagger} a_{\mathbf{k}} + b_{\mathbf{k}}{}^{\dagger} b_{\mathbf{k}})$$

and

$$\lambda' = -\frac{\pi}{2}$$

Combining results, we have

$$\mathscr{P}_s = e^{-i\frac{\pi}{2}S'} e^{i\frac{\pi}{2}S}$$

$$= \exp\left[-\frac{i\pi}{2}\sum_{\mathbf{k}}(a_{\mathbf{k}}{}^{\dagger}a_{\mathbf{k}} - a_{\mathbf{k}}{}^{\dagger}a_{-\mathbf{k}} + b_{\mathbf{k}}{}^{\dagger}b_{\mathbf{k}} - b_{\mathbf{k}}{}^{\dagger}b_{-\mathbf{k}})\right] \qquad 3(62a)$$

Our analysis also indicates that

$$\mathscr{P}_{ps} = e^{i\frac{\pi}{2}S'} e^{i\frac{\pi}{2}S}$$

$$= \exp\left[\frac{i\pi}{2}\sum_{\mathbf{k}}(a_{\mathbf{k}}{}^{\dagger}a_{\mathbf{k}} + a_{\mathbf{k}}{}^{\dagger}a_{-\mathbf{k}} + b_{\mathbf{k}}{}^{\dagger}b_{\mathbf{k}} + b_{\mathbf{k}}{}^{\dagger}b_{-\mathbf{k}})\right] \qquad 3(62b)$$

Both $\mathscr{P}_s$ and $\mathscr{P}_{ps}$ are unitary. Moreover, they satisfy

$$\mathscr{P}_s|0\rangle = \mathscr{P}_{ps}|0\rangle = |0\rangle \qquad 3(63)$$

as can be seen by expanding the exponentials in 3(62a) and 3(62b). Thus the parity of the vacuum is always even. As a consequence of 3(59a) and 3(59b), the momentum operator

$$\mathbf{P} = \sum_{\mathbf{k}}(a_{\mathbf{k}}{}^{\dagger}a_{\mathbf{k}} + b_{\mathbf{k}}{}^{\dagger}b_{\mathbf{k}})\mathbf{k}$$

transforms correctly according to

$$\mathscr{P}_s\mathbf{P}\mathscr{P}_s^{-1} = \mathscr{P}_{ps}\mathbf{P}\mathscr{P}_{ps}^{-1} = -\mathbf{P}$$

On the other hand, the parity operator commutes with the Hamiltonian and the angular momentum M_{ij}. [See Problem 4.]

We turn now to the charge-conjugation or particle-antiparticle conjugation operation. We seek an operator $\mathscr{C}$ which transforms the particle operators $a_{\mathbf{k}}$ and $a_{\mathbf{k}}{}^{\dagger}$ into the antiparticle operators $b_{\mathbf{k}}$ and $b_{\mathbf{k}}{}^{\dagger}$ respectively, i.e.

$$\mathscr{C}a_{\mathbf{k}}\mathscr{C}^{-1} = b_{\mathbf{k}} \qquad 3(64a)$$

$$\mathscr{C}b_{\mathbf{k}}\mathscr{C}^{-1} = a_{\mathbf{k}} \qquad 3(64b)$$

or, in terms of the field operators

$$\mathscr{C}\phi(x)\mathscr{C}^{-1} = \phi^{\dagger}(x) \qquad 3(65)$$

This operation is clearly a symmetry of the charged scalar theory. The explicit construction of $\mathscr{C}$ can be performed by the same procedure as for $\mathscr{P}$. The result is

$$\mathscr{C} = \exp\left[-\frac{i\pi}{2}\sum_{\mathbf{k}}(a_{\mathbf{k}}{}^{\dagger}a_{\mathbf{k}} - a_{\mathbf{k}}{}^{\dagger}b_{\mathbf{k}} + b_{\mathbf{k}}{}^{\dagger}b_{\mathbf{k}} - b_{\mathbf{k}}{}^{\dagger}a_{\mathbf{k}})\right] \qquad 3(66)$$

[See Problem 5.] $\mathscr{C}$ is unitary and satisfies $\mathscr{C}|0\rangle = |0\rangle$. From 3(65) we see that

$$\mathscr{C}j_{\mu}\mathscr{C}^{-1} = -j_{\mu} \qquad 3(67)$$

where j_{μ} is the current operator for the charged scalar field

$$j_{\mu} = i[(\partial_{\mu}\phi^{\dagger})\phi - (\partial_{\mu}\phi)\phi^{\dagger}] \qquad 3(68)$$

Finally we consider time-reflection. This operation differs from the other two in that it cannot be represented by a unitary transformation in Hilbert space. The Klein–Gordon equation is invariant under the transformation

$$\phi(\mathbf{x}, t) \to \pm\phi(\mathbf{x}, -t) \qquad 3(69)$$

but the equal-time commutation rules, for example

$$[\phi(\mathbf{x}, t), \dot{\phi}(\mathbf{x}', t)] = i\delta^{(3)}(\mathbf{x}-\mathbf{x}') \qquad 3(70)$$

do not exhibit this invariance unless 3(69) is accompanied by the change $i \to -i$. Therefore, if we seek a time-reversal operator $\mathscr{T}$ such that the transformation

$$\mathscr{T}\phi(\mathbf{x}, t)\mathscr{T}^{-1} = \pm\phi(\mathbf{x}, -t) \qquad 3(71)$$

is a symmetry operation of the theory, we must include in $\mathscr{T}$ the complex conjugation operation K satisfying

$$K\lambda = \lambda^{*}K \qquad 3(72)$$

where λ is a c-number. Multiplying 3(70) by $\mathscr{T}$ on the left and $\mathscr{T}^{-1}$ on the right, we have

$$\mathscr{T}[\phi(\mathbf{x}, t), \dot{\phi}(\mathbf{x}', t)]\mathscr{T}^{-1} = \mathscr{T}i\mathscr{T}^{-1}\delta^{(3)}(\mathbf{x}-\mathbf{x}')$$

or

$$[\phi(\mathbf{x}, -t), \dot{\phi}(\mathbf{x}', -t)] = -i\delta^{(3)}(\mathbf{x}-\mathbf{x}')$$

so that the equal-time commutation rules are now invariant under 3(71).

Let us seek a unitary operator $\mathscr{U}$ such that

$$\mathscr{T} = \mathscr{U}K$$

satisfies 3(71). From the expansion 3(36a) we see that

$$\mathscr{U} a_{\mathbf{k}} \mathscr{U}^{-1} = \pm a_{-\mathbf{k}} \tag{3(73a)}$$

$$\mathscr{U} b_{\mathbf{k}} \mathscr{U}^{-1} = \pm b_{-\mathbf{k}} \tag{3(73b)}$$

Hence $\mathscr{U}$ is just the parity operator $\mathscr{P}_s$ or $\mathscr{P}_{ps}$ depending on the choice of sign in 3(71). From 3(73a) and 3(73b) it follows that

$$\mathscr{T}\mathbf{P}\mathscr{T}^{-1} = -\mathbf{P}$$

in accord with physical intuition. Moreover, from 3(68) and 3(71) we infer that the current operator transforms as

$$\mathscr{T}\mathbf{j}(\mathbf{x}, t)\mathscr{T}^{-1} = -\mathbf{j}(\mathbf{x}, -t)$$

$$\mathscr{T}j_0(\mathbf{x}, t)\mathscr{T}^{-1} = j_0(\mathbf{x}, -t)$$

which is also in accord with what one expects on physical grounds.

The time reversal operator $\mathscr{T} = \mathscr{U}K$ is not a unitary operator owing to the non-linear character of the complex-conjugation operation. K is known as an *anti-unitary* (or unitary antilinear) operator; it satisfies

$$\langle K\alpha | K\beta \rangle = \langle \alpha | \beta \rangle^* \tag{3(74)}$$

where $|\alpha\rangle$ and $|\beta\rangle$ are arbitrary states in Hilbert space. By contrast, an ordinary linear unitary operator satisfies

$$\langle \mathscr{U}\alpha | \mathscr{U}\beta \rangle = \langle \alpha | \beta \rangle$$

To establish 3(74) we use the defining properties of K, namely

$$K(\lambda_1|\alpha_1\rangle + \lambda_2|\alpha_2\rangle) = \lambda_1{}^*K|\alpha_1\rangle + \lambda_2{}^*K|\alpha_2\rangle \tag{3(75a)}$$

where λ_1 and λ_2 are *c*-numbers, and

$$\langle K\alpha | K\alpha \rangle = \langle \alpha | \alpha \rangle \tag{3(75b)}$$

The second property simply expresses the fact that the norm of a state is unaltered by complex conjugation. Setting

$$|\alpha\rangle = |\alpha_1\rangle + \lambda|\alpha_2\rangle$$

we have, by 3(75a) and 3(75b),

$$\langle K\alpha_1|K\alpha_1\rangle + |\lambda|^2\langle K\alpha_2|K\alpha_2\rangle + \lambda\langle K\alpha_2|K\alpha_1\rangle + \lambda^*\langle K\alpha_1|K\alpha_2\rangle$$
$$= \langle\alpha_1|\alpha_1\rangle + |\lambda|^2\langle\alpha_2|\alpha_2\rangle + \lambda^*\langle\alpha_2|\alpha_1\rangle + \lambda\langle\alpha_1|\alpha_2\rangle$$

or

$$\lambda\langle K\alpha_2|K\alpha_1\rangle + \lambda^*\langle K\alpha_1|K\alpha_2\rangle = \lambda^*\langle\alpha_2|\alpha_1\rangle + \lambda\langle\alpha_1|\alpha_2\rangle$$

If, in this relation, we successively set $\lambda = 1$ and $\lambda = i$ and add the two resulting equalities, we get

$$\langle K\alpha_2|K\alpha_1\rangle = \langle\alpha_1|\alpha_2\rangle \qquad 3(76)$$

confirming that K is antiunitary. Moreover $\mathscr{T} = \mathscr{U}K$ is also antiunitary, since

$$\begin{aligned}\langle\mathscr{T}\alpha|\mathscr{T}\beta\rangle &= \langle\mathscr{U}K\alpha|\mathscr{U}K\beta\rangle \\ &= \langle K\alpha|K\beta\rangle \\ &= \langle\alpha|\beta\rangle^*\end{aligned}$$

Thus time reversal is represented in Hilbert space by an anti-unitary operator.

Invariant Commutation Rules Since $\phi(\mathbf{x}, t)$ is a Lorentz scalar, we expect the commutator

$$i\Delta(x_1 - x_2) = [\phi(x_1), \phi^\dagger(x_2)] \qquad 3(77)$$

to be a Lorentz-invariant function. To confirm this by explicit calculation we use 3(36a) and 3(36b) and write

$$[\phi(x_1), \phi^\dagger(x_2)] = [\phi^{(+)}(x_1), \phi^{\dagger(-)}(x_2)] + [\phi^{(-)}(x_1), \phi^{\dagger(+)}(x_2)]$$

where the superscripts (+) and (−) denote positive frequency (destruction) and negative frequency (creation) parts respectively,

$$\begin{aligned}\phi^{(+)}(x) &= \frac{1}{\sqrt{V}}\sum_{\mathbf{k}}\frac{1}{\sqrt{2\omega_{\mathbf{k}}}}a_{\mathbf{k}}\,\mathrm{e}^{ik.x} \qquad & \phi^{\dagger(-)}(x) &= \frac{1}{\sqrt{V}}\sum_{\mathbf{k}}\frac{1}{\sqrt{2\omega_{\mathbf{k}}}}a_{\mathbf{k}}^\dagger\,\mathrm{e}^{-ik.x} \\ \phi^{(-)}(x) &= \frac{1}{\sqrt{V}}\sum_{\mathbf{k}}\frac{1}{\sqrt{2\omega_{\mathbf{k}}}}b_{\mathbf{k}}^\dagger\,\mathrm{e}^{-ik.x} \qquad & \phi^{\dagger(+)}(x) &= \frac{1}{\sqrt{V}}\sum_{\mathbf{k}}\frac{1}{\sqrt{2\omega_{\mathbf{k}}}}b_{\mathbf{k}}\,\mathrm{e}^{ik.x}\end{aligned} \qquad 3(78)$$

Using the commutation rules 3(38) we find

$$i\Delta(x_1 - x_2) = i\Delta^{(+)}(x_1 - x_2) + i\Delta^{(-)}(x_1 - x_2) \qquad 3(79)$$

where

$$i\Delta^{(+)}(x_1 - x_2) = [\phi^{(+)}(x_1), \phi^{\dagger(-)}(x_2)] = \frac{1}{V}\sum_{\mathbf{k}}\frac{1}{2\omega_{\mathbf{k}}}\mathrm{e}^{ik.(x_1 - x_2)} \qquad 3(80a)$$

$$i\Delta^{(-)}(x_1 - x_2) = [\phi^{(-)}(x_1), \phi^{\dagger(+)}(x_2)] = -\frac{1}{V}\sum_{\mathbf{k}}\frac{1}{2\omega_{\mathbf{k}}}\mathrm{e}^{-ik.(x_1 - x_2)} \qquad 3(80b)$$

Note that

$$\Delta^{(+)}(x) = -\Delta^{(-)}(-x) \qquad 3(81)$$

It is convenient at this stage to go over from discrete to continuum normalization. In accordance with the well-known prescription, we replace $V^{-1}\Sigma_{\mathbf{k}}$ by $(2\pi)^{-3}\int d^3k$ to get

$$i\Delta^{(+)}(x_1-x_2) = \frac{1}{(2\pi)^3}\int d^3k\frac{e^{ik.(x_1-x_2)}}{2\omega_{\mathbf{k}}} \qquad 3(82a)$$

$$i\Delta^{(-)}(x_1-x_2) = -\frac{1}{(2\pi)^3}\int d^3k\frac{e^{-ik.(x_1-x_2)}}{2\omega_{\mathbf{k}}} \qquad 3(82b)$$

and

$$\Delta(x_1-x_2) = \frac{1}{(2\pi)^3}\int \frac{d^3k}{\omega_{\mathbf{k}}}\sin k\,.\,(x_1-x_2) \qquad 3(83)$$

We can now exhibit $\Delta^{(+)}$, $\Delta^{(-)}$ and Δ in the 4-covariant form

$$\Delta^{(+)}(x) = \frac{-i}{(2\pi)^3}\int d^4k\theta(k_0)\delta(k^2+\mu^2)\,e^{ik.x} \qquad 3(84a)$$

$$\Delta^{(-)}(x) = \frac{i}{(2\pi)^3}\int d^4k\theta(-k_0)\delta(k^2+\mu^2)\,e^{ik.x} \qquad 3(84b)$$

$$\Delta(x) = \frac{-i}{(2\pi)^3}\int d^4k\epsilon(k_0)\delta(k^2+\mu^2)\,e^{ik.x} \qquad 3(85)$$

where

$$\theta(k_0) = \begin{cases} 1 & \text{for} \quad k_0 > 0 \\ 0 & \text{for} \quad k_0 < 0 \end{cases} \qquad 3(86)$$

and $\epsilon(k_0) = \theta(k_0)-\theta(-k_0)$. The relativistic invariance of the commutator function $\Delta(x)$ follows from the fact that $\epsilon(k_0)$ is invariant under Lorentz transformations for time-like k $(k^2 < 0)$.

Using 3(83) it is a simple matter to check that 3(77) yields the canonical commutation relations 3(29a) and 3(29b) in the special case $t_1 = t_2$. [See Problem 6.] The remaining commutation relations for the charged scalar field are

$$[\phi(x_1), \phi(x_2)] = [\phi^\dagger(x_1), \phi^\dagger(x_2)] = 0 \qquad 3(87)$$

We can also write the invariant commutation rules in terms of the hermitian fields ϕ_1 and ϕ_2. We have then

$$[\phi_i(x_1), \phi_j(x_2)] = i\delta_{ij}\Delta(x_1 - x_2) \tag{3(88)}$$

On the other hand, for a *neutral* scalar field $\phi(x)$ we have

$$[\phi(x_1), \phi(x_2)] = i\Delta(x_1 - x_2) \tag{3(89)}$$

An important consequence of these commutation relations is the fact that field observables attached to points of space-like separation commute*. This follows from the observation that the commutator function $\Delta(x_1 - x_2)$ vanishes for $(x_1 - x_2)^2 > 0$. Indeed, since by 3(83), $\Delta(x)$ vanishes for equal times,

$$\Delta(\mathbf{x}_1 - \mathbf{x}_2, 0) = 0$$

it must, being a Lorentz-invariant function, vanish for all space-like intervals. Physically, the commutativity of field observables at points of spacelike separation can be understood as a requirement imposed by relativistic causality. Since two points separated by a spacelike interval cannot be connected by means of a light signal, measurements performed at these points cannot interfere with each other and we expect physical observables attached to the two points to commute with each other. We refer to this requirement as the *principle of microcausality.*

Feynman Propagator In the theory of interacting quantum fields, a key role will be played by the *Feynman propagator function* $\Delta_F(x_1 - x_2)$. This is defined to be the vacuum expectation value

$$\Delta_F(x_1 - x_2) = \langle 0| T\phi(x_1)\phi^\dagger(x_2)|0\rangle \tag{3(90)}$$

of the so-called *time-ordered* product

$$T\phi(x_1)\phi^\dagger(x_2) = \begin{cases} \phi(x_1)\phi^\dagger(x_2) & t_1 > t_2 \\ \phi^\dagger(x_2)\phi(x_1) & t_2 > t_1 \end{cases} \tag{3(91)}$$

or for a neutral field,

$$\Delta_F(x_1 - x_2) = \langle 0| T\phi(x_1)\phi(x_2)|0\rangle \tag{3(92)}$$

Note that although the T-symbol 3(91) does not prescribe the ordering of operator factors for equal times, there is no ambiguity here since $\phi(x_1)$ and $\phi^\dagger(x_2)$ commute for $t_1 = t_2$. More generally, the fact that $\phi(x_1)$ and

* The hermitian fields $\phi(x_1)$ and $\phi(x_2)$ are regarded as observables in this context, but whether quantities attached to an infinitely well localized point x may be regarded as *physically* observable is open to question.

$\phi^\dagger(x_2)$ commute for space-like separations $(x_1-x_2)^2 > 0$ ensures that the time ordering in 3(91) remains invariant under a Lorentz transformation that reverses the sign of t_1-t_2. Hence the Feynman propagator is Lorentz-invariant.

To evaluate the vacuum expectation value 3(90) we note that for $t_1 > t_2$

$$\begin{aligned}\langle 0|T\phi(x_1)\phi^\dagger(x_2)|0\rangle &= \langle 0|\phi(x_1)\phi^\dagger(x_2)|0\rangle \\ &= \langle 0|\phi^{(+)}(x_1)\phi^{\dagger(-)}(x_2)|0\rangle \\ &= \langle 0|[\phi^{(+)}(x_1), \phi^{\dagger(-)}(x_2)]|0\rangle \\ &= i\Delta^{(+)}(x_1-x_2) \qquad (t_1 > t_2) \qquad 3(93a)\end{aligned}$$

where we have used 3(78) and the fact that destruction operators annihilate the vacuum. Similarly, for $t_2 > t_1$

$$\langle 0|T\phi(x_1)\phi^\dagger(x_2)|0\rangle = -i\Delta^{(-)}(x_1-x_2) \qquad (t_2 > t_1) \qquad 3(93b)$$

We now use the explicit expressions 3(80a) and 3(80b) for $\Delta^{(+)}$ and $\Delta^{(-)}$ to write

$$\begin{aligned}\Delta_F(x) &= \theta(x_0)i\Delta^{(+)}(x)-\theta(-x_0)i\Delta^{(-)}(x) \\ &= \frac{1}{(2\pi)^3}\int\frac{d^3k}{2\omega_{\mathbf{k}}}[\theta(x_0)\,e^{i\mathbf{k}.\mathbf{x}-i\omega_{\mathbf{k}}x_0}+\theta(-x_0)\,e^{i\mathbf{k}.\mathbf{x}+i\omega_{\mathbf{k}}x_0}] \qquad 3(94)\end{aligned}$$

Using Cauchy's residue theorem, we can convert the integrand into a contour integral in the complex k_0-plane, giving

$$\begin{aligned}\Delta_F(x) &= -\frac{1}{(2\pi)^3}\int d^3k\int_{C_F}\frac{dk_0}{2\pi i}\frac{e^{i\mathbf{k}.\mathbf{x}-ik_0x_0}}{(k_0-\omega_{\mathbf{k}})(k_0+\omega_{\mathbf{k}})} \\ &= -\frac{i}{(2\pi)^4}\int d^3k\int_{C_F} dk_0\frac{e^{ik.x}}{k^2+\mu^2} \qquad 3(95)\end{aligned}$$

where C_F is the contour shown in Fig. 3.1. To recover 3(94), we observe that when $x_0 > 0$ the contour C_F can be closed in the lower half plane,

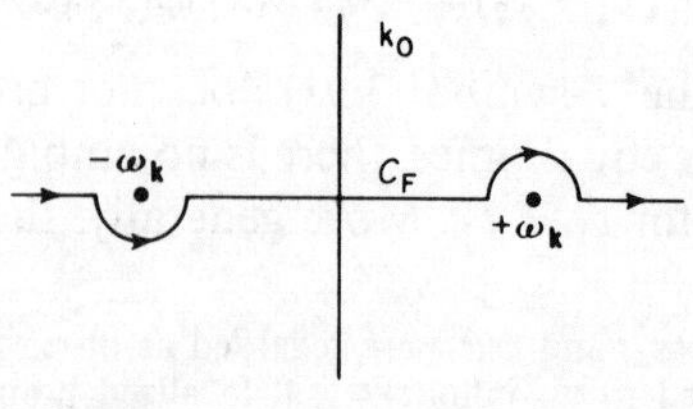

FIG. 3.1. Contour C_F for Feynman Propagator Δ_F.

giving a residue $-2\pi i\, e^{i\mathbf{k}.\mathbf{x}-i\omega_{\mathbf{k}}t}/2\omega_{\mathbf{k}}$ at the pole $k_0 = \omega_{\mathbf{k}}$. When $x_0 < 0$, the contour is closed in the upper half plane, giving a residue $-2\pi i\, e^{i\mathbf{k}.\mathbf{x}+i\omega_{\mathbf{k}}t}/2\omega_{\mathbf{k}}$ at the pole $k_0 = -\omega_{\mathbf{k}}$. The result 3(95) can be further transformed to

$$\Delta_F(x) = \frac{1}{(2\pi)^4}\int d^4k\, e^{ik.x}\Delta_F(k) \qquad 3(96a)$$

$$\Delta_F(k) = \frac{-i}{k^2+\mu^2-i\varepsilon} \qquad 3(96b)$$

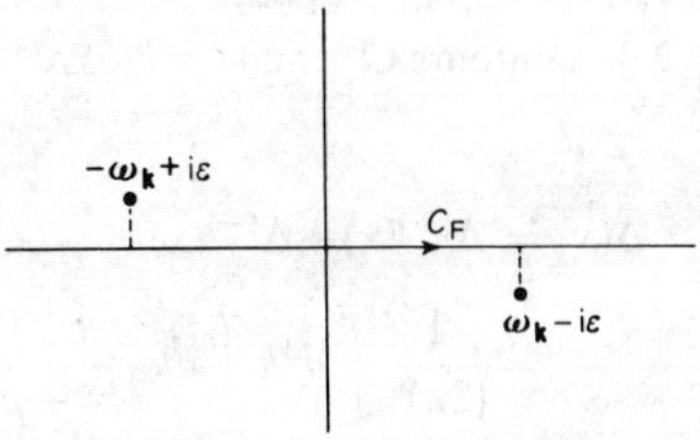

FIG. 3.2. Displacement of Poles and Contour for Δ_F.

by displacing the poles to $k_0 = \pm\omega_{\mathbf{k}}\mp i\varepsilon$. Equation 3(96b) is the expression of the Feynman propagator in momentum space. Notice that $\Delta_F(x)$ satisfies the inhomogeneous equation

$$(\Box-\mu^2)\Delta_F(x) = -\frac{1}{(2\pi)^4}\int d^4k\, e^{ik.x}(k^2+\mu^2)\Delta_F(k)$$

$$= i\delta^{(4)}(x) \qquad 3(97)$$

so that the Feynman propagator is a Green's function for the Klein–Gordon equation. [Feynman (1949), Stückelberg (1949).]

Invariant Functions The Feynman propagator Δ_F and the commutator functions $\Delta^{(\pm)}$ and Δ are members of a family of so-called invariant functions which can be represented in terms of contour integrals in the complex k_0 plane. To exhibit $\Delta^{(\pm)}$ and Δ in this form we note that 3(82a), for instance, can be written as

$$i\Delta^{(+)}(x) = \frac{1}{(2\pi)^3}\int d^3k \int_{C^{(+)}} \frac{dk_0}{2\pi i}\,\frac{e^{ik.x}}{(k_0+\omega_{\mathbf{k}})(k_0-\omega_{\mathbf{k}})}$$

or

$$\Delta^{(+)}(x) = \frac{1}{(2\pi)^4}\int d^3k \int_{C^{(+)}} dk_0 \frac{e^{ik.x}}{k^2+\mu^2} \qquad 3(98a)$$

where the contour $C^{(+)}$ is shown in Fig. 3.3. Similarly,

$$\Delta^{(-)}(x) = \frac{1}{(2\pi)^4}\int d^3k \int_{C^{(-)}} dk_0 \frac{e^{ik.x}}{k^2+\mu^2} \qquad 3(98b)$$

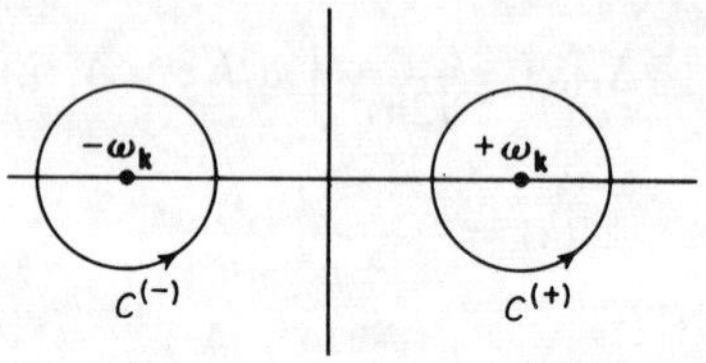

FIG. 3.3. Contours $C^{(+)}$ and $C^{(-)}$ for $\Delta^{(+)}$ and $\Delta^{(-)}$

and

$$\Delta(x) = \Delta^{(+)}(x)+\Delta^{(-)}(x)$$
$$= \frac{1}{(2\pi)^4}\int d^3k \int_{C} dk_0 \frac{e^{ik.x}}{k^2+\mu^2} \qquad 3(99)$$

where C is exhibited in Fig. 3.4. Other useful invariant functions are the *retarded* and *advanced* functions

$$\Delta_R(x) = \frac{1}{(2\pi)^4}\int d^3k \int_{C_R} dk_0 \frac{e^{ik.x}}{k^2+\mu^2} \qquad 3(100a)$$

$$\Delta_A(x) = \frac{1}{(2\pi)^4}\int d^3k \int_{C_A} dk_0 \frac{e^{ik.x}}{k^2+\mu^2} \qquad 3(100b)$$

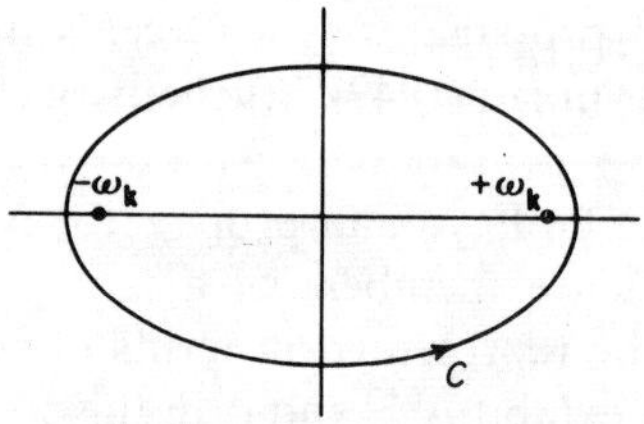

FIG. 3.4. Contour C for Δ.

where the contours C_R and C_A are shown in Fig. 3.5. The same argument used to establish 3(95) yields, in this case,

$$\Delta_R(x) = \begin{cases} -\Delta(x) & t > 0 \\ 0 & t < 0 \end{cases} \qquad 3(101a)$$

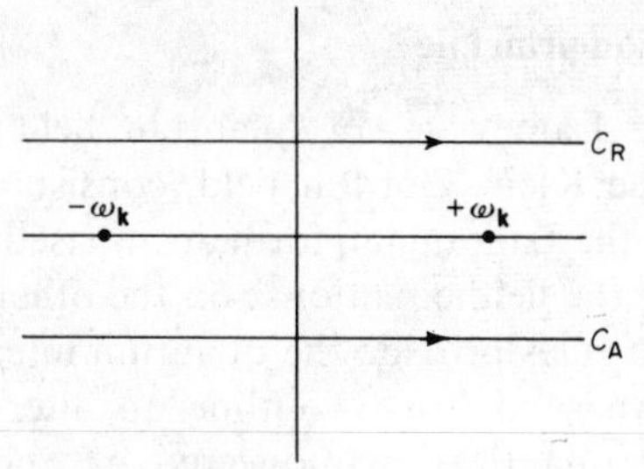

FIG. 3.5. Contours for Δ_A and Δ_R.

$$\Delta_A(x) = \begin{cases} 0 & t > 0 \\ \Delta(x) & t < 0 \end{cases} \qquad 3(101b)$$

Finally, the function Δ_1, defined by

$$\Delta_1(x) = i\Delta^{(+)}(x) - i\Delta^{(-)}(x) \qquad 3(102)$$

may be represented in the form

$$\Delta_1(x) = \frac{i}{(2\pi)^4} \int d^3k \int_{C_1} dk_0 \frac{e^{ik.x}}{k^2+\mu^2} \qquad 3(103)$$

where C_1 is the contour exhibited in Fig. 3.6.

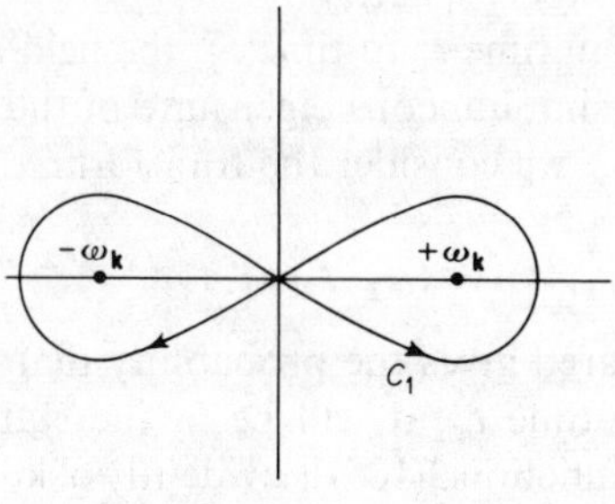

FIG. 3.6. Contour for Δ_1.

All invariant functions corresponding to integrals over *finite* contours, i.e. Δ, $\Delta^{(\pm)}$, Δ_1, satisfy the homogeneous Klein–Gordon equation

$$(\Box - \mu^2)\Delta^{(\ )} = 0 \qquad 3(104)$$

whereas the functions Δ_F, Δ_R, and Δ_A, which correspond to infinite contours, satisfy inhomogeneous Klein–Gordon equations with a $\delta^{(4)}(x)$ source function.

3-3 Quantum action principle

Introduction The framework of quantum field theory as set up in Section 3-2 for the Klein–Gordon field, consists of two distinct parts. On the one hand, the Lagrangian formalism based on Hamilton's action principle provides the field equations; on the other, the canonical equal-time commutation rules provide the quantum interpretation of the field. In this Section we shall briefly outline an alternative, more abstract approach to field quantization pioneered by Schwinger (1951) (1953). In this approach the two aspects of quantum field theory are unified in a single quantum action principle from which flow both the field equations and the commutation relations. We shall illustrate the procedure in the simplest case—that of a neutral scalar field.

Transformation Function In Section 3-2 we selected, as our complete set of commuting observables for the neutral scalar field, the number operators $N_{\mathbf{k}} = a_{\mathbf{k}}^{\dagger} a_{\mathbf{k}}$. Simultaneous eigenstates of this set were constructed by means of 3(22) and used as the basis for the Hilbert space in which the field operators are defined. For our present purposes, however, it is convenient to select the complete commuting set of operators formed by taking the field operator $\phi(\mathbf{x}, t_1)$ at each point of 3-space at time t_1. Denoting simultaneous eigenstates of this set by $|\varphi_1, t_1\rangle$ where φ_1 represents a set of eigenvalues of $\phi(\mathbf{x}, t_1)$, we adopt the set of all $|\varphi_1, t_1\rangle$ as the basis for the Hilbert space.

If we now pass from time t_1 to time t_2, the field $\phi(\mathbf{x}, t_1)$ goes over into $\phi(\mathbf{x}, t_2)$. Denoting a simultaneous eigenstate of the complete commuting set $\phi(\mathbf{x}, t_2)$ by $|\varphi_2, t_2\rangle$ we consider the transformation function or transition amplitude

$$\langle \varphi_2, t_2 | \varphi_1, t_1 \rangle \qquad 3(105)$$

whose modulus squared gives the probability that if the system is in the state $|\varphi_1, t_1\rangle$ at the time t_1, it will be in the state $|\varphi_2, t_2\rangle$ at time t_2. The transformation function 3(105) is evidently of key physical importance in describing the temporal development of the system.

Schwinger's Action Principle The quantum action principle formulated by Schwinger is a differential characterization of the transformation function 3(105). It states that under infinitesimal variations in the quantities upon which $\langle \varphi_2, t_2 | \varphi_1, t_1 \rangle$ depends—for example the field operators, or the times t_1 and t_2—the variation in the transformation function is given by

$$\delta \langle \varphi_2, t_2 | \varphi_1, t_1 \rangle = i \langle \varphi_2, t_2 | \delta W_{21} | \varphi_1, t_1 \rangle \qquad 3(106)$$

where δW_{21} is the corresponding variation in the action integral

$$\delta W_{21} = \delta \int_{t_1}^{t_2} dt \int \mathscr{L}(x)\, d^3x \tag{3(107)}$$

By choosing suitable classes of variations, the action principle 3(107) can be made to yield all the ingredients of quantum field theory,* in particular the field equations and the canonical commutation relations. To derive the field equations, for example, we consider 3(106) for arbitrary variations $\delta\phi(\mathbf{x}, t)$ which vanish for $t = t_1$ and $t = t_2$. For variations of this type we have $\delta\langle\varphi_2, t_2|\varphi_1, t_1\rangle = 0$ since the eigenvalues φ_1 and φ_2 at t_1 and t_2 are left unchanged. It follows from 3(106) that

$$\delta W_{21} = 0 \tag{3(108)}$$

which is just Hamilton's principle. From 3(108) one derives in the usual way the Euler–Lagrange equations

$$\frac{\partial \mathscr{L}}{\partial \phi} = \partial_\mu \frac{\partial \mathscr{L}}{\partial(\partial_\mu \phi)} \tag{3(109)}$$

To derive the equal-time commutation relations from 3(106) one must consider arbitrary variations of the fields which do *not* vanish at $t = t_1$ and t_2. We shall not carry through this programme explicitly but refer the reader to Schwinger's original paper for details. Instead, we shall consider 3(106) for variations $\delta\phi$ of the type 3(52) i.e.

$$\delta\phi(x) = -\omega_{\mu\nu} x_\nu \partial_\mu \phi(x) - \varepsilon_\mu \partial_\mu \phi(x) \tag{3(110)}$$

corresponding to infinitesimal space-time rotations and displacements. We shall show that as a consequence of the action principle, we recover the result

$$\delta\phi(\mathbf{x}, t) = \frac{1}{i}[\phi(\mathbf{x}, t), G(t)] \tag{3(111)}$$

where G is the generator

$$G = \varepsilon_\mu P_\mu - \omega_{\mu\nu} M_{\mu\nu} \tag{3(112)}$$

In the conventional formulation of Section 3-2, the result 3(111) is derived by application of the commutation relations. In Schwinger's approach, it is a direct consequence of the quantum action principle, which indicates that the latter does indeed contain information about the commutation rules.

* A similar action principle can be formulated for ordinary non-relativistic quantum mechanics, [Schwinger (1955)].

To derive 3(111) we recall that under the variation 3(110) we have

$$\delta W_{21} = G(t_2) - G(t_1)$$

with G given by 3(112). Hence, the action principle 3(106) takes the form

$$\delta\langle\varphi_2, t_2|\varphi_1, t_1\rangle = i\langle\varphi_2, t_2|G(t_2)|\varphi_1, t_1\rangle - i\langle\varphi_2, t_2|G(t_1)|\varphi_1, t_1\rangle$$

Since the variation of the transformation function $\langle\varphi_2, t_2|\varphi_1, t_1\rangle$ can only arise from alterations of the initial and final states, i.e.

$$\delta\langle\varphi_2, t_2|\varphi_1, t_1\rangle = \delta(\langle\varphi_2, t_2|)|\varphi_1, t_1\rangle + \langle\varphi_2, t_2|\delta(|\varphi_1, t_1\rangle)$$

we can identify the variations of the basis vectors according to

$$\delta|\varphi_1, t_1\rangle = -iG(t_1)|\varphi_1, t_1\rangle\,;\, \delta|\varphi_2, t_2\rangle = -iG(t_2)|\varphi_2, t_2\rangle$$

But, as indicated in the discussion following 3(48a), 3(48b), we can express this as the variation

$$\delta\phi(\mathbf{x}, t) = \frac{1}{i}[\phi(\mathbf{x}, t), G(t)]$$

of the field operator $\phi(x)$ for $t = t_1$ and t_2. Hence, we recover the result 3(111).

A particular consequence of 3(111) is the Hamiltonian equation of motion

$$\dot{\phi}(\mathbf{x}, t) = \frac{1}{i}[\phi(\mathbf{x}, t), H] \qquad 3(113)$$

This may be obtained from 3(111) by considering time displacements $\delta\phi(\mathbf{x}, t) = -\varepsilon_0\dot{\phi}(\mathbf{x}, t)$ for which the generator is $G = -\varepsilon_0 H$. In the conventional formulation of Section 3-2, it was necessary to check the consistency of the Lagrangian and Hamiltonian equations of motion explicitly by applying the equal-time commutation relations. In the present scheme, both 3(109) and 3(113) follow from the same postulate.

3-4 Nonrelativistic quantum field theory

Introduction In this Section we apply field quantization to the Schrödinger field. We have stressed that for relativistic quantum mechanics, one-particle wave theories are unsatisfactory and must be recast as quantum field theories to achieve a consistent formulation. The outlook is different in the case of nonrelativistic quantum mechanics. The one-particle theory itself is fully consistent and we shall see that the quantized field theory is equivalent to standard Schrödinger theory for an assembly of indistinguishable non-relativistic particles. Bose–Einstein statistics for the quanta are ensured by quantizing the field by means of commutators. To get Fermi–Dirac statistics, on the other hand, an important modification is necessary, namely the use of *anti*commutation rules rather than commutation rules. The Schrödinger field is the only field which can be quantized consistently using either commutators or anticommutators.

Bose–Einstein Quantization Let us quantize the Schrödinger field according to the canonical prescription, replacing the Poisson brackets in 2(98) by commutators in accordance with 3(1). This yields the commutation relations

$$[\psi(x), \psi^\dagger(x')]_{t=t'} = \delta^{(3)}(\mathbf{x}-\mathbf{x}') \qquad 3(114a)$$

$$[\psi(x), \psi(x')]_{t=t'} = [\psi^\dagger(x), \psi^\dagger(x')]_{t=t'} = 0 \qquad 3(114b)$$

From the Hamiltonian equation of motion

$$\dot{\psi}(\mathbf{x}, t) = \frac{1}{i}[\psi(\mathbf{x}, t), H(t)] \qquad 3(115)$$

with

$$H = \int d^3x \left[-\frac{1}{2m}\psi^\dagger \nabla^2 \psi + \psi^\dagger V(x)\psi \right] \qquad 3(116)$$

we recover the Schrödinger equation

$$i\dot{\psi}(x) = -\frac{1}{2m}\nabla^2\psi(x) + V(x)\psi(x) \qquad 3(117)$$

for the field operator $\psi(x)$.

To simplify matters we restrict our attention to time-independent potentials $V(x)$. A complete set of solutions to the classical time-dependent Schrödinger equation is then of the form

$$u_i(\mathbf{x})\, e^{-iE_i t} \qquad 3(118)$$

where the functions $u_i(\mathbf{x})$ satisfying

$$E_i u_i(\mathbf{x}) = \left[-\frac{1}{2m}\nabla^2 + V(\mathbf{x}) \right] u_i(\mathbf{x}) \qquad 3(119)$$

(no summation on i)

form a complete set of eigenfunctions of the Schrödinger Hamiltonian with the orthogonality, normalization, and completeness properties

$$\int u_i^*(\mathbf{x})u_{i'}(\mathbf{x})\,d^3x = \delta_{ii'} \qquad 3(120a)$$

$$\sum_i u_i(\mathbf{x})u_i^*(\mathbf{x}') = \delta^{(3)}(\mathbf{x}-\mathbf{x}') \qquad 3(120b)$$

We now expand the field operator $\psi(x)$ in terms of the solutions 3(118) according to

$$\psi(x) = \sum_i a_i u_i(\mathbf{x})\,\mathrm{e}^{-iE_it} \qquad 3(121)$$

The inverse relations expressing a_i in terms of $\psi(x)$ are easily obtained with the aid of 3(120a):

$$a_i = \int u_i^*(\mathbf{x})\,\mathrm{e}^{iE_it}\psi(x)\,d^3x \qquad 3(122)$$

From 3(114a) and 3(120a) we derive the commutation relations

$$\begin{aligned}[a_i, a_{i'}^\dagger] &= \int u_i^*(\mathbf{x})\,\mathrm{e}^{iE_it}u_{i'}(\mathbf{x}')\,\mathrm{e}^{-iE_{i'}t}[\psi(x), \psi^\dagger(x')]\,d^3x\,d^3x' \\ &= \int u_i^*(\mathbf{x})u_{i'}(\mathbf{x})\,\mathrm{e}^{iE_it}\,\mathrm{e}^{-iE_{i'}t}\,d^3x \\ &= \delta_{ii'} \end{aligned} \qquad 3(123a)$$

and similarly

$$[a_i, a_{i'}] = [a_i^\dagger, a_{i'}^\dagger] = 0 \qquad 3(123b)$$

Notice that in contrast to the Klein–Gordon case, the field operator ψ contains only destruction operators. The Hamiltonian 3(116), when expressed in terms of the a_n and $a_n^\dagger$ by means of 3(121), becomes simply

$$\begin{aligned} H &= \sum_{ii'} a_i^\dagger a_{i'}\,\mathrm{e}^{iE_it}\,\mathrm{e}^{-iE_{i'}t}\int d^3x u_i^*(\mathbf{x})E_{i'}u_{i'}(\mathbf{x}) \\ &= \sum_i E_i a_i^\dagger a_i \end{aligned} \qquad 3(124)$$

which we recognize as the energy of an assembly of particles. The eigenvalues of the number operators

$$N_i = a_i^\dagger a_i$$

give the number of particles with energy E_i. Since

$$[N_i, H] = 0 \qquad 3(125)$$

the number of particles in each eigenstate i is constant in time. The operators $N_1, N_2 \ldots$ form a complete commuting set and simultaneous eigenstates $|n_1, n_2, \ldots\rangle$ of this set are given by

$$|n_1, n_2, \ldots\rangle = \frac{1}{(n_1! n_2! \ldots)^{1/2}} (a_1{}^\dagger)^{n_1} (a_2{}^\dagger)^{n_2} \ldots |0\rangle \qquad 3(126)$$

As in the Klein–Gordon case, the state vectors 3(126) are characterized by specifying only the *number* of quanta in the various energy states E_i. The field quanta are therefore indistinguishable. They obey Bose–Einstein statistics, since the state vector 3(126) is symmetric under the interchange of any two creation operators.

At this point the physical equivalence of the quantized field with an assembly of identical Bose–Einstein particles is clear. To establish the formal equivalence of the theory with standard n-particle quantum mechanics, we must introduce configuration space wave functions. We first define the n-particle state

$$|\mathbf{x}_1, \mathbf{x}_2, \ldots \mathbf{x}_n; t\rangle = \frac{1}{\sqrt{n!}} \psi^\dagger(\mathbf{x}_1, t) \psi^\dagger(\mathbf{x}_2, t) \ldots \psi^\dagger(\mathbf{x}_n, t)|0\rangle \qquad 3(127)$$

This state represents n particles completely localized at time t at the positions $\mathbf{x}_1, \mathbf{x}_2, \ldots \mathbf{x}_n$. To see this we note that, with the aid of 3(122) and 3(120b), the total number operator

$$N = \sum_i a_i{}^\dagger a_i$$

can be written in the form*

$$N = \int \psi^\dagger(\mathbf{x}, t)\psi(\mathbf{x}, t)\, d^3x \qquad 3(128)$$

Defining a local number operator

$$N_v(t) = \int_v \psi^\dagger(\mathbf{x}, t)\psi(\mathbf{x}, t)\, d^3x \qquad 3(129)$$

whose eigenvalues give the number of particles in a spatial volume v, we note that by 3(114a) and 3(114b)

$$[N_v(t), \psi^\dagger(\mathbf{x}, t)] = \begin{cases} \psi^\dagger(\mathbf{x}, t) & \text{if } \mathbf{x} \text{ is in } v \\ 0 & \text{if } \mathbf{x} \text{ is not in } v \end{cases} \qquad 3(130)$$

* The total number operator is the analogue of the total probability of the one-particle theory. Note that N is time-independent, since $[H, N] = 0$. The same is not true of the local number operator 3(129), reflecting the fact that wave packets spread out with time.

We now consider *infinitesimally small* volumes $v_1, v_2, \ldots v_n$, centred at the points $\mathbf{x}_1, \mathbf{x}_2 \ldots \mathbf{x}_n$ featured in 3(127) and construct the corresponding operators $N_{v_1}, N_{v_2}, \ldots, N_{v_n}$. If the n-points are all distinct, then 3(130) and the property $N_v|0\rangle = 0$ indicate that $N_{v_1}, N_{v_2}, \ldots N_{v_n}$ all yield the eigenvalue $+1$ when operating on the state 3(127). On the other hand, if two points coincide, say $\mathbf{x}_1$ and $\mathbf{x}_2$, then the application of 3(130) yields the eigenvalue 2 for N_{v_1}. Either way, we are assured that 3(127) represents a state of n particles completely localized at $\mathbf{x}_1, \mathbf{x}_2, \ldots \mathbf{x}_n$. We now define the probability amplitude

$$\Phi^{(n)}_{n_1 n_2 \ldots}(\mathbf{x}_1, \mathbf{x}_2 \ldots, \mathbf{x}_n; t) = \langle \mathbf{x}_1, \mathbf{x}_2, \ldots \mathbf{x}_n; t | n_1, n_2, \ldots \rangle \qquad 3(131)$$

$$(n = n_1 + n_2 + \ldots)$$

whose modulus squared gives the probability* of finding the $n = n_1 + n_2 + \ldots$ particles at positions $\mathbf{x}_1, \mathbf{x}_2, \ldots, \mathbf{x}_n$ at time t when n_1 particles are in the state $u_1(\mathbf{x})$, n_2 in the state $u_2(\mathbf{x})$, etc. The amplitude 3(131) is thus the configuration space wave function of the n-particle system. Using 3(127) and 3(117) we see that it satisfies the usual n-body Schrödinger equation

$$i\partial_t \Phi^{(n)}_{n_1 n_2 \ldots}(\mathbf{x}_1, \mathbf{x}_2, \ldots \mathbf{x}_n; t) = \sum_{r=1}^{n} \left[-\frac{1}{2m} \nabla_r^2 + V(\mathbf{x}_r) \right] \Phi^{(n)}_{n_1 n_2 \ldots}(\mathbf{x}_1, \mathbf{x}_2, \ldots \mathbf{x}_n; t) \qquad 3(132)$$

Moreover, $\Phi^{(n)}_{n_1 n_2 \ldots}(\mathbf{x}_1, \mathbf{x}_2, \ldots \mathbf{x}_n; t)$ is symmetric under permutation of any pair of points $\mathbf{x}_i$ and $\mathbf{x}_j$. This follows from the symmetry of $|\mathbf{x}_1, \mathbf{x}_2, \ldots \mathbf{x}_n; t\rangle$ as defined by 3(127), since all creation operators $\psi^\dagger(\mathbf{x}_i, t)$ mutually commute. We have thus confirmed the formal equivalence of our theory with standard non-relativistic quantum mechanics for an assembly of n Bose–Einstein particles.

Fermi–Dirac Quantization Quantization of the Schrödinger field by means of the commutation relations 3(114a), 3(114b) necessarily leads to Bose–Einstein statistics, as we have just seen. To obtain Fermi–Dirac statistics, a new departure is needed involving a modification of the fundamental commutation relations. We recall that the wave function of an assembly of n Fermi–Dirac particles is completely antisymmetric under permutation of the positions of any two of the particles. This leads, in the well known manner, to the *exclusion principle* that not more than one particle may occupy a given state†. Hence the number eigenvalues $n_1, n_2 \ldots$ can take on only the values 0 and 1.

* See Problem 14.

† See for example E. Merzbacher, *Quantum Mechanics*, J. Wiley & Sons, 1961, Chap. 18.

The requirement that the wave function

$$\Phi^{(n)}_{n_1 n_2 \ldots}(\mathbf{x}_1, \mathbf{x}_2, \ldots \mathbf{x}_n; t) = \langle \mathbf{x}_1, \mathbf{x}_2, \ldots \mathbf{x}_n; t | n_1, n_2, \ldots \rangle \qquad 3(133)$$

be completely antisymmetric will be satisfied if the state

$$|\mathbf{x}_1, \mathbf{x}_2, \ldots \mathbf{x}_n; t\rangle = \frac{1}{\sqrt{n!}} \psi^\dagger(\mathbf{x}_1, t)\psi^\dagger(\mathbf{x}_2, t) \ldots \psi^\dagger(\mathbf{x}_n, t)|0\rangle \qquad 3(134)$$

is constructed with the aid of creation operators $\psi^\dagger(\mathbf{x}_1, t) \ldots \psi^\dagger(\mathbf{x}_2, t)$ which mutually *anti*commute

$$\{\psi^\dagger(x), \psi^\dagger(x')\}_{t=t'} = 0 \qquad 3(135a)$$

From 3(135a) it follows that

$$\{\psi(x), \psi(x')\}_{t=t'} = 0 \qquad 3(135b)$$

Moreover, to show that 3(134) represents a state of n particles localized at $\mathbf{x}_1, \mathbf{x}_2, \ldots, \mathbf{x}_n$, we need a relation of the type 3(130). Applying the formula

$$[AB, C] = A\{B, C\} - \{A, C\}B \qquad 3(136)$$

valid for arbitrary operators A, B, and C, we have

$$[N_v(t), \psi^\dagger(\mathbf{x}, t)] = \int_v [\psi^\dagger(\mathbf{x}', t)\psi(\mathbf{x}', t), \psi^\dagger(\mathbf{x}, t)] \, d^3x'$$

$$= \int_v \psi^\dagger(\mathbf{x}', t) \{\psi(\mathbf{x}', t), \psi^\dagger(\mathbf{x}, t)\} \, d^3x' \qquad 3(137)$$

where we have used 3(135a). Since 3(130) should hold for arbitrary $\mathbf{x}$ and v we get the remaining relation

$$\{\psi(x), \psi^\dagger(x')\}_{t=t'} = \delta^{(3)}(\mathbf{x} - \mathbf{x}') \qquad 3(135c)$$

The relations 3(135a), 3(135b), and 3(135c) are known as Jordan–Wigner anticommutation rules [Jordan (1928)].

The Hamiltonian equations of motion remain

$$\dot{\psi}(\mathbf{x}, t) = \frac{1}{i}[\psi(\mathbf{x}, t), H(t)]$$

with H given by 3(116). Evaluating the commutator on the right-hand side with the aid of 3(136) and the anticommutation rules 3(135a), 3(135b), and 3(135c), it is a simple matter to check that $\psi(\mathbf{x}, t)$ still satisfies the Schrödinger equation 3(117). Hence we are assured that the wave function 3(133) will satisfy the n-body Schrödinger equation 3(132). Also, we can again expand

$\psi(x)$ in terms of the solutions 3(118) according to

$$\psi(x) = \sum_i b_i u_i(\mathbf{x})\, e^{-iE_i t} \qquad 3(138)$$

with the inverse relations

$$b_i = \int u_i^*(\mathbf{x})\, e^{iE_i t} \psi(x)\, d^3x \qquad 3(139)$$

As a result of 3(135a), 3(135b) and 3(135c) the creation and destruction operators satisfy the *anti*commutation rules

$$\{b_i, b_{i'}^\dagger\} = \delta_{ii'} \qquad 3(140a)$$

$$\{b_i, b_{i'}\} = \{b_i^\dagger, b_{i'}^\dagger\} = 0 \qquad 3(140b)$$

and the Hamiltonian 3(116) is expressed in terms of the b_i and $b_i^\dagger$ by

$$H = \sum_i E_i b_i^\dagger b_i \qquad 3(141)$$

The anticommutation relations 3(140a) and 3(140b) have the consequence that the eigenvalues of the number operators

$$N_i = b_i^\dagger b_i$$

are restricted to the values 0 and 1, in accordance with the exclusion principle. Indeed

$$N_i^2 = b_i^\dagger b_i b_i^\dagger b_i = b_i^\dagger(1 - b_i^\dagger b_i) b_i = N_i \qquad 3(142)$$

since $b_i^2 = 0$ by 3(140b). From 3(142) it is a simple matter to show that the eigenvalues n_i must satisfy $n_i^2 = n_i$, which proves the result. The number operators $b_i^\dagger b_i$ are easily seen to form a complete commuting set and simultaneous eigenstates of this set can be constructed by taking

$$|n_1, n_2, \ldots\rangle = (b_1^\dagger)^{n_1}(b_2^\dagger)^{n_2} \ldots |0\rangle \qquad 3(143)$$

with $n_1, n_2 \ldots$ restricted to the values 0 or 1. The proof is trivial. Using the anticommutation rules 3(140a) and 3(140b) we have, for example

$$b_i^\dagger b_i |n_1, \ldots, n_i, \ldots\rangle = (b_1^\dagger)^{n_1} \ldots b_i^\dagger b_i (b_i^\dagger)^{n_i} \ldots |0\rangle$$

If $n_i = 0$, we can anticommute b_i through the remaining factors until it acts on $|0\rangle$. Since, by definition $b_i|0\rangle = 0$, the number operator $b_i^\dagger b_i$ yields the eigenvalue 0. If $n_i = 1$, we can replace $b_i^\dagger b_i b_i^\dagger$ by $b_i^\dagger\{b_i, b_i^\dagger\} = b_i^\dagger$ and recover the eigenvalue 1.

In conclusion, the Schrödinger field can be quantized consistently using either commutation or anticommutation relations. In the latter case, the theory is equivalent to ordinary quantum mechanics for an assembly of n Fermi–Dirac particles.

3-5 Localizability of field quanta

Relativistic and non-relativistic quantum fields exhibit a striking difference in regard to the localizability of their respective field quanta. In the non-relativistic case, there is in principle no limitation on the accuracy with which the position of a particle can be determined. This is evidenced by our ability to construct completely localized n-particle states of the type 3(127) and 3(134). In the case of relativistic field theory, however, we shall show that the localizability of field quanta is subject to certain intrinsic limitations.

Let us consider the case of a neutral scalar field, for which the total number operator is given by

$$N = \sum_{\mathbf{k}} a_{\mathbf{k}}{}^{\dagger} a_{\mathbf{k}}$$

To express this in terms of the field operators, we recall that according to 3(9) the positive and negative frequency parts of $\phi(x)$ are given by

$$\phi^{(+)}(x) = \frac{1}{\sqrt{V}} \sum_{\mathbf{k}} \frac{1}{\sqrt{2\omega_{\mathbf{k}}}} a_{\mathbf{k}} \, \mathrm{e}^{i\mathbf{k}.\mathbf{x} - i\omega_{\mathbf{k}} t}$$

$$\phi^{(-)}(x) = \frac{1}{\sqrt{V}} \sum_{\mathbf{k}} \frac{1}{\sqrt{2\omega_{\mathbf{k}}}} a_{\mathbf{k}}{}^{\dagger} \, \mathrm{e}^{-i\mathbf{k}.\mathbf{x} + i\omega_{\mathbf{k}} t}$$

It is then a simple matter to check that

$$\begin{aligned} N &= \sum_{\mathbf{k}} a_{\mathbf{k}}{}^{\dagger} a_{\mathbf{k}} \\ &= -i \int [\dot{\phi}^{(-)}(x)\phi^{(+)}(x) - \phi^{(-)}(x)\dot{\phi}^{(+)}(x)] \, d^3x \end{aligned} \qquad 3(144)$$

Let us define the local number operator

$$N_v(t) = \int_v N(x) \, d^3x \qquad 3(145a)$$

with*

$$N(x) = -i[\dot{\phi}^{(-)}(x)\phi^{(+)}(x) - \phi^{(-)}(x)\dot{\phi}^{(+)}(x)] \qquad 3(145b)$$

One can now readily convince oneself that the construction of perfectly localized states—eigenstates of N_v for infinitesimally small v—is impossible in this theory. For example, if we imitate the non-relativistic prescription and form the state

$$\phi^{(-)}(x)|0\rangle \qquad 3(146)$$

* Both $-2i\dot{\phi}^{(-)}\phi^{(+)}$ and $2i\phi^{(-)}\dot{\phi}^{(+)}$ give the total number operator N when integrated over all space, but to get a *hermitian* local operator N_v we must take the average.

by applying the creation operator $\phi^{(-)}(x)$ to the vacuum, we find, using the commutation relations

$$[\phi^{(+)}(x), \dot{\phi}^{(-)}(x')]_{t=t'} = [\phi^{(-)}(x), \dot{\phi}^{(+)}(x')]_{t=t'} = \tfrac{1}{2}i\delta^{(3)}(\mathbf{x}-\mathbf{x}')$$

$$[\phi^{(\pm)}(x), \phi^{(\mp)}(x')]_{t=t'} = i\Delta^{(\pm)}(\mathbf{x}-\mathbf{x}', 0)$$

that

$$\begin{aligned}[N_v(t), \phi^{(-)}(\mathbf{x}, t)] &= \tfrac{1}{2}\phi^{(-)}(\mathbf{x}, t)+R(\mathbf{x}, t) \quad \text{if } \mathbf{x} \text{ is in } v \\ &= R(\mathbf{x}, t) \qquad\qquad\qquad\quad \text{if } \mathbf{x} \text{ is not in } v\end{aligned}$$

where

$$R(\mathbf{x}, t) = \int_v \dot{\phi}^{(-)}(\mathbf{x}', t)\Delta^{(+)}(\mathbf{x}-\mathbf{x}', 0)\, d^3x'$$

so that $\phi^{(-)}(x)|0\rangle$ is clearly not an eigenstate of N_v. To assure ourselves that this is not simply due to our special choice, 3(146), and that the construction of a state localized at $\mathbf{x}$ is actually impossible, we observe that whereas for the non-relativistic case, we have

$$[N_v(t), N_{v'}(t)] = 0 \tag{3(147a)}$$

for arbitrary volumes v and v', in the relativistic case

$$[N_v(t), N_{v'}(t)] \neq 0 \tag{3(147b)}$$

even when the volumes v and v' do not overlap. In fact, a detailed analysis* reveals that the commutator $[N(\mathbf{x}, t), N(\mathbf{x}', t)]$ becomes infinite for $|\mathbf{x}-\mathbf{x}'| \to 0$ and tends rapidly to zero for separations $|\mathbf{x}-\mathbf{x}'|$ greater than the Compton wave length μ^{-1}. For π mesons for example, this is a distance of the order 10^{-13} cm. This means that the number of particles in two spatial volumes v and v' cannot be measured simultaneously unless v and v' are separated by distances greater than μ^{-1}. In particular, one cannot fix the particle number in a volume smaller than μ^{-3}.

Physically, the meaning of this result is that any attempt to determine the position of a particle with an accuracy greater than μ^{-1} requires the use, as a measuring agency, of an external field involving wavelengths less than μ^{-1} or frequencies greater than μ. The external agency can then create additional particles which are indistinguishable from the particle whose position is being measured. The merging of quantum theory and relativity provides an intrinsic limitation on the measurability of position†.

These results generalize directly to the case of a charged scalar field. In this connection we note that although we cannot form one-particle

* E. M. Henley and W. Thirring, *Elementary Quantum Field Theory*, McGraw Hill, 1962, Chap. 5.

† In the one-particle theory, this relativistic position uncertainty is described in terms of a zigzag motion of the particle over a region $\sim \mu^{-3}$, known as the 'zitterbewegung.'

states which are localized at **x**, we *can* construct point-localized states which have one unit of charge. Consider the total charge operator Q given by 3(35) and define the volume charge operator by

$$Q_v = -i \int_v (\dot{\phi}^\dagger \phi - \dot{\phi}\phi^\dagger)\, d^3x \qquad 3(148)$$

Then

$$\begin{aligned} [Q_v, \phi^\dagger(x)] &= \phi^\dagger(x) \qquad \text{if } \mathbf{x} \text{ is in } v \\ &= 0 \qquad \text{if } \mathbf{x} \text{ is not in } v \end{aligned} \qquad 3(149)$$

Hence the state

$$\phi^\dagger(x)|\ \rangle \qquad 3(150)$$

with $Q_v|\ \rangle = 0$, is an eigenstate of Q_v with eigenvalue 1, if **x** is in v, and eigenvalue 0, if **x** is not in v*. Specializing to an infinitesimally small volume $v(\mathbf{x})$ centred on **x**, we deduce that $\phi^\dagger(x)|\ \rangle$ is a point-localized state with unit charge. The localizability of charge quanta within an arbitrarily small volume may also be inferred from the relation

$$[Q_v(t), Q_{v'}(t)] = 0 \qquad 3(151)$$

which holds for arbitrary v and v'. Note however that

$$[Q_v, N] \neq 0 \qquad 3(152)$$

so that the eigenstates of Q_v do not have sharp particle number.

3-6 Quantization of the Dirac field

Introduction We now turn to the quantization of the spin $\frac{1}{2}$ field. Here we shall be *forced* to adopt Jordan–Wigner anticommutation rules of the type 3(135a), 3(135b) and 3(135c) to ensure that we get a positive-definite field energy. To convince ourselves, we first pursue the canonical approach and replace 2(77) by the equal-time commutation relations†

$$\begin{aligned} [\psi_\alpha(x), \psi_\beta{}^\dagger(x')]_{t=t'} &= \delta_{\alpha\beta}\delta^{(3)}(\mathbf{x}-\mathbf{x}') \\ [\psi_\alpha(x), \psi_\beta(x')]_{t=t'} &= [\psi_\alpha{}^\dagger(x), \psi_\beta{}^\dagger(x')]_{t=t'} = 0 \end{aligned}$$

From the quantum equations of motion

$$\dot{\psi}_\alpha(\mathbf{x}, t) = \frac{1}{i}[\psi_\alpha(\mathbf{x}, t), H(t)]$$

* Henley and Thirring, loc. cit. pp. 62–63. Note that $|\ \rangle$ is *not* the vacuum state $|0\rangle$.

† For an operator field, the dagger denotes the adjoint both with respect to the spinor indices and with respect to Hilbert space.

with H given by the quantum analogue of 2(75)

$$H = \int \psi^\dagger(\gamma_4 \boldsymbol{\gamma} \cdot \boldsymbol{\nabla} + m)\psi \, d^3x \qquad 3(153)$$

we easily recover the Dirac equation

$$(\gamma_\mu \partial_\mu + m)\psi = 0$$

for the quantum field. Hence $\psi(x)$ may be expanded in terms of the complete set of solutions 1(53a) and 1(53b) with operator coefficients $c_{\mathbf{k}\sigma}$ and $d_{\mathbf{k}\sigma}{}^\dagger$:

$$\psi(x) = \frac{1}{\sqrt{V}} \sum_{\mathbf{k}} \sum_{\sigma=1}^{2} \sqrt{\frac{m}{E_{\mathbf{k}}}} (u_{\mathbf{k}\sigma} c_{\mathbf{k}\sigma} \, e^{ik.x} + v_{\mathbf{k}\sigma} \, d_{\mathbf{k}\sigma}{}^\dagger \, e^{-ik.x}) \qquad 3(154a)$$

with $k_0 = E_{\mathbf{k}} = +(\mathbf{k}^2 + m^2)^{1/2}$. For $\bar{\psi} = \psi^\dagger \gamma_4$ we have the corresponding expansion

$$\bar{\psi}(x) = \frac{1}{\sqrt{V}} \sum_{\mathbf{k}} \sum_{\sigma=1}^{2} \sqrt{\frac{m}{E_{\mathbf{k}}}} (\bar{u}_{\mathbf{k}\sigma} c_{\mathbf{k}\sigma}{}^\dagger \, e^{-ik.x} + \bar{v}_{\mathbf{k}\sigma} \, d_{\mathbf{k}\sigma} \, e^{ik.x}) \qquad 3(154b)$$

Inverting these expansions with the aid of 1(58a), 1(58b) and 1(56b) we have

$$c_{\mathbf{k}\sigma} = \frac{1}{\sqrt{V}} \sqrt{\frac{m}{E_{\mathbf{k}}}} \int d^3x \, e^{-ik.x} \bar{u}_{\mathbf{k}\sigma} \gamma_4 \psi(x) \qquad 3(155a)$$

$$c_{\mathbf{k}\sigma}{}^\dagger = \frac{1}{\sqrt{V}} \sqrt{\frac{m}{E_{\mathbf{k}}}} \int d^3x \bar{\psi}(x) \gamma_4 u_{\mathbf{k}\sigma} \, e^{ik.x} \qquad 3(155b)$$

$$d_{\mathbf{k}\sigma} = \frac{1}{\sqrt{V}} \sqrt{\frac{m}{E_{\mathbf{k}}}} \int d^3x \bar{\psi}(x) \gamma_4 v_{\mathbf{k}\sigma} \, e^{-ik.x} \qquad 3(155c)$$

$$d_{\mathbf{k}\sigma}{}^\dagger = \frac{1}{\sqrt{V}} \sqrt{\frac{m}{E_{\mathbf{k}}}} \int d^3x \, e^{ik.x} \bar{v}_{\mathbf{k}\sigma} \gamma_4 \psi(x) \qquad 3(155d)$$

from which we derive the characteristic commutation relations for creation and destruction operators

$$\left.\begin{aligned} [c_{\mathbf{k}\sigma}, c_{\mathbf{k}'\sigma'}{}^\dagger] &= \delta_{\sigma\sigma'} \delta_{\mathbf{k}\mathbf{k}'} \\ [d_{\mathbf{k}\sigma}, d_{\mathbf{k}'\sigma'}{}^\dagger] &= \delta_{\sigma\sigma'} \delta_{\mathbf{k}\mathbf{k}'} \end{aligned}\right\} \qquad 3(156)$$

$$[c_{\mathbf{k}\sigma}, c_{\mathbf{k}'\sigma'}] = [d_{\mathbf{k}\sigma}, d_{\mathbf{k}'\sigma'}] = [c_{\mathbf{k}\sigma}, d_{\mathbf{k}'\sigma'}{}^\dagger] = [c_{\mathbf{k}\sigma}, d_{\mathbf{k}'\sigma'}] = 0$$

We now insert 3(154a) and 3(154b) into the expression

$$H = \int \psi^\dagger(\gamma_4 \boldsymbol{\gamma} \,.\, \boldsymbol{\nabla} + m)\psi \, d^3x$$

$$= \int \bar{\psi}(x)\gamma_4 i\partial_t \psi(x) \, d^3x \qquad 3(157)$$

for the Hamiltonian and apply 1(58a), 1(58b) and 1(56b) to find

$$H = \sum_{\mathbf{k}} \sum_{\sigma=1}^{2} E_{\mathbf{k}}(c_{\mathbf{k}\sigma}{}^\dagger c_{\mathbf{k}\sigma} - d_{\mathbf{k}\sigma}\, d_{\mathbf{k}\sigma}{}^\dagger) \qquad 3(158)$$

Using the commutation rules 3(156) and dropping the zero-point energy term as in the scalar field case, we get the expression

$$H = \sum_{\mathbf{k}} \sum_{\sigma=1}^{2} E_{\mathbf{k}}(c_{\mathbf{k}\sigma}{}^\dagger c_{\mathbf{k}\sigma} - d_{\mathbf{k}\sigma}{}^\dagger\, d_{\mathbf{k}\sigma}) \qquad 3(159)$$

which can take on both positive and negative values.

Fermi–Dirac Quantization To avoid the negative energy difficulty we must modify the passage from 3(158) to 3(159). This can be done by replacing the equal-time commutation relations by

$$\{\psi_\alpha(x), \psi_\beta{}^\dagger(x')\}_{t=t'} = \delta_{\alpha\beta}\delta^{(3)}(\mathbf{x}-\mathbf{x}') \qquad 3(160a)$$

$$\{\psi_\alpha(x), \psi_\beta(x')\}_{t=t'} = \{\psi_\alpha{}^\dagger(x), \psi_\beta{}^\dagger(x')\}_{t=t'} = 0 \qquad 3(160b)$$

which, as we have seen in Section 3-4, imply Fermi–Dirac statistics for the field quanta. The quantum equation of motion is still

$$\dot{\psi}(\mathbf{x}, t) = \frac{1}{i}[\psi(\mathbf{x}, t), H(t)] \qquad 3(161)$$

where H is given by 3(153). To evaluate the commutator on the right-hand side in terms of the anticommutators $\{\psi, \psi^\dagger\}$ and $\{\psi, \psi\}$ we apply the identity

$$[AB, C] = A\{B, C\} - \{A, C\}B \qquad 3(162)$$

It is then a simple matter to check that 3(161) still reproduces the Dirac equation for ψ. Hence the expansions 3(154a) and 3(154b) remain valid, as do the inverse relations 3(155a), 3(155b), 3(155c) and 3(155d). However, the creation and destruction operators now satisfy the anticommutation rules

$$\begin{aligned}\{c_{\mathbf{k}\sigma}, c_{\mathbf{k}'\sigma'}{}^{\dagger}\} &= \{d_{\mathbf{k}\sigma}, d_{\mathbf{k}'\sigma'}{}^{\dagger}\} = \delta_{\mathbf{k}\mathbf{k}'}\delta_{\sigma\sigma'} \\ \{c_{\mathbf{k}\sigma}, c_{\mathbf{k}'\sigma'}\} &= \{d_{\mathbf{k}\sigma}, d_{\mathbf{k}'\sigma'}\} = \{c_{\mathbf{k}\sigma}, d_{\mathbf{k}'\sigma'}{}^{\dagger}\} = \{c_{\mathbf{k}\sigma}, d_{\mathbf{k}'\sigma'}\} = 0\end{aligned} \tag{3(163)}$$

The Hamiltonian is still given by

$$H = \sum_{\mathbf{k}} \sum_{\sigma=1}^{2} E_{\mathbf{k}}(c_{\mathbf{k}\sigma}{}^{\dagger} c_{\mathbf{k}\sigma} - d_{\mathbf{k}\sigma}\, d_{\mathbf{k}\sigma}{}^{\dagger}) \tag{3(164a)}$$

as the commutation rules were not used in deriving 3(158). The momentum

$$\mathbf{P} = -i \int \psi^{\dagger} \nabla \psi \, d^3x$$

is readily found to be

$$\mathbf{P} = \sum_{\mathbf{k}} \sum_{\sigma=1}^{2} \mathbf{k}(c_{\mathbf{k}\sigma}{}^{\dagger} c_{\mathbf{k}\sigma} - d_{\mathbf{k}\sigma}\, d_{\mathbf{k}\sigma}{}^{\dagger}) \tag{3(164b)}$$

and the total charge—the quantum analogue of the total probability—given by

$$Q = \int \psi^{\dagger} \psi \, d^3x$$

is found to be

$$Q = \sum_{\mathbf{k}} \sum_{\sigma=1}^{2} (c_{\mathbf{k}\sigma}{}^{\dagger} c_{\mathbf{k}\sigma} + d_{\mathbf{k}\sigma}\, d_{\mathbf{k}\sigma}{}^{\dagger}) \tag{3(164c)}$$

It is now straightforward to rewrite H, P and Q in the standard form

$$H = \sum_{\mathbf{k}} \sum_{\sigma=1}^{2} E_{\mathbf{k}}(c_{\mathbf{k}\sigma}{}^{\dagger} c_{\mathbf{k}\sigma} + d_{\mathbf{k}\sigma}{}^{\dagger}\, d_{\mathbf{k}\sigma}) \tag{3(165a)}$$

$$\mathbf{P} = \sum_{\mathbf{k}} \sum_{\sigma=1}^{2} \mathbf{k}(c_{\mathbf{k}\sigma}{}^{\dagger} c_{\mathbf{k}\sigma} + d_{\mathbf{k}\sigma}{}^{\dagger}\, d_{\mathbf{k}\sigma}) \tag{3(165b)}$$

$$Q = \sum_{\mathbf{k}} \sum_{\sigma=1}^{2} (c_{\mathbf{k}\sigma}{}^{\dagger} c_{\mathbf{k}\sigma} - d_{\mathbf{k}\sigma}{}^{\dagger}\, d_{\mathbf{k}\sigma}) \tag{3(165c)}$$

in terms of the number operators $c_{\mathbf{k}\sigma}{}^{\dagger} c_{\mathbf{k}\sigma}$ and $d_{\mathbf{k}\sigma}{}^{\dagger} d_{\mathbf{k}\sigma}$. To get 3(165a), 3(165b) and 3(165c) from 3(164a), 3(164b) and 3(164c) we simply apply the anticommutation relations and subtract out the zero point energy, momentum* and charge

$$E_0 = -\sum_{\mathbf{k}} \sum_{\sigma=1}^{2} E_{\mathbf{k}} \qquad 3(166a)$$

$$P_0 = -\sum_{\mathbf{k}} \sum_{\sigma=1}^{2} \mathbf{k} \qquad 3(166b)$$

$$Q_0 = +\sum_{\mathbf{k}} \sum_{\sigma=1}^{2} 1 \qquad 3(166c)$$

The energy is now positive-definite. The total charge, on the other hand, *is no longer positive-definite*. In resolving the negative energy difficulty of the Dirac theory, we have converted the latter into a theory describing particles and antiparticles (e.g. electrons and positrons), with equal and opposite charges. The creation operators $c_{\mathbf{k}\sigma}{}^{\dagger}$ and $d_{\mathbf{k}\sigma}{}^{\dagger}$ create particles of positive and negative charge respectively.

As in the Klein–Gordon case, the subtraction of the zero-point energy and charge can be performed by redefining H and Q as normal ordered products. However, the definition of normal ordering must be modified in the case of Fermi–Dirac fields. To the rearrangement of all creation operators to the left of all destruction operators, we add the prescription of a change in sign each time two Fermi–Dirac operators are interchanged in the process of normal ordering. Thus, for example,

$$:\bar{\psi}_\alpha \psi_\beta: = \bar{\psi}_\alpha^{(+)}\psi_\beta^{(+)} + \bar{\psi}_\alpha^{(-)}\psi_\beta^{(+)} + \bar{\psi}_\alpha^{(-)}\psi_\beta^{(-)} - \psi_\beta^{(-)}\bar{\psi}_\alpha^{(+)} \qquad 3(167)$$

where we have denoted positive and negative frequency parts by (+) and (−) respectively. With this definition, the expressions 3(165a), 3(165b) and 3(165c) are just the normal ordered forms of 3(164a), 3(164b) and 3(164c). The redefined Hamiltonian and charge are given by

$$H = \int :\psi^{\dagger}(\gamma_4 \boldsymbol{\gamma} \cdot \boldsymbol{\nabla} + m)\psi : d^3x \qquad 3(168a)$$

$$Q = \int :\psi^{\dagger}\psi : d^3x \qquad 3(168b)$$

Hole Theory In the case of the Dirac theory, the operation of subtracting out the zero point energy and charge can be given a simple physical interpretation. This is based on the so-called ‘ hole theory ’, [Dirac (1930)], which postulates that the vacuum state is such that all positive energy levels are empty but all negative energy levels are *filled*. This concept of a

* The zero-point momentum is actually zero, by cancellation of **k** and −**k**.

filled Dirac 'sea' naturally requires an exclusion principle limiting the number of particles in a given state and is therefore applicable only when quantization is performed with anticommutators. Now if we return to the expression 3(164a) for the field energy, we see that $c_{\mathbf{k}\sigma}{}^{\dagger}c_{\mathbf{k}\sigma}$ and $d_{\mathbf{k}\sigma}\,d_{\mathbf{k}\sigma}{}^{\dagger}$ are number operators for positive and negative energy particles respectively*. By 3(164c), both positive and negative energy particles have positive charge. Applying the hole theory, we define the vacuum state to be the state in which all negative energy levels are filled, that is in which all $c_{\mathbf{k}\sigma}{}^{\dagger}c_{\mathbf{k}\sigma}$ have eigenvalue 0 and all $d_{\mathbf{k}\sigma}\,d_{\mathbf{k}\sigma}{}^{\dagger}$ have eigenvalue $+1$. By 3(164a) and 3(164c), this state has energy $E_0 = -\Sigma_{\mathbf{k}}\Sigma_{\sigma}E_{\mathbf{k}}$ and charge $Q_0 = \Sigma_{\mathbf{k}}\Sigma_{\sigma}1$. We must therefore subtract the latter quantities from the energy and charge of the field since all physical quantities are measured with respect to the vacuum state.

In this picture anti-particles are viewed as *holes* in the filled Dirac sea, whence the name 'hole theory'. For example, the antiparticle state

$$d_{\mathbf{k}\sigma}{}^{\dagger}|0\rangle \tag{3(169)}$$

yields the eigenvalue $+1$ for the antiparticle number operator $d_{\mathbf{k}\sigma}{}^{\dagger}\,d_{\mathbf{k}\sigma}$ and 0 for the negative energy particle number operator

$$d_{\mathbf{k}\sigma}\,d_{\mathbf{k}\sigma}{}^{\dagger} = 1 - d_{\mathbf{k}\sigma}{}^{\dagger}\,d_{\mathbf{k}\sigma} \tag{3(170)}$$

Hence, presence of an antiparticle signifies absence of a negative energy particle in the sea. We may also deduce this result from the fact that $d_{\mathbf{k}\sigma}{}^{\dagger}$ is at the same time a creation operator for an antiparticle of momentum $+\mathbf{k}$ and a destruction operator for a negative energy particle of momentum $-\mathbf{k}$.

Spin We have seen in Section 1-3 that the spin orientations of the spinors $u_{\mathbf{k}\sigma}$ and $v_{\mathbf{k}\sigma}$ are defined by the eigenvalues of the covariant operator $\omega(k).e$ as in 1(63a) and 1(63b). Generalization of this quantity to field theory is rather unrewarding however, and it is simplest to work with the integral

$$S_z = \tfrac{1}{2}\int :\psi^{\dagger}\Sigma_z\psi:\,d^3x \tag{3(171)}$$

representing the spin part of the angular momentum of the field along the z-axis, as given classically by 2(124). The normal product ensures the

* Note that by 3(164b), $d_{\mathbf{k}\sigma}\,d_{\mathbf{k}\sigma}{}^{\dagger}$ is the number operator for negative energy particles of momentum $-\mathbf{k}$. We are free to interpret $d_{\mathbf{k}\sigma}\,d_{\mathbf{k}\sigma}{}^{\dagger}$ as a number operator owing to the symmetry of anticommutation relations under interchange of creation and destruction operators ($d_{\mathbf{k}\sigma}$ may be interpreted as a creation operator and $d_{\mathbf{k}\sigma}{}^{\dagger}$ as a destruction operator).

automatic subtraction of the vacuum spin. Insertion of 3(154a) and 3(154b) into 3(171) reveals that S_z is not diagonal in the basis spanned by the eigenstates of the number operators $c_{\mathbf{k}\sigma}{}^\dagger c_{\mathbf{k}\sigma}$ and $d_{\mathbf{k}\sigma}{}^\dagger d_{\mathbf{k}\sigma}$. This is not surprising, as the $u_{\mathbf{k}\sigma}$ and $v_{\mathbf{k}\sigma}$ are themselves not eigenstates of $\frac{1}{2}\Sigma_z$. However, it is a simple matter to check that

$$\langle 1 \text{ electron}, \mathbf{k} = 0, \sigma | S_z | 1 \text{ electron}, \mathbf{k} = 0, \sigma \rangle = \begin{cases} +\frac{1}{2} & (\sigma = 1) \\ -\frac{1}{2} & (\sigma = 2) \end{cases}$$

for a one-electron state at rest

$$|1 \text{ electron}, \mathbf{k} = 0, \sigma\rangle = c_{0\sigma}{}^\dagger|0\rangle$$

On the other hand, for a one-positron state at rest we have

$$\langle 1 \text{ positron}, \mathbf{k} = 0, \sigma | S_z | 1 \text{ positron}, \mathbf{k} = 0, \sigma \rangle = \begin{cases} -\frac{1}{2} & (\sigma = 1) \\ +\frac{1}{2} & (\sigma = 2) \end{cases}$$

owing to the minus sign arising from the normal ordering of the positron operators in S_z. Thus the positron spin is opposite to that of the missing negative energy electron. Positrons with spin up (down) are described by $v_{\mathbf{k}_2}(v_{\mathbf{k}_1})$ respectively.

Generators Although we have quantized the Dirac field by means of anticommutation rules, basic quantum relations such as the Hamiltonian equation of motion 3(161) remain valid. The same applies to the relations expressing infinitesimal symmetry transformations of $\psi(x)$ in terms of the corresponding generators. For space translations, for example, we have

$$\delta\psi(\mathbf{x}, t) = -\boldsymbol{\epsilon} \,.\, \nabla\psi(\mathbf{x}, t) = \frac{1}{i}[\psi(\mathbf{x}, t), \boldsymbol{\epsilon} \,.\, \mathbf{P}(t)] \qquad 3(172)$$

where $\mathbf{P}$ is the field momentum

$$\mathbf{P} = -i \int \psi^\dagger \nabla \psi \, d^3x$$

To check 3(172), we evaluate the commutator with the aid of 3(160a), 3(160b) and 3(162) with the result

$$\left[\psi(\mathbf{x}, t), -i \int \psi^\dagger(\mathbf{x}', t)\boldsymbol{\epsilon} \,.\, \nabla\psi(\mathbf{x}', t) \, d^3x'\right] = -i\boldsymbol{\epsilon} \,.\, \nabla\psi(\mathbf{x}, t)$$

As another example, the charge operator

$$Q = \int \psi^\dagger \psi \, d^3x$$

is the generator of infinitesimal phase transformations

$$\delta\psi = i\alpha\psi = \frac{1}{i}[\psi, -\alpha Q]$$
$$\delta\psi^\dagger = -i\alpha\psi^\dagger = \frac{1}{i}[\psi^\dagger, -\alpha Q] \quad 3(173)$$

from which one can recover the statement that $\psi(\bar{\psi})$ destroys (creates) one unit of positive charge by following the same steps as in 3(48a) and 3(48b).

As in the Klein–Gordon case the validity of relations of the type 3(172) and 3(173) ensures that the symmetry transformations of the fields may be represented by means of unitary operators in Hilbert space. Under a general infinitesimal space-time rotation and displacement

$$x'_\mu = a_{\mu\nu}x_\nu + \varepsilon_\mu$$

with $a_{\mu\nu} = \delta_{\mu\nu} + \omega_{\mu\nu}$, the field variation

$$\delta\psi(x) = -\omega_{\mu\nu}x_\nu\partial_\mu\psi(x) + \frac{i}{4}\omega_{\mu\nu}\Sigma_{\mu\nu}\psi(x) - \varepsilon_\mu\partial_\mu\psi(x) \quad 3(174)$$

is generated by

$$G(a, \varepsilon) = \varepsilon_\mu P_\mu - \omega_{\mu\nu}M_{\mu\nu} \quad 3(175)$$

according to

$$\delta\psi = \frac{1}{i}[\psi, G(a, \varepsilon)] \quad 3(176)$$

P_μ and $M_{\mu\nu}$ in 3(175) are respectively the 4-momentum and angular momentum of the Dirac field. Setting $\psi' = \psi + \delta\psi$ and $U(a, \varepsilon) = 1 - iG(a, \varepsilon)$ as in the corresponding discussion for the Klein–Gordon case we can write*

$$\psi'(x) = L(a)\psi(a^{-1}x - \varepsilon)$$
$$= U(a, \varepsilon)^\dagger\psi(x)U(a, \varepsilon) \quad 3(177a)$$

where

$$L(a) = 1 + \frac{i}{4}\omega_{\mu\nu}\Sigma_{\mu\nu}$$

Substituting $x \to ax + \varepsilon$ in 3(177a), we can present the Lorentz transformation law for the Dirac field in the standard form

$$U(a, \varepsilon)\psi(x)U(a, \varepsilon)^\dagger = L^{-1}(a)\psi(ax + \varepsilon) \quad 3(177b)$$

* See Equation 1(76).

We also note that the explicit form of the transformation law for finite space-time displacements $x'_\mu = x_\mu + b_\mu$ is

$$e^{-ib.P}\psi(x)\,e^{ib.P} = \psi(x+b)$$

as in 3(55).

The discussion of the connection between symmetries and conservation laws for the Dirac field parallels the corresponding discussion in Section 3-2 and is left as an exercise for the reader.

Discrete Symmetries We now turn to the discrete symmetry transformations **P**, $\mathscr{C}$, and $\mathscr{T}$. Treating first the case of space reflections, we observe that the Dirac equation and the equal-time anticommutation rules 3(160a) and 3(160b) are invariant under the active parity transformation

$$\psi(\mathbf{x}, t) \to \psi'(\mathbf{x}, t) \qquad 3(178a)$$

with

$$\psi'(\mathbf{x}, t) = \gamma_4 \psi(-\mathbf{x}, t) \qquad 3(178b)$$

We therefore seek a unitary operator $\mathscr{P}$ such that

$$\mathscr{P}\psi(\mathbf{x}, t)\mathscr{P}^{-1} = \gamma_4 \psi(-\mathbf{x}, t) \qquad 3(179)$$

Using the expansion 3(154a) and the properties

$$\begin{aligned} \gamma_4 u_{\mathbf{k}\sigma} &= u_{-\mathbf{k}\sigma} \\ \gamma_4 v_{\mathbf{k}\sigma} &= -v_{-\mathbf{k}\sigma} \end{aligned} \qquad 3(180a)$$

which can be read off directly from 1(28) and 1(51a) and 1(51b), we can write 3(179) as

$$\begin{aligned} \mathscr{P} c_{\mathbf{k}\sigma} \mathscr{P}^{-1} &= c_{-\mathbf{k}\sigma} \\ \mathscr{P}\, d_{\mathbf{k}\sigma}{}^{\dagger} \mathscr{P}^{-1} &= -d_{-\mathbf{k}\sigma}{}^{\dagger} \end{aligned} \qquad 3(180b)$$

The explicit form of $\mathscr{P}$ can be obtained by following the same procedure as in the corresponding calculation of $\mathscr{P}_s$ and $\mathscr{P}_{ps}$, with the result

$$\mathscr{P} = \exp\left[-\frac{i\pi}{2}\sum_{\mathbf{k}\sigma}(c_{\mathbf{k}\sigma}{}^{\dagger}c_{\mathbf{k}\sigma} - c_{\mathbf{k}\sigma}{}^{\dagger}c_{-\mathbf{k}\sigma} - d_{\mathbf{k}\sigma}{}^{\dagger}\, d_{\mathbf{k}\sigma} - d_{\mathbf{k}\sigma}{}^{\dagger}\, d_{-\mathbf{k}\sigma})\right] \qquad 3(181)$$

In the case of charge conjugation one seeks a unitary operator $\mathscr{C}$ such that

$$\mathscr{C}\psi\mathscr{C}^{-1} = \psi^c \qquad 3(182a)$$

where ψ^c is the charge conjugate field

$$\psi^c = C\bar{\psi}^T \tag{3(182b)}$$

defined in Section 1-3. The charge conjugation matrix C is defined by the properties 1(64a), 1(64b) and 1(64c). We have seen in Section 1-3 that the substitution $\psi \to \psi^c$ is a symmetry operation of the Dirac equation. To see that the equal-time anticommutation rules are also left invariant by this substitution, we write 3(160a) for example in the form

$$\{\psi_\alpha(x), \bar{\psi}_\beta(x')\}_{t=t'} = (\gamma_4)_{\alpha\beta}\delta^{(3)}(\mathbf{x}-\mathbf{x}')$$

and contract it with $-C_{\rho\beta}(C^{-1})_{\alpha\sigma}$. Since, as follows easily from 3(182b),

$$\overline{\psi^c} = -\psi^T C^{-1} \tag{3(183)}$$

we get

$$\begin{aligned}\{\overline{\psi^c_\sigma}(x), \psi^c_\rho(x')\}_{t=t'} &= -(C\gamma_4^T C^{-1})_{\rho\sigma}\delta^{(3)}(\mathbf{x}-\mathbf{x}') \\ &= (\gamma_4)_{\rho\sigma}\delta^{(3)}(\mathbf{x}-\mathbf{x}')\end{aligned}$$

which establishes the invariance upon substituting $\mathbf{x} \rightleftarrows \mathbf{x}'$.

The explicit construction of the unitary operator $\mathscr{C}$ in terms of creation and destruction operators follows the usual pattern and is left as an exercise.

Finally we turn to time reflection. As in the Klein–Gordon case, we expect this operation to be represented by an antiunitary operator $\mathscr{T}$ of the form

$$\mathscr{T} = \mathscr{U}K$$

where K is the complex conjugation operator satisfying 3(72). To determine $\mathscr{T}$ we multiply the Dirac equation

$$(\gamma\,.\,\partial+m)\psi(\mathbf{x}, t) = 0 \tag{3(184)}$$

on the left by $\mathscr{T}$ and on the right by $\mathscr{T}^{-1}$. This gives, using 3(72)

$$(\gamma^*\,.\,\partial^*+m)\mathscr{T}\psi(\mathbf{x}, t)\mathscr{T}^{-1} = 0$$

To recover 3(184), it suffices to set

$$\mathscr{T}\psi(\mathbf{x}, t)\mathscr{T}^{-1} = T^{-1}\psi(\mathbf{x}, -t) \tag{3(185)}$$

where T^{-1} is a non-singular matrix with the property $\gamma_\mu{}^*T^{-1} = T^{-1}\gamma_\mu$ or

$$T^{-1}\gamma_\mu T = \gamma_\mu^T \tag{3(186a)}$$

Hence the operation 3(185) is a symmetry transformation of the Dirac equation. To ensure the invariance of the equal-time anticommutation rules 3(160a) and 3(160b) one must impose the further restriction that T be a unitary matrix

$$T^\dagger = T^{-1} \qquad 3(186b)$$

A matrix T with the properties 3(186a) and 3(186b) can be constructed with the aid of the charge conjugation matrix C satisfying 1(64a), 1(64b) and 1(64c) and the γ_5 matrix satisfying 1(45a) and 1(45b). We easily see that

$$T = \gamma_5 C \qquad 3(187)$$

has the required properties 3(186a) and 3(186b) and in addition satisfies

$$T^T = -T \qquad 3(186c)$$

Repeating 3(185) and using the property $T^* = -T^{-1}$ we get

$$\begin{aligned}\mathscr{T}\mathscr{T}\psi(\mathbf{x},t)\mathscr{T}^{-1}\mathscr{T}^{-1} &= \mathscr{T}T^{-1}\psi(\mathbf{x},-t)\mathscr{T}^{-1}\\ &= -T\mathscr{T}\psi(\mathbf{x},-t)\mathscr{T}^{-1}\\ &= -\psi(\mathbf{x},t)\end{aligned}$$

so that $\mathscr{T}$ is a reasonable reflection operator. Moreover, the current operator

$$j_\mu(\mathbf{x},t) = i\bar{\psi}(\mathbf{x},t)\gamma_\mu\psi(\mathbf{x},t)$$

transforms according to

$$\begin{aligned}\mathscr{T}j_\mu(\mathbf{x},t)\mathscr{T}^{-1} &= -i\mathscr{T}\psi^\dagger(\mathbf{x},t)\mathscr{T}^{-1}\gamma_4{}^*\gamma_\mu{}^*\mathscr{T}\psi(\mathbf{x},t)\mathscr{T}^{-1}\\ &= -i\psi^\dagger(\mathbf{x},-t)T\gamma_4{}^*\gamma_\mu{}^*T^{-1}\psi(\mathbf{x},-t)\\ &= -i\bar{\psi}(\mathbf{x},-t)\gamma_\mu\psi(\mathbf{x},-t)\\ &= -j_\mu(\mathbf{x},-t)\end{aligned} \qquad 3(188)$$

The explicit construction of $\mathscr{U}$ such that $\mathscr{T} = \mathscr{U}K$ satisfies 3(185) is left to the reader.

General Commutation Rules Our next task is to calculate the anticommutators $\{\psi_\alpha(x), \bar{\psi}_\beta(x')\}$ for arbitrary times x_0 and x_0'. Applying the anticommutation rules 3(163) we have

$$\begin{aligned}\{\psi_\alpha(x), \bar{\psi}_\beta(x')\} &= \{\psi_\alpha^{(+)}(x), \bar{\psi}_\beta^{(-)}(x')\} + \{\psi_\alpha^{(-)}(x), \bar{\psi}_\beta^{(+)}(x')\}\\ &= \frac{1}{V}\sum_{\mathbf{k},k_0=E_\mathbf{k}} \frac{m}{E_\mathbf{k}}\Big\{\sum_{\sigma=1}^{2} u^\alpha_{\mathbf{k}\sigma}\bar{u}^\beta_{\mathbf{k}\sigma}\, e^{ik.(x-x')}\\ &\quad + \sum_{\sigma=1}^{2} v^\alpha_{\mathbf{k}\sigma}\bar{v}^\beta_{\mathbf{k}\sigma}\, e^{-ik.(x-x')}\Big\}\end{aligned} \qquad 3(189)$$

The spin sums which appear on the right-hand side have been evaluated in Section 1-3. Using the results 1(83a) and 1(83b) we find

$$\{\psi_\alpha(x), \bar{\psi}_\beta(x')\} = \frac{1}{V}\sum_{\mathbf{k},k_0=E_\mathbf{k}} \frac{1}{2iE_\mathbf{k}}[(\gamma\,.\,k+im)_{\alpha\beta}\,\mathrm{e}^{ik.(x-x')}$$
$$+(\gamma\,.\,k-im)_{\alpha\beta}\,\mathrm{e}^{-ik.(x-x')}]$$
$$= -(\gamma\,.\,\partial-m)_{\alpha\beta}\frac{1}{V}\sum_{\mathbf{k},k_0=E_\mathbf{k}}\left(\frac{1}{2E_\mathbf{k}}\mathrm{e}^{ik.(x-x')}-\frac{1}{2E_\mathbf{k}}\mathrm{e}^{-ik.(x-x')}\right)$$

Recalling 3(80a) and 3(80b) we see that

$$\{\psi_\alpha^{(+)}(x), \bar{\psi}_\beta^{(-)}(x')\} = -i(\gamma\,.\,\partial-m)_{\alpha\beta}\Delta^{(+)}(x-x') \qquad 3(190a)$$

$$\{\psi_\alpha^{(-)}(x), \bar{\psi}_\beta^{(+)}(x')\} = -i(\gamma\,.\,\partial-m)_{\alpha\beta}\Delta^{(-)}(x-x') \qquad 3(190b)$$

and

$$\{\psi_\alpha(x), \bar{\psi}_\beta(x')\} = -i(\gamma\,.\,\partial-m)_{\alpha\beta}[\Delta^{(+)}(x-x')+\Delta^{(-)}(x-x')]$$
$$= -i(\gamma\,.\,\partial-m)_{\alpha\beta}\Delta(x-x') \qquad 3(191)$$

All other anticommutators vanish:

$$\{\psi_\alpha(x), \psi_\beta(x')\} = \{\bar{\psi}_\alpha(x), \bar{\psi}_\beta(x')\} = 0 \qquad 3(192)$$

It is customary to define anticommutator functions $S^{(+)}$, $S^{(-)}$, and S corresponding to $\Delta^{(+)}$, $\Delta^{(-)}$, and Δ by

$$S_{\alpha\beta}^{(\ )}(x-x') = (\gamma\,.\,\partial-m)_{\alpha\beta}\Delta^{(\ )}(x-x') \qquad 3(193)$$

in terms of which the anticommutation relations take the form

$$\{\psi_\alpha^{(+)}(x), \bar{\psi}_\beta^{(-)}(x')\} = -iS_{\alpha\beta}^{(+)}(x-x') \qquad 3(194a)$$

$$\{\psi_\alpha^{(-)}(x), \bar{\psi}_\beta^{(+)}(x')\} = -iS_{\alpha\beta}^{(-)}(x-x') \qquad 3(194b)$$

and

$$\{\psi_\alpha(x), \bar{\psi}_\beta(x')\} = -iS_{\alpha\beta}(x-x') \qquad 3(195)$$

These functions all satisfy the Dirac equation

$$(\gamma\,.\,\partial+m)_{\alpha\gamma}S_{\gamma\beta}^{(\ )}(x) = 0 \qquad 3(196)$$

In the Dirac theory, local observable quantities are associated with bilinear expressions in $\psi(x)$ and $\bar{\psi}(x)$. The anticommutation relations 3(160a) and 3(160b) ensure that these commute at spacelike separations.

Indeed, using 3(162)

$$\begin{aligned}[\bar{\psi}_\alpha(x)\psi_\beta(x), \bar{\psi}_\rho(x')\psi_\sigma(x')] \\ &= \bar{\psi}_\rho(x')[\bar{\psi}_\alpha(x)\psi_\beta(x), \psi_\sigma(x')] + [\bar{\psi}_\alpha(x)\psi_\beta(x), \bar{\psi}_\rho(x')]\psi_\sigma(x') \\ &= i\bar{\psi}_\rho(x')S_{\sigma\alpha}(x'-x)\psi_\beta(x) - i\bar{\psi}_\alpha(x)S_{\beta\rho}(x-x')\psi_\sigma(x') \qquad 3(197)\end{aligned}$$

Since $\Delta(x-x')$, and hence $S(x-x')$, vanishes for $(x-x')^2 > 0$ the commutator 3(197) vanishes and the principle of microcausality is assured.

Feynman Propagator The concept of time-ordering was introduced for Bose–Einstein fields by 3(91). To extend this to Fermi–Dirac fields we modify the definition to include a minus sign for each interchange of Fermi–Dirac field operators. Thus, for example,

$$T\psi(x_1)\bar{\psi}(x_2) = \begin{cases} \psi(x_1)\bar{\psi}(x_2) & t_1 > t_2 \\ -\bar{\psi}(x_2)\psi(x_1) & t_2 > t_1 \end{cases} \qquad 3(198)$$

The change of sign is necessary to ensure a Lorentz-invariant definition of time ordering. For spacelike separations the sign of $t_1 - t_2$ can be reversed by means of a Lorentz transformation. However, since $\psi(x_1)$ and $\bar{\psi}(x_2)$ anticommute for spacelike separations, the time ordering 3(198) has a Lorentz-invariant meaning.

We now define the Feynman propagator for the Dirac field to be

$$S_{F\alpha\beta}(x_1 - x_2) = \langle 0|T\psi_\alpha(x_1)\bar{\psi}_\beta(x_2)|0\rangle \qquad 3(199)$$

For $t_1 > t_2$ we have

$$\begin{aligned}\langle 0|T\psi_\alpha(\chi_1)\bar{\psi}_\beta(\chi_2)|0\rangle &= \langle 0|\psi_\alpha(\chi_1)\bar{\psi}_\beta(\chi_2)|0\rangle \\ &= \langle 0|\{\psi_\alpha^{(+)}(x_1), \bar{\psi}_\beta^{(-)}(x_2)\}|0\rangle \\ &= -iS_{\alpha\beta}^{(+)}(x_1 - x_2)\end{aligned}$$

Similarly, for $t_2 > t_1$

$$\langle 0|T\psi_\alpha(x_1)\,\bar{\psi}_\beta(x_2)|0\rangle = +iS_{\alpha\beta}^{(-)}(x_1 - x_2)$$

Thus the Feynman propagator is equal to

$$\begin{aligned}S_{F\alpha\beta}(x_1 - x_2) = &-i\theta(t_1 - t_2)(\gamma\,.\,\partial_1 - m)_{\alpha\beta}\Delta^{(+)}(x_1 - x_2) \\ &+ i\theta(t_2 - t_1)(\gamma\,.\,\partial_1 - m)_{\alpha\beta}\Delta^{(-)}(x_1 - x_2)\end{aligned} \qquad 3(200)$$

where we have recalled the definition 3(193). The θ functions can be commuted past the $\gamma . \partial - m$ factors since

$$\gamma_4[\partial_{t_1}\theta(t_1-t_2)]\Delta^{(+)}(x_1-x_2)-\gamma_4[\partial_{t_1}\theta(t_2-t_1)]\Delta^{(-)}(x_1-x_2)$$
$$= \gamma_4\delta(t_1-t_2)\Delta(x_1-x_2) = 0$$

where we have used the fact that $\partial_t\theta(t) = \delta(t)$ and the relation 3(79). Hence we get, suppressing spinor indices,

$$S_F(x) = -(\gamma . \partial - m)[i\theta(x_0)\Delta^{(+)}(x) - i\theta(-x_0)\Delta^{(-)}(x)]$$
$$= -(\gamma . \partial - m)\Delta_F(x)$$

where Δ_F is the Feynman propagator 3(94) for the spin 0 field. Using 3(96a) and 3(96b) we obtain the final result

$$S_F(x) = \frac{1}{(2\pi)^4}\int d^4k\, e^{ik.x} S_F(k) \qquad 3(201a)$$

$$S_F(k) = \frac{-1}{\gamma . k - im - i\varepsilon} \qquad 3(201b)$$

We note that the definition of the T-product 3(198), and hence the Feynman propagator 3(199), suffers from ambiguity when $t_1 = t_2$ and $\mathbf{x}_1 = \mathbf{x}_2$, since the ordering of the operator factors $\psi(x_1)$ and $\bar{\psi}(x_2)$ is not prescribed in that case*. The ambiguity is frequently resolved by adopting the convention that $T\psi_\alpha(x)\bar{\psi}_\beta(x)$ is to equal the antisymmetrized product

$$\tfrac{1}{2}(\psi_\alpha(x)\bar{\psi}_\beta(x) - \bar{\psi}_\beta(x)\psi_\alpha(x)) \qquad 3(202)$$

Antisymmetrized Current One can also use the product 3(202) to define an antisymmetrized current operator

$$j_\mu = \tfrac{1}{2}i\gamma_\mu^{\beta\alpha}(\bar{\psi}_\beta(x)\psi_\alpha(x) - \psi_\alpha(x)\bar{\psi}_\beta(x))$$
$$= \tfrac{1}{2}i(\bar{\psi}(x)\gamma_\mu\psi(x) - \psi(x)\gamma_\mu^T\bar{\psi}(x))$$
$$= \tfrac{1}{2}i[\bar{\psi}(x), \gamma_\mu\psi(x)] \qquad 3(203)$$

which will be of importance later. In contrast to the un-antisymmetrized current, 3(203) has the simple transformation property

$$\mathscr{C}j_\mu\mathscr{C}^{-1} = -j_\mu \qquad 3(204)$$

under the charge conjugation operation 3(182)†.

* For $t_1 = t_2$ and $\mathbf{x}_1 \neq \mathbf{x}_2$ the field operators $\psi(x_1)$ and $\bar{\psi}(x_2)$ anticommute so that the T-product is well defined.

† See Problem 11. Note that 3(204) is of the same form as 3(67).

We now show that antisymmetrization of the current is equivalent to normal ordering, i.e. that

$$\tfrac{1}{2}[\bar{\psi}(x), \gamma_\mu \psi(x)] = :\bar{\psi}(x)\gamma_\mu \psi(x): \qquad 3(205)$$

Indeed, evaluating the difference with the aid of 3(167), we find

$$:\bar{\psi}_\alpha(x)\psi_\beta(x): - \tfrac{1}{2}[\bar{\psi}_\alpha(x), \psi_\beta(x)] = \tfrac{1}{2}\{\bar{\psi}_\alpha^{(-)}(x), \psi_\beta^{(+)}(x)\} - \tfrac{1}{2}\{\psi_\beta^{(-)}(x), \bar{\psi}_\alpha^{(+)}(x)\}$$

The anticommutators appearing on the right-hand side can be computed as in 3(189) with the result

$$\frac{1}{2V}\sum_{\mathbf{k}} \frac{m}{E_{\mathbf{k}}} \sum_{\sigma=1}^{2} (u_{\mathbf{k}\sigma}^\beta \bar{u}_{\mathbf{k}\sigma}^\alpha - v_{\mathbf{k}\sigma}^\beta \bar{v}_{\mathbf{k}\sigma}^\alpha) = \frac{1}{2V}\sum_{\mathbf{k}} \frac{m}{E_{\mathbf{k}}} \delta_{\beta\alpha}$$

Hence

$$:\bar{\psi}\gamma_\mu \psi: - \tfrac{1}{2}[\bar{\psi}, \gamma_\mu \psi] \sim T_r \gamma_\mu = 0$$

The same argument also shows that

$$:\bar{\psi}(x)\gamma_5 \psi(x): = \tfrac{1}{2}[\bar{\psi}(x), \gamma_5 \psi(x)] \qquad 3(206a)$$

but

$$:\bar{\psi}(x)\psi(x): \neq \tfrac{1}{2}[\bar{\psi}(x), \psi(x)] \qquad 3(206b)$$

since $Tr\ 1 \neq 0$. [See Problem 12.]

Owing to 3(205), the vacuum expectation value of the antisymmetrized current vanishes:

$$\langle 0|j_\mu(x)|0\rangle = \langle 0|\tfrac{1}{2}[\bar{\psi}(x), \gamma_\mu \psi(x)]|0\rangle = 0 \qquad 3(207)$$

3-7 Connection between spin and statistics

We recall from our work on classical fields in Chapter 2 that for integer spin fields like the Klein–Gordon and vector fields, the field energy or Hamiltonian is positive-definite, whereas the charge can take on either sign. The situation is just the reverse in the case of half-integer spin fields. There the total charge is positive-definite, while the field energy is indefinite. As a result, the quantization of half-integer spin fields follows a different pattern from that of integer spin fields. Whereas integer spin fields are quantized by means of the canonical prescription 3(1), the quantization of half integer spin fields requires the use of anticommutators. For example, the quantization of the charged scalar field, as performed in Section 3-2, ensured that the quantized field energy and charge retained their positive-definite and indefinite character respectively, in accordance with the interpretation of the field as an assembly of positively and negatively charged quanta. Correspondingly, the quantization of the Dirac field by means of anticommutation relations converts the indefinite field energy of the classical field into a positive-definite form and at the same time destroys the positive-definite character of the charge.

This connection between the spin and the character of the commutation relations, or equivalently the statistics of the field quanta, is one of the most significant predictions of local relativistic quantum field theory [Pauli (1940)]. It is specifically a relativistic effect; we have seen that in the case of non-relativistic field theory, both commutation and anticommutation relations yield a consistent quantization scheme. Experimentally, the connection between spin and statistics is verified for all particles which one studies in large assemblies. Spin $\frac{1}{2}$ particles such as protons, neutrons and electrons obey Fermi–Dirac statistics, whereas spin 1 photons obey Bose–Einstein statistics [Bethe (1936)]. The statistics of other known particles have not yet been determined with certainty, but there is strong evidence that the spin 0 π's obey Bose–Einstein statistics [Messiah (1964)].

The connection between spin and statistics can also be recovered through the microcausality requirement that local observables must commute for space-like separations. Let us impose this requirement on bilinear local observables of the form

$$0(x) = \varphi_a(x)\varphi_b(x)$$

We therefore require

$$[0(x), 0(x')] = 0 \qquad \text{for} \quad (x-x')^2 > 0 \tag{3(208)}$$

and it is easy to see that 3(208) in turn requires either

$$[\varphi_a(x), \varphi_b(x')] = 0 \qquad \text{for} \quad (x-x')^2 > 0 \tag{3(209a)}$$

or

$$\{\varphi_a(x), \varphi_b(x')\} = 0 \qquad \text{for} \quad (x-x')^2 > 0 \tag{3(209b)}$$

depending on whether the φ_a obey Bose–Einstein or Fermi–Dirac commutation relations. Now we have seen that the requirement 3(209a) is satisfied when the Klein–Gordon field is quantized according to Bose–Einstein statistics. Similarly, the Dirac field, when quantized according to Fermi–Dirac statistics satisfies 3(209b). What happens if we quantize, say, the neutral Klein–Gordon field with *anti*commutators? Referring to the derivation of the invariant commutation rules in Section 3-2, one finds, instead of 3(89),

$$\{\phi(x_1), \phi(x_2)\} = i[\Delta^{(+)}(x_1-x_2)-\Delta^{(-)}(x_1-x_2)]$$
$$= \Delta_1(x_1-x_2)$$

where Δ_1 is given by 3(103). Since, as can readily be checked, $\Delta_1(x)$ does *not* vanish for $x^2 > 0$, the microcausality requirement is violated. Similarly, if we attempt to quantize the Dirac field by means of commutation relations, we find, repeating the steps which led to 3(191),

$$[\psi_\alpha(x), \bar{\psi}_\beta(x')] = -S_1(x-x')$$

with

$$S_1(x-x') = (\gamma \,.\, \partial - m)\Delta_1(x-x')$$

Since $S_1(x)$ fails to vanish for $x^2 > 0$ we are again in contradiction with microcausality.

PROBLEMS

1. Check the relations 3(10a) and 3(10b).
2. Show that the convention 3(26) implies 3(27).
3. Complete the proof of 3(46a) and 3(46b) for time translations and space-time rotations. Confirm the Lorentz invariance of the quantized scalar theory by checking the relations 3(57a) and 3(57b).
4. Show that $\mathscr{P}_{ps}$ and $\mathscr{P}_s$ commute with the angular momentum operator M_{ij} for the scalar field.
5. Check that 3(66) is the charge conjugation operator $\mathscr{C}$ for the charged scalar theory.
6. Check that for equal times 3(77) yields the canonical equal-time commutation rules 3(29a) and 3(29b).
7. Verify 3(151) and 3(152).
8. Check the results 3(164a), 3(164b) and 3(164c) for the energy, momentum and charge of the Dirac field.
9. Show that the energy momentum of the Dirac field transforms as a four-vector under Lorentz transformations.
10. Construct the charge conjugation and time reflection operators $\mathscr{C}$ and $\mathscr{T}$ for the Dirac theory.
11. Establish the transformation property 3(204) for the antisymmetrized current.
12. Establish the equalities

$$\bar{\psi}(x)\gamma_5\psi(x) = \tfrac{1}{2}[\bar{\psi}(x), \gamma_5\psi(x)]$$

$$\bar{\psi}(x)\gamma_\mu\psi(x) - \tfrac{1}{2}[\bar{\psi}(x), \gamma_\mu\psi(x)] = \delta_{\mu 4}\frac{1}{V}\sum_{\mathbf{k}} 2$$

where, in the second relation, the right-hand side represents the divergent vacuum charge density.

13. Show explicitly that the invariance of the vacuum state $|0\rangle$ under a Lorentz transformation requires the vanishing of zero-point energy and momentum. [Hint: Use the requirement that $(\mathbf{P}, iH)$ transform as a 4-vector under Lorentz transformations.]
14. Use the definition 3(131) with 3(127), and the normalization

$$\langle n_1', n_2', \ldots | n_1, n_2, \ldots \rangle = \delta_{n_1 n_1'} \delta_{n_2 n_2'} \cdots$$

of the basis vectors to show that the probability amplitude 3(131) is properly normalized:

$$\int d^3x_1 \ldots d^3x_n \Phi^{(n)*}_{n_1 n_2} \ldots (\mathbf{x}_1, \ldots \mathbf{x}_n; t) \Phi^{(n)}_{n_1' n_2'} \ldots (\mathbf{x}_1, \ldots \mathbf{x}_n; t) = \delta_{n_1 n_1'} \delta_{n_2 n_2'} \cdots$$

REFERENCES

Bethe, H. A. (1936) (with R. F. Bacher) *Rev. Mod. Phys.* **8**, 82
Dirac, P. A. M. (1930) *Proc. Cambridge Phil. Soc.* **26**, 376
Federbush, P. (1958) (with M. Grisaru) *Nuovo Cimento* **9**, 890
Feynman, R. P. (1949) *Phys. Rev.* **76**, 749, 769
Gärding, L. (1954) (with A. S. Wightman) *Proc. Nat. Acad. Sci* (*USA*) **40**, 612
Haag, R. (1955) *Kgl. Danske Vidensk. Selsk., Mat.-Fys. Medd.* **29**, No. 12
Jordan, P. (1928) (with E. P. Wigner) *Z., Phys.* **47**, 631
Messiah, A. M. L. (1964) (with O. W. Greenberg) *Phys. Rev.* **136**, B 248
Pauli, W. (1940) *Phys. Rev.* **58**, 716
Schwinger, J. (1951) *Phys. Rev.* **82**, 914
Schwinger, J. (1953) *Phys. Rev.* **91**, 713
Schwinger, J. (1955) Lectures on Quantum Dynamics, Les Houches Summer School (1955)
Stückelberg, E. C. G. (1949) (with D. Rivier), *Helv. Phys. Acta* **13**, 215
Welton, T. A. (1948) *Phys. Rev.* **74**, 1157
Wightman, A. S. (1955) (with S. Schweber) *Phys. Rev.* **98**, 812

4

Quantization of Spin 1 and Spin $\frac{3}{2}$ Fields

4-1 Quantization of the massive vector field

Neutral Vector Field To quantize the neutral vector field we apply the standard quantization prescription 3(1) to the Poisson-bracket relations 2(83). We thereby obtain the equal-time commutation relations

$$\begin{aligned} [A_i(x), -E_j(x')]_{t=t'} &= i\delta_{ij}\delta^{(3)}(\mathbf{x}-\mathbf{x}') \\ [A_i(x), A_j(x')]_{t=t'} &= [E_i(x), E_j(x')]_{t=t'} = 0 \end{aligned} \tag{4(1)}$$

for the canonically conjugate fields $\mathbf{A}$ and $-\mathbf{E}$ which characterize the neutral massive vector field. From 4(1) we deduce the commutation relation

$$[A_i(x), A_0(x')]_{t=t'} = i\frac{1}{\mu^2}\partial_i\delta^{(3)}(\mathbf{x}-\mathbf{x}') \tag{4(2)}$$

where A_0 is the fourth component of the vector field, given by

$$A_0 = -\frac{1}{\mu^2}\,\text{div}\,\mathbf{E} \tag{4(3)}$$

as in 1(122d). The Hamiltonian is given by the quantum analogue of 2(81), namely

$$H = \tfrac{1}{2}\int\left[\mathbf{E}^2 + \mu^2\mathbf{A}^2 + (\mathbf{curl}\,\mathbf{A})^2 + \frac{1}{\mu^2}(\text{div}\,\mathbf{E})^2\right]d^3x \tag{4(4)}$$

and it is a matter of straightforward calculation to check that the quantum equations of motion

$$\dot{\mathbf{A}}(\mathbf{x}, t) = \frac{1}{i}[\mathbf{A}(\mathbf{x}, t), H] \tag{4(5a)}$$

$$\dot{\mathbf{E}}(\mathbf{x}, t) = \frac{1}{i}[\mathbf{E}(\mathbf{x}, t), H] \tag{4(5b)}$$

reduce to

$$\dot{\mathbf{A}} = -\mathbf{E}+\frac{1}{\mu^2}\,\mathbf{grad}\,(\mathrm{div}\,\mathbf{E}) \qquad 4(6a)$$

$$\dot{\mathbf{E}} = \mu^2\mathbf{A}+\mathbf{curl}\,(\mathbf{curl}\,A) \qquad 4(6b)$$

in agreement with the corresponding classical equations 1(123a) and 1(123b). To recover the set 1(122a), 1(122b), 1(122c) and 1(122d) it suffices to define

$$\mathbf{B} = \mathbf{curl}\,\mathbf{A} \qquad 4(7)$$

and substitute 4(3) and 4(7) into 4(6a) and 4(6b) respectively. For convenience we now revert to four-dimensional notation, defining the anti-symmetric tensor $F_{\lambda\nu} = -F_{\nu\lambda}$ by

$$F_{\lambda\nu} = \begin{cases} F_{ij} = \varepsilon_{ijk}B_k \\ F_{k4} = -F_{4k} = -iE_k \end{cases} \qquad 4(8)$$

In this notation the set of equations 4(3), 4(6a), 4(6b), 4(7) and 4(8) become simply

$$\partial_\lambda F_{\lambda\nu} = \mu^2 A_\nu \qquad 4(9a)$$

$$F_{\lambda\nu} = \partial_\lambda A_\nu - \partial_\nu A_\lambda \qquad 4(9b)$$

from which we easily deduce that

$$\partial_\lambda A_\lambda = 0 \qquad 4(10a)$$

and

$$\Box A_\nu - \mu^2 A_\nu = 0 \qquad 4(10b)$$

exactly as in the classical case. Finally, we note that the Lagrangian density for the quantized neutral vector field is

$$\mathscr{L} = -\tfrac{1}{4}F_{\mu\nu}F_{\mu\nu}-\tfrac{1}{2}\mu^2 A_\mu A_\mu \qquad 4(11)$$

as in 2(21).

Fourier Analysis Since the quantum and classical equations of motion are the same, we can expand $A_\mu(x)$ in terms of the complete set of plane wave solutions 1(135a) and 1(135b) according to

$$A^\mu(x) = \frac{1}{\sqrt{V}}\sum_{\mathbf{k}}\frac{1}{\sqrt{2\omega_{\mathbf{k}}}}\sum_{\lambda=1}^{3}\varepsilon^\mu_{\mathbf{k}\lambda}(a_{\mathbf{k}\lambda}\,\mathrm{e}^{ik.x}+a_{\mathbf{k}\lambda}{}^\dagger\,\mathrm{e}^{-ik.x}) \qquad 4(12)$$

where we have taken into account the fact that $\mathbf{A}$ and $A_0 = -iA_4$ are hermitian operators. The triad of polarization 4-vectors $\varepsilon_{\mathbf{k}\lambda}$ ($\lambda = 1, 2, 3$), appearing in 4(12) are given by 1(133) with the convenient choice 1(136a) and 1(136b) for the 3-vectors $\boldsymbol{\varepsilon}_{\mathbf{k}\lambda}$. The expansions of $\mathbf{A}(x)$ and $\mathbf{E}(x)$ corresponding to 4(12) are then

$$\mathbf{A}(x) = \frac{1}{\sqrt{V}}\sum_{\mathbf{k}}\frac{1}{\sqrt{2\omega_{\mathbf{k}}}}\left[\sum_{\lambda=1}^{2}\boldsymbol{\varepsilon}_{\mathbf{k}\lambda}(a_{\mathbf{k}\lambda}\,\mathrm{e}^{ik.x}+a_{\mathbf{k}\lambda}{}^{\dagger}\,\mathrm{e}^{-ik.x}) + \frac{\omega_{\mathbf{k}}}{\mu}\frac{\mathbf{k}}{|\mathbf{k}|}(a_{\mathbf{k}3}\,\mathrm{e}^{ik.x}+a_{\mathbf{k}3}{}^{\dagger}\,\mathrm{e}^{-ik.x})\right] \qquad 4(13a)$$

$$\mathbf{E}(x) = \frac{i}{\sqrt{V}}\sum_{\mathbf{k}}\sqrt{\frac{\omega_{\mathbf{k}}}{2}}\left[\sum_{\lambda=1}^{2}\boldsymbol{\varepsilon}_{\mathbf{k}\lambda}(a_{\mathbf{k}\lambda}\,\mathrm{e}^{ik.x}-a_{\mathbf{k}\lambda}{}^{\dagger}\,\mathrm{e}^{-ik.x}) + \frac{\mu}{\omega_{\mathbf{k}}}\frac{\mathbf{k}}{|\mathbf{k}|}(a_{\mathbf{k}3}\,\mathrm{e}^{ik.x}-a_{\mathbf{k}3}{}^{\dagger}\,\mathrm{e}^{-ik.x})\right] \qquad 4(13b)$$

as is easily seen by recalling the result 1(142) and 1(143). To derive the commutation rules for the $a_{\mathbf{k}\lambda}$ ($\lambda = 1, 2, 3$) we must invert 4(13a) and 4(13b). Using 1(136a) we find [Problem 2]

$$a_{\mathbf{k}\lambda} = \frac{1}{\sqrt{V}}\int d^3x\,\mathrm{e}^{-ik.x}\left(\sqrt{\frac{\omega_{\mathbf{k}}}{2}}\boldsymbol{\varepsilon}_{\mathbf{k}\lambda}\,.\,\mathbf{A}(x) - \frac{i}{\sqrt{2\omega_{\mathbf{k}}}}\boldsymbol{\varepsilon}_{\mathbf{k}\lambda}\,.\,\mathbf{E}(x)\right) \qquad 4(14a)$$

$(\lambda = 1, 2)$

$$a_{\mathbf{k}3} = \frac{1}{\sqrt{V}}\int d^3x\,\mathrm{e}^{-ik.x}\left(\sqrt{\frac{\omega_{\mathbf{k}}}{2}}\frac{\mu}{\omega_{\mathbf{k}}}\boldsymbol{\varepsilon}_{\mathbf{k}3}\,.\,\mathbf{A}(x) - \frac{i}{\sqrt{2\omega_{\mathbf{k}}}}\frac{\omega_{\mathbf{k}}}{\mu}\boldsymbol{\varepsilon}_{\mathbf{k}3}\,.\,\mathbf{E}(x)\right) \qquad 4(14b)$$

with $\boldsymbol{\varepsilon}_{\mathbf{k}3} = \mathbf{k}/|\mathbf{k}|$. From 4(14a) and 4(14b) follows

$$a_{\mathbf{k}\lambda}{}^{\dagger} = \frac{1}{\sqrt{V}}\int d^3x\,\mathrm{e}^{ik.x}\left(\sqrt{\frac{\omega_{\mathbf{k}}}{2}}\boldsymbol{\varepsilon}_{\mathbf{k}\lambda}\,.\,\mathbf{A}(x) + \frac{i}{\sqrt{2\omega_{\mathbf{k}}}}\boldsymbol{\varepsilon}_{\mathbf{k}\lambda}\,.\,\mathbf{E}(x)\right) \qquad 4(15a)$$

$(\lambda = 1, 2)$

$$a_{\mathbf{k}3}{}^{\dagger} = \frac{1}{\sqrt{V}}\int d^3x\,e^{ik.x}\left(\sqrt{\frac{\omega_{\mathbf{k}}}{2}}\frac{\mu}{\omega_{\mathbf{k}}}\boldsymbol{\varepsilon}_{\mathbf{k}3}\,.\,\mathbf{A}(x) + \frac{i}{\sqrt{2\omega_{\mathbf{k}}}}\frac{\omega_{\mathbf{k}}}{\mu}\boldsymbol{\varepsilon}_{\mathbf{k}3}\,.\,\mathbf{E}(x)\right) \qquad 4(15b)$$

Thus, for $\lambda, \lambda' = (1, 2)$, for example

$$[a_{\mathbf{k}\lambda}, a_{\mathbf{k}'\lambda'}{}^{\dagger}] = \frac{i}{2V}\int d^3x\, d^3x'\, \mathrm{e}^{-ik.x}\, \mathrm{e}^{ik'.x'} \varepsilon^i_{\mathbf{k}\lambda}\varepsilon^j_{\mathbf{k}'\lambda'}$$
$$\times([A^i(\mathbf{x}, t), E^j(\mathbf{x}', t)] - [E^i(\mathbf{x}, t), A^j(\mathbf{x}', t)])$$
$$= \frac{1}{V}\int d^3x\, d^3x'\, \mathrm{e}^{-ik.x}\, \mathrm{e}^{ik'.x'} \boldsymbol{\varepsilon}_{\mathbf{k}\lambda}\cdot\boldsymbol{\varepsilon}_{\mathbf{k}'\lambda'}\delta^{(3)}(\mathbf{x}-\mathbf{x}')$$
$$= \delta_{\mathbf{k}\mathbf{k}'}\delta_{\lambda\lambda'}$$

Similarly, for $\lambda, \lambda' = (1, 2, 3)$ we easily verify the standard relations

$$[a_{\mathbf{k}\lambda}, a_{\mathbf{k}'\lambda'}{}^{\dagger}] = \delta_{\mathbf{k}\mathbf{k}'}\delta_{\lambda\lambda'} \tag{4(16a)}$$

$$[a_{\mathbf{k}\lambda}, a_{\mathbf{k}'\lambda'}] = [a_{\mathbf{k}\lambda}{}^{\dagger}, a_{\mathbf{k}'\lambda'}{}^{\dagger}] = 0 \tag{4(16b)}$$

A straightforward calculation for the Hamiltonian yields, upon inserting 4(13a) and 4(13b) into 4(4),

$$H = \sum_{\mathbf{k}}\sum_{\lambda=1}^{3} \omega_{\mathbf{k}} a_{\mathbf{k}\lambda}{}^{\dagger} a_{\mathbf{k}\lambda} \tag{4(17)}$$

where we have dropped the zero-point energy term. This is of course equivalent to redefining 4(4) as a normal product, as in the Klein–Gordon case. The verification of 4(17) is left as an exercise.

Spin From the discussion following 1(137b) we anticipate that linear combinations of the operators $a_{\mathbf{k}1}{}^{\dagger}$ and $a_{\mathbf{k}2}{}^{\dagger}$ will create states with definite spin ± 1 along $\mathbf{k}$ while $a_{\mathbf{k}3}$ creates states of zero helicity. To confirm this we consider the spin angular momentum of the field. From the general expression for the angular momentum 2(123), we get, for the spin of the vector field, the expression

$$S^{ij} = i\int d^3x \sum_{r,s=1}^{3} E_r(s^{ij})_{rs}A_s \tag{4(18)}$$
$$i, j = (1, 2, 3)$$

where the spin matrices s^{ij} are given by 1(120a) or equivalently

$$(s^{ij})_{rs} = -i(\delta^i_r\delta^j_s - \delta^i_s\delta^j_r) \tag{4(19)}$$

Inserting 4(19) into 4(18) and remembering to take the normal product to ensure the subtraction of the vacuum spin, we have

$$S^{ij} = \int d^3x : (E^iA^j - E^jA^i): \tag{4(20)}$$

We now use 4(13a) and 4(13b) to compute the contribution to S_{12} from a single Fourier wave $\mathbf{k}$ where $\mathbf{k}$ points along the positive z-axis. In that case $\boldsymbol{\varepsilon}_{\mathbf{k}3} = \mathbf{k}/|\mathbf{k}|$ has no components along the x or y axes and we find

$$S^{12}[\mathbf{k} = (0, 0, k)] = i \sum_{\lambda, \lambda' = 1}^{2} \varepsilon_{\mathbf{k}\lambda}{}^{1} \varepsilon_{\mathbf{k}\lambda'}{}^{2} (a_{\mathbf{k}\lambda'}{}^{\dagger} a_{\mathbf{k}\lambda} - a_{\mathbf{k}\lambda}{}^{\dagger} a_{\mathbf{k}\lambda'}) \qquad 4(21)$$

after a brief manipulation. The right-hand side of 4(21) is equal to

$$i(\varepsilon_{\mathbf{k}1}{}^{1} \varepsilon_{\mathbf{k}2}{}^{2} - \varepsilon_{\mathbf{k}1}{}^{2} \varepsilon_{\mathbf{k}2}{}^{1})(a_{\mathbf{k}2}{}^{\dagger} a_{\mathbf{k}1} - a_{\mathbf{k}1}{}^{\dagger} a_{\mathbf{k}2})$$

If we select the polarization vectors $\boldsymbol{\varepsilon}_{\mathbf{k}1}$ and $\boldsymbol{\varepsilon}_{\mathbf{k}2}$ to lie along the x and y axes respectively, as in Fig. 4.1, then the above expression is simply

$$i(a_{\mathbf{k}2}{}^{\dagger} a_{\mathbf{k}1} - a_{\mathbf{k}1}{}^{\dagger} a_{\mathbf{k}2}) \qquad 4(22)$$

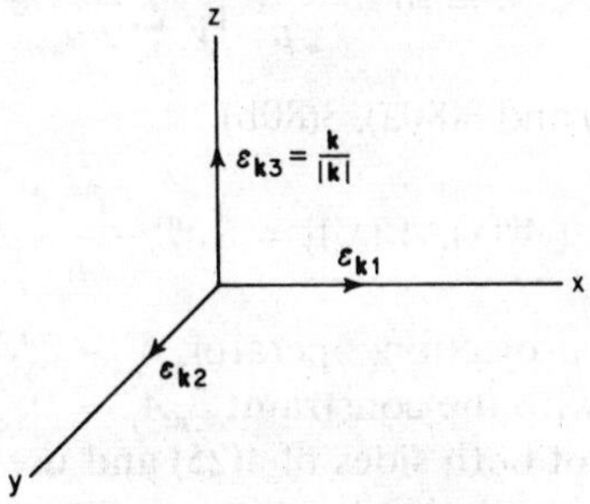

FIG. 4.1. Triad of polarization vectors for photon of momentum $\mathbf{k}$ directed along z-axis.

To diagonalize 4(22) we introduce the linear combinations

$$a_{\mathbf{k},+1} = \frac{1}{\sqrt{2}}(a_{\mathbf{k}1} - i a_{\mathbf{k}2}) \qquad 4(23a)$$

$$a_{\mathbf{k},-1} = \frac{1}{\sqrt{2}}(a_{\mathbf{k}1} + i a_{\mathbf{k}2}) \qquad 4(23b)$$

4(23a) and 4(23b) represents a transformation from plane to circularly polarized states. In terms of the $a_{\mathbf{k},\pm 1}$ and their adjoints, 4(22) becomes simply

$$a_{\mathbf{k},+1}{}^{\dagger} a_{\mathbf{k},+1} - a_{\mathbf{k},-1}{}^{\dagger} a_{\mathbf{k},-1}$$

so that $a_{\mathbf{k},+1}{}^{\dagger}$ and $a_{\mathbf{k},-1}{}^{\dagger}$ are respectively creation operators for spin $+1$ and -1 along z. The $\lambda = 3$ mode gives no contribution to S_{12} as it corresponds to the eigenvalue 0.

General Commutation Rules To compute the commutator

$$[A_\mu(x), A_\nu(x')]$$

for arbitrary times t and t' we apply the expansion 4(12) and the commutation rules 4(16a) and 4(16b). This yields

$$[A^\mu(x), A^\nu(x')] = \frac{1}{V}\sum_{\mathbf{k}} \frac{1}{2\omega_{\mathbf{k}}} \sum_{\lambda=1}^{3} \varepsilon^\mu_{\mathbf{k}\lambda}\varepsilon^\nu_{\mathbf{k}\lambda}(\mathrm{e}^{ik.(x-x')} - \mathrm{e}^{-ik.(x-x')}) \tag{4(24)}$$

The spin sum appearing in 4(24) is given explicitly by 1(141). Applying this result we get

$$[A^\mu(x), A^\nu(x')] = \frac{1}{V}\sum_{\mathbf{k}} \frac{1}{2\omega_{\mathbf{k}}}\left(\delta^{\mu\nu} + \frac{k^\mu k^\nu}{\mu^2}\right)(\mathrm{e}^{ik.(x-x')} - \mathrm{e}^{-ik.(x-x')})$$

$$= \left(\delta^{\mu\nu} - \frac{\partial^\mu\partial^\nu}{\mu^2}\right)\frac{1}{V}\sum_{\mathbf{k}} \frac{1}{2\omega_{\mathbf{k}}}(\mathrm{e}^{ik.(x-x')} - \mathrm{e}^{-ik.(x-x')})$$

or, recalling 3(79) and 3(80a), 3(80b)

$$[A^\mu(x), A^\nu(x')] = i\left(\delta^{\mu\nu} - \frac{\partial^\mu\partial^\nu}{\mu^2}\right)\Delta(x-x') \tag{4(25)}$$

Notice that the projection operator $\delta^{\mu\nu} - \partial^\mu\partial^\nu/\mu^2$ guarantees the consistency of 4(25) with the constraint $\partial_\mu A_\mu = 0$. To see this we simply take the 4-divergence of both sides of 4(25) and use the fact that $\Delta(x)$ satisfies the Klein–Gordon equation.

Feynman Propagator The Feynman propagator $\Delta_{F\mu\nu}$ for the neutral vector field is defined by

$$\Delta_{F\mu\nu}(x_1 - x_2) = \langle 0|TA_\mu(x_1)A_\nu(x_2)|0\rangle \tag{4(26)}$$

where T is the time-ordered product

$$TA_\mu(x_1)A_\nu(x_2) = \begin{cases} A_\mu(x_1)A_\nu(x_2) & t_1 > t_2 \\ A_\nu(x_2)A_\mu(x_1) & t_2 > t_1 \end{cases} \tag{4(27)}$$

Proceeding in the usual fashion we easily find

$$\Delta_{F\mu\nu}(x_1 - x_2) = i\theta(t_1 - t_2)\left(\delta^{\mu\nu} - \frac{\partial^\mu\partial^\nu}{\mu^2}\right)\Delta^{(+)}(x_1 - x_2)$$

$$- i\theta(t_2 - t_1)\left(\delta^{\mu\nu} - \frac{\partial^\mu\partial^\nu}{\mu^2}\right)\Delta^{(-)}(x_1 - x_2)$$

as in the corresponding relation 3(200) for Dirac fields. However, in the present case the θ functions cannot simply be commuted past the $\delta^{\mu\nu} - \partial^\mu\partial^\nu/\mu^2$ factor. We find, using the relation $\partial_t\theta(t) = \delta(t)$,

$$\Delta_{F\mu\nu}(x_1 - x_2) = \left(\delta^{\mu\nu} - \frac{\partial^\mu\partial^\nu}{\mu^2}\right)[i\theta(t_1 - t_2)\Delta^{(+)}(x_1 - x_2) - i\theta(t_2 - t_1)\Delta^{(-)}(x_1 - x_2)]$$
$$+ \frac{i}{\mu^2}\delta^{\mu 4}\delta^{\nu 4}\delta^{(3)}(\mathbf{x}_1 - \mathbf{x}_2)\delta(t_1 - t_2)$$

or, by 3(94)

$$\Delta_{F\mu\nu}(x) = \left(\delta^{\mu\nu} - \frac{\partial^\mu\partial^\nu}{\mu^2}\right)\Delta_F(x) + \frac{i}{\mu^2}\delta^{\mu 4}\delta^{\nu 4}\delta^{(4)}(x_1 - x_2) \qquad 4(28)$$

Actually the additional term in 4(28) can be neglected; in practical calculations for interacting fields, the effect of this term is cancelled by a so-called 'normal dependent' term in the interaction Hamiltonian. [See Eq. 5(86).] Using 3(96a) and 3(96b) we get the Fourier representation for $\Delta_{F\mu\nu}(x)$ in the form

$$\Delta_{F\mu\nu}(x) = \frac{1}{(2\pi)^4}\int d^4k\, e^{ik.x}\Delta_{F\mu\nu}(k)$$

$$\Delta_{F\mu\nu}(k) = \left(\delta_{\mu\nu} + \frac{k_\mu k_\nu}{\mu^2}\right)\frac{-i}{k^2 + \mu^2 - i\varepsilon} + \frac{i}{\mu^2}\delta^{\mu 4}\delta^{\nu 4} \qquad 4(29)$$

Charged Vector Field To quantize the complex vector field theory described by the classical Lagrangian density 2(22) we perform the substitution 3(1) in the Poisson bracket relations 2(86a) and 2(86b). This yields the equal-time commutation relations

$$[A_i(x), -E_j^\dagger(x')]_{t=t'} = i\delta_{ij}\delta^{(3)}(\mathbf{x} - \mathbf{x}') \qquad 4(30a)$$
$$[A_i^\dagger(x), -E_j(x')]_{t=t'} = i\delta_{ij}\delta^{(3)}(\mathbf{x} - \mathbf{x}') \qquad 4(30b)$$

with all other independent equal-time commutators being equal to zero. Since the reality, or hermiticity, constraint on $\mathbf{A}$ and $A_0 = -iA_4$ has been lifted, the expansion of $A_\mu(x)$ is now of the form

$$A^\mu(x) = \frac{1}{\sqrt{V}}\sum_{\mathbf{k}}\frac{1}{\sqrt{2\omega_{\mathbf{k}}}}\sum_{\lambda=1}^{3}\varepsilon^\mu_{\mathbf{k}\lambda}(a_{\mathbf{k}\lambda}\, e^{ik.x} + b_{\mathbf{k}\lambda}{}^\dagger\, e^{-ik.x}) \qquad 4(31)$$

where the $a_{\mathbf{k}}$ and $b_{\mathbf{k}}$ are independent operators, as in the corresponding scalar field expansions 3(36a) and 3(36b). A corresponding change occurs in the expansions 4(13a) and 4(13b). The commutation relations for the $a_{\mathbf{k}}$, $a_{\mathbf{k}}{}^\dagger$, $b_{\mathbf{k}}$ and $b_{\mathbf{k}}{}^\dagger$ are easily found to be

$$[a_{\mathbf{k}\lambda}, a_{\mathbf{k}'\lambda'}{}^\dagger] = [b_{\mathbf{k}\lambda}, b_{\mathbf{k}'\lambda'}{}^\dagger] = \delta_{\mathbf{k}\mathbf{k}'}\delta_{\lambda\lambda'} \qquad 4(32a)$$
$$[a_{\mathbf{k}\lambda}, a_{\mathbf{k}'\lambda'}] = [b_{\mathbf{k}\lambda}, b_{\mathbf{k}'\lambda'}] = [a_{\mathbf{k}\lambda}, b_{\mathbf{k}'\lambda'}] = [a_{\mathbf{k}\lambda}, b_{\mathbf{k}'\lambda'}{}^\dagger] = 0 \qquad 4(32b)$$

and the Hamiltonian operator, given classically by 2(85),

$$H = \int \left[\mathbf{E}^\dagger \mathbf{E} + \mu^2 \mathbf{A}^\dagger \mathbf{A} + (\mathbf{curl}\ \mathbf{A}^\dagger)(\mathbf{curl}\ \mathbf{A}) + \frac{1}{\mu^2}(\mathrm{div}\ \mathbf{E}^\dagger)(\mathrm{div}\ \mathbf{E}) \right] d^3x \qquad 4(33)$$

becomes

$$H = \sum_{\mathbf{k}} \sum_{\lambda=1}^{3} (a_{\mathbf{k}\lambda}{}^\dagger a_{\mathbf{k}\lambda} + b_{\mathbf{k}\lambda}{}^\dagger b_{\mathbf{k}\lambda})\omega_{\mathbf{k}} \qquad 4(34)$$

As in the case of the charged scalar field, the distinction between the a and b operators shows up in the expression for the charge. From 2(131c) we have

$$Q = -i \int j_4(x)\, d^3x = \int (F^\dagger{}_{4i} A_i - F_{4i} A_i{}^\dagger)\, d^3x$$

or, since $F_{4i} = iE_i$ and $F^\dagger{}_{4i} = iE_i{}^\dagger$

$$Q = i \int (\mathbf{E}^\dagger . \mathbf{A} - \mathbf{A}^\dagger . \mathbf{E})\, d^3x \qquad 4(35)$$

If the Fourier expansions for $\mathbf{A}$ and $\mathbf{E}$ are inserted into 4(35), we find [see Problem 4]

$$Q = \sum_{\mathbf{k}} \sum_{\lambda=1}^{3} (a_{\mathbf{k}\lambda}{}^\dagger a_{\mathbf{k}\lambda} - b_{\mathbf{k}\lambda}{}^\dagger b_{\mathbf{k}\lambda}) \qquad 4(36)$$

so that the a and b operators refer to vector quanta of charge $+1$ and -1 respectively.

For the general covariant commutation relations and the Feynman propagator one finds*

$$[A_\mu(x), A^\dagger{}_\nu(x')] = i\left(\delta^{\mu\nu} - \frac{\partial^\mu \partial^\nu}{\mu^2}\right)\Delta(x-x') \qquad 4(37)$$

and

$$\Delta_{F\mu\nu}(x_1 - x_2) = \langle 0| T A_\mu(x_1) A^\dagger{}_\nu(x_2)|0\rangle$$
$$= \left(\delta^{\mu\nu} - \frac{\partial^\mu \partial^\nu}{\mu^2}\right)\Delta_F(x_1 - x_2) + \frac{i}{\mu^2}\delta^{\mu 4}\delta^{\nu 4}\delta^{(4)}(x_1 - x_2) \qquad 4(38)$$

by following the same steps as in the neutral case.

4-2 Quantization of the Maxwell field

Introduction In one respect, the Maxwell field is the most difficult to quantize for, as a result of the gauge invariance of the vector potential A_μ, different quantization procedures are required for different gauges.

* Note that, as in 1(114), $A^\dagger{}_\mu = (\mathbf{A}^\dagger, iA_0{}^\dagger)$.

We begin by carrying out the quantization in the radiation gauge, where the physical nature of the electromagnetic field is most apparent.

Radiation Gauge Quantization Recalling that the independent canonical fields in the radiation gauge are the transverse vector potential $\mathbf{A}^t$ and $\boldsymbol{\pi}^t = -\mathbf{E}^t$, we seek canonical equal-time commutation relations which are consistent with the transversality conditions $\operatorname{div} \mathbf{A}^t = \operatorname{div} \mathbf{E}^t = 0$. The usual form

$$[A_i^t(x), -E_j^t(x')]_{t=t'} = i\delta_{ij}\delta^{(3)}(\mathbf{x}-\mathbf{x}') \qquad 4(39)$$

is inconsistent with these constraints, since, taking the divergence of both sides we get

$$[\operatorname{div} \mathbf{A}^t(x), -E_j^t(x')]_{t=t'} = i\delta_j\delta^{(3)}(\mathbf{x}-\mathbf{x}') \neq 0$$

To find the correct relations we must recognize that canonical equal-time commutation relations can only be applied to truly independent field variables. To isolate the latter, we decompose $\mathbf{A}^t$ and $\mathbf{E}^t$ into normal modes, according to

$$\mathbf{A}^t(x) = \frac{1}{\sqrt{V}} \sum_{\mathbf{k}} \mathbf{q}_{\mathbf{k}}(t)\, \mathrm{e}^{i\mathbf{k}.\mathbf{x}} \qquad 4(40a)$$

$$-\mathbf{E}^t(x) = \frac{1}{\sqrt{V}} \sum_{\mathbf{k}} \mathbf{p}_{\mathbf{k}}(t)\, \mathrm{e}^{-i\mathbf{k}.\mathbf{x}} \qquad 4(40b)$$

as in 2(35) and 2(51). Since, for each value of $\mathbf{k}$, the coefficients $\mathbf{q}_{\mathbf{k}}$ and $\mathbf{p}_{\mathbf{k}}$ satisfy the transversality conditions

$$\mathbf{k}\,.\,\mathbf{q}_{\mathbf{k}} = \mathbf{k}\,.\,\mathbf{p}_{\mathbf{k}} = 0 \qquad 4(41)$$

they can be decomposed in terms of two independent transverse polarization vectors $\boldsymbol{\varepsilon}_{\mathbf{k}\lambda}$ $(\lambda = 1, 2)$ according to

$$\mathbf{q}_{\mathbf{k}} = \sum_{\lambda=1}^{2} q_{\mathbf{k}\lambda}\boldsymbol{\varepsilon}_{\mathbf{k}\lambda} \qquad \mathbf{p}_{\mathbf{k}} = \sum_{\lambda=1}^{2} p_{\mathbf{k}\lambda}\boldsymbol{\varepsilon}_{\mathbf{k}\lambda} \qquad 4(42)$$

The set of all coefficients $q_{\mathbf{k}\lambda}$ and $p_{\mathbf{k}\lambda}$ constitute the independent canonical coordinates and momenta of the field. We can therefore impose the canonical commutation rules

$$\begin{aligned} [q_{\mathbf{k}\lambda}, p_{\mathbf{k}'\lambda'}] &= i\delta_{\mathbf{k}\mathbf{k}'}\delta_{\lambda\lambda'} \\ [q_{\mathbf{k}\lambda}, q_{\mathbf{k}'\lambda'}] &= [p_{\mathbf{k}\lambda}, p_{\mathbf{k}'\lambda'}] = 0 \end{aligned} \qquad 4(43)$$

for λ and λ' varying from 1 to 2. From 4(40a), 4(40b) and 4(43) we get

$$[A_i^t(x), -E_j^t(x')]_{t=t'}$$

$$= \frac{1}{V}\sum_{\mathbf{k},\mathbf{k}'}\sum_{\lambda,\lambda'=1}^{2} \varepsilon_{\mathbf{k}\lambda}^{j}\varepsilon_{\mathbf{k}'\lambda'}^{j}\, \mathrm{e}^{i\mathbf{k}.\mathbf{x}}\, \mathrm{e}^{-i\mathbf{k}'.\mathbf{x}'}[q_{\mathbf{k}\lambda}(t), p_{\mathbf{k}'\lambda'}(t)]$$

$$= \frac{i}{V}\sum_{\mathbf{k}}\sum_{\lambda=1}^{2} \varepsilon_{\mathbf{k}\lambda}^{i}\varepsilon_{\mathbf{k}\lambda}^{j}\, \mathrm{e}^{i\mathbf{k}.(\mathbf{x}-\mathbf{x}')}$$

or, by 1(166)

$$[A_i^t(x), -E_j^t(x')]_{t=t'} = \frac{i}{V}\sum_{\mathbf{k}}\left(\delta_{ij} - \frac{k_i k_j}{\mathbf{k}^2}\right)\mathrm{e}^{i\mathbf{k}.(\mathbf{x}-\mathbf{x}')}$$

$$= i\left(\delta_{ij} - \frac{\partial_i\partial_j}{\boldsymbol{\nabla}^2}\right)\delta^{(3)}(\mathbf{x}-\mathbf{x}') \qquad 4(44a)$$

where the mathematical definition of the symbol $\boldsymbol{\nabla}^{-2}$ is given by 1(158). The relation 4(44a) is obviously consistent with the constraints div $\mathbf{A}^t =$ div $\mathbf{E}^t = 0$. For the remaining equal-time commutation rules we have

$$[A_i^t(x), A_j^t(x')]_{t=t'} = [E_i^t(x), E_j^t(x')]_{t=t'} = 0 \qquad 4(44b)$$

One can now proceed in the usual fashion to check that the Hamiltonian equations of motion

$$\dot{\mathbf{A}}^t = \frac{1}{i}[\mathbf{A}^t, H] \qquad 4(45a)$$

$$\dot{\mathbf{E}}^t = \frac{1}{i}[\mathbf{E}^t, H] \qquad 4(45b)$$

with H given by the quantum analogue of 2(91)

$$H = \tfrac{1}{2}\sum_{i=1}^{3}\int (E_i^t E_i^t + \boldsymbol{\nabla} A_i^t\,.\,\boldsymbol{\nabla} A_i^t)\, d^3x \qquad 4(46)$$

yield the equation of motion

$$\Box \mathbf{A}^t = 0 \qquad 4(47)$$

identical to the classical equation 1(162). The proof is left to the reader as an exercise. From 4(47) we infer that $\mathbf{A}^t(x)$ may be expanded in terms of the complete set of solutions 1(164a) and 1(164b) according to

$$\mathbf{A}^t(x) = \frac{1}{\sqrt{V}}\sum_{\mathbf{k}}\frac{1}{\sqrt{2|\mathbf{k}|}}\sum_{\lambda=1}^{2} \boldsymbol{\varepsilon}_{\mathbf{k}\lambda}(a_{\mathbf{k}\lambda}\,\mathrm{e}^{ik.x} + a_{\mathbf{k}\lambda}{}^{\dagger}\,\mathrm{e}^{-ik.x}) \qquad 4(48)$$

Since it follows from 4(45a) that

$$\mathbf{E}^t = -\dot{\mathbf{A}}^t \tag{4(49)}$$

as in 1(163), we get the corresponding expansion for $\mathbf{E}^t$ in the form

$$\mathbf{E}^t(x) = \frac{i}{\sqrt{V}} \sum_{\mathbf{k}} \sqrt{\frac{|\mathbf{k}|}{2}} \sum_{\lambda=1}^{2} \boldsymbol{\varepsilon}_{\mathbf{k}\lambda}(a_{\mathbf{k}\lambda}\,\mathrm{e}^{ik.x} - a_{\mathbf{k}\lambda}{}^{\dagger}\,\mathrm{e}^{-ik.x}) \tag{4(50)}$$

The expansions 4(48) and 4(50) are similar to 4(13a) and 4(13b) with the important difference that the longitudinal mode associated with $\boldsymbol{\varepsilon}_{\mathbf{k}3}$ is now absent*. Inverting 4(48) and 4(50) we find

$$a_{\mathbf{k}\lambda} = \frac{1}{\sqrt{V}} \int d^3x\, \mathrm{e}^{-ikx}\left(\sqrt{\frac{\omega_{\mathbf{k}}}{2}}\boldsymbol{\varepsilon}_{\mathbf{k}\lambda}\,.\,\mathbf{A}^t(x) - \frac{i}{\sqrt{2\omega_{\mathbf{k}}}}\boldsymbol{\varepsilon}_{\mathbf{k}\lambda}\,.\,\mathbf{E}^t(x)\right) \tag{4(51a)}$$

$$a_{\mathbf{k}\lambda}{}^{\dagger} = \frac{1}{\sqrt{V}} \int d^3x\, \mathrm{e}^{ikx}\left(\sqrt{\frac{\omega_{\mathbf{k}}}{2}}\boldsymbol{\varepsilon}_{\mathbf{k}\lambda}\,.\,\mathbf{A}^t(x) + \frac{i}{\sqrt{2\omega_{\mathbf{k}}}}\boldsymbol{\varepsilon}_{\mathbf{k}\lambda}\,.\,\mathbf{E}^t(x)\right) \tag{4(51b)}$$

as in 4(14a) and 4(15a). From 4(51a), 4(51b) and 4(44a), 4(44b) we get, by following the same steps as in the massive vector case

$$[a_{\mathbf{k}\lambda}, a_{\mathbf{k}'\lambda'}{}^{\dagger}]$$
$$= \frac{1}{V}\int d^3x\, d^3x'\, \mathrm{e}^{-ik.x}\,\mathrm{e}^{ik'.x'}\varepsilon^i_{\mathbf{k}\lambda}\varepsilon^j_{\mathbf{k}'\lambda'}\left(\delta^{ij} - \frac{\partial^i\partial^j}{\nabla^2}\right)\delta^{(3)}(\mathbf{x}-\mathbf{x}')$$

The right-hand side can be simplified by means of the relation

$$\left(\delta^{ij} - \frac{\partial^i\partial^j}{\nabla^2}\right)\delta^{(3)}(\mathbf{x}-\mathbf{x}') = \frac{1}{V}\sum_{\mathbf{p}}\left(\delta^{ij} - \frac{p^ip^j}{\mathbf{p}^2}\right)e^{i\mathbf{p}.(\mathbf{x}-\mathbf{x}')}$$

and we find

$$[a_{\mathbf{k}\lambda}, a_{\mathbf{k}'\lambda'}{}^{\dagger}] = \delta_{\mathbf{k}\mathbf{k}'}\delta_{\lambda\lambda'} \tag{4(52a)}$$

by virtue of the orthogonality relations 1(136a). Similarly

$$[a_{\mathbf{k}\lambda}, a_{\mathbf{k}'\lambda'}] = [a_{\mathbf{k}\lambda}{}^{\dagger}, a_{\mathbf{k}'\lambda'}{}^{\dagger}] = 0 \tag{4(52b)}$$

The Hamiltonian 4(46) when expressed in terms of the creation and destruction operators $a_{\mathbf{k}\lambda}{}^{\dagger}$ and $a_{\mathbf{k}\lambda}$ takes the form

$$H = \sum_{\mathbf{k}}\sum_{\lambda=1}^{2} |\mathbf{k}| a_{\mathbf{k}\lambda}{}^{\dagger} a_{\mathbf{k}\lambda} \tag{4(53)}$$

* Accordingly, the zero helicity state is suppressed for electromagnetic quanta, and the spin can only take on the values ± 1 along the direction of motion.

after subtraction of the zero-point energy. The field momentum, given classically by 2(117a) or*

$$\mathbf{P} = \sum_{i=1}^{3} \int E_i^t \nabla A_i^t \, d^3x \tag{4(54)}$$

becomes

$$\mathbf{P} = \sum_{\mathbf{k}} \sum_{\lambda=1}^{2} \mathbf{k} a_{\mathbf{k}\lambda}{}^{\dagger} a_{\mathbf{k}\lambda} \tag{4(55)}$$

These equations exhibit H and $\mathbf{P}$ as the energy and momentum of an assembly of transverse massless quanta and thereby provide the particle interpretation of the electromagnetic field as an assembly of *photons.*

General Commutation Rules The commutator

$$[A_i^t(x), A_j^t(x')]$$

for $t \neq t'$ can be computed in the usual way by applying 4(48) and 4(52a), 4(52b). We find

$$\begin{aligned}[A_i^t(x), A_j^t(x')] &= \frac{1}{V}\sum_{\mathbf{k}} \frac{1}{2|\mathbf{k}|} \sum_{\lambda=1}^{2} \varepsilon_{\mathbf{k}\lambda}^i \varepsilon_{\mathbf{k}\lambda}^j (e^{ik(x-x')} - e^{-ik(x-x')}) \\ &= \frac{1}{V}\sum_{\mathbf{k}} \frac{1}{2|\mathbf{k}|}\left(\delta^{ij} - \frac{k^i k^j}{\mathbf{k}^2}\right)(e^{ik(x-x')} - e^{-ik(x-x')})\end{aligned}$$

where we have used 1(166). Recalling 3(79) and 3(80a), 3(80b) we get

$$[A_i^t(x), A_j^t(x')] = i\left(\delta_{ij} - \frac{\partial_i \partial_j}{\nabla^2}\right) D(x-x') \tag{4(56)}$$

where $D(x)$ is just $\Delta(x)$ in the special case of zero rest mass. Explicitly $D(x)$ is given by [Problem 6]

$$D(x) = \frac{1}{4\pi|\mathbf{x}|}[\delta(t+|\mathbf{x}|) - \delta(t-|\mathbf{x}|)] \tag{4(57)}$$

* The expression 4(54) is equivalent to the more familiar form

$$\mathbf{P} = \int (\mathbf{E} \times \mathbf{B})\, d^3x = \int (\mathbf{E}^t \times \mathbf{curl}\, \mathbf{A}^t)\, d^3x$$

in terms of the Poynting vector. To see this, write

$$E_i^t \partial_r A_i^t = \delta_{rs}\delta_{ij} E_i^t \partial_s A_j^t = (\varepsilon_{rik}\varepsilon_{sjk} - \delta_{rj}\delta_{si}) E_i^t \partial_s A_j^t$$

where summation over repeated indices is understood. The last form is just $(\mathbf{E}^t \times \mathbf{curl}\, \mathbf{A}^t)_r - \mathbf{E} \,.\, \nabla A_r^t$ where the last term can be disposed of by integrating by parts and using $\operatorname{div} \mathbf{E}^t = 0$.

We can rewrite the result 4(56) in 4-dimensional notation by setting $A_\mu = (\mathbf{A}^t, 0)$ and extending the polarization vectors $\boldsymbol{\varepsilon}_{\mathbf{k}\lambda}$ $(\lambda = 1, 2)$ into 4-vectors $e_{\mathbf{k}\lambda} = (\boldsymbol{\varepsilon}_{\mathbf{k}\lambda}, 0)$ as in 1(167) and 1(168). We have then, applying 1(173)

$$[A_\mu(x), A_\nu(x')] = i\left[\delta_{\mu\nu} - \frac{\partial_\mu \partial_\nu}{(\partial \,.\, \eta)^2} - \frac{\eta_\mu \partial_\nu + \partial_\mu \eta_\nu}{(\partial \,.\, \eta)}\right]\Delta(x - x') \qquad 4(58)$$

The result is of course non-covariant, owing to the dependence of the right-hand side on the timelike 4-vector η. In the Lorentz frame in which we have carried out the quantization, η is given by $(0, 0, 0, i)$.

Lorentz Gauge Quantization The radiation gauge formalism has the merit that only the real dynamical degrees of freedom of the Maxwell field are quantized, but this is achieved at the expense of manifest Lorentz covariance. On the other hand, for practical calculations in quantum electrodynamics, it is convenient to have at our disposal a covariant formalism for the electromagnetic field. Here we must confront the fact that owing to the structure of the Maxwell field, canonical quantization—which necessarily bears on the real dynamical degrees of freedom—and manifest covariance are mutually incompatible. Accordingly, we must modify the structure of Maxwell's equations to enable us to treat all four components of the vector potential as independent canonical variables.

The modified Lagrangian for the electromagnetic field is taken to be [Fermi (1930)]

$$\mathscr{L} = -\tfrac{1}{2}(\partial_\nu A_\mu)(\partial_\nu A_\mu) \qquad 4(59)$$

for which the Euler–Lagrange equations of motion are

$$\Box A_\mu = 0 \qquad 4(60)$$

The four components of the vector potential may now be treated as independent degrees of freedom. We stress that the equation of motion 4(60) is *not* equivalent to Maxwell's equation; equivalence would require the additional constraint $\partial_\mu A_\mu = 0$. To recover the description of the electromagnetic field, we shall eventually have to suppress the additional non-Maxwellian degrees of freedom by imposing a constraint on the admissible states of the quantum system.

The quantization of the field is performed by imposing the canonical equal-time commutation relations

$$\begin{aligned} [A_\mu(x), \pi_\nu(x')]_{t=t'} &= i\delta_{\mu\nu}\delta^{(3)}(\mathbf{x} - \mathbf{x}') \\ [A_\mu(x), \pi_\nu(x')]_{t=t'} &= [\pi_\mu(x), \pi_\nu(x')]_{t=t'} = 0 \end{aligned} \qquad 4(61)$$

where

$$\pi_\mu = \frac{\partial \mathscr{L}}{\partial \dot{A}_\mu} = \dot{A}_\mu \tag{4(62)}$$

Since the electromagnetic field carries no charge, $\mathbf{A}(x)$ and $A_0(x) = -iA_4(x)$ are hermitian field operators after quantization.

It is a simple matter to check that the quantum equations of motion

$$\dot{A}_\mu = \frac{1}{i}[A_\mu, H] \tag{4(63a)}$$

$$\dot{\pi}_\mu = \frac{1}{i}[\pi_\mu, H] \tag{4(63b)}$$

with

$$H = \int (\dot{A}_\mu \pi_\mu - \mathscr{L})\, d^3x = \tfrac{1}{2}\int (\pi_\mu \pi_\mu + \nabla A_\mu \nabla A_\mu)\, d^3x \tag{4(64)}$$

reproduce the field equation

$$\Box A_\mu = 0 \tag{4(65)}$$

We can therefore expand $A_\mu(x)$ in terms of a complete set of solutions

$$\frac{1}{\sqrt{V}}\frac{1}{\sqrt{2|\mathbf{k}|}} e_{\mathbf{k}\lambda}\, \mathrm{e}^{\pm ik.x} \qquad (\lambda = 1, 2, 3, 4) \tag{4(66)}$$

with $k_0 = |\mathbf{k}|$. For each $\mathbf{k}$ the $e_{\mathbf{k}\lambda}$ ($\lambda = 1, \ldots 4$) form a quartet of linearly independent 4-vectors, which for convenience we take to be the set 1(167), 1(168), 1(169) and 1(170) satisfying the orthonormality and completeness relations 1(171) and 1(172). Taking into account the fact that $\mathbf{A}$ and $A_0 = -iA_4$ are hermitian operators, the expansion of A_μ reads

$$A^\mu(x) = \frac{1}{\sqrt{V}}\sum_{\mathbf{k}} \frac{1}{\sqrt{2|\mathbf{k}|}} \sum_{\lambda=1}^{4} e^\mu_{\mathbf{k}\lambda}(a_{\mathbf{k}\lambda}\, \mathrm{e}^{ik.x} + \bar{a}_{\mathbf{k}\lambda}\, \mathrm{e}^{-ik.x}) \tag{4(67)}$$

where we have defined

$$\bar{a}_{\mathbf{k}\lambda} = \begin{cases} a_{\mathbf{k}\lambda}{}^\dagger & (\lambda = 1, 2, 3) \\ -a_{\mathbf{k}\lambda}{}^\dagger & (\lambda = 4) \end{cases} \tag{4(68)}$$

The minus sign for $\lambda = 4$ is traceable to the factor i in the definition 1(170) of $e_{\mathbf{k}4}$. Correspondingly, we have for $\pi_\mu = \dot{A}_\mu$

$$\pi^\mu(x) = \frac{-i}{\sqrt{V}}\sum_{\mathbf{k}} \sqrt{\frac{|\mathbf{k}|}{2}} \sum_{\lambda=1}^{4} e^\mu_{\mathbf{k}\lambda}(a_{\mathbf{k}\lambda}\, \mathrm{e}^{ik.x} - \bar{a}_{\mathbf{k}\lambda}\, \mathrm{e}^{-ikx}) \tag{4(69)}$$

Inverting these expansions we get

$$a_{\mathbf{k}\lambda} = \frac{i}{\sqrt{V}} \frac{1}{\sqrt{2|\mathbf{k}|}} \int d^3x \, \mathrm{e}^{-ik.x} \overleftrightarrow{\partial}_t e^{\mu}_{\mathbf{k}\lambda} A_\mu(x) \qquad 4(70a)$$

and

$$\bar{a}_{\mathbf{k}\lambda} = -\frac{i}{\sqrt{V}} \frac{1}{\sqrt{2|\mathbf{k}|}} \int d^3x e^{\mu}_{\mathbf{k}\lambda} A_\mu(x) \overleftrightarrow{\partial}_t \, \mathrm{e}^{ik.x} \qquad 4(70b)$$

as in 3(10a) and 3(10b). From 4(70a), 4(70b) and 4(61) we derive, in the usual manner

$$[a_{\mathbf{k}\lambda}, \bar{a}_{\mathbf{k}'\lambda'}] = \delta_{\mathbf{k}\mathbf{k}'}\delta_{\lambda\lambda'} \qquad 4(71a)$$

$$[a_{\mathbf{k}\lambda}, a_{\mathbf{k}'\lambda'}] = [\bar{a}_{\mathbf{k}\lambda}, \bar{a}_{\mathbf{k}'\lambda'}] = 0 \qquad 4(71b)$$

with the aid of the orthonormality relations 1(171). Armed with 4(67) and 4(71a), 4(71b) it is now a simple matter to calculate the covariant commutator

$$[A_\mu(x), A_\nu(x')] = i\delta_{\mu\nu} D(x-x') \qquad 4(72)$$

with $D(x)$ given by 4(57), and the Feynman propagator

$$\langle 0| T A_\mu(x_1) A_\nu(x_2)|0\rangle = \delta_{\mu\nu} D_F(x_1 - x_2) \qquad 4(73)$$

where D_F is just Δ_F for zero rest mass, i.e.

$$D_F(x_1 - x_2) = \frac{1}{(2\pi)^4} \int d^4k \, \mathrm{e}^{ik.(x_1 - x_2)} D_F(k^2)$$

with

$$D_F(k^2) = \frac{-i}{k^2 - i\varepsilon}$$

Indefinite Metric The commutation relations 4(71a) and 4(71b) for λ and $\lambda' \neq 4$ are the usual ones for a set of creation and destruction operators. However the relation for $\lambda = \lambda' = 4$,

$$[a_{\mathbf{k}4}, a_{\mathbf{k}'4}{}^\dagger] = -\delta_{\mathbf{k}\mathbf{k}'} \qquad 4(74)$$

exhibits an unfamiliar minus sign on the right-hand side. The appearance of this minus sign is a consequence of the indefinite character of the Minkowski metric and is unavoidable*. It has the important consequence

* If, for example, $e_{\mathbf{k}4}$ were defined without the factor i in 1(170), then $\bar{a}_{\mathbf{k}\lambda}$ would be replaced by $a_{\mathbf{k}\lambda}{}^\dagger$ in 4(67), but the minus sign would be reintroduced in 4(74) by a corresponding modification of the orthonormality relations 1(171).

that the space in which the field operators are defined carries an *indefinite metric*—that is the space contains states of negative norm. For instance, the state $a_{\mathbf{k}4}{}^{\dagger}|0\rangle$ has the norm

$$\langle 0|a_{\mathbf{k}4}a_{\mathbf{k}4}{}^{\dagger}|0\rangle$$

Since $a_{\mathbf{k}4}$ is a destruction operator, we have $a_{\mathbf{k}4}|0\rangle = 0$ and hence

$$\langle 0|a_{\mathbf{k}4}a_{\mathbf{k}4}{}^{\dagger}|0\rangle = \langle 0|[a_{\mathbf{k}4}, a_{\mathbf{k}4}{}^{\dagger}]|0\rangle = -1$$

More generally, one can show that the state $|n_{\mathbf{k}4}\rangle$ containing $n_{\mathbf{k}4}$ 'timelike' photons has the norm

$$\langle n_{\mathbf{k}4}|n_{\mathbf{k}4}\rangle = (-1)^{n_{\mathbf{k}4}} \qquad 4(75)$$

The appearance of these negatively normed states has the result that expectation value of the Hamiltonian 4(64) can take on negative values. Computing H in terms of creation and destruction operators, we get

$$H = \sum_{\mathbf{k}} \sum_{\lambda=1}^{4} |\mathbf{k}|\bar{a}_{\mathbf{k}\lambda}a_{\mathbf{k}\lambda} \qquad 4(76a)$$

$$= \sum_{\mathbf{k}} |\mathbf{k}| \left(\sum_{\lambda=1}^{3} a_{\mathbf{k}\lambda}{}^{\dagger}a_{\mathbf{k}\lambda} - a_{\mathbf{k}4}{}^{\dagger}a_{\mathbf{k}4} \right) \qquad 4(76b)$$

The presence of the minus sign in front of $a_{\mathbf{k}4}{}^{\dagger}a_{\mathbf{k}4}$ seems to indicate that the timelike photons give a negative contribution to the energy, but in fact all eigenvalues of H are non-negative. This is a result of the fact that the number operator $N_{\mathbf{k}4}$ for timelike photons is not $a_{\mathbf{k}4}{}^{\dagger}a_{\mathbf{k}4}$ but

$$N_{\mathbf{k}4} = -a_{\mathbf{k}4}{}^{\dagger}a_{\mathbf{k}4} \qquad 4(77)$$

since, applying 4(74) we have

$$\begin{aligned} N_{\mathbf{k}4}a_{\mathbf{k}4}{}^{\dagger}|0\rangle &= -a_{\mathbf{k}4}{}^{\dagger}[a_{\mathbf{k}4}, a_{\mathbf{k}4}{}^{\dagger}]|0\rangle \\ &= 1a_{\mathbf{k}4}{}^{\dagger}|0\rangle \\ N_{\mathbf{k}4}(a_{\mathbf{k}4}{}^{\dagger})^2|0\rangle &= 2(a_{k4}{}^{\dagger})^2|0\rangle \\ &\vdots \end{aligned}$$

as required. However, due to the indefinite character of the metric, the *expectation value* of H can be negative. For example, we have by virtue of 4(75)

$$\begin{aligned} \langle n_{\mathbf{k}4}|H|n_{\mathbf{k}4}\rangle &= |\mathbf{k}|\langle n_{\mathbf{k}4}|N_{\mathbf{k}4}|n_{\mathbf{k}4}\rangle \\ &= |\mathbf{k}|(-1)^{n_{\mathbf{k}4}}n_{\mathbf{k}4} \end{aligned}$$

Fortunately the indefinite metric does not cause a breakdown in the interpretation of the theory, for, as we shall show presently, the equation of constraint which is needed to suppress the non-Maxwellian degrees of freedom also suppresses the states of negative norm.

Subsidiary Condition The commutation relations 4(63a) and 4(63b) and the Hamiltonian 4(64) do not by themselves constitute a mathematical description of the electromagnetic field. Additional, non-Maxwellian, degrees of freedom are present which we must now eliminate by imposing a constraint on the admissible physical states. Classically, the additional degrees of freedom would be eliminated by imposing the Lorentz condition $\partial_\mu A_\mu = 0$; this would restore the equivalence of 4(60) with the Maxwell equation 1(146). Since we have chosen to work with 4(60), we can attempt to reintroduce the necessary constraint by imposing the condition

$$\partial_\mu A_\mu(x)|a\rangle = 0 \tag{4(78)}$$

on the state vectors $|a\rangle$. This is too stringent a constraint however, as it would rule out the vacuum state. Indeed, the vacuum state $|0\rangle$ cannot satisfy 4(78) for, if it did, we would have

$$\partial_\mu A_\mu(x)|0\rangle = \partial_\mu A_\mu^{(-)}(x)|0\rangle = 0 \tag{4(79)}$$

and

$$\begin{aligned} 0 = A_\nu^{(+)}(y)\frac{\partial}{\partial x_\mu}A_\mu^{(-)}(x)|0\rangle &= \frac{\partial}{\partial x_\mu}[A_\nu^{(+)}(y), A_\mu^{(-)}(x)]|0\rangle \\ &= \frac{\partial}{\partial x_\mu} i\delta_{\mu\nu} D^{(+)}(y-x)|0\rangle \\ &= i\frac{\partial}{\partial x_\nu} D^{(+)}(y-x)|0\rangle \end{aligned}$$

Since the right-hand side is non-vanishing, 4(79) cannot hold*.

The correct formulation of the subsidiary condition is due to Gupta (1950) and Bleuler (1950). Instead of 4(78), we adopt the more relaxed condition

$$\partial_\mu A_\mu^{(+)}(x)|a\rangle = 0 \tag{4(80)}$$

involving only the positive frequency part of the field. Note that the vacuum state $|0\rangle$ is *automatically* a solution of 4(80). Moreover, 4(80)

* See also Federbush (1960) for a general proof that if $O(x)$ is a local operator such that $O(x)|0\rangle = 0$, then $O(x) = 0$.

is sufficient to ensure that the classical constraint 1(152) is satisfied by the *expectation values* of A_μ in allowed states. Indeed

$$\begin{aligned}\langle a|\partial_\mu A_\mu|a\rangle &= \langle a|\partial_\mu(A_\mu^{(+)}+A_\mu^{(-)})|a\rangle \\ &= \langle a|\partial_\mu A_\mu^{(-)}|a\rangle \\ &= \langle a|\partial_\mu A_\mu^{(+)}|a\rangle^* = 0\end{aligned}$$

Expressing the subsidiary condition 4(80) in terms of creation and destruction operators with the aid of 4(67) we find

$$L(\mathbf{k})|a\rangle = 0 \qquad (\text{all } \mathbf{k}) \tag{4(81)}$$

where we have defined

$$L(\mathbf{k}) = \sum_{\lambda=1}^{4} k \,.\, e_{\mathbf{k}\lambda}\, a_{\mathbf{k}\lambda} \tag{4(82a)}$$

Using the explicit expressions 1(167)–1(170) for the polarization vectors $e_{\mathbf{k}\lambda}$, and noting that $k \,.\, e_{\mathbf{k}1} = k \,.\, e_{\mathbf{k}2} = 0$, we can write $L(\mathbf{k})$ as

$$L(\mathbf{k}) = -(k \,.\, \eta)(a_{\mathbf{k}3} - i a_{\mathbf{k}4}) \tag{4(82b)}$$

which exhibits the subsidiary condition 4(81) as a restriction on the possible admixtures of longitudinal and timelike photons in physical states. As an immediate consequence of 4(81) we have

$$\langle a|a_{\mathbf{k}3}{}^\dagger a_{\mathbf{k}3}|a\rangle - \langle a|a_{\mathbf{k}4}{}^\dagger a_{\mathbf{k}4}|a\rangle = 0 \tag{4(83)}$$

for all $\mathbf{k}$, so that the total contribution of the longitudinal and timelike photons to the expectation value of the Hamiltonian 4(76b) is zero. Thus the subsidiary condition 4(81) ensures that only the transverse photons—which represent the real dynamical degrees of freedom of the field—contribute to the expectation value of H.

To show that the subsidiary condition effectively eliminates states with negative norms, we now construct the most general state vector satisfying 4(81). This is given by

$$|a_f\rangle = R_f|a_0\rangle \tag{4(84)}$$

where $|a_0\rangle$ is an arbitrary state containing only *transverse* photons and where

$$\begin{aligned}R_f &= 1 + \sum_{\mathbf{k}} f(\mathbf{k}_1)L(\mathbf{k}_1)^\dagger + \ldots \\ &\qquad \ldots + \sum_{\mathbf{k}_1\ldots\mathbf{k}_n} f(\mathbf{k}_1\ldots\mathbf{k}_n)L(\mathbf{k}_1)^\dagger\ldots L(\mathbf{k}_n)^\dagger + \ldots\end{aligned} \tag{4(85)}$$

To prove that $|a_f\rangle$ satisfies 4(81) we simply note that

$$\begin{aligned}[L(\mathbf{k}), L(\mathbf{k}')^\dagger] &= (k\,.\,\eta)^2[(a_{\mathbf{k}3} - ia_{\mathbf{k}4}), (a_{\mathbf{k}'3}{}^\dagger + ia_{\mathbf{k}'4}{}^\dagger)] \\ &= (k\,.\,\eta)^2\{[a_{\mathbf{k}3}, a_{\mathbf{k}'3}{}^\dagger] + [a_{\mathbf{k}4}, a_{\mathbf{k}'4}{}^\dagger]\} \\ &= 0 \end{aligned} \qquad 4(86)$$

and that the states $|a_0\rangle$ satisfy 4(81) trivially. For an arbitrary 'transverse' state $|a_0\rangle$, 4(85) represents the most general admixture of longitudinal and timelike photons compatible with 4(81). Let us compute the scalar product of $\langle a_f|b_g\rangle$ where $a_f = R_f|a_0\rangle$ and $b_g = R_g|b_0\rangle$ are two states satisfying 4(81). Using the property

$$[R_f{}^\dagger, R_g] = 0 \qquad 4(87)$$

which follows directly from 4(86), we have

$$\begin{aligned}\langle a_f|b_g\rangle &= \langle a_0|R_f{}^\dagger R_g|b_0\rangle \\ &= \langle a_0|R_g R_f{}^\dagger|b_0\rangle\end{aligned}$$

But

$$\begin{aligned}R_f{}^\dagger|b_0\rangle &= |b_0\rangle \\ R_g{}^\dagger|a_0\rangle &= |a_0\rangle\end{aligned} \qquad 4(88)$$

by virtue of 4(81), so that

$$\langle a_f|b_g\rangle = \langle a_0|b_0\rangle \qquad 4(89)$$

This important result states that the allowed admixtures of longitudinal and timelike photons do not affect the scalar products (and, in particular, the norms) of state vectors, the latter being determined only by the transverse photons. Thus we conclude that all allowed state vectors have positive norm.

Essentially, the function of the subsidiary condition is to ensure that the longitudinal and timelike degrees of freedom are physically unobservable. From a physical point of view, the states $|a_f\rangle$ and $|a_0\rangle$ are completely equivalent. Mathematically, the difference between them corresponds to the gauge freedom of the classical theory [see Problem 7].

4-3 Quantization of the Rarita–Schwinger field

Introduction For spin $\frac{3}{2}$ (and in general for all spins greater than 1), the canonical quantization procedure becomes rather awkward, owing to the difficulty of isolating the independent dynamical degrees of freedom.

In particular, the spinor–vector ψ_μ introduced in Section 1-7 would have to be decomposed into its irreducible spin $\frac{1}{2}$ and spin $\frac{3}{2}$ parts and only the latter part subjected to canonical quantization. To avoid the difficult calculations which this entails [Moldauer (1956)] we shall bypass the canonical procedure altogether and work directly with the creation and destruction operators for the normal modes.

Fourier Decomposition A consistent quantization scheme for the Rarita–Schwinger field must ensure that the quantum equations of motion

$$\dot{\psi}_\mu = \frac{1}{i}[\psi_\mu, H] \tag{4(90)}$$

are equivalent to the Rarita–Schwinger equations 1(186a), 1(186b) or 1(188). The field can then be expanded in terms of the complete set of solutions 1(202a) and 1(202b) according to

$$\psi^\mu(x) = \frac{1}{\sqrt{V}}\sum_{\mathbf{k}}\sqrt{\frac{m}{E_{\mathbf{k}}}}\sum_{\sigma=1}^{4}(c_{\mathbf{k}\sigma}u^\mu_{\mathbf{k}\sigma}\,\mathrm{e}^{ik.x} + d_{\mathbf{k}\sigma}{}^\dagger v^\mu_{\mathbf{k}\sigma}\,\mathrm{e}^{-ik.x}) \tag{4(91)}$$

In the normal canonical approach the commutation relations for the $c_{\mathbf{k}\sigma}$ and $d_{\mathbf{k}\sigma}$ are deduced from the equal-time commutators of the fields. Here we shall reverse the usual order and impose the commutation relations directly on the creation and destruction operators. In keeping with the connection between spin and statistics we set

$$\{c_{\mathbf{k}\sigma}, c_{\mathbf{k}'\sigma'}{}^\dagger\} = \{d_{\mathbf{k}\sigma}, d_{\mathbf{k}'\sigma'}{}^\dagger\} = \delta_{\mathbf{k}\mathbf{k}'}\delta_{\sigma\sigma'} \tag{4(92a)}$$

$$\{c_{\mathbf{k}\sigma}, c_{\mathbf{k}'\sigma'}\} = \{d_{\mathbf{k}\sigma}, d_{\mathbf{k}'\sigma'}\} = \{c_{\mathbf{k}\sigma}, d_{\mathbf{k}'\sigma'}{}^\dagger\} = \{c_{\mathbf{k}\sigma}, d_{\mathbf{k}'\sigma'}\} = 0 \tag{4(92b)}$$

We must now check the validity of this procedure by showing that 4(90) is consistent with 4(91) and 4(92a), 4(92b). Let us insert 4(91) and the corresponding expansion for $\bar{\psi}^\mu$

$$\bar{\psi}^\mu(x) = \frac{1}{\sqrt{V}}\sum_{\mathbf{k}}\sqrt{\frac{m}{E_{\mathbf{k}}}}\sum_{\sigma=1}^{4}(c_{\mathbf{k}\sigma}{}^\dagger\bar{u}^\mu_{\mathbf{k}\sigma}\,\mathrm{e}^{-ikx} + d_{\mathbf{k}\sigma}\bar{v}^\mu_{\mathbf{k}\sigma}\,\mathrm{e}^{ikx}) \tag{4(93)}$$

into the expression for the Hamiltonian, given classically by 2(94)

$$H = \int \psi^\dagger{}_\mu i\partial_t\psi_\mu\,d^3x = \int \bar{\psi}_\mu\gamma_4 i\partial_t\psi_\mu\,d^3x \tag{4(94)}$$

Applying 1(202a)–1(202d) and subtracting the zero point energy, we obtain the usual formula

$$H = \sum_{\mathbf{k}}\sum_{\sigma=1}^{4} E_{\mathbf{k}}(c_{\mathbf{k}\sigma}{}^\dagger c_{\mathbf{k}\sigma} + d_{\mathbf{k}\sigma}{}^\dagger d_{\mathbf{k}\sigma}) \tag{4(95)}$$

expressing the energy of the field as the sum of the energies carried by the field quanta. The commutator $[\psi_\mu, H]$ is now easily evaluated with the aid of 4(92a) and 4(92b) and the formula

$$[A, BC] = \{A, B\}C - B\{A, C\} \qquad 4(96)$$

We find

$$\begin{aligned}[\psi_\mu, H] &= \frac{1}{\sqrt{V}} \sum_{\mathbf{k}} \sqrt{\frac{m}{E_{\mathbf{k}}}} \sum_{\sigma=1}^{4} E_{\mathbf{k}}(c_{\mathbf{k}\sigma} u^\mu_{\mathbf{k}\sigma}\, e^{ikx} - d_{\mathbf{k}\sigma}{}^\dagger v^\mu_{\mathbf{k}\sigma}\, e^{-ikx}) \\ &= i\dot{\psi}_\mu\end{aligned}$$

in agreement with 4(90).

Commutation Rules and Feynman Propagator Using 4(92a) and 4(92b) we can now compute the commutator of $\psi^\mu(x)$ and $\bar{\psi}^\mu(x')$ for arbitrary times t and t'. We get

$$\{\psi^\mu(x), \bar{\psi}^\nu(x')\} = \frac{1}{V} \sum_{\mathbf{k}} \frac{m}{E_{\mathbf{k}}} \left\{ \sum_{\sigma=1}^{4} u^\mu_{\mathbf{k}\sigma} \bar{u}^\nu_{\mathbf{k}\sigma}\, e^{ik(x-x')} + \sum_{\sigma=1}^{4} v^\mu_{\mathbf{k}\sigma} \bar{v}^\nu_{\mathbf{k}\sigma}\, e^{-ik(x-x')} \right\}$$

or, applying 1(206a) and 1(206b)

$$\{\psi^\mu(x), \bar{\psi}^\nu(x')\} = i\partial_{\mu\nu}\Delta(x-x') \qquad 4(97a)$$

where $\partial_{\mu\nu}$ is the differential operator

$$\partial_{\mu\nu} = -(\gamma . \partial - m)\left[\delta_{\mu\nu} - \tfrac{1}{3}\gamma_\mu\gamma_\nu + \frac{1}{3m}(\gamma_\mu\partial_\nu - \gamma_\nu\partial_\mu) - \frac{2}{3m^2}\partial_\mu\partial_\nu \right] \qquad 4(97b)$$

For the Feynman propagator

$$S_{F\mu\nu}(x_1 - x_2) = \langle 0| T\psi_\mu(x_1)\bar{\psi}_\nu(x_2)|0\rangle \qquad 4(98)$$

we obtain, by the usual procedure

$$\begin{aligned}S_{F\mu\nu}(x_1 - x_2) &= \theta(t_1 - t_2)\partial_{\mu\nu} i\Delta^{(+)}(x_1 - x_2) - \theta(t_1 - t_2)\partial_{\mu\nu} i\Delta^{(-)}(x_1 - x_2) \\ &= \partial_{\mu\nu}\Delta_F(x_1 - x_2) + \ldots\end{aligned}$$

where the dots denote additional non-covariant terms proportional to $\delta^{(4)}(x_1 - x_2)$ and its derivatives which arise when the θ functions are commuted past the differential operator $\partial_{\mu\nu}$. We have encountered an earlier instance of this in the expression 4(28) for the massive vector meson propagator. As in the vector case, these additional terms can be dropped in practical calculations for interacting fields.*

* There is an additional complication in the spin $\frac{3}{2}$ case, in that the cancellation of the non-covariant terms introduces an additional *covariant* term proportional to $\delta^{(4)}(x_1-x_2)$ and its derivatives. The effective propagator is in fact the function

$$\partial_{\mu\nu}\Delta_F(x_1-x_2)-\frac{2i}{3m^2}[(\gamma_\mu\partial_\nu-\gamma_\nu\partial_\mu)+(\gamma\,.\,\partial-m)\gamma_\mu\gamma_\nu]\delta^{(4)}(x_1-x_2)$$

which satisfies the inhomogeneous Green's function equation corresponding to 1(188) [Takahashi (1953). See also H. Umezawa, *Quantum Field Theory*, North Holland (1956), p. 151].

PROBLEMS

1. Derive the operator equations of motion 4(6a) and 4(6b) from 4(5a) and 4(5b).
2. Check the relations 4(14a) and 4(14b).
3. Derive the expression 4(17).
4. Verify Eq. 4(36).
5. Applying the equal-time commutation relations 4(44a) and 4(44b), show that the quantum equations of motion 4(45a) and 4(45b) reduce to the single equation 4(47) for the transverse vector potential.
6. Derive the expression 4(57) for the photon commutator function.
7. Show that the difference between the expectation values $\langle a_f|A_\mu(x)|a_f\rangle$ and $\langle a_0|A_\mu(x)|a_0\rangle$ is of the form $\partial_\mu\Lambda(x)$ where

$$\Lambda(x)=\frac{-i}{\sqrt{V}}\sum_{\mathbf{k}}\frac{1}{\sqrt{2|\mathbf{k}|}}[f(\mathbf{k})\,e^{ik.x}-f^*(\mathbf{k})\,e^{-ik.x}]$$

satisfies $\Box\Lambda(x)=0$.

REFERENCES

Bleuler, K. (1950) *Helv. Phys. Acta* **23**, 567
Fermi, E. (1930) *Atti. Accad. Lincei* **12**, 431
Federbush, P. (1960) (with K. Johnson) *Phys. Rev.* **120**, 1926
Gupta, S. N. (1950) *Proc. Phys. Soc.* **A 63**, 681
Moldauer, P. A. (1956) (with K. M. Case) *Phys. Rev.* **102**, 279
Takahashi, Y. (1953) (with H. Umezawa) *Prog. Theoret. Phys.* **9**, 1

5

Interacting Quantum Fields

5-1 Introduction

Free field theories are completely soluble since, in each theory, the eigenstates and eigenvalues of the Hamiltonian can be calculated exactly. To apply field theory to the physical world we must turn to the far more difficult and as yet unsolved problem of coupled fields.

In this chapter we examine various types of couplings between fields and establish the general framework for the theory of interacting quantum fields. We begin by considering the electromagnetic coupling of Dirac fields.

5-2 The electromagnetic interaction

Minimal Coupling for Spin $\frac{1}{2}$ Fields Our choice for the form of the electromagnetic coupling is motivated by the correspondence principle. In classical physics the electromagnetic interaction is introduced by the so-called *minimal* substitution $k_\mu \to k_\mu - eA_\mu$ in the free field equations of motion. Let us extend this prescription to quantum field theory and perform the substitution

$$\partial_\mu \to \partial_\mu - ieA_\mu(x) \qquad 5(1)$$

in the free Dirac Lagrangian density

$$\mathscr{L}_{\text{Dirac}} = -\bar{\psi}(\gamma \,.\, \partial + m)\psi \qquad 5(2)$$

and add the Lagrangian density for the free electromagnetic field

$$\mathscr{L}_{\text{e.m.}} = -\tfrac{1}{2}\partial_\nu A_\mu \partial_\nu A_\mu \qquad 5(3)$$

where we have adopted the Lorentz gauge for convenience. By this process we obtain the total Lagrangian density for the interacting system

$$\mathscr{L} = \mathscr{L}_{\text{Dirac}} + \mathscr{L}_{\text{e.m.}} + \mathscr{L}_I \qquad 5(4)$$

where $\mathscr{L}_I$, the *interaction Lagrangian density*, is given by

$$\mathscr{L}_I = ie\bar{\psi}(x)\gamma_\mu\psi(x)A_\mu(x) \tag{5(5)}$$

So far, our experimental knowledge of the electron-positron-photon interaction is perfectly consistent with the minimal coupling 5(5); there is no evidence for any additional couplings*. This is satisfactory from a theoretical point of view, since the theory based on 5(5) has several appealing features, notably *renormalizibility*, (see Section 6-4).

The Euler–Lagrange equations for ψ and A_μ which follow from 5(4) are

$$(\gamma_\mu\partial_\mu + m)\psi = ie\gamma_\mu\psi A_\mu \tag{5(6a)}$$

$$\Box A_\mu = -ie\bar{\psi}\gamma_\mu\psi \tag{5(6b)}$$

Note that by virtue of 5(6a) and the adjoint equation

$$\bar{\psi}(\gamma_\mu\overleftarrow{\partial}_\mu - m) = -ie\bar{\psi}\gamma_\mu A_\mu$$

the current

$$j_\mu(x) = i\bar{\psi}(x)\gamma_\mu\psi(x) \tag{5(7)}$$

appearing on the right-hand side of 5(6b) obeys the conservation law

$$\begin{aligned}\partial_\mu j_\mu &= \bar{\psi}(\overleftarrow{\partial}_\mu + \partial_\mu)\gamma_\mu\psi \\ &= 0\end{aligned} \tag{5(8)}$$

To the equations of motion 5(6a) and 5(6b) we must add the Lorentz gauge condition

$$(\partial_\mu A_\mu(x))^{(+)}|a\rangle = 0 \tag{5(9)}$$

as a restriction on the realizable state vectors $|a\rangle$ of the quantum system†. The canonical equal-time commutation and anticommutation relations for the fields ψ, A_μ and their conjugate momenta

$$\pi_\psi = \frac{\partial\mathscr{L}}{\partial\dot{\psi}} = i\psi^\dagger \tag{5(10a)}$$

$$\pi_\mu = \frac{\partial\mathscr{L}}{\partial\dot{A}_\mu} = \dot{A}_\mu \tag{5(10b)}$$

* For example, the so-called Pauli term $f\bar{\psi}\Sigma_{\mu\nu}\psi F_{\mu\nu}$ where $F_{\mu\nu}$ is the field strength tensor, is a priori possible, but experimentally absent.

† Note that by virtue of 5(6b) and 5(8), $\partial_\mu A_\mu(x)$ satisfies the free Klein–Gordon equation. Hence a decomposition of $\partial_\mu A_\mu(x)$ into positive and negative frequency waves is meaningful.

are

$$\{\psi_\alpha(x), \psi_\beta{}^\dagger(x')\}_{t=t'} = \delta_{\alpha\beta}\delta^{(3)}(\mathbf{x}-\mathbf{x}') \qquad 5(11a)$$

$$\{\psi_\alpha(x), \psi_\beta(x')\}_{t=t'} = \{\psi_\alpha{}^\dagger(x), \psi_\beta{}^\dagger(x')\}_{t=t'} = 0 \qquad 5(11b)$$

$$[A_\mu(x), \dot{A}_\nu(x')]_{t=t'} = i\delta_{\mu\nu}\delta^{(3)}(\mathbf{x}-\mathbf{x}') \qquad 5(11c)$$

$$[A_\mu(x), A_\nu(x')]_{t=t'} = [\dot{A}_\mu(x), \dot{A}_\nu(x')]_{t=t'} = 0 \qquad 5(11d)$$

In addition, we require that Dirac field operators commute with electromagnetic field operators at equal times:

$$[\psi_\alpha(x), A_\mu(x')]_{t=t'} = [\psi_\alpha(x), \dot{A}_\mu(x')]_{t=t'} = 0 \qquad 5(11e)$$

The Hamiltonian density $\mathscr{H}(x)$ is defined by the usual prescription

$$\mathscr{H} = \pi_\psi\dot{\psi} + \pi_\mu\dot{A}_\mu - \mathscr{L}$$

and we find

$$\mathscr{H} = \mathscr{H}_{\text{Dirac}} + \mathscr{H}_{\text{e.m.}} + \mathscr{H}_I \qquad 5(12)$$

where $\mathscr{H}_{\text{Dirac}}$ and $\mathscr{H}_{\text{e.m.}}$ are the free field Hamiltonian densities

$$\mathscr{H}_{\text{Dirac}} = \bar{\psi}(\boldsymbol{\gamma}\,.\,\boldsymbol{\nabla}+m)\psi \qquad 5(13a)$$

$$\mathscr{H}_{\text{e.m.}} = \tfrac{1}{2}[\pi_\mu\pi_\mu + (\boldsymbol{\nabla}A_\mu)\,.\,(\boldsymbol{\nabla}A_\mu)] \qquad 5(13b)$$

and

$$\mathscr{H}_I = -\mathscr{L}_I = -ie\bar{\psi}\gamma_\mu\psi A_\mu \qquad 5(13c)$$

The temporal evolution of the interacting field system is determined, as in the free field case, by the quantum equations of motion

$$\dot{\psi} = \frac{1}{i}[\psi, H] \qquad 5(14a)$$

$$\dot{A}_\mu = \frac{1}{i}[A_\mu, H] \qquad \dot{\pi}_\mu = \frac{1}{i}[\pi_\mu, H] \qquad 5(14b)$$

where H is the total Hamiltonian

$$H = \int \mathscr{H}(x)\, d^3x$$

$$= H_{\text{Dirac}} + H_{\text{e.m.}} + H_I \qquad 5(15)$$

The proof of the consistency of 5(15a) and 5(15b) with the Euler–Lagrange equations 5(6a) and 5(6b) is left to the reader as an exercise. Since the Hamiltonian depends on time only through the fields ψ, $\bar{\psi}$, A_μ and π_μ, we have

$$\dot{H} = \frac{1}{i}[H, H]$$
$$= 0$$

so that H is conserved in time. Equations 5(15a) and 5(15b) can be formally integrated to read

$$\psi(\mathbf{x}, t) = e^{iHt}\psi(\mathbf{x}, 0)\, e^{-iHt} \qquad 5(16a)$$

$$A_\mu(\mathbf{x}, t) = e^{iHt}A_\mu(\mathbf{x}, 0)\, e^{-iHt} \qquad 5(16b)$$

As in the free field case, the accompanying relations

$$\psi(\mathbf{x}, t) = e^{-i\mathbf{P}.\mathbf{x}}\psi(0, t)\, e^{i\mathbf{P}.\mathbf{x}} \qquad 5(16c)$$

$$A_\mu(\mathbf{x}, t) = e^{-i\mathbf{P}.\mathbf{x}}A_\mu(0, t)\, e^{i\mathbf{P}.\mathbf{x}} \qquad 5(16d)$$

with

$$\mathbf{P} = -\int (\pi_\psi \nabla\psi + \pi_\mu \nabla A_\mu)\, d^3x \qquad 5(17)$$

follow from the equalities

$$\nabla\psi(\mathbf{x}, t) = i[\psi(\mathbf{x}, t), \mathbf{P}] \qquad 5(18a)$$

$$\nabla A_\mu(\mathbf{x}, t) = i[A_\mu(\mathbf{x}, t), \mathbf{P}] \qquad 5(18b)$$

exhibiting the momentum $\mathbf{P}$ as the generator of infinitesimal spatial displacements. Note that $\mathbf{P}$ coincides with the sum of the free field momentum operators. [See Problem 1.]

As the field operators ψ and A_μ obey coupled non-linear equations of motion, we can no longer expand them in terms of free field solutions as in 3(154a), 3(154b) and 4(67). We can of course still expand the operators at a given time, say $t = 0$, in terms of their Fourier components:

$$\psi(\mathbf{x}, 0) = \frac{1}{\sqrt{V}}\sum_{\mathbf{k}}\sum_{\sigma=1}^{2}\sqrt{\frac{m}{E_{\mathbf{k}}}}(u_{\mathbf{k}\sigma}c_{\mathbf{k}\sigma}\, e^{i\mathbf{k}\mathbf{x}} + v_{\mathbf{k}\sigma}\, d_{\mathbf{k}\sigma}{}^\dagger\, e^{-i\mathbf{k}\mathbf{x}}) \qquad 5(19a)$$

$$\bar{\psi}(\mathbf{x}, 0) = \frac{1}{\sqrt{V}}\sum_{\mathbf{k}}\sum_{\sigma=1}^{2}\sqrt{\frac{m}{E_{\mathbf{k}}}}(\bar{u}_{\mathbf{k}\sigma}c_{\mathbf{k}\sigma}{}^\dagger\, e^{-i\mathbf{k}\mathbf{x}} + \bar{v}_{\mathbf{k}\sigma}\, d_{\mathbf{k}\sigma}\, e^{i\mathbf{k}\mathbf{x}}) \qquad 5(19b)$$

$$A_\mu(\mathbf{x}, 0) = \frac{1}{\sqrt{V}}\sum_{\mathbf{k}}\sum_{\lambda=1}^{4}\frac{1}{\sqrt{2k_0}}e^\mu_{\mathbf{k}\lambda}(a_{\mathbf{k}\lambda}\, e^{i\mathbf{k}\mathbf{x}} + a_{\mathbf{k}\lambda}{}^\dagger\, e^{-i\mathbf{k}\mathbf{x}}) \qquad 5(19c)$$

$$\dot{A}_\mu(\mathbf{x}, 0) = \frac{-i}{\sqrt{V}}\sum_{\mathbf{k}}\sum_{\lambda=1}^{4}\sqrt{\frac{k_0}{2}}e^\mu_{\mathbf{k}\lambda}(a_{\mathbf{k}\lambda}\, e^{i\mathbf{k}\mathbf{x}} - a_{\mathbf{k}\lambda}{}^\dagger\, e^{-i\mathbf{k}\mathbf{x}}) \qquad 5(19d)$$

where, for convenience, we have identified the form of the expansions 5(19a), 5(19b), 5(19c) and 5(19d) with that of the corresponding free field expansions at $t = 0$. The operators $c_{\mathbf{k}\sigma}$, $d_{\mathbf{k}\sigma}$, $a_{\mathbf{k}\lambda}$ and their adjoints obey the same algebra as the corresponding destruction and creation operators of the free field theories, since the equal-time commutation relations 5(11a), 5(11b), 5(11c) and 5(11d) are formally the same as for free fields. In addition, owing to 5(11e), the electromagnetic operators $a_{\mathbf{k}\lambda}$ and $a_{\mathbf{k}\lambda}{}^\dagger$ commute with the Dirac operators $c_{\mathbf{k}\sigma}$, $d_{\mathbf{k}\sigma}$ and their adjoints. However, we are not at liberty to interpret $a_{\mathbf{k}\lambda}{}^\dagger$, $a_{\mathbf{k}\lambda}$, $c_{\mathbf{k}\sigma}{}^\dagger$, $c_{\mathbf{k}\sigma}$, etc., as single-particle creation and destruction operators, since they are no longer associated with eigenmodes of the Hamiltonian. Passage from time $t = 0$ to time t involves the replacements

$$a_{\mathbf{k}\lambda} \rightarrow e^{iHt} a_{\mathbf{k}\lambda}\, \mathrm{e}^{-iHt} \qquad 5(20a)$$

$$a_{\mathbf{k}\lambda}{}^\dagger \rightarrow \mathrm{e}^{iHt} a_{\mathbf{k}\lambda}{}^\dagger\, \mathrm{e}^{-iHt} \qquad 5(20b)$$

$$c_{\mathbf{k}\lambda} \rightarrow e^{iHt} c_{\mathbf{k}\lambda}\, \mathrm{e}^{-iHt} \qquad 5(20c)$$

etc.

in the expansions 5(19a), 5(19b), 5(19c) and 5(19d). For free fields, the right-hand side of 5(20a), 5(20b) and 5(20c) reduces respectively to $\mathrm{e}^{-i|\mathbf{k}|t} a_{\mathbf{k}\lambda}$, $\mathrm{e}^{i|\mathbf{k}|t} a_{\mathbf{k}\lambda}{}^\dagger$, and $\mathrm{e}^{-iE_{\mathbf{k}}t} c_{\mathbf{k}\lambda}$, but in the interacting field case additional terms appear, due to the interaction Hamiltonian H_I. In fact the entire apparatus developed in the free field case for constructing the eigenstates of the Hamiltonian breaks down and the exact solution to the coupled field problem is unknown. The same applies to all other coupled field theories which we shall consider. Interacting quantum field theories are too complex to solve exactly and we must either apply perturbative methods or fall back on assumptions about the nature of the solutions. The perturbative approach will be developed and applied in Chapter 6.

On occasion it will be useful to consider the much simpler problem of the coupling of the electron field to an externally applied c-number electromagnetic field $A_\mu^{\text{ext.}}(x)$. The interaction Lagrangian for this coupling is of the same form as 5(5), i.e.

$$\mathscr{L}_I = ie\bar{\psi}\gamma_\mu \psi A_\mu^{\text{ext.}} \qquad 5(21)$$

but there is in this case no electromagnetic Lagrangian $\mathscr{L}_{\text{e.m.}}$ and no equation of motion for $A_\mu^{\text{ext.}}$, since the latter is an *applied* field.

Minimal Coupling for Spin 0 *Fields* If we perform the minimal substitutions

$$\partial_\mu\phi \to \partial_\mu\phi - ieA_\mu\phi$$
$$\partial_\mu\phi^\dagger \to \partial_\mu\phi^\dagger + ieA_\mu\phi^\dagger$$

in the Lagrangian density

$$\mathscr{L}_{\text{KG}} = -\partial_\mu\phi^\dagger\partial_\mu\phi - \mu^2\phi^\dagger\phi \qquad 5(22)$$

for the free charged scalar field, and add the electromagnetic field Lagrangian 5(3), we obtain the Lagrangian density

$$\mathscr{L} = \mathscr{L}_{\text{KG}} + \mathscr{L}_{\text{e.m.}} + \mathscr{L}_I \qquad 5(23)$$

with

$$\mathscr{L}_I = ie[(\partial_\mu\phi^\dagger)\phi - \phi^\dagger(\partial_\mu\phi)]A_\mu - e^2\phi^\dagger\phi A_\mu A_\mu \qquad 5(24)$$

describing the interaction of charged scalar bosons (e.g. charged pions or charged kaons) with the electromagnetic field. The coupling Lagrangian 5(24) is of the so-called *derivative* type, since it features the field derivatives $\partial_\mu\phi$ and $\partial_\mu\phi^\dagger$.

The Euler–Lagrange equations of motion for ϕ, $\phi^\dagger$ and A_μ are easily derived from 5(23). We find

$$(\partial_\mu - ieA_\mu)^2\phi - \mu^2\phi = 0 \qquad 5(25a)$$
$$(\partial_\mu + ieA_\mu)^2\phi^\dagger - \mu^2\phi^\dagger = 0 \qquad 5(25b)$$

and

$$\Box A_\mu = -ie[(\partial_\mu\phi^\dagger)\phi - \phi^\dagger(\partial_\mu\phi)] + 2e^2\phi^\dagger\phi A_\mu \qquad 5(25c)$$

where the current appearing on the right-hand side, i.e.

$$j_\mu(x) = i[(\partial_\mu\phi^\dagger)\phi - \phi^\dagger(\partial_\mu\phi)] - 2e\phi^\dagger\phi A_\mu \qquad 5(26)$$

is conserved by virtue of the equations of motion:

$$\partial_\mu j_\mu = 0 \qquad 5(27)$$

[See Problem 2.] Note that j_μ is an explicit function of the electromagnetic field in the present case, and that owing to the factor 2 in the last term of 5(26), the interaction Lagrangian density is not simply $ej_\mu A_\mu$ as it is in the Dirac case. As we are working in the Lorentz gauge, physically realizable states $|a\rangle$ are subject to the subsidiary condition

$$(\partial_\mu A_\mu(x))^{(+)}|a\rangle = 0 \qquad 5(28)$$

as in 5(9). For the conjugate momenta, we find

$$\pi = \frac{\partial \mathscr{L}}{\partial \dot{\phi}} = \dot{\phi}^\dagger - e\phi^\dagger A_4 \qquad 5(29a)$$

$$\pi^\dagger = \frac{\partial \mathscr{L}}{\partial \dot{\phi}^\dagger} = \dot{\phi} + e\phi A_4 \qquad 5(29b)$$

and

$$\pi_\mu = \frac{\partial \mathscr{L}}{\partial \dot{A}_\mu} = \dot{A}_\mu \qquad 5(30)$$

where we note that 5(29a) and 5(29b) differ from the corresponding free field expressions, owing to the derivative coupling term in 5(24). The canonical equal-time commutation relations are, as usual

$$[\phi(x), \pi(x')]_{t=t'} = i\delta^{(3)}(\mathbf{x}-\mathbf{x}') \qquad 5(31a)$$

$$[\phi^\dagger(x), \pi^\dagger(x')]_{t=t'} = i\delta^{(3)}(\mathbf{x}-\mathbf{x}') \qquad 5(31b)$$

$$[A_\mu(x), \pi_\nu(x')]_{t=t'} = i\delta_{\mu\nu}\delta^{(3)}(\mathbf{x}-\mathbf{x}') \qquad 5(31c)$$

etc.

with the additional requirement that all Klein–Gordon fields commute with all Maxwell fields at equal times. Defining the Hamiltonian density

$$\mathscr{H} = \pi\dot{\phi} + \dot{\phi}^\dagger\pi^\dagger + \pi_\mu\dot{A}_\mu - \mathscr{L} \qquad 5(32)$$

and using 5(29a), 5(29b) and 5(30) to express $\dot{\phi}$, $\dot{\phi}^\dagger$ and $\dot{A}_\mu$ in terms of π, $\pi^\dagger$ and π_μ, we find

$$\mathscr{H} = \mathscr{H}_{\text{KG}} + \mathscr{H}_{\text{e.m.}} + \mathscr{H}_I \qquad 5(33)$$

where

$$\mathscr{H}_{\text{KG}} = \pi^\dagger\pi + \nabla\phi^\dagger \cdot \nabla\phi + \mu^2\phi^\dagger\phi \qquad 5(34a)$$

$$\mathscr{H}_{\text{e.m.}} = \tfrac{1}{2}[\pi_\mu\pi_\mu + (\nabla A_\mu) \cdot (\nabla A_\mu)] \qquad 5(34b)$$

and

$$\mathscr{H}_I = -\mathscr{L}_I - e^2\phi^\dagger\phi A_0^2 \qquad 5(35)$$

with $\mathscr{L}_I$ given by 5(24). [See Problem 3.]

We draw the reader's attention to the second term in 5(35) and to the fact that $\mathscr{H}_I \neq -\mathscr{L}_I$ in contrast to 5(13c). The additional term in 5(35) is often referred to as a *normal-dependent* term to underline the fact that it singles out the fourth component of the vector potential, i.e. the component along the normal to 3-space as viewed in the four-dimensional continuum.

The proof of the consistency of the Hamiltonian equations of motion

$$\dot{\phi} = \frac{1}{i}[\phi, H] \qquad \dot{\phi}^\dagger = \frac{1}{i}[\phi^\dagger, H] \tag{5(36a)}$$

$$\dot{A}_\mu = \frac{1}{i}[A_\mu, H] \tag{5(36b)}$$

with the Euler–Lagrange equations 5(25a), 5(25b) and 5(25c) is left to the reader. The total Hamiltonian H, given by

$$\begin{aligned} H &= \int \mathscr{H}(x)\, d^3x \\ &= H_{\text{KG}} + H_{\text{e.m.}} + H_I \end{aligned} \tag{5(37)}$$

is a constant of the motion, and we have

$$\phi(\mathbf{x}, t) = \mathrm{e}^{iHt}\phi(\mathbf{x}, 0)\,\mathrm{e}^{-iHt} \tag{5(38a)}$$

$$A_\mu(\mathbf{x}, t) = \mathrm{e}^{iHt}A_\mu(\mathbf{x}, 0)\,\mathrm{e}^{-iHt} \tag{5(38b)}$$

as in 5(16a) and 5(16b). The accompanying relations

$$\phi(\mathbf{x}, t) = \mathrm{e}^{-i\mathbf{P}.\mathbf{x}}\phi(0, t)\,\mathrm{e}^{i\mathbf{P}.\mathbf{x}} \tag{5(38c)}$$

$$A_\mu(\mathbf{x}, t) = \mathrm{e}^{-i\mathbf{P}.\mathbf{x}}A_\mu(0, t)\,\mathrm{e}^{i\mathbf{P}.\mathbf{x}} \tag{5(38d)}$$

with

$$\mathbf{P} = -\int (\pi\nabla\phi + \pi_\mu\nabla A_\mu)\, d^3x \tag{5(39)}$$

are derived in the same way as 5(16c) and 5(16d).

Minimal Coupling and Current Conservation By effecting the minimal substitution 5(1) in the Klein–Gordon and Dirac Lagrangians, we have been led, in each case, to an equation of motion of the form

$$\Box A_\mu = -ej_\mu \tag{5(40)}$$

where the current j_μ is conserved by virtue of the equations of motion

$$\partial_\mu j_\mu = 0$$

The same applies to the minimal electromagnetic coupling of charged spin 1 and spin $\frac{3}{2}$ particles, as can be seen by effecting the substitution

5(1) in 2(22) and 2(26). [Problem 4.] The important point is that, in each case, the conserved current appearing on the left-hand side of 5(39) is just the current

$$j_\mu = -i\sum_A \left(\frac{\partial \mathscr{L}}{\partial(\partial_\mu \phi_A)}\phi_A - \frac{\partial \mathscr{L}}{\partial(\partial_\mu {\phi_A}^\dagger)}{\phi_A}^\dagger\right) \tag{5(41)}$$

whose conservation is assured by the invariance of the total Lagrangian under infinitesimal phase transformations of the type 2(129). For the charged scalar field, for example, 5(41) yields

$$j_\mu = i(\partial_\mu \phi^\dagger + ieA_\mu \phi^\dagger)\phi - i(\partial_\mu \phi - ieA_\mu \phi)\phi^\dagger \tag{5(42)}$$

in agreement with 5(26)*. This point can be understood by noting that the Euler–Lagrange equation of motion for the electromagnetic field reads

$$-\partial_\nu \frac{\partial \mathscr{L}}{\partial(\partial_\nu A_\mu)} = \Box A_\mu = -\frac{\partial \mathscr{L}}{\partial A_\mu}$$

Hence, comparing with 5(40), we see that

$$ej_\mu = \frac{\partial \mathscr{L}}{\partial A_\mu} \tag{5(43)}$$

or, by virtue of the minimal coupling hypothesis

$$\begin{aligned} ej_\mu &= \sum_A \left[\frac{\partial \mathscr{L}}{\partial(A_\mu \phi_A)}\phi_A + \frac{\partial \mathscr{L}}{\partial(A_\mu {\phi_A}^\dagger)}{\phi_A}^\dagger\right] \\ &= -ie\sum_A \left[\frac{\partial \mathscr{L}}{\partial(-ieA_\mu \phi_A)}\phi_A - \frac{\partial \mathscr{L}}{\partial(ieA_\mu {\phi_A}^\dagger)}{\phi_A}^\dagger\right] \\ &= -ie\sum_A \left[\frac{\partial \mathscr{L}}{\partial(\partial_\mu \phi_A)}\phi_A - \frac{\partial \mathscr{L}}{\partial(\partial_\mu {\phi_A}^\dagger)}{\phi_A}^\dagger\right] \end{aligned}$$

in agreement with 5(41).

The conservation of the current to which A_μ is coupled is actually a direct result of the structure of the Maxwell field. If instead of 5(3) we adopt the 'true' Lagrangian 2(24), or

$$\mathscr{L}_{\text{e.m.}} = -\tfrac{1}{4}F_{\mu\nu}F_{\mu\nu} \tag{5(44)}$$

* 5(42) actually differs from 5(26) in the ordering of the operator factors in the second term. This difference may be ignored since, as we shall see later, the current operator for a Bose–Einstein field must be symmetrized with respect to its noncommuting factors.

then the equation of motion 5(39) is replaced by the true Maxwell equation

$$\partial_\nu F_{\nu\mu} = -ej_\mu \qquad 5(45)$$

The current j_μ is unaffected by this change and we find directly,

$$\partial_\mu j_\mu = 0 \qquad 5(46)$$

by virtue of the antisymmetric character of the field tensor $F_{\mu\nu}$. The fact that A_μ is necessarily coupled to a conserved current is a crucial feature of the electromagnetic interaction. We shall see in Section 8-1 that it ensures that the electromagnetic coupling is *universal*, i.e. that all charged particles couple to the electromagnetic field with the same strength. This is in agreement with experimental observation.

Radiation Gauge In our discussion of electrodynamic phenomena, we shall generally adhere to the Lorentz gauge in order to maintain manifest Lorentz covariance. Nevertheless, it is instructive to examine, say, the interacting electron–photon system in the radiation gauge. The total Lagrangian of the system is then

$$\mathscr{L} = \mathscr{L}_{\text{Dirac}} + \mathscr{L}_{\text{e.m.}} + \mathscr{L}_I \qquad 5(47)$$

where $\mathscr{L}_{\text{Dirac}}$ and $\mathscr{L}_I$ are still given by

$$\mathscr{L}_{\text{Dirac}} = -\bar{\psi}(\gamma\,.\,\partial + m)\psi \qquad 5(48a)$$

$$\mathscr{L}_I = ie\bar{\psi}\gamma_\mu\psi A_\mu \qquad 5(48b)$$

but

$$\mathscr{L}_{\text{e.m.}} = -\tfrac{1}{4}F_{\mu\nu}F_{\mu\nu} \qquad 5(49)$$

as in 5(44). The equations of motion are

$$(\gamma\,.\,\partial + m)\psi = ie\gamma_\mu\psi A_\mu \qquad 5(50a)$$

and

$$\partial_\lambda F_{\lambda\mu} = -ie\bar{\psi}\gamma_\mu\psi \qquad 5(50b)$$

The radiation gauge is characterized by the condition div $\mathbf{A} = 0$, or

$$\mathbf{A} = \mathbf{A}^t \qquad 5(51)$$

For the electric field $\mathbf{E}$ on the other hand we have now

$$\mathbf{E} = \mathbf{E}^t + \mathbf{E}^l \qquad 5(52)$$

where, by virtue of Maxwell's equation 1(147b), i.e.

$$\mathbf{E} = -\mathbf{grad}\, A_0 - \partial_t \dot{\mathbf{A}}^t$$

we can identify $\mathbf{E}^t$ as

$$\mathbf{E}^t = -\dot{\mathbf{A}}^t \tag{5(53a)}$$

since $\mathbf{grad}^t \equiv 0$. [See Eq. 1(157a).] For $\mathbf{E}^l$ we have

$$\mathbf{E}^l = -\mathbf{grad}\, A_0 \tag{5(53b)}$$

where we note that A_0 does not vanish in the interacting case but is determined in terms of the charge density $\psi^\dagger\psi$ by 5(50b), or

$$\nabla^2 A_0 = -e\psi^\dagger\psi \tag{5(54)}$$

Explicitly

$$A_0(\mathbf{x}, t) = \frac{e}{4\pi}\int \frac{d^3x'}{|\mathbf{x}-\mathbf{x}'|}\psi^\dagger(\mathbf{x}', t)\psi(\mathbf{x}', t) \tag{5(55)}$$

Let us write the Lagrangian density 5(49) in the form

$$\begin{aligned}\mathscr{L}_{\text{e.m.}} &= \tfrac{1}{2}(\mathbf{E}^2-\mathbf{B}^2)\\ &= \tfrac{1}{2}[(\dot{\mathbf{A}}^t+\nabla A_0)^2-(\mathbf{curl}\,\mathbf{A}^t)^2]\\ &= \tfrac{1}{2}[(\dot{\mathbf{A}}^t)^2+(\nabla A_0)^2-(\mathbf{curl}\,\mathbf{A}^t)^2]\end{aligned} \tag{5(56)}$$

where we have dropped the cross term

$$\begin{aligned}\dot{A}^t\,.\,\nabla A_0 &= \nabla(\dot{\mathbf{A}}^t A_0)-(\nabla\,.\,\dot{\mathbf{A}}^t)A_0\\ &= \nabla(\dot{\mathbf{A}}^t A_0)\end{aligned}$$

on the grounds that a 3-divergence in $\mathscr{L}$ does not affect the total integrated Lagrangian

$$L_{\text{e.m.}} = \int \mathscr{L}_{\text{e.m.}}\, d^3x$$

The canonically conjugate momentum to $\mathbf{A}^t$ is therefore

$$\boldsymbol{\pi}^t = \frac{\partial\mathscr{L}}{\partial\dot{\mathbf{A}}^t} = \dot{\mathbf{A}}^t = -\mathbf{E}^t \tag{5(57)}$$

The field $A_0(x)$, though non-vanishing, does not represent an independent degree of freedom, since it is determined by the charge density $\psi^\dagger\psi$ according to 5(55). The same remark applies to the longitudinal component $\mathbf{E}^l$ of the electric field given by 5(53b). Thus, the independent canonical variables for the Maxwell field are again $\mathbf{A}^t$ and $-\mathbf{E}^t$ as in the free field case, and we can impose the usual canonical equal-time commutation relations

$$[A_i^t(x), -E_j^t(x')]_{t=t'} = i\left(\delta_{ij} - \frac{\partial_i \partial_j}{\nabla^2}\right)\delta^{(3)}(\mathbf{x}-\mathbf{x}') \qquad 5(58a)$$

$$\{\psi_\alpha(x), \psi_\beta^\dagger(x')\}_{t=t'} = \delta_{\alpha\beta}\delta^{(3)}(\mathbf{x}-\mathbf{x}') \qquad 5(58b)$$

etc.

In addition, the Maxwell fields $\mathbf{A}^t(x)$ and $\mathbf{E}^t(x)$ are assumed to commute with the Dirac field ψ at equal times. [See Problem 5.]

Defining the Hamiltonian density

$$\mathscr{H} = \pi_\psi \dot{\psi} - \mathbf{E}^t . \dot{\mathbf{A}}^t - \mathscr{L} \qquad 5(59)$$

we find, using 5(48a), 5(48b) and 5(56)

$$\mathscr{H} = \mathscr{H}_{\text{Dirac}} + \mathscr{H}_{\text{rad.}} + \mathscr{H}_I \qquad 5(60)$$

where $\mathscr{H}_{\text{Dirac}}$ and $\mathscr{H}_{\text{rad.}}$ are the usual free field densities

$$\mathscr{H}_{\text{Dirac}} = \bar{\psi}(\boldsymbol{\gamma} . \nabla + m)\psi \qquad 5(60a)$$

$$\mathscr{H}_{\text{rad.}} = \tfrac{1}{2}(\dot{\mathbf{A}}^t)^2 + \tfrac{1}{2}(\mathbf{curl}\ \mathbf{A}^t)^2 \qquad 5(60b)$$

and $\mathscr{H}_I$ is given by

$$\mathscr{H}_I = -ie\bar{\psi}\gamma_\mu \psi A_\mu - \tfrac{1}{2}(\nabla A_0)^2 \qquad 5(60c)$$

Note that $\mathscr{H}_I \neq -\mathscr{L}_I$. The total integrated interaction Hamiltonian is given by

$$H_I = \int (-ie\bar{\psi}\gamma_\mu \psi A_\mu + A_0 \nabla^2 A_0 - \tfrac{1}{2} A_0 \nabla^2 A_0)\, d^3x$$

$$= -ie \int \bar{\psi}(x)\boldsymbol{\gamma}\psi(x)\mathbf{A}(x)\, d^3x$$

$$+ \frac{e^2}{8\pi}\int \psi^\dagger(\mathbf{x}, t)\psi(\mathbf{x}, t)\frac{1}{|\mathbf{x}-\mathbf{x}'|}\psi^\dagger(\mathbf{x}', t)\psi(\mathbf{x}', t)\, d^3x\, d^3x' \qquad 5(61)$$

where we have performed an integration by parts and used 5(54) and 5(55). Equation 5(61) exhibits the interaction Hamiltonian as the sum of two distinct contributions. The first represents the coupling of the electron current to the transverse vector potential. The second term represents the ordinary Coulomb interaction energy.

5-3 Nonelectromagnetic couplings

Introduction When we turn to the nonelectromagnetic interactions of elementary particles, the correspondence principle can no longer serve to determine the interaction Lagrangian. In the case of the weak interactions, direct appeal to experiment yields information on the form of the coupling. For strong interactions, however, the coupling constants are too large to allow the use of perturbative techniques, with the result that there is no reliable way of testing theoretical coupling schemes against experiment. As the exact form of the interaction is unknown, we shall be guided in our choice of coupling schemes by simplicity and by the symmetries exhibited by strong interactions. The latter include exact symmetries like Lorentz invariance, as well as approximate ones like SU_2 and SU_3. We shall assume throughout our discussion that the reader is familiar with the phenomenology of elementary particles and with their group theoretical classification.

Coupling of Neutral Pions to Nucleons We begin with the simplest example of a nonelectromagnetic interaction—the interaction of π^0 mesons with protons and neutrons. We represent this interaction in its simplest form as a Lorentz-invariant coupling

$$\mathscr{L}_I = iG\bar{\psi}\gamma_5\psi\phi \qquad 5(62)$$

of a Dirac field operator ψ with a neutral (hermitian) Klein–Gordon field ϕ. The coupling constant G is assumed to be real and the factor i in 5(62) ensures the hermicity of $\mathscr{L}_I$.

The trilinear interaction 5(62) is known as a *Yukawa coupling*. Notice that we have coupled ϕ to the pseudoscalar density $i\bar{\psi}\gamma_5\psi$ rather than the scalar $\bar{\psi}\psi$, our choice being determined by the observed negative parity of the π^0 and the fact that the strong interactions are parity-conserving. The Euler–Lagrange equations of motion which follow from the Lagrangian density

$$\mathscr{L} = \mathscr{L}_{\text{KG}} + \mathscr{L}_{\text{Dirac}} + \mathscr{L}_I$$

are

$$(\gamma\,.\,\partial + m)\psi = iG\gamma_5\psi\phi \qquad 5(63a)$$

$$\bar{\psi}(\gamma\,.\,\overleftarrow{\partial} - m) = -iG\bar{\psi}\gamma_5\phi \qquad 5(63b)$$

and

$$(\Box - \mu^2)\phi = -iG\bar{\psi}\gamma_5\psi \qquad 5(64)$$

The remainder of the discussion parallels that of the electromagnetic coupling theory based on 5(5). The conjugate momenta π_ψ and π are equal to $i\psi^\dagger$ and $\dot{\phi}$ respectively and the Hamiltonian density is given by

$$\begin{aligned}\mathscr{H} &= \pi\dot{\phi}+\pi_\psi\dot{\psi}-\mathscr{L} \\ &= \mathscr{H}_{\text{Dirac}}+\mathscr{H}_{\text{KG}}+\mathscr{H}_I \qquad 5(65)\end{aligned}$$

where $\mathscr{H}_{\text{KG}}$ is the Hamiltonian density for the neutral scalar field

$$\mathscr{H}_{\text{KG}} = \tfrac{1}{2}(\pi^2+(\nabla\phi)^2+\mu^2\phi^2) \qquad 5(66)$$

and where

$$\mathscr{H}_I = -\mathscr{L}_I = -iG\bar{\psi}\gamma_5\psi\phi \qquad 5(67)$$

Actually, the two couplings 5(5) and 5(62) have a number of features in common. Both couplings are linear in the boson field, bilinear in the fermion field variables, and contain no derivatives. Moreover, both couplings yield renormalizable field theories [see Section 6-4]. Also note that both e and G are dimensionless coupling constants.

A rather more complicated choice for the π^0-nucleon interaction is the Lorentz-invariant derivative coupling Lagrangian

$$\mathscr{L}_I = if\bar{\psi}\gamma_\mu\gamma_5\psi\partial_\mu\phi \qquad 5(68)$$

where the coupling constant f has the dimension of length. The coupling 5(68) is a less attractive candidate than 5(62), owing to the fact that it is *non*renormalizable. The Euler–Lagrange equations of motion are in this case

$$(\gamma\,.\,\partial+m)\psi = if\gamma_\mu\gamma_5\psi\partial_\mu\phi \qquad 5(69a)$$

$$\bar{\psi}(\gamma\,.\,\overleftarrow{\partial}-m) = -if\bar{\psi}\gamma_\mu\gamma_5\partial_\mu\phi \qquad 5(69b)$$

and

$$(\Box-\mu^2)\phi = if\partial_\mu(\bar{\psi}\gamma_\mu\gamma_5\psi) \qquad 5(70)$$

The momentum conjugate to ψ is still $\pi_\psi = i\psi^\dagger$ but we now have

$$\pi = \frac{\partial\mathscr{L}}{\partial\dot{\phi}} = \dot{\phi}+f\bar{\psi}\gamma_4\gamma_5\psi \qquad 5(71)$$

We also get an additional ‘normal-dependent’ term in the expression for the interaction Hamiltonian density. We find [Problem 6]

$$\begin{aligned}\mathscr{H} &= \pi_\psi\dot{\psi}+\pi\dot{\phi}-\mathscr{L} \\ &= \mathscr{H}_{\text{Dirac}}+\mathscr{H}_{\text{KG}}+\mathscr{H}_I \qquad 5(72)\end{aligned}$$

where $\mathscr{H}_{KG}$ is given by 5(66) and

$$\mathscr{H}_I = -\mathscr{L}_I + \tfrac{1}{2}f^2(\bar{\psi}\gamma_4\gamma_5\psi)^2 \qquad 5(73)$$

depends on the fourth component of the axial vector current $i\bar{\psi}\gamma_\mu\gamma_5\psi$.

Obviously there is no limit to the number of couplings which one can construct. For example, one can allow progressively higher powers of ϕ to appear in the interaction Lagrangian, say

$$f_1\bar{\psi}\psi\phi^2 \qquad \text{or} \qquad f_2\bar{\psi}\psi\phi^4 \qquad 5(74)$$

or even*

$$\bar{\psi}(e^{ig\gamma_5\phi} - 1)\psi \qquad 5(75)$$

but we shall restrict our attention to the simpler couplings and more particularly to the renormalizable coupling 5(62).

Coupling of Neutral Vector Mesons The simplest coupling of a massive neutral vector field to a Dirac field is the trilinear form

$$\mathscr{L}_I = ig\bar{\psi}\gamma_\mu\psi A_\mu \qquad 5(76)$$

similar to the electrodynamic coupling 5(5). This can be viewed as a model for the interaction of, say, ρ^0 mesons with protons or neutrons. The total Lagrangian

$$\mathscr{L} = \mathscr{L}_{\text{Dirac}} + \mathscr{L}_{\text{Vector}} + \mathscr{L}_I \qquad 5(77)$$

with

$$\mathscr{L}_{\text{Vector}} = -\tfrac{1}{4}F_{\mu\nu}F_{\mu\nu} - \frac{\mu^2}{2}A_\mu A_\mu \qquad 5(78)$$

yields the Euler–Lagrange equations of motion

$$(\gamma\,.\,\partial + m)\psi = ig\gamma_\mu\psi A_\mu \qquad 5(79a)$$

$$\bar{\psi}(\gamma\,.\,\overleftarrow{\partial} - m) = -ig\bar{\psi}\gamma_\mu A_\mu \qquad 5(79b)$$

$$\partial_\nu F_{\nu\mu} - \mu^2 A_\mu = -ig\bar{\psi}\gamma_\mu\psi \qquad 5(80)$$

Note that, in contrast with the corresponding equation in quantum electrodynamics, the *structure* of 5(80) for $\mu \neq 0$ does not demand that the current on the right-hand side be conserved†. Nevertheless, by virtue

* The coupling 5(75) has been considered by Nishijima (1959) in an entirely different context. See also Nambu (1962).

† Unless one also requires that $\partial_\mu A_\mu = 0$.

of 5(79a) and 5(79b) we find that

$$\partial_\mu(\bar{\psi}\gamma_\mu\psi) = 0$$

This in turn ensures that

$$\partial_\mu A_\mu = 0 \qquad 5(81)$$

by taking the divergence of both sides of 5(80). The canonically conjugate momenta are $\pi_\psi = i\psi^\dagger$ and

$$\pi_i = -iF_{i4} = -E_i \qquad 5(82)$$

as in the free field case. Defining the Hamiltonian density by

$$\mathscr{H} = i\psi^\dagger\dot{\psi} - \mathbf{E}\,.\,\dot{\mathbf{A}} - \mathscr{L} \qquad 5(83)$$

we find, by repeating the steps followed in passing from 2(79) to 2(81),

$$\mathscr{H} = \mathscr{H}_{\text{Dirac}} + \mathscr{H}_{\text{Vector}} + \mathscr{H}_I \qquad 5(84)$$

where $\mathscr{H}_{\text{Vector}}$ is the free field Hamiltonian

$$\mathscr{H}_{\text{Vector}} = \tfrac{1}{2}\left[\mathbf{E}^2 + \mu^2\mathbf{A}^2 + (\textbf{curl}\,\mathbf{A})^2 + \frac{1}{\mu^2}(\text{div}\,\mathbf{E})^2\right] \qquad 5(85)$$

expressed in terms of the canonically conjugate fields **A** and $-\mathbf{E}$ as in 2(81), and where $\mathscr{H}_I$ is given by

$$\mathscr{H}_I = -\mathscr{L}_I - \frac{1}{2\mu^2}g^2(\bar{\psi}\gamma_4\psi)^2 \qquad 5(86)$$

Note that $\mathscr{H}_I$ exhibits a normal dependent term. The latter can be traced to the terms $\mathbf{E}\,.\,\nabla A_0 - \frac{1}{2}\mu^2 A_0^2$ appearing in 2(79). [See Problem 7.]

The simplest choice for the coupling of a neutral vector field A_μ with a neutral scalar field ϕ is of the form

$$g_1(\partial_\mu\phi)\phi A_\mu$$

but if A_μ and ϕ respectively represent the ρ^0 and π^0 the above coupling is forbidden by SU_2-invariance (see below). The coupling

$$ig_2[(\partial_\mu\phi^\dagger)\phi - \phi^\dagger(\partial_\mu\phi)]A_\mu \qquad 5(87)$$

representing the $\rho^0\pi^+\pi^-$ interaction is allowed, on the other hand. If we wish the ρ^0 to be coupled to a conserved current, we must add to 5(87) a term of the form $-g_2^2\phi^\dagger\phi A_\mu A_\mu$, as in the corresponding electrodynamic interaction 5(24).

We shall not consider the interactions of spin $\frac{3}{2}$ particles in any detail. We simply note that the simplest trilinear couplings of a spin $\frac{3}{2}$ field ψ_μ

to neutral pseudoscalar and vector fields ϕ and A_μ are

$$ih_1\bar{\psi}_\mu\gamma_5\psi_\mu\phi \qquad 5(88a)$$

and

$$ih_2\bar{\psi}_\mu\gamma_\lambda\psi_\mu A_\lambda \qquad 5(88b)$$

Other couplings like

$$h_3\bar{\psi}\psi_\mu A_\mu + \text{hermitian conjugate}$$

or

$$ih_4\bar{\psi}\gamma_5\psi_\mu\partial_\mu\phi + \text{hermitian conjugate}$$

can be used to describe the decay of a spin $\frac{3}{2}$ baryon into a spin $\frac{1}{2}$ baryon and neutral vector or pseudoscalar meson, respectively.

SU_2-invariant Couplings Let us extend our treatment of the π^0-nucleon interaction to include the three charge states of the π-meson. In the absence of electromagnetic couplings, the $\pi - N$ interaction conserves isotopic spin and we must therefore build isotopic spin or SU_2 symmetry into the interaction Lagrangian. Under infinitesimal SU_2 transformations the nucleon doublet*

$$N(x) = \begin{pmatrix} p(x) \\ n(x) \end{pmatrix} \qquad 5(89)$$

transforms according to the basic doublet representation

$$N(x) \to \left(1 + \frac{i}{2}\boldsymbol{\alpha} \cdot \boldsymbol{\tau}\right) N(x) \qquad 5(90)$$

where α_i ($i = 1, 2, 3$) are a set of three infinitesimal parameters and the τ_i ($i = 1, 2, 3$) are a set of three Pauli matrices satisfying

$$[\tau_i, \tau_j] = 2i\varepsilon_{ijk}\tau_k \qquad 5(91)$$

where ε_{ijk} is the 3-dimensional Levi–Civita tensor. The π-meson triplet

$$\boldsymbol{\phi} = (\phi_1, \phi_2, \phi_3)$$

* For convenience, we denote proton and neutron fields by $p(x)$ and $n(x)$ respectively. π^+, π^- and π^0 fields will be denoted by ϕ_{π^+}, ϕ_{π^-} and ϕ_{π^0} respectively. In the terminology of Section 3-2, ϕ_{π^0} is a hermitian field operator, while $\phi_{\pi^+} = \phi$ and $\phi_{\pi^-} = \phi^\dagger$ are hermitian conjugates of each other.

with

$$\phi_1 = \frac{1}{\sqrt{2}}(\phi_{\pi^+} + \phi_{\pi^-}) \tag{5(92a)}$$

$$\phi_2 = \frac{i}{\sqrt{2}}(\phi_{\pi^+} - \phi_{\pi^-}) \tag{5(92b)}$$

$$\phi_3 = \phi_{\pi^0} \tag{5(92c)}$$

transforms according to the adjoint representation of SU_2, that is, the traceless 2×2 matrix

$$\phi_{ij} = (\boldsymbol{\tau} . \boldsymbol{\phi})_{ij} = \begin{pmatrix} \phi_{\pi^0} & \sqrt{2}\phi_{\pi^+} \\ \sqrt{2}\phi_{\pi^-} & -\phi_{\pi^0} \end{pmatrix} \tag{5(93)}$$

transforms as a second-rank SU_2 tensor. An equivalent statement is that $\boldsymbol{\phi}$ transforms like $\bar{N}\boldsymbol{\tau}N$ or explicitly, using 5(90) and 5(91),

$$\phi_i \to \phi_i + \varepsilon_{ijk}\alpha_k\phi_j \tag{5(94)}$$

The transformation law 5(94) is also expressed as the statement that $\boldsymbol{\phi}$ transforms as a vector in 3-dimensional isotopic spin space. The simplest SU_2 invariant and Lorentz-invariant coupling is clearly

$$\mathscr{L}_I = iG\bar{N}_i\gamma_5 N_j\phi_{ij} \tag{5(95a)}$$

or equivalently

$$\mathscr{L}_I = iG\bar{N}\gamma_5\boldsymbol{\tau}N\boldsymbol{\phi} \tag{5(95b)}$$

An alternative Lagrangian density is provided by the derivative coupling

$$\mathscr{L}_I = if\bar{N}\gamma_\mu\gamma_5\boldsymbol{\tau}N\partial_\mu\boldsymbol{\phi} \tag{5(96)}$$

generalizing 5(68).

To recover the experimentally observed conservation law for isotopic spin we note the Lagrangian densities for the nucleon and π-meson fields

$$\mathscr{L}_N = -\bar{p}(\gamma . \partial + m)p - \bar{n}(\gamma . \partial + m)n \tag{5(97a)}$$

and

$$\mathscr{L}_\pi = -\tfrac{1}{2}\sum_{i=1}^{3}(\partial_\mu\phi_i\partial_\mu\phi_i + \mu^2\phi_i\phi_i) \tag{5(97b)}$$

are invariant under the SU_2 transformations 5(90) and 5(94). Since the coupling 5(95) or 5(96) is also invariant under the combined effect of 5(90) and 5(94), we can apply the remarks immediately following 2(132) to derive a conserved current. For the coupling 5(95), the conserved current is given by

$$\mathbf{j}_\mu = \frac{i}{2}\overline{N}\gamma_\mu\boldsymbol{\tau}N+(\partial_\mu\boldsymbol{\phi})\times\boldsymbol{\phi} \qquad 5(98)$$

and for the coupling 5(96) by

$$\mathbf{j}_\mu = \frac{i}{2}\overline{N}\gamma_\mu\boldsymbol{\tau}N+(\partial_\mu\boldsymbol{\phi})\times\boldsymbol{\phi}-if\,\overline{N}\gamma_\mu\gamma_5\boldsymbol{\tau}N\times\boldsymbol{\phi} \qquad 5(99)$$

In either case, the three components of the isotopic spin vector

$$\mathbf{I} = -i\int \mathbf{j}_4\,d^3x = \int [\tfrac{1}{2}N^\dagger\boldsymbol{\tau}N-\boldsymbol{\pi}\times\boldsymbol{\phi}]\,d^3x \qquad 5(100)$$

are constants of the motion where $\boldsymbol{\pi}$, the canonically conjugate field vector, is equal to either $\dot{\boldsymbol{\phi}}$ or $\dot{\boldsymbol{\phi}}+f\,\overline{N}\gamma_4\gamma_5\boldsymbol{\tau}N$, depending on whether the coupling is given by 5(95) or 5(96). Thus we recover isotopic spin conservation. Although we have derived this result on the basis of the classical discussion of Section 2-3, the argument is equally valid when $p(x)$, $n(x)$ and $\boldsymbol{\phi}(x)$ are quantum fields. The equal-time commutation relations are the usual ones

$$\{p(x), p^\dagger(x')\}_{t=t'} = \{n(x), n^\dagger(x')\}_{t=t'} = \delta^{(3)}(\mathbf{x}-\mathbf{x}')$$
$$[\phi_i(x), \pi_j(x')]_{t=t'} = i\delta_{ij}\delta^{(3)}(\mathbf{x}-\mathbf{x}') \qquad (i,j = 1,2,3) \qquad 5(101a)$$

etc.

in addition to which we require that*

$$\{p(x), n(x')\}_{t=t'} = \{p(x), n^\dagger(x')\}_{t=t'} = 0 \qquad 5(101b)$$

and that $p(x)$, $p^\dagger(x)$, $n(x)$ and $n^\dagger(x)$ commute with $\boldsymbol{\phi}(x)$ and $\boldsymbol{\pi}(x) = \dot{\boldsymbol{\phi}}(x)$ at equal times. Using the commutation relations 5(101a) and 5(101b) it is a simple matter to check that the components of the isotopic spin vector 5(100) satisfy the usual angular momentum commutation relations

$$[I_i, I_j] = i\varepsilon_{ijk}I_k \qquad 5(102)$$

The construction of SU_2 invariant couplings for other strongly interacting particles follows the same lines as 5(95). For example, the Σ-hyperon triplet, represented by the fields

$$\Sigma_1(x) = \frac{1}{\sqrt{2}}(\Sigma^+(x)+\Sigma^-(x)) \qquad 5(103a)$$

* See Problem 8.

$$\Sigma_2(x) = \frac{i}{\sqrt{2}}(\Sigma^+(x) - \Sigma^-(x)) \qquad 5(103b)$$

$$\Sigma_3(x) = \Sigma^0(x) \qquad 5(103c)$$

transforms according to the adjoint representation of SU_2

$$\Sigma_i \to \Sigma_i + \varepsilon_{ijk}\alpha_k\Sigma_j \qquad 5(104)$$

Hence a suitable SU_2-invariant $\Sigma - \pi$ coupling Lagrangian is given by

$$\mathscr{L}_I = G'\bar{\Sigma}_{ij}\gamma_5\Sigma_{ki}\phi_{jk} \qquad 5(105a)$$

$$= -G'\bar{\Sigma}_{ij}\gamma_5\Sigma_{jk}\phi_{ki} \qquad 5(105b)$$

where Σ_{ij} is the traceless 2×2 matrix

$$\Sigma_{ij} = (\boldsymbol{\tau} . \boldsymbol{\Sigma})_{ij} = \begin{pmatrix} \Sigma^0 & \sqrt{2}\Sigma^+ \\ \sqrt{2}\Sigma^- & -\Sigma^0 \end{pmatrix} \qquad 5(106)$$

analogous to 5(93)*. The equivalence of 5(105a) with 5(105b) is easily verified with the aid of the definition 5(106) and the wellknown properties of the τ-matrices. An alternate form for 5(105a) and 5(105b) is [see Problem 9]

$$\mathscr{L}_I = -2iG'\bar{\boldsymbol{\Sigma}}\gamma_5 \times \boldsymbol{\Sigma} . \boldsymbol{\phi} \qquad 5(107)$$

The addition of the coupling 5(105) and the free Σ Lagrangian

$$\mathscr{L}_\Sigma = -\bar{\Sigma}^+(\gamma . \partial + m)\Sigma^+ - \bar{\Sigma}^-(\gamma . \partial + m)\Sigma^- - \bar{\Sigma}^0(\gamma . \partial + m)\Sigma^0 \qquad 5(108)$$

to the Lagrangian for the interacting $\pi - N$ system results in the addition of the term

$$\mathbf{j}_\mu^\Sigma = \bar{\boldsymbol{\Sigma}}\gamma_\mu \times \boldsymbol{\Sigma} \qquad 5(109)$$

to the isotopic spin current 5(98) or 5(99). [See Problem 10.]

SU_2 invariance is only exact in the absence of electromagnetic couplings, as the latter introduce a physical distinction between the charged and neutral members of an SU_2 multiplet, thereby breaking the SU_2-symmetry. SU_2 remains an approximate symmetry, however, as electromagnetic couplings are only a small correction to the strong interactions.

* The adjoint matrix is

$$\bar{\Sigma}_{ij} = (\boldsymbol{\tau} . \bar{\boldsymbol{\Sigma}})_{ij} = \begin{pmatrix} \bar{\Sigma}^0 & \sqrt{2}\bar{\Sigma}^- \\ \sqrt{2}\bar{\Sigma}^+ & -\bar{\Sigma}^0 \end{pmatrix}$$

as is seen by applying 5(103a), 5(103b) and 5(103c). Note that the Σ^- hyperon differs from the anti-Σ^+ hyperon, so that we must distinguish $\bar{\Sigma}^+(x)$ from $\Sigma^-(x)$. Similarly $\bar{\Sigma}^- \neq \Sigma^+$.

Electromagnetic effects are also held to be responsible for the small mass difference within an SU_2 multiplet which are neglected in the SU_2-invariant limit, as, for example, in 5(97a), 5(97b) and 5(108).

Charge and Baryon Number Conservation In addition to isotopic spin, strong coupling schemes must conserve charge and baryon number which represent absolute conservation laws, valid for all interactions. To check charge conservation for the $\pi - N$ coupling 5(95) or 5(96), we subject all *charged* fields to a common phase transformation.

$$p \to p + i\alpha p \tag{5(110)}$$

$$\phi_{\pi^+} \to \phi_{\pi^+} + i\alpha\phi_{\pi^+} \tag{5(111a)}$$

$$\phi_{\pi^-} \to \phi_{\pi^-} - i\alpha\phi_{\pi^-} \tag{5(111b)}$$

where the phase is taken proportional to the charge. The Lagrangian densities $\mathscr{L}_N$ and $\mathscr{L}_\pi$ are clearly invariant under this variation; to check this invariance for $\mathscr{L}_\pi$ it suffices to rewrite 5(97b) in the form

$$\mathscr{L}_\pi = -\partial_\mu\phi_{\pi^-}\partial_\mu\phi_{\pi^+} - \mu^2\phi_{\pi^-}\phi_{\pi^+} - \tfrac{1}{2}\partial_\mu\phi_{\pi^0}\partial_\mu\phi_{\pi^0} - \tfrac{1}{2}\mu^2\phi_{\pi^0}\phi_{\pi^0} \tag{5(112)}$$

using 5(92a), 5(92b) and 5(92c). Moreover, the couplings 5(95) and 5(96) are also invariant under the phase transformation 5(110), 5(111a) and 5(111b). For example, 5(95) may be rewritten in the form

$$\mathscr{L}_I = iG(\sqrt{2}\bar{p}\gamma_5 n\phi_{\pi^+} + \sqrt{2}\bar{n}\gamma_5 p\phi_{\pi^-} + \bar{p}\gamma_5 p\phi_{\pi^0} - \bar{n}\gamma_5 n\phi_{\pi^0}) \tag{5(113)}$$

where the phase invariance is apparent. We deduce that the current 5(41) is conserved. If the coupling is the nonderivative form 5(95), the conserved current is given by

$$j_\mu = i\bar{p}\gamma_\mu p - i(\partial_\mu\phi_{\pi^+})\phi_{\pi^-} + i(\partial_\mu\phi_{\pi^-})\phi_{\pi^+} = \frac{i}{2}\bar{N}\gamma_\mu(1+\tau_3)N + (\partial_\mu\boldsymbol{\phi}\times\boldsymbol{\phi})_3 \tag{5(114)}$$

In case the coupling is of the derivative type 5(96), the conserved current is

$$j_\mu = \frac{i}{2}\bar{N}\gamma_\mu(1+\tau_3)N + (\partial_\mu\boldsymbol{\phi}\times\boldsymbol{\phi})_3 - if(\bar{N}\gamma_\mu\gamma_5\boldsymbol{\tau}N\times\boldsymbol{\phi})_3 \tag{5(115)}$$

but in either case the conserved charge is given by

$$Q = -i\int j_4\, d^3x = \int (p^\dagger p - \pi_1\phi_2 + \pi_2\phi_1)\, d^3x \qquad 5(116)$$

where the pion contribution, represented by the second and third terms, is of the same form as 2(138a).

The conservation of baryon number can be verified by applying a similar procedure. By subjecting the baryon fields $p(x)$ and $n(x)$ to a common phase transformation, one deduces from the invariance of the Lagrangian that the baryonic charge

$$N = \int (p^\dagger p + n^\dagger n)\, d^3x \qquad 5(117)$$

is conserved. Notice that from 5(100), 5(116) and 5(117) we derive the well-known relation

$$Q = I_3 + \frac{N}{2} \qquad 5(118)$$

Using the commutation rules 5(101a) and 5(101b) one can verify that the charge 5(116) is the generator of the infinitesimal phase transformations 5(110), 5(111a) and 5(111b). Thus, for example,

$$\delta p = i\alpha p = \frac{1}{i}[p, -\alpha Q] \qquad 5(119)$$

Note that as Q is conserved, its time label may be taken equal to that of p in the commutator; this allows us to apply the equal-time commutation relations 5(101a) and 5(101b). From 5(119) and its adjoint we infer, by the same reasoning as in 3(48a) and 3(48b), that $p(p^\dagger)$ destroys (creates) one unit of charge. This is an important result. If $p(x)$ is a free field, the corresponding statement may be deduced directly from the expansion of the field in terms of creation and destruction operators, but the latter are no longer available in the interacting case.

Similarly, N given by 5(117) is the generator of infinitesimal phase transformations on p and n. By the same argument it follows that p and n ($p^\dagger$ and $n^\dagger$) destroy (create) one unit of baryonic charge.

The $\rho\pi N$ Interaction—Yang–Mills Coupling We now turn to the interaction of the ρ-triplet, ρ^+, ρ^-, and ρ^0 with nucleons and pions. We represent the ρ-triplet by the massive vector isovector field*

$$\mathbf{A}^\mu = (A_1^\mu, A_2^\mu, A_3^\mu)$$

* As in the π-meson case, $A^\mu_{\rho^+}$ and $A^\mu_{\rho^-}$ denote charged vector meson fields for ρ^+ and ρ^-. In terms of the notation of Section 4-1, $A^\mu_{\rho^+} = A^\mu$ and $A^\mu_{\rho^-} = A^{\dagger\mu}$. $A^\mu_{\rho o}$ is a neutral vector meson field.

with

$$A_1^\mu = \frac{1}{\sqrt{2}}(A_{\rho^+}^\mu + A_{\rho^-}^\mu) \qquad 5(120a)$$

$$A_2^\mu = \frac{i}{\sqrt{2}}(A_{\rho^+}^\mu - A_{\rho^-}^\mu) \qquad 5(120b)$$

$$A_3^\mu = A_{\rho^0} \qquad 5(120c)$$

and the free Lagrangian density

$$\mathscr{L}_\rho = -\sum_{i=1}^{3}\left(\tfrac{1}{4}F_i^{\mu\nu}F_i^{\mu\nu} + \frac{\mu^2}{2}A_i^\mu A_i^\mu\right) \qquad 5(121)$$

where we assume equal masses for the $\rho^\pm$ and ρ^0 in the SU_2-invariant limit.

In the simplest case of neutral vector mesons coupled to nucleons, the simplest possible coupling 5(76) automatically ensures that the current appearing on the right-hand side of 5(80) is conserved. We shall attempt to preserve this feature for the coupling of the ρ-triplet to nucleons and pions. As in electrodynamics, the coupling of $\mathbf{A}_\mu$ to a conserved current—in this case the conserved isotopic spin current $\mathbf{j}_\mu$—would ensure the universality of vector meson couplings, as first proposed by Sakurai (1960). In the case at hand, it would guarantee that nucleons and pions are coupled to ρ's with the same strength, which appears to be consistent with the experimental data [Gell-Mann (1962)].

The problem of constructing the coupling of a vector-isovector field to the isotopic spin current has been solved by Yang and Mills [Yang (1954)]. The chief difficulty to be faced is the fact that the current $\mathbf{j}$ must have a contribution coming from the vector field itself, since the latter carries unit isotopic spin. To recover the Yang–Mills coupling scheme we proceed as follows. Setting

$$\mathbf{j}^\mu = \mathbf{j}_{(N)}^\mu + \mathbf{j}_{(\pi)}^\mu + \mathbf{j}_{(\rho)}^\mu \qquad 5(122)$$

where $\mathbf{j}_{(\rho)}^\mu$ is the isotopic spin current of the ρ meson, we wish to assign to the ρ-field an equation of motion of the form

$$-\partial_\nu \mathbf{F}_{\nu\mu} = -\mu^2 \mathbf{A}_\mu + 2g\mathbf{j}_\mu \qquad 5(123)$$

in analogy with 5(80). Here $\mathbf{F}_{\mu\nu}$ is the field strength tensor

$$\mathbf{F}_{\mu\nu} = \partial_\mu \mathbf{A}_\nu - \partial_\nu \mathbf{A}_\mu \qquad 5(124)$$

Actually, 5(123) will require modification in one respect, as we shall see.

We seek to determine the explicit form of $\mathbf{j}^\mu$ and of $\mathscr{L}_I$ such that $\mathbf{A}_\mu$ satisfies an equation of the form 5(123). Now the isotopic spin current of the $\mathbf{A}_\mu$ field is obtained by applying 2(134) or

$$\mathbf{j}^\mu_{(\rho)} = -\frac{\partial \mathscr{L}}{\partial(\partial^\mu \mathbf{A}^\nu)} \times \mathbf{A}^\nu \tag{5(125)}$$

where we have used the fact that $\mathbf{A}^\mu$ transforms according to the adjoint representation of SU_2

$$\mathbf{A}^\nu \rightarrow \mathbf{A}^\nu + \mathbf{A}^\nu \times \boldsymbol{\alpha} \tag{5(126)}$$

With the aid of 5(125) we identify the contribution of the free vector field Lagrangian 5(121) to $\mathbf{j}^\mu_{(\rho)}$ as

$$(\mathbf{F}^{\mu\nu} \times \mathbf{A}^\nu) \tag{5(127)}$$

This term (multiplied by $2g$) will therefore appear on the right-hand side of 5(123). Since, by the general form of the Euler–Lagrange equations, the right-hand side of 5(123) must equal $\partial\mathscr{L}/\partial\mathbf{A}_\mu$, we can identify a piece of the interaction Lagrangian density as

$$g(\mathbf{F}^{\mu\nu} \times \mathbf{A}^\nu) \,.\, \mathbf{A}^\mu = g\mathbf{F}^{\mu\nu} \,.\, (\mathbf{A}^\nu \times \mathbf{A}^\mu) \tag{5(128)}$$

We now note that this term depends explicitly on $\partial_\nu \mathbf{A}_\mu$. It will therefore generate an additional contribution to the current $\mathbf{j}^\mu_{(\rho)}$, which by 5(125) is seen to be

$$-2g(\mathbf{A}^\nu \times \mathbf{A}^\mu) \times \mathbf{A}^\nu \tag{5(129)}$$

But by the same argument as before this implies that the interaction Lagrangian contains the additional piece

$$-g^2(\mathbf{A}^\nu \times \mathbf{A}^\mu) \times \mathbf{A}^\nu \,.\, \mathbf{A}^\mu = -g^2(\mathbf{A}^\nu \times \mathbf{A}^\mu) \,.\, (\mathbf{A}^\nu \times \mathbf{A}^\mu) \tag{5(130)}$$

whose derivative with respect to $\mathbf{A}^\mu$ is easily seen to yield 5(129) times $2g$. Note that 5(128) and 5(130) are *self-coupling* terms. They represent interactions of the ρ-triplet with itself.

Pausing to collect our results so far, we have obtained the total ρ-contribution to the conserved isotopic spin current in the form

$$\mathbf{j}^\mu_{(\rho)} = \mathbf{f}^{\mu\nu} \times \mathbf{A}^\nu \tag{5(131)}$$

with

$$\mathbf{f}^{\mu\nu} = \mathbf{F}^{\mu\nu} + 2g\mathbf{A}^\mu \times \mathbf{A}^\nu \tag{5(132)}$$

The total self-coupling Lagrangian for the ρ is*

$$\mathscr{L}_I^{(\rho)} = g\mathbf{F}^{\mu\nu} \,.\, (\mathbf{A}^\nu \times \mathbf{A}^\mu) - g^2(\mathbf{A}^\nu \times \mathbf{A}^\mu) \,.\, (\mathbf{A}^\nu \times \mathbf{A}^\mu) \tag{5(133)}$$

* The sum of 5(121) and 5(133) is frequently written in the simple form $-\frac{1}{4}\mathbf{f}_{\mu\nu}\mathbf{f}_{\mu\nu}$.

We now observe that since the first term in 5(133) involves the derivatives $\partial_\nu \mathbf{A}_\mu$ it will also modify the left-hand side of the Euler–Lagrange equation 5(123). The true equation of motion for $\mathbf{A}_\mu$ is therefore not 5(123) but

$$-\partial_\nu[\mathbf{F}_{\nu\mu} - 2g\mathbf{A}_\mu \times \mathbf{A}_\nu] = -\mu^2 \mathbf{A}_\mu + 2g\mathbf{j}_\mu$$

or

$$\partial_\nu \mathbf{f}_{\nu\mu} - \mu^2 \mathbf{A}_\mu = -2g\mathbf{j}_\mu \qquad 5(134)$$

Because we have chosen $\mathbf{j}_\mu$ to be a *conserved* current, we deduce from 5(134) that $\mathbf{A}_\mu$ satisfies the subsidiary condition

$$\partial_\mu \mathbf{A}_\mu = 0 \qquad 5(135)$$

$\mathbf{f}_{\nu\mu}$ being antisymmetric.

The current $\mathbf{j}_\mu$ is given by 5(122) as the sum of 5(131) and of the nucleon and pion contributions. The nucleon part $\mathbf{j}^\mu_{(N)}$, is given by $i\bar{N}\gamma^\mu \boldsymbol{\tau} N/2$ as in 5(98). However, the π contribution $\mathbf{j}^\mu_{(\pi)}$ differs from its expression in 5(98) or 5(99) by an additional term proportional to $\mathbf{A}_\mu$. Indeed, by the same argument as above, the appearance on the right-hand side of 5(134) of the term

$$(\partial_\mu \boldsymbol{\phi}) \times \boldsymbol{\phi}$$

in $\mathbf{j}^\mu_{(\pi)}$ must correspond to a contribution of the form

$$2g(\partial^\mu \boldsymbol{\phi}) \times \boldsymbol{\phi} \,.\, \mathbf{A}^\mu \qquad 5(136)$$

in the Lagrangian density. Since 5(136) depends explicitly on $\partial_\mu \boldsymbol{\phi}$, it gives rise to the additional term

$$-2g(\boldsymbol{\phi} \times \mathbf{A}^\mu) \times \boldsymbol{\phi}$$

in the expression for $\mathbf{j}^\mu_{(\pi)}$. This in turn gives rise to a further contribution of the form

$$-g^2(\boldsymbol{\phi} \times \mathbf{A}^\mu) \times \boldsymbol{\phi} \,.\, \mathbf{A}^\mu$$

in the Lagrangian density. Thus the total $\mathbf{j}^\mu_{(\pi)}$ is given by

$$\mathbf{j}^\mu_{(\pi)} = (\partial^\mu \boldsymbol{\phi}) \times \boldsymbol{\phi} - 2g(\boldsymbol{\phi} \times \mathbf{A}^\mu) \times \boldsymbol{\phi} \qquad 5(137)$$

(assuming nonderivative πN coupling as in 5(95)) and the total $\pi\rho$ coupling Lagrangian is given by

$$\mathscr{L}_I^{(\pi\rho)} = 2g(\partial^\mu \boldsymbol{\phi}) \times \boldsymbol{\phi} \,.\, \mathbf{A}_\mu - g^2(\boldsymbol{\phi} \times \mathbf{A}^\mu) \times \boldsymbol{\phi} \,.\, \mathbf{A}^\mu \qquad 5(138)$$

The additional term in the current 5(137) is analogous to the additional A_μ-dependent term which appears in 5(26) for the case of minimal electromagnetic coupling.

Thus a rather complicated scheme is necessary if we wish the ρ to be coupled to a conserved current. In return we are assured that the coupling is universal, as mentioned earlier, and that $\mathbf{A}_\mu$ satisfies the subsidiary condition 5(135).

SU_3-Invariant Couplings To construct model interactions for strong couplings which are invariant under SU_3 transformations*, we begin by recalling that the fundamental triplet representation

$$b = \begin{pmatrix} b_1 \\ b_2 \\ b_3 \end{pmatrix} \tag{5(139)}$$

analogous to 5(89), transforms as

$$b \to \left(1 + i \sum_{r=1}^{8} \alpha_r \lambda_r\right) b \tag{5(140)}$$

under infinitesimal SU_3 transformations. Here the λ_r $(r = 1, \dots 8)$ are the following set of traceless hermitian 3×3 matrices, [Gell-Mann (1961)]

$$\lambda_1 = \begin{pmatrix} 0 & 1 & 0 \\ 1 & 0 & 0 \\ 0 & 0 & 0 \end{pmatrix} \qquad \lambda_2 = \begin{pmatrix} 0 & -i & 0 \\ i & 0 & 0 \\ 0 & 0 & 0 \end{pmatrix} \qquad \lambda_3 = \begin{pmatrix} 1 & 0 & 0 \\ 0 & -1 & 0 \\ 0 & 0 & 0 \end{pmatrix}$$

$$\lambda_4 = \begin{pmatrix} 0 & 0 & 1 \\ 0 & 0 & 0 \\ 1 & 0 & 0 \end{pmatrix} \qquad \lambda_5 = \begin{pmatrix} 0 & 0 & -i \\ 0 & 0 & 0 \\ i & 0 & 0 \end{pmatrix} \qquad \lambda_6 = \begin{pmatrix} 0 & 0 & 0 \\ 0 & 0 & 1 \\ 0 & 1 & 0 \end{pmatrix}$$

$$\lambda_7 = \begin{pmatrix} 0 & 0 & 0 \\ 0 & 0 & -i \\ 0 & i & 0 \end{pmatrix} \qquad \lambda_8 = \frac{1}{\sqrt{3}} \begin{pmatrix} 1 & 0 & 0 \\ 0 & 1 & 0 \\ 0 & 0 & -2 \end{pmatrix} \tag{5(141)}$$

satisfying the commutation rules

$$[\lambda_r, \lambda_s] = 2i f_{rst} \lambda_t \tag{5(142)}$$

with f_{rst} real and totally antisymmetric, like the Levi–Civita tensor ε_{ijk} which appears in the corresponding formula 5(91) for SU_2. Explicitly,

* For details regarding SU_3 symmetry, we refer the reader to M. Gell-Mann and Y. Ne'eman, *The Eightfold Way*, Benjamin, New York, 1964.

the f_{rst} are given by [Gell-Mann (1961)]

$$f_{123} = 1 \qquad f_{147} = \tfrac{1}{2} \qquad f_{156} = -\tfrac{1}{2}$$

$$f_{246} = \tfrac{1}{2} \qquad f_{257} = \tfrac{1}{2} \qquad f_{345} = \tfrac{1}{2} \qquad f_{367} = -\tfrac{1}{2} \qquad 5(143)$$

$$f_{458} = \frac{\sqrt{3}}{2} \qquad f_{678} = \frac{\sqrt{3}}{2}$$

All other components f_{rst} that cannot be obtained from 5(143) by a permutation of (r, s, t) are equal to zero.

So far, no particles associated with the triplet representation 5(139) have been discovered.† The octet of pseudoscalar mesons (π, η, K, $\bar{K}$) represented by the eight hermitian fields

$$M_1 = \frac{1}{\sqrt{2}}(\phi_{\pi^+} + \phi_{\pi^-}) \qquad M_5 = \frac{i}{\sqrt{2}}(\phi_{K^+} - \phi_{K^-})$$

$$M_2 = \frac{i}{\sqrt{2}}(\phi_{\pi^+} - \phi_{\pi^-}) \qquad M_6 = \frac{1}{\sqrt{2}}(\phi_{K^0} + \phi_{\bar{K}^0})$$

$$M_3 = \phi_{\pi^0} \qquad M_7 = \frac{i}{\sqrt{2}}(\phi_{K^0} - \phi_{\bar{K}^0})$$

$$M_4 = \frac{1}{\sqrt{2}}(\phi_{K^+} + \phi_{K^-}) \qquad M_8 = \phi_\eta$$

5(142)

transforms according to the adjoint representation of SU_3 [Gell-Mann (1961), Ne'eman (1961)]. Thus the traceless 3×3 matrix

$$M_{ij} = \sum_{r=1}^{8} (\lambda^r)_{ij} M_r = \begin{pmatrix} \phi_{\pi^0} + \frac{1}{\sqrt{3}}\phi_\eta & \sqrt{2}\phi_{\pi^+} & \sqrt{2}\phi_{K^+} \\ \sqrt{2}\phi_{\pi^-} & -\phi_{\pi^0} + \frac{1}{\sqrt{3}}\phi_\eta & \sqrt{2}\phi_{K^0} \\ \sqrt{2}\phi_{K^-} & \sqrt{2}\phi_{\bar{K}^0} & -\frac{2}{\sqrt{3}}\phi_\eta \end{pmatrix}$$

$(i, j = 1, 2, 3)$

5(143)

† The original version of SU_3 symmetry [Ikeda (1960)] identified (p, n, λ) as the basic triplet representation, in accordance with the Sakata model [Sakata (1956)]. This model later gave way to the more successful 'eightfold way' described in the text.

transforms as a second-rank tensor under 5(140) or, equivalently, M_r transforms as $\bar{b}\lambda_r b$. Explicitly

$$M_r \to M_r + f_{rst}\alpha_t M_s \tag{5(144)}$$

in analogy with 5(94). The baryon octet (Σ, Λ, N, Ξ) transforms in the same way and has a similar 3×3 matrix representation

$$B_{ij} = \sum_{r=1}^{8} (\lambda^r)_{ij} B_r = \begin{pmatrix} \Sigma^0 + \frac{1}{\sqrt{3}}\Lambda & \sqrt{2}\Sigma^+ & \sqrt{2}p \\ \sqrt{2}\Sigma^- & -\Sigma^0 + \frac{1}{\sqrt{3}}\Lambda & \sqrt{2}n \\ \sqrt{2}\Xi^- & \sqrt{2}\Xi^0 & -\frac{2}{\sqrt{3}}\Lambda \end{pmatrix} \tag{5(145)}$$

$$(i, j = 1, 2, 3)$$

The simplest trilinear baryon-meson couplings which are invariant under SU_3 are then given by [Okubo (1962)]

$$\mathscr{L}_I = ig\bar{B}_{ij}\gamma_5 B_{ki} M_{jk} \tag{5(146a)}$$

and

$$\mathscr{L}_{I'} = ig\bar{B}_{ij}\gamma_5 B_{jk} M_{ki} \tag{5(146b)}$$

in analogy with 5(105a) and 5(105b). Here $\bar{B}_{ij}$ denotes the 3×3 matrix

$$\bar{B}_{ij} = \sum_{r=1}^{8} (\lambda^r)_{ij} \bar{B}_r = \begin{pmatrix} \bar{\Sigma}^0 + \frac{1}{\sqrt{3}}\bar{\Lambda} & \sqrt{2}\bar{\Sigma}^- & \sqrt{2}\bar{\Xi}^- \\ \sqrt{2}\bar{\Sigma}^+ & -\bar{\Sigma}^0 + \frac{1}{\sqrt{3}}\bar{\Lambda} & \sqrt{2}\bar{\Xi}^0 \\ \sqrt{2}\bar{p} & \sqrt{2}\bar{n} & -\frac{2}{\sqrt{3}}\bar{\Lambda} \end{pmatrix} \tag{5(147)}$$

Note that in the SU_3 case the two couplings 5(146a) and 5(146b) are no longer equivalent but represent two distinct possibilities for the baryon-meson interaction [see Problem 11]. The linear combinations $\mathscr{L}_I + \mathscr{L}_{I'}$ and $\mathscr{L}_I - \mathscr{L}_{I'}$ are known as D- and F-type couplings respectively [Gell-Mann (1961)].

SU_3-invariant couplings for the nonet of vector mesons ($\rho, \omega, K^*, \bar{K}^*, \varphi$) can be constructed in a similar way. For the generalization of the Yang–Mills coupling scheme to SU_3 we refer the reader to Ne'eman's 1961 paper [see also Glashow (1961)].

In the exact symmetry limit which we are considering here, the masses of the baryons within each multiplet are assumed to be identical and the invariance of the total Lagrangian under 5(144) and the corresponding transformation for B_r yields an octet of conserved 'unitary spin' currents, namely

$$J_t^\mu = \bar{B}_r \gamma^\mu f_{rst} B_s + (\partial^\mu M_r) f_{rst} M_t \qquad (t = 1 \ldots 8) \tag{5(148)}$$

Experimentally, of course, SU_3 symmetry is badly broken by the large mass differences within the baryon and meson octets, and symmetry-breaking terms transforming like the 8th component of unitary spin must be added to 5(146a) and 5(146b) to take this into account [Gell-Mann (1961), Okubo (1962)]. Of the eight currents 5(148), only J_i^μ $(i = 1, 2, 3)$ and J_8^μ remain conserved once the symmetry-breaking interaction is switched on. Using 5(143) and setting $\mathbf{J}^\mu = (J_1^\mu, J_2^\mu, J_3^\mu)$ we find, by straightforward calculation

$$\mathbf{J}^\mu = \bar{\mathbf{\Sigma}} \gamma^\mu \times \mathbf{\Sigma} + \frac{i}{2} \bar{N} \gamma^\mu \boldsymbol{\tau} N + \frac{i}{2} \bar{\Xi} \gamma^\mu \boldsymbol{\tau} \Xi + (\partial^\mu \boldsymbol{\phi}_\pi) \times \boldsymbol{\phi}_\pi + \frac{i}{2} [(\partial^\mu \phi_K^\dagger) \boldsymbol{\tau} \phi_K - (\partial^\mu \phi_K^T) \boldsymbol{\tau} \phi_K^*] \tag{5(149)}$$

and

$$\frac{2}{\sqrt{3}} J_8^\mu = i \bar{N} \gamma^\mu N - i \bar{\Xi} \gamma^\mu \Xi + i[(\partial^\mu \phi_K^\dagger) \phi_K - (\partial^\mu \phi_K^T) \phi_K^*] \tag{5(150)}$$

where we have used the notation

$$N = \begin{pmatrix} p \\ n \end{pmatrix} \qquad \phi_K = \begin{pmatrix} \phi_{K^+} \\ \phi_{K^0} \end{pmatrix} \qquad \Xi = \begin{pmatrix} \Xi^0 \\ \Xi^- \end{pmatrix} \tag{5(151)}$$

The currents $\mathbf{J}^\mu$ and $(2/\sqrt{3}) J_8^\mu$ are just the total isotopic spin and hypercharge currents for the interacting baryon meson system.

General Remarks At present the detailed form of the strong interaction is unknown, and the couplings considered here must be viewed simply as models. It should also be borne in mind that the observed strongly interacting particles might well be bound states or 'excitations' either of more fundamental, as yet undiscovered, particles associated with the triplet representation of SU_3, or bound states of each other, as in the so-called 'bootstrap' hypothesis. If so, the procedure of associating to each observed particle a separate quantum field is rather doubtful. On the other hand, attempts to actually exhibit certain particles as bound states of others, (i.e. π's as $N\bar{N}$ bound states), within the framework of quantum field theory have met with considerable technical and, in some cases, fundamental difficulties. One such case, featuring a self-interacting Dirac field with four-fermion coupling:

$$\mathscr{L} = -\bar{\psi}(\gamma\,.\,\partial+m)\psi-g_0\bar{\psi}\gamma_5\psi\bar{\psi}\gamma_5\psi \qquad 5(152)$$

will be considered in Section 9-5. A more complicated theory of this general type has been extensively studied by Heisenberg and his collaborators*.

Finally we note that we have restricted our attention to *local* couplings, that is, couplings in which all field operators appearing in the interaction Lagrangian density refer to the same point in space time. Possibly, the correct description of particle interactions involves a certain degree of nonlocality. An example of a nonlocal coupling is

$$\mathscr{L}_I(x) = \int d^4x'\, d^4x''\, d^4x'''\bar{\psi}(x')\gamma_5\psi(x'')\phi(x''')F(x-x', x-x'', x-x''') \quad 5(153)$$

in which the interaction is ‘ smeared ’ over a finite space–time region by means of some function F. For an account of the special difficulties that are associated with such theories, see Kristensen (1952), Bloch (1952), Chrétien (1954), Arnous (1959), (1960), and O'Raifeartaigh (1960a, b).

Weak Couplings In the case of weak interactions, reasonable confidence can be placed in the results of lowest order perturbation theory and these can be confronted with experiment. As a result, the form of the weak coupling Lagrangian is now fairly well determined, at least for the purely leptonic and semileptonic processes. The simplest weak process is μ-decay,

$$\mu^- \rightarrow e^- + \bar{\nu}_e + \nu_\mu \qquad 5(154)$$

This is a purely leptonic process, unaffected by the strong interactions. The experimental data for 5(154) are well described by a local four-fermion interaction of the form

$$\mathscr{L}_I = \frac{1}{\sqrt{2}}G[\bar{\nu}_\mu\gamma_\lambda(1+\gamma_5)\mu]\,[\bar{e}\gamma_\lambda(1+\gamma_5)\nu_e]+\text{hermitian conjugate} \qquad 5(155)$$

where the coupling constant G has the dimension of length squared. This interaction violates both parity and charge conjugation invariance. For

* See for example, Dürr (1965). This paper contains references to earlier work.

example, under a parity transformation, represented by simultaneous transformations of the type 3(178) on the four Dirac field operators e, ν_e, ν_μ and μ, the interaction Lagrangian 5(155) goes over into

$$\frac{1}{\sqrt{2}}G[\bar{\nu}_\mu\gamma_\lambda(1-\gamma_5)\mu][\bar{e}\gamma_\lambda(1-\gamma_5)\nu_e]+\text{hermitian conjugate}$$

However, 5(155) remains invariant under the combined effect of space reflection and charge-conjugation [see Problem 12].

The general form of the interaction Lagrangian for strangeness-conserving semileptonic processes such as

$$\beta\text{-decay:} \quad n \to p+e^-+\bar{\nu}_e \qquad 5(156a)$$

$$\pi\text{-decay:} \quad \pi^- \to \begin{cases} \mu^-+\bar{\nu}_\mu \\ e^-+\bar{\nu}_e \end{cases} \qquad 5(156b)$$

$$\mu\text{-capture:} \ \mu^-+p \to n+\nu_\mu \qquad 5(156c)$$

is now rather well established.* The coupling is of the current-current form

$$\mathscr{L}_I = G\cos\theta\frac{1}{\sqrt{2}}h_\lambda l_\lambda+\text{hermitian conjugate} \qquad 5(157)$$

where G is the weak coupling constant appearing in 5(155), $\theta = 0{\cdot}26$ is the so-called Cabibbo angle [Cabibbo (1963)], l_λ is the lepton current

$$l_\lambda = i\bar{e}\gamma_\lambda(1+\gamma_5)\nu_e+i\bar{\mu}\gamma_\lambda(1+\gamma_5)\nu_\mu \qquad 5(158)$$

and h^λ is the weak current of the hadrons. The latter is composed of a vector part J_+^λ and of an axial vector (or pseudovector) part A_+^λ:

$$h^\lambda = J_+^\lambda + A_+^\lambda \qquad 5(159)$$

as originally proposed by Feynman and Gell-Mann [Feynman (1958)] and by Sudarshan and Marshak [Sudarshan (1958)].

The vector part J_+^λ has been identified with the component

$$J_+^\lambda = J_1^\lambda+iJ_2^\lambda \qquad 5(160a)$$

of the conserved isotopic spin current of the strong interactions [Feynman (1958)]. The component 5(160a) carries the isotopic spin selection rule $\Delta I_3 = +1$. Accordingly, it has non zero matrix elements corresponding

* For a detailed account see, for example, R. E. Marshak, Riazuddin and C. Ryan, *Weak Interactions and Elementary Particles*, Wiley, New York, to be published.

to β-decay. The component

$$J_-^\lambda = J_1^\lambda - iJ_2^\lambda \qquad 5(160b)$$

appears in the hermitian conjugate term in 5(157); since 5(160b) carries the selection rule $\Delta I_3 = -1$ it will have nonzero matrix elements for μ-capture. Note that if the strong interaction Lagrangian is identified as a *nonderivative* Yukawa coupling among the observed baryons (as in 5(146a) and 5(146b)), then the explicit form of the current $\mathbf{J}^\lambda$ is known; it is simply 5(149). However, there is no compelling reason to suppose that the strong interaction has such a simple form, and in any case, the observed baryons may not be 'fundamental'. Thus the tendency has been to avoid the explicit construction of J_i^λ in terms of particle fields but to rely more on *general* properties of the current. In Chapter 8 we shall see how useful, quantitative results can be derived without recourse to the explicit form of the current.

The axial vector current A_+^λ in 5(159) is also the component

$$A_+^\lambda = A_1^\lambda + iA_2^\lambda \qquad 5(161)$$

of an isotopic vector $\mathbf{A}^\lambda$. This component carries $\Delta I_3 = 1$ and possesses nonzero matrix elements both for β-decay and for π-decay. In contrast to the vector current, however, the axial vector current is *not* conserved. Instead, its 4-divergence is usually assumed to be proportional to the π^--field operator

$$\partial_\lambda A_+^\lambda(x) \sim \phi_{\pi^-}(x) \qquad 5(162a)$$

The exact relation will be stated and applied in Chapter 8. For the component

$$A_-^\lambda = A_1^\lambda - iA_2^\lambda$$

which enters into the hermitian conjugate term in 5(157), the relation 5(162a) is replaced by

$$\partial_\lambda A_-^\lambda(x) \sim \phi_{\pi^+}(x) \qquad 5(162b)$$

The process of π-decay, 5(156b), can also be described in terms of a phenomenological derivative coupling Lagrangian

$$\mathscr{L}_I = ig_\pi[\bar{\mu}\gamma_\lambda(1+\gamma_5)\nu_\mu + \bar{e}\gamma_\lambda(1+\gamma_5)\nu_e]\partial_\lambda\phi_{\pi^-} + \text{hermitian conjugate} \qquad 5(163)$$

in which the pion field is coupled directly to the lepton current. In Section 6-3 we shall use 5(163) to compute the π-decay lifetime. The more fundamental treatment based on 5(157) will be applied in Section 8-2.

Strangeness-violating semileptonic processes such as

$$K^- \to \mu^- + \bar{\nu}_\mu$$

$$\Lambda \to p + e^- + \bar{\nu}_e$$

are presently described in terms of a current–current coupling of the form

$$\mathscr{L}_I = G \sin\theta \frac{1}{\sqrt{2}} h^\lambda_{\Delta S} l_\lambda \qquad 5(164a)$$

where the strangeness-changing hadronic current is of the form

$$h^\lambda_{\Delta S} = J^\lambda_4 + iJ^\lambda_5 + A^\lambda_4 + iA^\lambda_5 \qquad 5(164b)$$

Here J^λ_4 and J^λ_5 belong to the same SU_3 octet as J^λ_i $(i = 1, 2, 3)$ while A^λ_4 and A^λ_5 belong to the same octet as A^λ_i $(i = 1, 2, 3)$.

5-4 Discrete symmetries

$\mathscr{T}$, $\mathscr{C}$, $\mathscr{P}$ Transformations for Interacting Fields As a general rule we formally assign to interacting fields the same $\mathscr{P}$, $\mathscr{C}$, and $\mathscr{T}$ transformation laws as for free fields. This assures the $\mathscr{P}$, $\mathscr{C}$, and $\mathscr{T}$ invariance of the equal-time commutation relations and of the free part of the equations of motion. One must then verify whether or not the interaction terms remain invariant. For both strong and electromagnetic interactions, the couplings must be such as to conserve $\mathscr{P}$, $\mathscr{C}$ and $\mathscr{T}$.

As an example, let us consider the π^0-nucleon interaction 5(62). We define the parity transformation for ψ and ϕ to be

$$\mathscr{P}\psi(\mathbf{x}, t)\mathscr{P}^{-1} = \gamma_4\psi(-\mathbf{x}, t) \qquad 5(165a)$$

and

$$\mathscr{P}\phi(\mathbf{x}, t)\mathscr{P}^{-1} = -\phi(-\mathbf{x}, t) \qquad 5(165b)$$

as in the free field case. Note the choice of sign in 5(165b) which is dictated by the fact that, in keeping with the observed negative parity of the π relative to the nucleon, we have coupled ϕ to the pseudoscalar density $\bar{\psi}\gamma_5\psi$. The latter transforms according to

$$\begin{aligned}\mathscr{P}\bar{\psi}(\mathbf{x}, t)\gamma_5\psi(\mathbf{x}, t)\mathscr{P}^{-1} &= \mathscr{P}\bar{\psi}(\mathbf{x}, t)\mathscr{P}^{-1}\gamma_5\mathscr{P}\psi(\mathbf{x}, t)\mathscr{P}^{-1} \\ &= \bar{\psi}(-\mathbf{x}, t)\gamma_4\gamma_5\gamma_4\psi(-\mathbf{x}, t) \\ &= -\bar{\psi}(-\mathbf{x}, t)\gamma_5\psi(-\mathbf{x}, t) \qquad 5(166)\end{aligned}$$

With the definitions 5(165a) and 5(165b) for the parity transformation, both the equal-time commutation relations, which are formally the same as for free fields, and the equations of motion 5(63a), 5(63b) and 5(64) are invariant under application of $\mathscr{P}$. For example, 5(64) is transformed into the equivalent equation

$$(\square - \mu^2)\phi(-\mathbf{x}, t) = -iG\bar{\psi}(-\mathbf{x}, t)\gamma_5\psi(-\mathbf{x}, t)$$

To recover the original equation of motion, it suffices to relabel $-\mathbf{x} \to \mathbf{x}$. We emphasize that the presence of the interaction resolves the ambiguity of sign in the definition of $\mathscr{P}$ for the Klein–Gordon field. The parity transformation is a symmetry operation for one choice of sign only.

The time reversal transformation for ψ is defined by 3(185), i.e.

$$\mathscr{T}\psi(\mathbf{x}, t)\mathscr{T}^{-1} = T^{-1}\psi(\mathbf{x}, -t) \qquad 5(167a)$$

and for ϕ by

$$\mathscr{T}\phi(\mathbf{x}, t)\mathscr{T}^{-1} = +\phi(\mathbf{x}, -t) \qquad 5(167b)$$

where the choice of the plus sign in 3(71) is dictated by the transformation property

$$\begin{aligned}\mathscr{T}i\bar{\psi}(\mathbf{x}, t)\gamma_5\psi(\mathbf{x}, t)\mathscr{T}^{-1} &= -i\mathscr{T}\psi^\dagger(\mathbf{x}, t)\mathscr{T}^{-1}\gamma_4^T\gamma_5^T\mathscr{T}\psi(\mathbf{x}, t)\mathscr{T}^{-1} \\ &= -i\psi^\dagger(\mathbf{x}, -t)T\gamma_4^T\gamma_5^T T^{-1}\psi(\mathbf{x}, -t) \\ &= -i\bar{\psi}(\mathbf{x}, -t)\gamma_5\psi(\mathbf{x}, -t) \qquad 5(168)\end{aligned}$$

of the pseudoscalar density to which ϕ is coupled. With the definitions 5(167a) and 5(167b) it is a simple matter to check that both the equal-time commutation relations and the equations of motion are $\mathscr{T}$-invariant. Finally, charge conjugation is defined by

$$\mathscr{C}\psi(\mathbf{x}, t)\mathscr{C}^{-1} = \psi^c = C\bar{\psi} \qquad 5(169a)$$

and

$$\mathscr{C}\phi(\mathbf{x}, t)\mathscr{C}^{-1} = \phi(\mathbf{x}, t) \qquad 5(169b)$$

since ϕ is assumed to be hermitian in the present case. To check $\mathscr{C}$ invariance, it is simplest to write the pseudoscalar density $i\bar{\psi}\gamma_5\psi$ in its antisymmetrized form*

$$\tfrac{1}{2}[\bar{\psi}, \gamma_5\psi] = \tfrac{1}{2}(\bar{\psi}\gamma_5\psi - \psi\gamma_5^T\bar{\psi}) \qquad 5(170)$$

Then, noting that

$$\mathscr{C}[\bar{\psi}, \gamma_5\psi]\mathscr{C}^{-1} = [\bar{\psi}, \gamma_5\psi] \qquad 5(171)$$

* See Chapter 3, Problem 12.

[Problem 13], we see that the equation of motion for ϕ

$$\Box\phi - \mu^2\phi = -\frac{iG}{2}[\bar{\psi}, \gamma_5\psi] \qquad 5(172)$$

transforms into itself under 5(169b) and 5(171). The $\mathscr{C}$-invariance of the remaining equations of motion and of the commutation relations is readily verified.

Let us turn to the minimal electromagnetic coupling 5(24) for a charged scalar field. The $\mathscr{P}$ and $\mathscr{T}$ transformations for ϕ are defined as above but since ϕ is now a *charged* field, we replace 5(169b) by

$$\mathscr{C}\phi(\mathbf{x}, t)\mathscr{C}^{-1} = \phi^\dagger(\mathbf{x}, t) \qquad 5(173)$$

The $\mathscr{P}$, $\mathscr{T}$ and $\mathscr{C}$ transformation laws for the electromagnetic field will be inferred by demanding the invariance of the coupled equations of motion 5(25a), 5(25b) and 5(25c).

To ensure the invariance of 5(25a), 5(25b) and 5(25c) under $\mathscr{P}$ we must define the parity transformation law for A_μ to be

$$\mathscr{P}\mathbf{A}(\mathbf{x}, t)\mathscr{P}^{-1} = -\mathbf{A}(-\mathbf{x}, t) \qquad 5(174a)$$

$$\mathscr{P}A_4(\mathbf{x}, t)\mathscr{P}^{-1} = +A_4(-\mathbf{x}, t) \qquad 5(174b)$$

With the above choice of signs, the current 5(26) to which A_μ is coupled transforms according to

$$\mathscr{P}\mathbf{j}(\mathbf{x}, t)\mathscr{P}^{-1} = -\mathbf{j}(-\mathbf{x}, t) \qquad 5(175a)$$

$$\mathscr{P}j_4(\mathbf{x}, t)\mathscr{P}^{-1} = +j_4(-\mathbf{x}, t) \qquad 5(175b)$$

and the equation of motion 5(25c) is left invariant. The invariance of the remaining equations 5(25a) and 5(25b) and of the commutation relations 5(31a), 5(31b) and 5(31c) is readily verified. As in the previous example, the interaction term imposes a special choice of signs in the definition of the parity transformation for the electromagnetic field; in the free field theory, the choice of signs in 5(174a) and 5(174b) is arbitrary. For time reversal we take

$$\mathscr{T}A_\mu(\mathbf{x}, t)\mathscr{T}^{-1} = -A_\mu(\mathbf{x}, -t) \qquad 5(176)$$

Taking into account the antiunitary character of $\mathscr{T}$ we then have

$$\mathscr{T}j_\mu(\mathbf{x}, t)\mathscr{T}^{-1} = -j_\mu(\mathbf{x}, -t) \qquad 5(177)$$

and the equations of motion and commutation relations are easily seen to be invariant. For the case of charge conjugation, comparison of

5(25a) and 5(25b) indicates that A_μ must transform according to

$$\mathscr{C} A_\mu \mathscr{C}^{-1} = -A_\mu \tag{5(178)}$$

and, moreover, that the current 5(26) must be *symmetrized* in ϕ and $\phi^\dagger$ to ensure that

$$\mathscr{C} j_\mu \mathscr{C}^{-1} = -j_\mu \tag{5(179)}$$

in analogy with 5(171). If A_μ is coupled to the symmetrized current, the charge conjugation invariance of the coupled equations of motion is assured.

The treatment of the minimal electromagnetic coupling 5(5) for spin $\frac{1}{2}$ fields follows the same pattern. The $\mathscr{P}$, $\mathscr{T}$ and $\mathscr{C}$ transformation laws for ψ and A_μ are defined as above and the current 5(7) must be replaced by the *antisymmetrized* current 3(203)

$$j_\mu = \tfrac{1}{2} i[\bar{\psi}, \gamma_\mu \psi] \tag{5(180)}$$

to ensure invariance of the coupled field equations under charge conjugation. We note that in contrast to the pseudoscalar density $i\bar{\psi}\gamma_5\psi$, the electric current $i\bar{\psi}\gamma_\mu\psi$ is not equal to its antisymmetrized form but differs from it by a term proportional to the infinite zero point charge [see Chapter 3, Problem 12]. Thus, in addition to ensuring charge conjugation invariance, antisymmetrization of the current, (or symmetrization, in the case of Bose–Einstein fields), is also necessary in order to remove a physically meaningless infinity which would otherwise be present in the theory. Also note that if we perform the minimal substitution 5(1) in the free Dirac Lagrangian in its normal ordered form

$$\begin{aligned}\mathscr{L}_{\text{Dirac}} &= :\bar{\psi}\gamma_\mu \partial_\mu \psi: + m:\bar{\psi}\psi: \\ &= \tfrac{1}{2}[\bar{\psi}, \gamma_\mu \partial_\mu \psi] + m:\bar{\psi}\psi:\end{aligned}$$

we automatically generate the coupling term $\mathscr{L}_I = ie[\bar{\psi}, \gamma_\mu \psi] A_\mu$ featuring the antisymmetrized current 5(180).

We conclude with a remark concerning the explicit construction of the operators $\mathscr{P}$, $\mathscr{C}$ and $\mathscr{T}$ in the interacting case. Since we have defined the operators $\mathscr{P}$, $\mathscr{C}$ and $\mathscr{T}$ so as to guarantee the invariance of the coupled equations of motion under the corresponding symmetry operations, these operators must commute with the total Hamiltonian. Accordingly $\mathscr{P}$, $\mathscr{C}$ and $\mathscr{T}$ are time-independent. We may therefore select the time $t = 0$ and use momentum expansions of the type 5(19a), 5(19b), 5(19c) and 5(19d) in exactly the same way as in the free field case to construct the operators $\mathscr{P}$, $\mathscr{C}$ and $\mathscr{T}$ explicitly.

TCP Theorem An important symmetry property exhibited by a wide class of interacting field theories is the invariance under the combined operation *TCP*, the product of space reflection, charge conjugation and time reflection. This invariance holds, even though, individually, *T*, *C* or *P* invariance may be violated.

As a simple illustration of this property, consider the parity-violating coupling

$$\mathscr{L}_I = g_1[\bar{\psi}, \psi]\phi + g_2 i[\bar{\psi}, \gamma_5\psi]\phi \tag{5(181)}$$

of a Dirac field with a neutral scalar field ϕ. The coupling constants g_1 and g_2 are assumed to be real, to ensure the hermiticity of $\mathscr{L}_I$. The interaction 5(181) is charge conjugation invariant, since we have, as in 5(171)

$$\begin{aligned} \mathscr{C}[\bar{\psi}, \psi]\mathscr{C}^{-1} &= [\bar{\psi}, \psi] \\ \mathscr{C}[\bar{\psi}, \gamma_5\psi]\mathscr{C}^{-1} &= [\bar{\psi}, \gamma_5\psi] \end{aligned} \tag{5(182)}$$

However, the coupling 5(181) is not invariant under space-reflection. The parity transformation

$$\begin{aligned} \mathscr{P}\psi(\mathbf{x}, t)\mathscr{P}^{-1} &= \gamma_4\psi(-\mathbf{x}, t) \\ \mathscr{P}\phi(\mathbf{x}, t)\mathscr{P}^{-1} &= \phi(-\mathbf{x}, t) \end{aligned} \tag{5(183)}$$

takes the equation of motion

$$(\Box - \mu^2)\phi(x) = -g_1[\bar{\psi}(x), \psi(x)] - ig_2[\bar{\psi}(x), \gamma_5\psi(x)] \tag{5(184)}$$

into

$$(\Box - \mu^2)\phi(x) = -g_1[\bar{\psi}(x), \psi(x)] + ig_2[\bar{\psi}(x), \gamma_5\psi(x)] \tag{5(185)}$$

upon relabelling $-\mathbf{x} \to +\mathbf{x}$ in the transformed equation. On the other hand, the coupling 5(181) also violates time reflection invariance, since the transformation

$$\begin{aligned} \mathscr{T}\psi(\mathbf{x}, t)\mathscr{T}^{-1} &= T^{-1}\psi(\mathbf{x}, t) \\ \mathscr{T}\phi(\mathbf{x}, t)\mathscr{T}^{-1} &= \phi(\mathbf{x}, -t) \end{aligned} \tag{5(186)}$$

also takes 5(184) into 5(185), (upon relabelling $-t \to +t$). Indeed by following the same steps as in 5(168) we have

$$\begin{aligned} \mathscr{T}\bar{\psi}(\mathbf{x}, t)\psi(\mathbf{x}, t)\mathscr{T}^{-1} &= \bar{\psi}(\mathbf{x}, -t)\psi(\mathbf{x}, -t) \\ \mathscr{T}i\bar{\psi}(\mathbf{x}, t)\gamma_5\psi(\mathbf{x}, t)\mathscr{T}^{-1} &= -i\bar{\psi}(\mathbf{x}, -t)\gamma_5\psi(\mathbf{x}, -t) \end{aligned} \tag{5(187)}$$

Thus the product of 5(183) and 5(186) restores the original equation of motion 5(184). It is a simple matter to check that the equation of motion for ψ is also C and PT invariant.

Another example is provided by the coupling

$$\mathscr{L}_I = g_1 i[\bar{\psi}, \gamma_\mu \psi] A_\mu + g_2 i[\bar{\psi}, \gamma_\mu \gamma_5 \psi] A_\mu \qquad 5(188)$$

which is T-invariant but violates P and C. This is similar to the case of the weak coupling Lagrangian 5(155). It is easy to verify that the product CP is conserved, so that TCP invariance is again assured.

The basic mechanism responsible for the TCP invariance of these (and other similar trilinear couplings) is the fact that, under TCP, neutral spin 0 and spin 1 fields transform in exactly the same way as the corresponding (symmetrized or antisymmetrized) densities to which they are coupled, an essential element being the fact that neutral fields are always coupled to *hermitian* densities so as to ensure hermiticity of $\mathscr{L}_I$. Thus, for example, we find, combining 5(165b), 5(167b), and 5(169b),

$$\mathscr{T}\mathscr{C}\mathscr{P}\phi(x)\mathscr{P}^{-1}\mathscr{C}^{-1}\mathscr{T}^{-1} = \phi(-x) \qquad 5(189)$$

which is to be compared with the transformation law for hermitian Dirac densities

$$\mathscr{T}\mathscr{C}\mathscr{P}[\bar{\psi}(x), \psi(x)]\mathscr{P}^{-1}\mathscr{C}^{-1}\mathscr{T}^{-1} = [\bar{\psi}(-x), \psi(-x)] \qquad 5(190a)$$

or

$$\mathscr{T}\mathscr{C}\mathscr{P}i[\bar{\psi}(x), \gamma_5\psi(x)]\mathscr{P}^{-1}\mathscr{C}^{-1}\mathscr{T}^{-1} = i[\bar{\psi}(-x), \gamma_5\psi(-x)] \qquad 5(190b)$$

obtained by combining 5(165a), 5(167a) and 5(169a). The proof of the corresponding property for spin 1 fields and densities is left to the reader as an exercise, [see Problem 14].

For the general proof of the TCP theorem, the reader is referred to the original papers by Lüders (1957) and Pauli (1955). The only requirements for the validity of the theorem are Lorentz invariance, locality and hermiticity of $\mathscr{L}_I(x)$, and the correct connection between spin and statistics, that is, the quantization of integer (half-integer) spin fields by means of commutation (anticommutation) relations respectively.

PROBLEMS

1. Construct the generators of infinitesimal space-time rotations and establish relations of the type 3(177b) for interacting Dirac and Maxwell fields.
2. Use the equations of motion 5(25a), 5(25b) and 5(25c) to prove that the current 5(26) is conserved.
3. Verify the result 5(35).

4. Construct the minimal electromagnetic coupling Lagrangians for charged spin 1 and spin $\frac{3}{2}$ fields. Derive the equations of motion, and show in each case that the current to which the electromagnetic field is coupled obeys a conservation law.
5. Write down the equal-time commutation relations featuring $A_0(x)$ for the radiation gauge.
6. Derive the result 5(73).
7. Verify 5(86).
8. Write down the Euler–Lagrange equations of motion for the interacting π meson-nucleon system. Show that the conditions for the consistency of the Euler–Lagrange and Hamiltonian equations of motion strongly suggest the choice of anticommutation relations 5(101b).
9. Check the equivalence of 5(105a), 5(105b) and 5(107).
10. Assuming non-derivative Yukawa couplings of the type 5(95) or 5(105), write down the most general SU_2-invariant Lagrangian for the coupling of the baryon octet $(N, \Xi, \Sigma, \Lambda)$ to the pseudoscalar meson octet $(K, \bar{K}, \pi, \eta)$, and derive the expression for the total isotopic spin current.
11. Compute the explicit expressions for the trilinear couplings 5(146a) and 5(146b) in terms of the physical baryon and meson fields.
12. Verify that the interaction 5(155) is invariant under the product of parity and charge conjugation.
13. Check the $\mathscr{C}$ transformation property 5(171).
14. Verify the TCP transformation properties 5(189) and 5(190a), 5(190b). Show that neutral vector fields transform under TCP in the same way as the vector and axial vector currents $i\bar{\psi}\gamma_\mu\psi$ and $i\bar{\psi}\gamma_\mu\gamma_5\psi$.

REFERENCES

Arnous, E. (1959) (with W. Heitler) *Nuovo Cimento* **11**, 443

Arnous, E. (1960) (with W. Heitler and Y. Takahashi) *Nuovo Cimento* **16**, 671

Bloch, C. (1952) *Kgl. Danske Vidensk. Selsk., Mat.-Fys. Medd.* **27**, No. 8

Cabibbo, N. (1963) *Phys. Rev. Letters* **10**, 531

Chrétien, M. (1954) (with R. E. Peierls) *Proc. Roy. Soc.* **A 223**, 468

Dürr, H. P. (1965) (with W. Heisenberg, H. Yamamoto and K. Yamazaki) *Nuovo Cimento* **38**, 1220

Feynman, R. P. (1958) (with M. Gell-Mann) *Phys. Rev.* **109**, 193

Gell-Mann, M. (1961) *The Eightfold Way: A Theory of Strong Interaction Symmetry*, (reprinted in *The Eightfold Way*, M. Gell-Mann and Y. Ne'eman, Benjamin, New York, 1964)

Gell-Mann, M. (1962) *Phys. Rev.* **125**, 1067

Glashow, S. (1961) (with M. Gell-Mann) *Ann. Phys.* **15**, 437

Ikeda, M. (1960) (with S. Ogawa and Y. Ohnuki) *Progr. Theoret. Phys.* **23**, 1073

Kristensen, P. (1952) (with C. Møller) *Kgl. Danske Vidensk. Selsk., Mat.-Fys. Medd.* **27**, No. 7

Lüders, G. (1957) *Ann. Phys.* **2**, 1

Nambu, Y. (1962) (with D. Lurié) *Phys. Rev.* **125**, 1429

Ne'eman, Y. (1961) *Nucl. Phys.* **26**, 222

Nishijima, M. (1959) *Nuovo Cimento* **11**, 698

Okubo, S. (1962) *Prog. Theoret. Phys.* **27**, 949
O'Raifeartaigh, L. (1960a) *Helv. Phys. Acta.* **33**, 783
O'Raifeartaigh, L. (1960b) (with Y. Takahashi) *Helv. Phys. Acta* **34**, 554
Pauli, W. (1955) in *Niels Bohr and the Development of Physics*, McGraw-Hill, New York, 1955
Sakata, S. (1956) *Progr. Theoret. Phys.* **16**, 686
Sakurai, J. J. (1960) *Ann. Phys.* **11**, 1
Sudarshan, E. C. G. (1958) (with R. E. Marshak) *Phys. Rev.* **109**, 1860
Yang, C. N. (1954) (with R. L. Mills) *Phys. Rev.* **96**, 191

6

Perturbation Theory

6-1 The interaction picture

Formulation Our presentation so far has been based on the Heisenberg picture. In this picture the development in time of a quantum field system is carried by the field operators ϕ_A according to the equations of motion

$$\begin{aligned}\dot{\phi}_A(x) &= \frac{1}{i}[\phi_A(x), H] \\ \dot{\pi}_A(x) &= \frac{1}{i}[\pi_A(x), H]\end{aligned} \tag{6(1)}$$

or, since the total Hamiltonian H is a constant of the motion

$$\begin{aligned}\phi_A(\mathbf{x}, t) &= e^{iHt}\phi_A(\mathbf{x}, 0)\, e^{-iHt} \\ \pi_A(\mathbf{x}, t) &= e^{iHt}\pi_A(\mathbf{x}, 0)\, e^{-iHt}\end{aligned} \tag{6(2)}$$

State vectors $|a\rangle$, on the other hand, are time-independent. As is well known, the time evolution of the quantum system can also be described in terms of the Schrödinger picture. If we define field operators ϕ_A^S in the Schrödinger picture by means of the unitary transformation

$$\begin{aligned}\phi_A^S(\mathbf{x}) &= e^{-iHt}\phi_A(\mathbf{x}, t)\, e^{iHt} \\ \pi_A^S(\mathbf{x}) &= e^{-iHt}\pi_A(\mathbf{x}, t)\, e^{iHt}\end{aligned} \tag{6(3)}$$

then the operators ϕ_A^S and π_A^S are just the Heisenberg picture fields ϕ_A and π_A at time $t = 0$

$$\begin{aligned}\phi_A^S(\mathbf{x}) &= \phi_A(\mathbf{x}, 0) \\ \pi_A^S(\mathbf{x}) &= \pi_A(\mathbf{x}, 0)\end{aligned} \tag{6(4)}$$

and are *time-independent*. On the other hand, Schrödinger picture state vectors, defined by

$$|a; t\rangle^S = e^{-iHt}|a\rangle \tag{6(5)}$$

vary with time according to the Schrödinger equation

$$i\frac{\partial}{\partial t}|a; t\rangle^S = H^S|a; t\rangle^S \tag{6(6)}$$

where

$$H^S = H(t = 0) = H \tag{6(7)}$$

Note that the two pictures coincide at time $t = 0$ and that plane-wave expansions for Schrödinger picture operators are given by expressions of the type 5(19a), 5(19b), 5(19c) and 5(19d).

A perturbation scheme based on the use of the Schrödinger picture can be developed along the same lines as in ordinary nonrelativistic quantum mechanics*. However, the Schrödinger picture suffers from the drawback that the field operators depend only on the *spatial* coordinates $\mathbf{x}$; as a result the formalism fails to exhibit manifest Lorentz covariance. For this reason perturbative calculations are more conveniently performed in the *interaction picture* which is intermediate between the Schrödinger and Heisenberg pictures. Writing

$$H = H_0 + H_I \tag{6(8)}$$

where H_I is the interaction Hamiltonian, we define operators ϕ_A^{ip}, π_A^{ip} by

$$\begin{aligned} \phi_A^{ip}(\mathbf{x}, t) &= e^{iH_0^S t}\phi_A^S(\mathbf{x})\, e^{-iH_0^S t} \\ \pi_A^{ip}(\mathbf{x}, t) &= e^{iH_0^S t}\pi_A^S(\mathbf{x})\, e^{-iH_0^S t} \end{aligned} \tag{6(9)}$$

where H_0^S is the free field Hamiltonian in the Schrödinger picture

$$H_0^S = H_0(t = 0) \tag{6(10)}$$

Correspondingly, we define state vectors $|a; t\rangle^{ip}$ by

$$|a; t\rangle^{ip} = e^{iH_0^S t}|a; t\rangle^S \tag{6(11)}$$

The relation of ϕ_A^{ip} and $|a; t\rangle^{ip}$ to the field operators and states in the Heisenberg picture is given by combining 6(9) with 6(3) and 6(11) with 6(5). Thus

$$\begin{aligned} \phi_A^{ip}(\mathbf{x}, t) &= e^{iH_0^S t}\, e^{-iHt}\phi_A(\mathbf{x}, t)\, e^{iHt}\, e^{-iH_0^S t} \\ \pi_A^{ip}(\mathbf{x}, t) &= e^{iH_0^S t}\, e^{-iHt}\pi_A(\mathbf{x}, t)\, e^{iHt}\, e^{-iH_0^S t} \end{aligned} \tag{6(12)}$$

* For an account of perturbation theory in the Schrödinger picture see for example W. Heitler, *Quantum Theory of Radiation*, Oxford, 1954.

and

$$|a;t\rangle^{ip} = e^{iH_0^S t}\, e^{-iHt}|a\rangle \qquad 6(13)$$

Note that the Heisenberg, Schrödinger and interaction pictures all coincide at $t = 0$.

Let us derive the equation of motion for $|a;t\rangle^{ip}$. Using 6(6) we have

$$\frac{\partial}{\partial t}|a;t\rangle^{ip} = iH_0^S\, e^{iH_0^S t}|a;t\rangle^S - e^{iH_0^S t} iH|a;t\rangle^S$$

Since

$$\begin{aligned} H_0^{ip}(t) &= e^{iH_0^S t} H_0^S\, e^{-iH_0^S t} \\ &= H_0^S \end{aligned} \qquad 6(14)$$

and

$$H^{ip}(t) = e^{iH_0^S t} H^S\, e^{-iH_0^S t} \qquad 6(15)$$

we can write the equation for $|a;t\rangle^{ip}$ in the form

$$\frac{\partial}{\partial t}|a;t\rangle^{ip} = iH_0^{ip}(t)|a;t\rangle^{ip} - iH^{ip}(t)|a;t\rangle^{ip}$$

or

$$i\frac{\partial}{\partial t}|a;t\rangle^{ip} = H_I^{ip}(t)|a;t\rangle^{ip} \qquad 6(16)$$

We see that the time dependence of $|a;t\rangle^{ip}$ is determined solely by the *interaction* Hamiltonian $H_I^{ip}(t)$ in the interaction picture. On the other hand, the time dependence of the fields ϕ_A^{ip} and π_A^{ip} (and hence of $H_I^{ip}(t)$) is determined by the *free* Hamiltonian H_0^{ip}, since by 6(9) we have

$$\begin{aligned} \dot{\phi}_A^{ip}(\mathbf{x}, t) &= \frac{1}{i}[\phi_A^{ip}(\mathbf{x}, t), H_0^S] \\ &= \frac{1}{i}[\phi_A^{ip}(\mathbf{x}, t), H_0^{ip}] \end{aligned} \qquad 6(17a)$$

and

$$\dot{\pi}_A^{ip}(\mathbf{x}, t) = \frac{1}{i}[\pi_A^{ip}(\mathbf{x}, t), H_0^{ip}] \qquad 6(17b)$$

We stress that H_0^{ip} is time-independent.

Since ϕ_A^{ip} and π_A^{ip} are related to ϕ_A and π_A by means of unitary transformations, they obey the same equal-time commutation relations as ϕ_A and π_A. Thus *both the equations of motion and the equal-time commutation relations for ϕ_A^{ip} and π_A^{ip} are the same as in the free-field case.* We may therefore represent these fields by their free-field plane wave expansions. For example, for a real scalar field, we have

$$\phi^{ip}(\mathbf{x}, t) = \frac{1}{\sqrt{V}} \sum_{\mathbf{k}} \frac{1}{\sqrt{2\omega_{\mathbf{k}}}} (a_{\mathbf{k}}\, \mathrm{e}^{i\mathbf{k}.\mathbf{x} - i\omega_{\mathbf{k}} t} + a_{\mathbf{k}}{}^{\dagger}\, \mathrm{e}^{-i\mathbf{k}.\mathbf{x} + i\omega_{\mathbf{k}} t})$$

and

$$\pi^{ip}(\mathbf{x}, t) = \frac{-i}{\sqrt{V}} \sum_{\mathbf{k}} \sqrt{\frac{\omega_{\mathbf{k}}}{2}} (a_{\mathbf{k}}\, \mathrm{e}^{i\mathbf{k}.\mathbf{x} - i\omega_{\mathbf{k}} t} - a_{\mathbf{k}}{}^{\dagger}\, \mathrm{e}^{-i\mathbf{k}.\mathbf{x} + i\omega_{\mathbf{k}} t})$$

for all times t.

Time-Displacement Operator We investigate the temporal evolution of the state vectors $|a; t\rangle^{ip}$ by defining a time-displacement operator $U(t_1, t_0)$ which transforms $|a; t_0\rangle^{ip}$ into $|a; t_1\rangle^{ip}$

$$|a; t_1\rangle^{ip} = U(t_1, t_0)|a; t_0\rangle^{ip} \qquad 6(18)$$

Clearly

$$U(t_0, t_0) = 1 \qquad 6(19)$$

Furthermore, the operator U must satisfy the group property

$$U(t_1, t)U(t, t_0) = U(t_1, t_0) \qquad 6(20)$$

since

$$|a; t_1\rangle^{ip} = U(t_1, t)|a; t\rangle^{ip}$$

and

$$|a; t\rangle^{ip} = U(t, t_0)|a; t_0\rangle^{ip}$$

From the group property 6(20) we infer the existence of the inverse operator $U(t, t_0)^{-1}$. Indeed, setting $t_1 = t_0$ and using 6(19) we have

$$U(t_0, t)U(t, t_0) = 1 \qquad 6(21)$$

so that

$$U(t, t_0)^{-1} = U(t_0, t) \qquad 6(22)$$

Let us show that $U(t, t_0)$ is a unitary operator, i.e.

$$U(t, t_0)^{-1} = U(t, t_0)^\dagger \tag{6(23)}$$

Enlisting the aid of the equation of motion 6(16) we see that $U(t, t_0)$ satisfies the differential equation

$$i\frac{\partial}{\partial t}U(t, t_0) = H_I^{ip}(t)U(t, t_0) \tag{6(24a)}$$

From 6(24a) it follows that the adjoint operator $U(t, t_0)^\dagger$ satisfies

$$-i\frac{\partial}{\partial t}U(t, t_0)^\dagger = U(t, t_0)^\dagger H_I^{ip}(t) \tag{6(24b)}$$

since H_I^{ip} is hermitian. Combining 6(24a) with 6(24b) we get

$$\frac{\partial}{\partial t}[U(t, t_0)^\dagger U(t, t_0)] = 0$$

which, together with 6(19) implies

$$U(t, t_0)^\dagger U(t, t_0) = 1 \tag{6(25)}$$

The unitarity of $U(t, t_0)$ is a consequence of 6(25) and of the existence of the inverse operator $U(t, t_0)^{-1}$. Multiplying 6(25) on the right by $U(t, t_0)^{-1}$ we get

$$U(t, t_0)^\dagger = U(t, t_0)^{-1}$$

as stated.

Comparing 6(18) with 6(13) and noting that $|a; t\rangle^{ip}$ coincides with the Heisenberg picture state vector $|a\rangle$ at $t = 0$, we see that

$$U(t, 0) = e^{iH_0^S t} e^{-iHt} \tag{6(26)}$$

Thus the relation 6(12) between field operators in the interaction and Heisenberg pictures may be expressed in terms of the time displacement operator according to

$$\phi_A^{ip}(\mathbf{x}, t) = U(t, 0)\phi_A(\mathbf{x}, t)U(t, 0)^{-1} \tag{6(27)}$$

The Perturbation Expansion We obtain a perturbative expansion for $U(t, t_0)$ by converting the differential equation 6(24a) with the boundary condition 6(19) into the integral equation*

$$U(t, t_0) = 1 - i\int_{t_0}^{t} dt_1 H_I(t_1)U(t_1, t_0) \tag{6(28)}$$

* For the remainder of this chapter superscripts *ip* will be suppressed, as our discussion will be carried out entirely within the framework of the interaction picture.

and applying the standard iterative solution

$$U(t,t_0)=1-i\int_{t_0}^{t} dt_1 H_I(t_1)\,dt_1+(-i)^2\int_{t_0}^{t} dt_1\int_{t_0}^{t_1} dt_2 H_I(t_1)H_I(t_2)$$

$$+\ldots+(-i)^n\int_{t_0}^{t} dt_1\int_{t_0}^{t_1} dt_2\ldots\int_{t_0}^{t_{n-1}} dt_n H_I(t_1)\ldots H_I(t_n)+\ldots \qquad 6(29)$$

To obtain a more convenient form, we consider the $n = 2$ term. Relabelling $t_1 \to t_2$ and $t_2 \to t_1$ we have

$$\int_{t_0}^{t} dt_1\int_{t_0}^{t_1} dt_2 H_I(t_1)H_I(t_2)=\int_{t_0}^{t} dt_2\int_{t_0}^{t_2} dt_1 H_I(t_2)H_I(t_1) \qquad 6(30)$$

Next we invert the order of integration on the right-hand side by writing

$$\int_{t_0}^{t} dt_2\int_{t_0}^{t_2} dt_1=\int_{t_0}^{t} dt_2\int_{t_0}^{t} dt_1\theta(t_2-t_1)=\int_{t_0}^{t} dt_1\int_{t_0}^{t} dt_2\theta(t_2-t_1)$$

$$=\int_{t_0}^{t} dt_1\int_{t_1}^{t} dt_2$$

so that 6(30) becomes

$$\int_{t_0}^{t} dt_1\int_{t_0}^{t_1} dt_2 H_I(t_1)H_I(t_2)=\int_{t_0}^{t} dt_1\int_{t_1}^{t} dt_2 H_I(t_2)H_I(t_1)$$

or equivalently

$$\int_{t_0}^{t} dt_1\int_{t_0}^{t_1} dt_2 H_I(t_1)H_I(t_2)$$

$$=\tfrac{1}{2}\int_{t_0}^{t} dt_1\int_{t_0}^{t_1} dt_2 H_I(t_1)H_I(t_2)+\tfrac{1}{2}\int_{t_0}^{t} dt_1\int_{t_1}^{t} dt_2 H_I(t_2)H_I(t_1) \qquad 6(31)$$

We now observe that the products $H_I(t_1)H_I(t_2)$ and $H_I(t_2)H_I(t_1)$ appearing on the right-hand side of 6(31) are in time-ordered form, since the integrations are over the regions $t_1 > t_2$ and $t_2 < t_1$ respectively. Hence

$$\int_{t_0}^{t} dt_1 \int_{t_0}^{t_1} dt_2 H_I(t_1) H_I(t_2)$$

$$= \tfrac{1}{2} \int_{t_0}^{t} dt_1 \int_{t_0}^{t_1} dt_2 T H_I(t_1) H_I(t_2) + \tfrac{1}{2} \int_{t_0}^{t} dt_1 \int_{t_1}^{t} dt_2 T H_I(t_2) H_I(t_1)$$

$$= \tfrac{1}{2} \int_{t_0}^{t} dt_1 \int_{t_0}^{t} dt_2 T H_I(t_1) H_I(t_2) \qquad 6(32)$$

Note that the T-symbol also prescribes a change of sign for each permutation of fermion factors. This presents no problem, however, since $H_I(t)$ always contains *pairs* of fermion factors. Thus the interchanges in the positions of the $H_I(t)$ prescribed by T will always involve an even number of minus signs.

The nth order term in the expansion can be treated in a similar way, with the result

$$\int_{t_0}^{t} dt_1 \int_{t_0}^{t_1} dt_2 \ldots \int_{t_0}^{t_{n-1}} dt_n H_I(t_1) \ldots H_I(t_n)$$

$$= \frac{1}{n!} \int_{t_0}^{t} dt_1 \int_{t_0}^{t} dt_2 \ldots \int_{t_0}^{t} dt_n T H_I(t_1) \ldots H_I(t_n) \qquad 6(33)$$

Hence the series expansion 6(29) may be written in the form

$$U(t, t_0) = 1 + \sum_{n=1}^{\infty} \frac{(-i)^n}{n!} \int_{t_0}^{t} dt_1 \ldots \int_{t_0}^{t} dt_n T H_I(t_1) \ldots H_I(t_n) \qquad 6(34)$$

or simply

$$U(t, t_0) = T \exp\left[-i \int_{t_0}^{t} H_I(t_1)\, dt_1 \right] \qquad 6(35)$$

Setting

$$H_I(t) = \int d^3x \mathscr{H}_I(x)$$

we obtain the final form

$$U(t, t_0) = 1 + \sum_{n=1}^{\infty} \frac{(-i)^n}{n!} \int_{t_0}^{t} d^4x_1 \int_{t_0}^{t} d^4x_2 \ldots \int_{t_0}^{t} d^4x_n T\mathscr{H}_I(x_1)\ldots\mathscr{H}_I(x_n) \qquad 6(36a)$$

or

$$U(t, t_0) = T\exp\left[-i\int d^4x \mathscr{H}_I(x)\right] \qquad 6(36b)$$

An important observation is that the time orderings

$$T\mathscr{H}_I(x_1)\ldots\mathscr{H}_I(x_n)$$

are Lorentz-invariant, since $\mathscr{H}_I(x_i)$ and $\mathscr{H}_I(x_j)$ commute at spacelike separations, i.e.

$$[\mathscr{H}_I(x), \mathscr{H}_I(x')] = 0 \quad \text{for} \quad (x-x')^2 > 0 \qquad 6(37)$$

as a consequence of microcausality.

The Adiabatic Hypothesis We are now equipped with a perturbative scheme for computing the time evolution of a given state vector $|a; t\rangle$. If we are given $|a; t_1\rangle$ we can construct $|a; t_2\rangle$ to any desired order in perturbation theory by means of the formula

$$|a; t_2\rangle = U(t_2, t_1)|a; t_1\rangle$$

with $U(t_2, t_1)$ given by 6(36a). We now ask, how are initial states $|a; t_1\rangle$ to be specified? In ordinary quantum mechanical scattering theory for potentials of finite range, initial and final states are specified at $t = -\infty$ and $t = \infty$ respectively, by means of unperturbed wave functions. The same procedure will be applied here. We shall assume that, as $t \to \pm\infty$, the interaction Hamiltonian $H_I(t)$ can be switched off adiabatically and that initial and final states can be represented by eigenstates of the free Hamiltonian H_0. We shall denote these unperturbed states by round brackets, i.e. $|a)$ or $|b)$. The probability amplitude for the system to make a transition from the state $|a)$ at $t = -\infty$ to the state $|b)$ at $t = +\infty$ is then just the scalar product $S_{ba} = (b|U(\infty, -\infty)|a)$. S_{ba} is known as the *scattering amplitude*, or *S-matrix element*. From the perturbation expansion 6(36a) we obtain the general formula

$$S_{ba} = (b|S|a) \qquad 6(38)$$

with

$$S = U(\infty, -\infty) = 1 + \sum_{n=1}^{\infty} \frac{(-i)^n}{n!} \int_{-\infty}^{\infty} d^4x_1 \ldots \int_{-\infty}^{\infty} d^4x_n T\mathscr{H}_I(x_1)\ldots\mathscr{H}_I(x_n) \qquad 6(39)$$

In a physical scattering process or production reaction, particles which are initially well-separated and noninteracting come together, interact, and produce reaction products which then separate back into noninteracting final states. This means that at any given time the various particles are *localized* and we must bear in mind that a quantum description of localized particles requires the use of wave packets. Mathematically, such a description of scattering is rather awkward and for this reason we usually represent initial and final states by plane waves, that is *we assign sharp momentum values to the states* $|a)$ *and* $|b)$. For example in neutral pseudoscalar meson theory with

$$\mathscr{H}_I(x) = -iG\bar{\psi}^{ip}(x)\gamma_5\psi^{ip}(x)\phi^{ip}(x)$$

we might consider a one-meson, one-nucleon initial state $|\mathbf{k}, \mathbf{p}\sigma) = a_{\mathbf{k}}^{\dagger}c_{\mathbf{p}\sigma}^{\dagger}|0)$ and a two-meson, one-nucleon final state $|\mathbf{k}'', \mathbf{k}', \mathbf{p}'\sigma') = a_{\mathbf{k}''}^{\dagger}a_{\mathbf{k}'}^{\dagger}c_{\mathbf{p}'\sigma'}^{\dagger}|0)$ with *sharp* initial and final momentum values. The S-matrix element

$$S_{ba} = (0|c_{\mathbf{p}'\sigma'}a_{\mathbf{k}''}a_{\mathbf{k}'}Sa_{\mathbf{k}}^{\dagger}c_{\mathbf{p}\sigma}^{\dagger}|0) \qquad 6(40)$$

would then give the probability amplitude for the process

$$\mathbf{k} + (\mathbf{p}, \sigma) \rightarrow \mathbf{k}' + \mathbf{k}'' + (\mathbf{p}', \sigma')$$

and can, in principle, be calculated to any desired order of perturbation theory by applying 6(39) and the free-field plane wave expansions for ψ^{ip} and ϕ^{ip}. The price that we pay for the mathematical convenience of using plane waves is that certain expressions become mathematically undefined* and a prescription is required for dealing with them. This prescription is provided by the *formal theory of scattering*† and it consists in associating to the interaction Hamiltonian $H_I(t)$ a factor $\exp \varepsilon t$ which serves to remove the mathematical ambiguities associated with the plane wave description; the limit $\varepsilon \rightarrow 0$ is taken after the mathematical ambiguities have been disposed of. Although for the sake of brevity we shall frequently omit the $\exp \varepsilon t$ factor, it will be implicit in all calculations based

* See 6(130) below.

† We shall assume that the reader is familiar with this description. A very clear presentation is given in E. Merzbacher, *Quantum Mechanics*, Wiley, New York, 1961, Chapter 21.

on 6(39) in which initial and final states are assigned sharp momentum values.

A further remark concerns the question of the applicability of the adiabatic hypothesis to field theoretic interactions. In contrast to the finite range potentials of ordinary quantum mechanics, field theoretic interactions cannot be completely switched off, even when the particles are far apart. We shall see that the S-matrix 6(39) contains so-called *self-energy* effects which are present even for an isolated single particle. Thus the formulation of the adiabatic hypothesis is not entirely correct as it stands. The error will be corrected by *mass renormalization* in Section 6-4.

Decomposition into Normal Products Our next task is an algebraic one; we must decompose the time-ordered products appearing in 6(39) into *normal*-ordered products in which all destruction operators stand to the right, all creation operators stand to the left. This will allow us to read off directly the expression for an arbitrary matrix element S_{ba}. For example, the pion-production amplitude 6(40) will clearly be given by the coefficient of the term $a_{\mathbf{k}'}{}^{\dagger}a_{\mathbf{k}''}{}^{\dagger}c_{\mathbf{p}'\sigma'}{}^{\dagger}a_{\mathbf{k}}c_{\mathbf{p}\sigma}$ in the normal-product decomposition of S.

To familiarize ourselves with the prescription for decomposing time-ordered products of field operators into normal-ordered products, we consider the case of two, three and four-fold products in detail. We begin with the trivial identity

$$T\phi_A(x_1) = :\phi_A(x_1): \qquad 6(41)$$

for a single field operator. To construct the decomposition formula for a time-ordered product of *two* fields we multiply 6(41) on the right by an arbitrary field operator $\phi_B(x_2)$, referring to a time $t_2 < t_1$. Then

$$(T\phi_A(x_1))\phi_B(x_2) = T\phi_A(x_1)\phi_B(x_2) = :\phi_A(x_1):\phi_B(x_2)$$

To absorb $\phi_B(x_2)$ into the normal product, we write

$$\begin{aligned} :\phi_A(x_1):\phi_B(x_2) &= \phi_A(x_1)\phi_B^{(+)}(x_2)+\phi_A(x_1)\phi_B^{(-)}(x_2) \\ &= \phi_A(x_1)\phi_B^{(+)}(x_2)+\phi_A^{(-)}(x_1)\phi_B^{(-)}(x_2)+\phi_A^{(+)}(x_1)\phi_B^{(-)}(x_2) \end{aligned} \qquad 6(42)$$

All terms on the right-hand side except $\phi_A^{(+)}\phi_B^{(-)}$ are in normal-ordered form. The latter term may be normal-ordered by commuting (or anti-commuting) $\phi_B^{(-)}$ through to the left, using the relation

$$\phi_A^{(+)}(x_1)\phi_B^{(-)}(x_2) = \pm\phi_A^{(-)}(x_2)\phi_B^{(+)}(x_1)+[\phi_A^{(+)}(x_1),\phi_B^{(-)}(x_2)]_{\mp} \qquad 6(43)$$

where the lower sign is to be used *only if* ϕ_A and ϕ_B are *both* Fermi–Dirac fields, and where

$$[\phi_A^{(+)}(x_1), \phi_B^{(-)}(x_2)]_{\mp} = \phi_A^{(+)}(x_1)\phi_B^{(-)}(x_2) \mp \phi_B^{(-)}(x_2)\phi_A^{(+)}(x_1) \qquad 6(44)$$

is necessarily a *c-number function*. Thus we have, for $t_2 < t_1$,

$$\begin{aligned}[\phi_A^{(+)}(x_1), \phi_B^{(-)}(x_2)]_{\mp} &= (0|[\phi_A^{(+)}(x_1), \phi_B^{(-)}(x_2)]_{\mp}|0) \\ &= (0|\phi_A^{(+)}(x_1)\phi_B^{(-)}(x_2)|0) \\ &= (0|\phi_A(x_1)\phi_B(x_2)|0) \\ &= (0|T\phi_A(x_1)\phi_B(x_2)|0) \end{aligned} \qquad 6(45)$$

and hence, combining results

$$T\phi_A(x_1)\phi_B(x_2) = :\phi_A(x_1)\phi_B(x_2): + (0|T\phi_A(x_1)\phi_B(x_2)|0) \qquad 6(46)$$

for $t_2 < t_1$. The latter restriction is removed by observing that $\phi_A(x_1)$ and $\phi_B(x_2)$ can now be interchanged on both sides of 6(46); by the definition of the normal-ordered and time-ordered products, the relation 6(46) remains invariant under this permutation. Therefore 6(46) is valid in general.

As a derivation of 6(46), the above procedure is longwinded and unnecessary*. It has the merit, however, that it is easily generalized to apply to higher order products. Thus, to obtain the normal-product decomposition of $T\phi_A(x_1)\phi_B(x_2)\phi_C(x_3)$, we multiply 6(46) on the right by $\phi_C(x_3)$ with $t_1, t_2 > t_3$:

$$\begin{aligned} T\phi_A(x_1)\phi_B(x_2)\phi_C(x_3) &= (T\phi_A(x_1)\phi_B(x_2))\phi_C(x_3) \\ &= :\phi_A(x_1)\phi_B(x_2):\phi_C(x_3) + (0|T\phi_A(x_1)\phi_B(x_2)|0)\phi_C(x_3) \end{aligned} \qquad 6(47)$$

Again $\phi_C(x_3)$ can be absorbed into the normal product by writing

$$:\phi_A(x_1)\phi_B(x_2):\phi_C(x_3) = :\phi_A(x_1)\phi_B(x_2):\phi_C^{(+)}(x_3) + :\phi_A(x_1)\phi_B(x_2):\phi_C^{(-)}(x_3) \qquad 6(48)$$

The first term on the right-hand side is already in normal-ordered form. The last term can be normal-ordered by successively commuting (or anticommuting) $\phi_C^{(-)}(x_3)$ through to the left with the aid of 6(45) and we

* We can derive the result 6(46) far more simply by noting that the difference between $T\phi_A(x_1)\phi_B(x_2)$ and $:\phi_A(x_1)\phi_B(x_2):$ can only be a *c*-number commutator (or anticommutator) function. This *c*-number function is then immediately identified as $(0|T\phi_A(x_1)\phi_B(x_2)|0)$ by taking the vacuum expectation value of both sides and noting that $(0|:\phi_A(x_1)\phi_B(x_2):|0) = 0$.

easily find

$$:\phi_A(x_1)\phi_B(x_2):\phi_C^{(-)}(x_3) = :\phi_A(x_1)\phi_B(x_2)\phi_C^{(-)}(x_3):$$
$$+\phi_A(x_1)(0|T\phi_B(x_2)\phi_C(x_3)|0)$$
$$\pm\phi_B(x_2)(0|T\phi_A(x_1)\phi_C(x_3)|0) \qquad 6(49)$$

Thus, combining 6(48) and 6(49) with 6(47) we get the result

$$T\phi_A(x_1)\phi_B(x_2)\phi_C(x_3) = :\phi_A(x_1)\phi_B(x_2)\phi_C(x_3):$$
$$+\phi_C(x_3)(0|T\phi_A(x_1)\phi_B(x_2)|0)$$
$$\pm\phi_B(x_2)(0|T\phi_A(x_1)\phi_C(x_3)|0)$$
$$+\phi_A(x_1)(0|T\phi_B(x_2)\phi_C(x_3)|0) \qquad 6(50)$$

for $t_1, t_2 > t_3$. The latter restriction is lifted in the same way as before by noting that 6(50) remains invariant under any permutation of $\phi_A(x_1)$, $\phi_B(x_2)$ and $\phi_C(x_3)$ as long as the same permutation is applied to both sides of the equation.

For higher order products it is useful to have a more compact notation. Accordingly we introduce the symbol

$$\underbracket{\phi_A(x_1)\phi_B}(x_2) = (0|T\phi_A(x_1)\phi_B(x_2)|0) \qquad 6(51)$$

for the vacuum expectation values which appear in the decompositions 6(46) and 6(50); this symbol will be referred to as the *contraction* of $\phi_A(x_1)$ with $\phi_B(x_2)$. We also define a generalized normal product containing one or more contractions by

$$:\underbracket{A\,\underbracket{B\,C\,\underbracket{D\,E\,F}\ldots J\,K}\,L\,M}\ldots: = \pm:CE\ldots JL\ldots:\underbracket{AK}\,\underbracket{BM}\,\underbracket{DF}\ldots \qquad 6(52)$$

where the + or − sign is assigned according to whether the number of interchanges of fermion factors required to go from the ordering $(ABCDEF\ldots.JKLM\ldots)$ to $(CE\ldots JL\ldots AKBMDF\ldots.)$ is even or odd. In this notation, 6(46) and 6(50) take the form

$$T\phi_A(x_1)\phi_B(x_2) = :\phi_A(x_1)\phi_B(x_2):+\underbracket{\phi_A(x_1)\phi_B}(x_2) \qquad 6(53)$$

and

$$T\phi_A(x_1)\phi_B(x_2)\phi_C(x_3) = :\phi_A(x_1)\phi_B(x_2)\phi_C(x_3):$$
$$+:\underbracket{\phi_A(x_1)\phi_B}(x_2)\phi_C(x_3):$$
$$+:\underbracket{\phi_A(x_1)\phi_B(x_2)\phi_C}(x_3):$$
$$+:\phi_A(x_1)\underbracket{\phi_B(x_2)\phi_C}(x_3): \qquad 6(54)$$

respectively. With the same notation, the decomposition formula for a *four*-fold product is given by

$$
\begin{aligned}
T\phi_A(x_1)\phi_B(x_2)\phi_C(x_3)\phi_D(x_4) = {} & :\phi_A(x_1)\phi_B(x_2)\phi_C(x_3)\phi_D(x_4): \\
& + :\underbrace{\phi_A(x_1)\phi_B(x_2)}\phi_C(x_3)\phi_D(x_4): \\
& + \text{all other singly contracted normal products} \\
& + :\underbrace{\phi_A(x_1)\phi_B(x_2)}\underbrace{\phi_C(x_3)\phi_D(x_4)}: \\
& + \text{all other doubly contracted normal products.}
\end{aligned}
\qquad 6(55)
$$

The derivation of 6(55) follows the same pattern as the derivation of 6(53) and 6(54) and is left to the reader as an exercise, [Problem 1]. Note that there are six singly contracted normal products in 6(55), namely

$$
:\underbrace{A\,B}\,C\,D:,\quad :\underbrace{A\,B\,C}\,D:,\quad :\underbrace{A\,B\,C\,D}:,\quad :A\,\underbrace{B\,C}\,D:,\quad :A\,\underbrace{B\,C\,D}:,\quad :A\,B\,\underbrace{C\,D}:
$$

and three doubly contracted products

$$
:\underbrace{A\,B}\,\underbrace{C\,D}: \qquad :\underbrace{A\,\overbrace{B\,C}\,}\!\!\overbrace{\,D}: \qquad :\underbrace{A\,\underbrace{B\,C}\,D}:
$$

Basing ourselves on the examples 6(53), 6(54) and 6(55), we can expect the decomposition formula for an n-fold product to read

$$
\begin{aligned}
T\phi_A(x_1)\phi_B(x_2)\ldots\phi_N(x_n) = {} & :\phi_A(x_1)\phi_B(x_2)\ldots\phi_N(x_n): \\
& + \text{all singly contracted normal products} \\
& + \text{all doubly contracted normal products} \\
& + \text{all triply contracted normal products} \\
& + \ldots
\end{aligned}
\qquad 6(56)
$$

the sum on the right-hand side comprising all possible sets of contractions between pairs of operators. The general formula 6(56) has been proved by Wick (1950). We shall not reproduce the proof which proceeds by induction, but refer the reader to Wick's original paper*.

The application of the Wick decomposition theorem 6(56) to the perturbation expansion 6(39) yields the set of all possible S-matrix elements $(b|S|a)$ for a given interaction Hamiltonian $\mathscr{H}_I(x)$. To each normal product there corresponds an S-matrix element which can be represented graphically in terms of a *Feynman diagram*. The explicit calculation of

* See also S. S. Schweber, *An Introduction to Relativistic Quantum Field Theory*, Harper and Row, New York, 1961, Section 13c.

these S-matrix elements or Feynman diagrams will be taken up in the following section.

6-2 Feynman diagrams

Spinor Electrodynamics We illustrate the Wick decomposition of the S-matrix by reference to the explicit example of spinor electrodynamics. The Hamiltonian for this theory is given by 5(13c), or rather

$$\mathscr{H}_I(x) = -ie\tfrac{1}{2}[\bar{\psi}(x), \gamma_\mu \psi(x)]A_\mu(x) \tag{6(57)}$$

where we have inserted the antisymmetrized spinor current 5(180) in accordance with the discussion in Section 5-4. We have adopted the Lorentz gauge formulation of electrodynamics, so that realizable state vectors are subject to the Lorentz condition 5(9).

To terms of order e^2, the scattering operator 6(39) is given by

$$\begin{aligned} S = 1 - e\int d^4x_1\, T[\tfrac{1}{2}[\bar{\psi}(x_1), \gamma_\mu\psi(x_1)]A_\mu(x_1)] \\ + \frac{e^2}{2!}\int d^4x_1 \int d^4x_2\, T[\tfrac{1}{2}[\bar{\psi}(x_1), \gamma_\mu\psi(x_1)]A_\mu(x_1)\tfrac{1}{2}[\bar{\psi}(x_2), \gamma_\nu\psi(x_2)]A_\nu(x_2)] \end{aligned} \tag{6(58)}$$

Let us decompose 6(58) into normal products. The Wick decomposition of the first order term is trivial. We get simply

$$S^{(1)} = -e\int d^4x_1 :\tfrac{1}{2}[\bar{\psi}(x_1), \gamma_\mu\psi(x_1)]A_\mu(x_1):$$

or, since $:\psi_\beta\bar{\psi}_\alpha: = -:\bar{\psi}_\alpha\psi_\beta:$,

$$S^{(1)} = -e\int d^4x_1 :\bar{\psi}(x_1)\gamma_\mu\psi(x_1)A_\mu(x_1): \tag{6(59)}$$

No contraction terms appear in this order. The contractions

$$\underbrace{\bar{\psi}(x_1)A_\mu}(x_1) \quad \text{and} \quad \underbrace{\psi(x_1)A_\mu}(x_1)$$

vanish identically and the antisymmetrized density $\frac{1}{2}[\bar{\psi}(x_1), \gamma_\mu\psi(x_1)]$ is already in normal ordered form by virtue of 3(205), so that the contraction

$$\begin{aligned} (0|T\tfrac{1}{2}[\bar{\psi}(x_1), \gamma_\mu\psi(x_1)]|0) &= (0|T:\bar{\psi}(x_1)\gamma_\mu\psi(x_1):|0) \\ &= (0|:\bar{\psi}(x_1)\gamma_\mu\psi(x_1):|0) \end{aligned} \tag{6(60)}$$

also vanishes identically. For the second order term in 6(58) the Wick decomposition gives the following array of normal products:

$$S^{(2)} = \frac{e^2}{2!}\int d^4x_1 \int d^4x_2 :\bar{\psi}(x_1)\gamma_\mu\psi(x_1)\bar{\psi}(x_2)\gamma_\nu\psi(x_2)A_\mu(x_1)A_\nu(x_2): \qquad 6(61a)$$

$$+\frac{e^2}{2!}\int d^4x_1 \int d^4x_2 :\bar{\psi}(x_1)\gamma_\mu\psi(x_1)\bar{\psi}(x_2)\gamma_\nu\psi(x_2)\underbrace{A_\mu(x_1)A_\nu(x_2)}: \qquad 6(61b)$$

$$+\frac{e^2}{2!}\int d^4x_1 \int d^4x_2 :\bar{\psi}(x_1)\gamma_\mu\underbrace{\psi(x_1)\bar{\psi}(x_2)}\gamma_\nu\psi(x_2)A_\mu(x_1)A_\nu(x_2): \qquad 6(61c)$$

$$+\frac{e^2}{2!}\int d^4x_1 \int d^4x_2 :\underbrace{\bar{\psi}(x_1)\gamma_\mu\psi(x_1)\bar{\psi}(x_2)\gamma_\nu\psi(x_2)}A_\mu(x_1)A_\nu(x_2): \qquad 6(61d)$$

$$+\frac{e^2}{2!}\int d^4x_1 \int d^4x_2 :\bar{\psi}(x_1)\gamma_\mu\underbrace{\psi(x_1)\bar{\psi}(x_2)}\gamma_\nu\psi(x_2)\underbrace{A_\mu(x_1)A_\nu(x_2)}: \qquad 6(61e)$$

$$+\frac{e^2}{2!}\int d^4x_1 \int d^4x_2 :\underbrace{\bar{\psi}(x_1)\gamma_\mu\psi(x_1)\bar{\psi}(x_2)\gamma_\nu\psi(x_2)}\underbrace{A_\mu(x_1)A_\nu(x_2)}: \qquad 6(61f)$$

$$+\frac{e^2}{2!}\int d^4x_1 \int d^4x_2 :\underbrace{\bar{\psi}(x_1)\gamma_\mu\underbrace{\psi(x_1)\bar{\psi}(x_2)}\gamma_\nu\psi(x_2)}A_\mu(x_1)A_\nu(x_2): \qquad 6(61g)$$

$$+\frac{e^2}{2!}\int d^4x_1 \int d^4x_2 :\underbrace{\bar{\psi}(x_1)\gamma_\mu\underbrace{\psi(x_1)\bar{\psi}(x_2)}\gamma_\nu\psi(x_2)}\underbrace{A_\mu(x_1)A_\nu(x_2)}: \qquad 6(61h)$$

Note that we have replaced the antisymmetrized product $\frac{1}{2}[\bar{\psi}, \gamma_\mu\psi]$ by $\bar{\psi}\gamma_\mu\psi$ throughout the generalized normal products 6(61a)–6(61h). This is justified by the fact that such normal products are antisymmetric under permutation of two neighbouring fermion fields. For example,

$$\tfrac{1}{4}:[\bar{\psi}, \gamma_\mu\underbrace{\psi][\bar{\psi}}, \gamma_\nu\psi]: = :\bar{\psi}\gamma_\mu\underbrace{\psi\bar{\psi}}\gamma_\nu\psi:$$

Contractions of two fermion fields at the same point, i.e. $\underbrace{\bar{\psi}(x_1)\psi(x_1)}$ and $\underbrace{\bar{\psi}(x_2)\psi(x_2)}$, do not appear, owing to 6(60). Moreover, there are no contractions of the type $\underbrace{\psi(x_1)\psi(x_2)}$ or $\underbrace{\bar{\psi}(x_1)\bar{\psi}(x_2)}$; we have*

$$(0|T\psi_\alpha(x_1)\psi_\beta(x_2)|0) = (0|T\bar{\psi}_\alpha(x_1)\bar{\psi}_\beta(x_2)|0) = 0 \qquad 6(62)$$

The only non-vanishing contractions to appear in 6(61a)–6(61h) are

$$\underbrace{A_\mu(x_1)A_\nu(x_2)} = (0|TA_\mu(x_1)A_\nu(x_2)|0) = \delta_{\mu\nu}D_F(x_1-x_2) \qquad 6(63)$$

$$\underbrace{\psi(x_1)\bar{\psi}(x_2)} = (0|T\psi(x_1)\bar{\psi}(x_2)|0) = S_F(x_1-x_2) \qquad 6(64)$$

* To check this statement, either use the explicit representations 3(154a) and 3(154b) or, alternatively, observe that $\psi(\bar{\psi})$ destroys (creates) one unit of electric charge. Since the vacuum state carries zero electric charge, the statement follows.

where we have used 4(73) and 3(199) to identify the contractions with the Feynman propagators

$$\delta_{\mu\nu} D_{\mathrm{F}}(x_1 - x_2) = \frac{1}{(2\pi)^4} \int d^4k \, e^{ik.(x_1 - x_2)} \delta_{\mu\nu} \frac{-i}{k^2 - i\varepsilon} \qquad 6(65)$$

$$S_{\mathrm{F}}(x_1 - x_2) = \frac{1}{(2\pi)^4} \int d^4p \, e^{ip.(x_1 - x_2)} \frac{-1}{\gamma \,.\, p - im - i\varepsilon} \qquad 6(66)$$

for electromagnetic and spinor fields respectively.

S-Matrix Elements All operator products in the Wick decomposition 6(59) and 6(61a)–6(61h) are in normal-ordered form—all destruction operators standing to the right, all creation operators standing to the left—so that we can read off the corresponding S-matrix elements directly. For the first order term 6(59), we list some of the normal products and their corresponding S-matrix elements in Table 6.1. We recall that $\psi^{(+)}$ and $A_\mu^{(+)}$ are the positive-frequency or *destruction* parts of ψ and A_μ; $\psi^{(-)}$ and $A_\mu^{(-)}$ are the *creation* parts.

TABLE 6.1

FIRST ORDER NORMAL PRODUCTS AND CORRESPONDING S-MATRIX ELEMENTS*

Normal Product	Initial States	Final States
$\bar{\psi}^{(-)}\gamma_\mu \psi^{(+)} A_\mu^{(+)}$	$c_{\mathbf{p}\sigma}{}^\dagger a_{\mathbf{k}\lambda}{}^\dagger\lvert 0)$	$c_{\mathbf{p}'\sigma'}{}^\dagger\lvert 0)$
$-\psi^{(-)}\gamma_\mu^T \bar{\psi}^{(+)} A_\mu^{(+)}$	$d_{\mathbf{p}\sigma}{}^\dagger a_{\mathbf{k}\lambda}{}^\dagger\lvert 0)$	$d_{\mathbf{p}'\sigma'}{}^\dagger\lvert 0)$
$\bar{\psi}^{(-)}\gamma_\mu \psi^{(-)} A_\mu^{(-)}$	$\lvert 0)$	$d_{\mathbf{p}\sigma}{}^\dagger c_{\mathbf{p}'\sigma'}{}^\dagger a_{\mathbf{k}\lambda}{}^\dagger\lvert 0)$
$\bar{\psi}^{(+)}\gamma_\mu \psi^{(+)} A_\mu^{(+)}$	$d_{\mathbf{p}\sigma}{}^\dagger c_{\mathbf{p}\sigma}{}^\dagger a_{\mathbf{k}\lambda}{}^\dagger\lvert 0)$	$\lvert 0)$
$\bar{\psi}^{(+)}\gamma_\mu \psi^{(+)} A_\mu^{(-)}$	$d_{\mathbf{p}\sigma}{}^\dagger c_{\mathbf{p}\sigma}{}^\dagger\lvert 0)$	$a_{\mathbf{k}\lambda}{}^\dagger\lvert 0)$
⋮	⋮	⋮

* Initial and final electron-photon states are designated by momentum **p** or **k** and by spin $\sigma = (1, 2)$ or $\lambda = (1, 2)$.

Altogether there are eight distinct normal products in the single term 6(59). The electron and positron creation operators $c_{\mathbf{p}\sigma}{}^\dagger$ and $d_{\mathbf{p}\sigma}{}^\dagger$ and the photon creation operators $a_{\mathbf{k}\lambda}{}^\dagger$ are of course identical to those which appear in the plane-wave expansions 3(154a), 3(154b) and 4(67). A word of explanation is required concerning the restriction of the photon spin index λ to the values 1 and 2. The latter refer to the states of *transverse* polarization. Since, in accordance with the adiabatic hypothesis, the initial and final states are assumed to be free, we can apply the analysis of

Section 4-2 to eliminate from them all longitudinal and timelike photons.* For the same reason, we can replace the creation operator $\bar{a}_{\mathbf{k}\lambda}$ appearing in 4(67) by $a_{\mathbf{k}\lambda}{}^{\dagger}$, [see Eq. 4(68)].

Actually, all eight normal products arising from 6(59) can be shown to vanish on grounds of energy-momentum conservation between initial and final free-particle states†. We must therefore turn to the second order terms 6(61a)–6(61h) to construct nonvanishing S-matrix elements. Just as in the case of 6(59), *each* term in 6(61) contains an array of normal products and possible S-matrix elements. Most of them vanish on grounds of energy-momentum conservation. This is true in particular of the 'pure' normal-product terms in 6(61a). We list some of the more important nonvanishing normal products in Table 6.2, together with the corresponding S-matrix elements.

Note that the position of the photon operators $A_\mu^{(+)}$ or $A_\nu^{(-)}$ relative to the Dirac operators is immaterial, since all photon operators commute with all Dirac operators.

The explicit evaluation of the S-matrix elements listed in Table 6.2 is straightforward. We first examine the matrix element for Møller scattering.

Møller Scattering The second order S-matrix element for Møller scattering is obtained from Table 6.2 as

$$(0|c_{\mathbf{q}'\tau'}c_{\mathbf{p}'\sigma'}:\bar{\psi}^{(-)}(x_1)\gamma_\mu\psi^{(+)}(x_1)\bar{\psi}^{(-)}(x_2)\gamma_\nu\psi^{(+)}(x_2):c_{\mathbf{p}\sigma}{}^{\dagger}c_{\mathbf{q}\tau}{}^{\dagger}|0) \\ \times \underbrace{A_\mu(x_1)A_\nu(x_2)} \qquad 6(67)$$

According to 6(61b), this must still be multiplied by $e^2/2$ and integrated over x_1 and x_2. We see from 6(67) that the particle $(\mathbf{p}, \sigma)$ can be destroyed either by $\psi^{(+)}(x_1)$ or by $\psi^{(+)}(x_2)$. In each case, the *remaining* operator destroys $(\mathbf{q}, \tau)$. Similarly, $(\mathbf{p}', \sigma')$ can be created either by $\bar{\psi}^{(-)}(x_1)$ or by $\bar{\psi}^{(-)}(x_2)$. Thus there is a total of four terms. Examination reveals that two

* However, this does not mean that the longitudinal and timelike photons do not play an important role *in the interaction*. In fact they generate the Coulomb force. [See the discussion of Møller scattering, below.]

† Consider for example the first matrix element in Table 6.1. Applying 3(154a), 3(154b) and 4(67) and integrating over x_1 in accordance with 6(59), we easily find

$$(\mathbf{p}'\sigma'|S^{(1)}|\mathbf{p}\sigma, \mathbf{k}\lambda) = \frac{e}{V^{3/2}}\frac{1}{\sqrt{2|\mathbf{k}|}}\sqrt{\frac{m^2}{E_{\mathbf{p}}E_{\mathbf{p}'}}}\, e^{\mu}_{\mathbf{k}\lambda}\bar{u}_{\mathbf{p}'\sigma'}\gamma_\mu u_{\mathbf{p}\sigma}\int d^4x_1\, e^{-i(p'-p-k).x_1}$$

The integral over x_1 yields a $\delta^{(4)}(p'-p-k)$ factor, ensuring energy-momentum conservation between the initial and final states. However, for finite k the constraint $p'-p = k$ is inconsistent with the mass shell restrictions $p^2 = p'^2 = -m^2, k^2 = 0$ for the initial and final particles. The same argument can be applied to show that to any order in perturbation theory, 'pure' normal products like 6(59) or 6(61a) vanish identically. This holds for all trilinear couplings.

TABLE 6.2
SECOND-ORDER NORMAL PRODUCTS

Normal Product	Initial States	Final States	Process
$:\bar{\psi}^{(-)}\gamma_\mu\psi^{(+)}\bar{\psi}^{(-)}\gamma_\nu\psi^{(+)}:\underbrace{A_\mu A_\nu}$	$c_{\mathbf{p}\sigma}{}^\dagger c_{\mathbf{q}\tau}{}^\dagger\|0)$	$c_{\mathbf{p}'\sigma'}{}^\dagger c_{\mathbf{q}'\tau'}{}^\dagger\|0)$	electron-electron scattering (Møller scattering)
$:\bar{\psi}^{(-)}\gamma_\mu\psi^{(+)}\bar{\psi}^{(+)}\gamma_\nu\psi^{(-)}:\underbrace{A_\mu A_\nu}$	$c_{\mathbf{p}\sigma}{}^\dagger d_{\mathbf{q}\tau}{}^\dagger\|0)$	$c_{\mathbf{p}'\sigma'}{}^\dagger d_{\mathbf{q}'\tau'}{}^\dagger\|0)$	electron-positron scattering (Bhabba scattering)
$\bar{\psi}^{(-)}\gamma_\mu\underbrace{\psi\bar{\psi}}\gamma_\nu\psi^{(+)}A_\mu^{(-)}A_\nu^{(+)}$	$c_{\mathbf{p}\sigma}{}^\dagger a_{\mathbf{k}\lambda}{}^\dagger\|0)$	$c_{\mathbf{p}'\sigma'}{}^\dagger a_{\mathbf{k}'\lambda'}{}^\dagger\|0)$	electron-photon scattering (Compton scattering)
$\bar{\psi}^{(+)}\gamma_\mu\underbrace{\psi\bar{\psi}}\gamma_\nu\psi^{(+)}A_\mu^{(-)}A_\nu^{(-)}$	$c_{\mathbf{p}\sigma}{}^\dagger d_{\mathbf{p}'\sigma'}{}^\dagger\|0)$	$a_{\mathbf{k}\lambda}{}^\dagger a_{\mathbf{k}'\lambda'}{}^\dagger\|0)$	electron-positron annihilation into 2 photons
$\bar{\psi}^{(-)}\gamma_\mu\underbrace{\psi\bar{\psi}}\gamma_\nu\psi^{(+)}\underbrace{A_\mu A_\nu}$	$c_{\mathbf{p}\sigma}{}^\dagger\|0)$	$c_{\mathbf{p}'\sigma'}{}^\dagger\|0)$	electron self-energy transition
$\underbrace{\bar{\psi}\gamma_\mu\underbrace{\psi\bar{\psi}}\gamma_\nu\psi}A_\mu^{(-)}A_\nu^{(+)}$	$a_{\mathbf{k}\lambda}{}^\dagger\|0)$	$a_{\mathbf{k}'\lambda'}{}^\dagger\|0)$	photon self-energy transition
$\underbrace{\bar{\psi}\gamma_\mu\underbrace{\psi\bar{\psi}}\gamma_\nu\psi}\underbrace{A_\mu A_\nu}$	$\|0)$	$\|0)$	vacuum-vacuum transition

of these differ only by the interchange of x_1 and x_2, and will therefore serve to cancel the factor $\frac{1}{2}$ in 6(61b) after integration over x_1 and x_2. There remain two distinct terms. We find, after a short calculation using 3(154a), 3(154b) and 6(65),

$$(\mathbf{p}'\sigma', \mathbf{q}'\tau'|S^{(2)}|\mathbf{p}\sigma, \mathbf{q}\tau) = A+B$$

where

$$\begin{aligned}
A &= e^2\int d^4x_1\, d^4x_2\, \mathrm{e}^{-ip'.x_1}\,\mathrm{e}^{ip.x_1}\,\mathrm{e}^{-iq'.x_2}\,\mathrm{e}^{iq.x_2} \\
&\quad\times\left[\frac{1}{\sqrt{V}}\sqrt{\frac{m}{E_{\mathbf{p}'}}}\bar{u}_{\mathbf{p}'\sigma'}\right]\gamma_\mu\left[\frac{1}{\sqrt{V}}\sqrt{\frac{m}{E_{\mathbf{p}}}}u_{\mathbf{p}\sigma}\right]\left[\frac{1}{\sqrt{V}}\sqrt{\frac{m}{E_{\mathbf{q}'}}}\bar{u}_{\mathbf{q}'\tau'}\right] \\
&\quad\times\gamma_\nu\left[\frac{1}{\sqrt{V}}\sqrt{\frac{m}{E_{\mathbf{q}}}}u_{\mathbf{q}\tau}\right]\frac{1}{(2\pi)^4}\int d^4k\,\mathrm{e}^{ik.(x_1-x_2)}\delta_{\mu\nu}\frac{-i}{k^2-i\varepsilon} \\
&= e^2\frac{1}{V^2}\left(\frac{m^4}{E_{\mathbf{p}'}E_{\mathbf{p}}E_{\mathbf{q}'}E_{\mathbf{q}}}\right)^{1/2}\int d^4k(2\pi)^4\delta^{(4)}(p'-p-k)(2\pi)^4\delta^{(4)}(q'-q+k) \\
&\quad\times\bar{u}_{\mathbf{p}'\sigma'}\gamma_\mu u_{\mathbf{p}\sigma}\left[\frac{1}{(2\pi)^4}\delta_{\mu\nu}\frac{-i}{k^2}\right]\bar{u}_{\mathbf{q}'\tau'}\gamma_\nu u_{\mathbf{q}\tau}
\end{aligned} \qquad 6(68a)$$

and

$$B = -e^2 \int d^4x_1\, d^4x_2\, e^{-iq'.x_1} e^{ip.x_1} e^{-ip'.x_2} e^{iq.x_2}$$

$$\times \left[\frac{1}{\sqrt{V}}\sqrt{\frac{m}{E_{\mathbf{q}'}}}\bar{u}_{\mathbf{q}'\tau'}\right]\gamma_\mu\left[\frac{1}{\sqrt{V}}\sqrt{\frac{m}{E_{\mathbf{p}}}}u_{\mathbf{p}\sigma}\right]\left[\frac{1}{\sqrt{V}}\sqrt{\frac{m}{E_{\mathbf{p}'}}}\bar{u}_{\mathbf{p}'\sigma'}\right]$$

$$\times \gamma_\nu\left[\frac{1}{\sqrt{V}}\sqrt{\frac{m}{E_{\mathbf{q}}}}u_{\mathbf{q}\tau}\right]\frac{1}{(2\pi)^4}\int d^4k\, e^{ik.(x_1-x_2)}\delta_{\mu\nu}\frac{-i}{k^2-i\varepsilon}$$

$$= -e^2\frac{1}{V^2}\left(\frac{m^4}{E_{\mathbf{p}'}E_{\mathbf{p}}E_{\mathbf{q}'}E_{\mathbf{q}}}\right)^{1/2}\int d^4k(2\pi)^4\delta^{(4)}(q'-p-k)(2\pi)^4\delta^{(4)}(p'-q+k)$$

$$\times \bar{u}_{\mathbf{q}'\tau'}\gamma_\mu u_{\mathbf{p}\sigma}\left[\frac{1}{(2\pi)^4}\delta_{\mu\nu}\frac{-i}{k^2}\right]\bar{u}_{\mathbf{p}'\sigma'}\gamma_\nu u_{\mathbf{q}\tau} \qquad 6(68b)$$

Note that B differs from A both by an overall change in sign and by the exchange of the wave functions of the two final electrons

$$(\mathbf{p}', \sigma') \rightleftarrows (\mathbf{q}', \tau')$$

Accordingly, B is known as the *exchange scattering* matrix element.

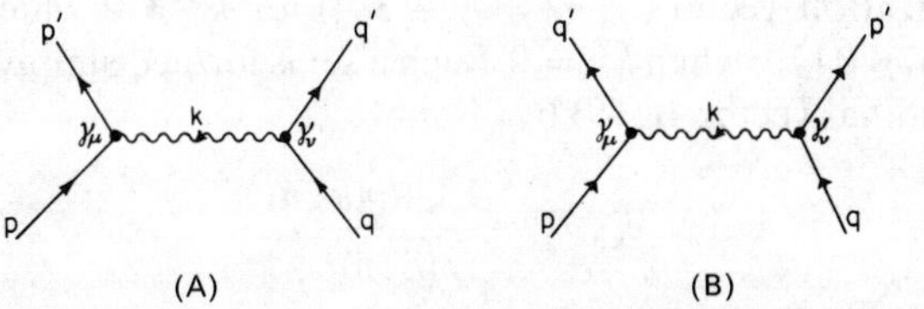

FIG. 6.1. Feynman diagrams for Direct (A) and exchange (B) Møller scattering. Internal wavy line represents Feynman propagator for photon given by 6(69). External lines represent spinor factors with $p^2 = p'^2 = q^2 = q'^2 = -m^2$.

A diagrammatic representation [Feynman (1949)] for the Møller scattering amplitudes A and B is shown in Fig. 6.1.* The factors γ_μ and γ_ν are represented by *vertices* joined by an internal wavy line representing the propagator factor

$$\delta_{\mu\nu}\frac{1}{(2\pi)^4}\frac{-i}{k^2-i\varepsilon} \qquad 6(69)$$

* These diagrams were introduced by Feynman (1949) on the basis of an entirely different, more intuitive approach. The procedure followed here is due to Dyson (1949). An extensive account of Feynman's original procedure is given in J. D. Bjorken and S. Drell, *Relativistic Quantum Mechanics*, McGraw-Hill, New York, (1964).

while the Dirac spinor factors are represented by external lines entering the diagram from the bottom and leaving at the top. Each vertex also carries an energy-momentum conservation delta function, in addition to the γ_μ factor, and the internal line carries the instruction to integrate over its momentum*.

The integrations over k in 6(68a) and 6(68b) are actually trivial and we find for A, for example,

$$e^2(2\pi)^4\delta^{(4)}(p'+q'-p-q)\bar{u}_{\mathbf{p}'\sigma'}\gamma_\mu u_{\mathbf{p}\sigma}\delta_{\mu\nu}\frac{-i}{(p'-p)^2}\bar{u}_{\mathbf{q}'\tau'}\gamma_\nu u_{\mathbf{q}\tau} \qquad 6(70)$$

where we have suppressed normalization factors for convenience. Note the appearance of the pole at $(p'-p)^2 = 0$ and the presence of the $\delta^{(4)}(p'+q'-p-q)$ factor ensuring overall conservation of energy and momentum. A further feature is the fact that the photon propagator 6(69) is taken for a ' photon mass ' $k^2 = (p'-p)^2$ which is in general nonzero. In this sense, Møller scattering can be viewed as proceeding through the exchange of a *virtual* (i.e. off the $k^2 = 0$ mass shell) photon between the two interacting electrons. All four types of virtual photons—transverse, longitudinal, and timelike—are exchanged. To exhibit the role of the longitudinal and timelike photons we introduce an orthonormal set of four polarization vectors $e_{\mathbf{k}\lambda}(\lambda = 1, 2, 3, 4)$ for $k^2 \neq 0$ which reduces to the set 1(167)–1(170) when $k^2 = 0$. Such a set is formed simply by replacing the longitudinal vector 1(169) by

$$e_{\mathbf{k}3} = -\frac{k+\eta(k\,.\,\eta)}{\sqrt{(k\,.\,\eta)^2+k^2}} \qquad 6(71)$$

We may then write

$$\begin{aligned}
&\bar{u}_{\mathbf{p}'\sigma'}\gamma_\mu u_{\mathbf{p}\sigma}\delta^{\mu\nu}\frac{-i}{k^2}\bar{u}_{\mathbf{q}'\tau'}\gamma_\nu u_{\mathbf{q}\tau} \\
&\quad = \sum_{\lambda=1}^{4} \bar{u}_{\mathbf{p}'\sigma'}\gamma_\mu u_{\mathbf{p}\sigma} e^\mu_{\mathbf{k}\lambda} e^\nu_{\mathbf{k}\lambda}\frac{-i}{k^2}\bar{u}_{\mathbf{q}'\tau'}\gamma_\nu u_{\mathbf{q}\tau} \\
&\quad = \sum_{\lambda=1}^{4} \bar{u}_{\mathbf{p}'\sigma'}\gamma\,.\,e_{\mathbf{k}\lambda}u_{\mathbf{p}\sigma}\frac{-i}{k^2}\bar{u}_{\mathbf{q}'\tau'}\gamma\,.\,e_{\mathbf{k}\lambda}u_{\mathbf{q}\tau}
\end{aligned} \qquad 6(72)$$

with $k = p'-p$. The $\lambda = 1$ and $\lambda = 2$ terms represent the contribution to the scattering amplitude due to the exchange of *transverse* photons. A simple manipulation using 6(71) and 1(170) allows us to rewrite the

*Detailed rules for associating Feynman diagrams to arbitrary S-matrix elements will be presented later.

contribution of the longitudinal and timelike photons in the form

$$\sum_{\lambda=3}^{4} \bar{u}_{\mathbf{p}'\sigma'}\gamma \,.\, e_{\mathbf{k}\lambda} u_{\mathbf{p}\sigma} \frac{-i}{k^2} \bar{u}_{\mathbf{q}'\tau'}\gamma \,.\, e_{\mathbf{k}\lambda} u_{\mathbf{q}\tau}$$

$$= \bar{u}_{\mathbf{p}'\sigma'}\gamma \,.\, \eta u_{\mathbf{p}\sigma} \bar{u}_{\mathbf{q}'\tau'}\gamma \,.\, \eta u_{\mathbf{q}\tau} \left[\frac{(k\,.\,\eta)^2}{(k\,.\,\eta)^2+k^2} - 1\right] \frac{-i}{k^2}$$

$$= \bar{u}_{\mathbf{p}'\sigma'}\gamma \,.\, \eta u_{\mathbf{p}\sigma} \bar{u}_{\mathbf{q}'\tau'}\gamma \,.\, \eta u_{\mathbf{q}\tau} \frac{-i}{(k\,.\,\eta)^2+k^2} \qquad 6(73)$$

where we have used the fact that

$$k_\mu \bar{u}_{\mathbf{p}'\sigma'}\gamma_\mu u_{\mathbf{p}\sigma} = \bar{u}_{\mathbf{p}'\sigma'}(\gamma \,.\, p' - \gamma \,.\, p) u_{\mathbf{p}\sigma}$$

$$= im\bar{u}_{\mathbf{p}'\sigma'} u_{\mathbf{p}\sigma} - im\bar{u}_{\mathbf{p}'\sigma'} u_{\mathbf{p}\sigma}$$

$$= 0 \qquad 6(74a)$$

and similarly,

$$k_\mu \bar{u}_{\mathbf{q}'\tau'}\gamma_\mu u_{\mathbf{q}\tau} = \bar{u}_{\mathbf{q}'\tau'}(\gamma \,.\, q - \gamma \,.\, q') u_{\mathbf{q}\tau}$$

$$= 0 \qquad 6(74b)$$

by virtue of the free Dirac equation. In the special Lorentz frame with $\eta = (0, 0, 0, i)$, 6(73) reduces to

$$\bar{u}_{\mathbf{p}'\sigma'}\gamma_4 u_{\mathbf{p}\sigma} \bar{u}_{\mathbf{q}'\tau'}\gamma_4 u_{\mathbf{q}\tau} \frac{i}{|\mathbf{k}|^2} \qquad 6(75)$$

which can easily be seen to represent the effect of the instantaneous Coulomb interaction*. A similar separation of radiation and Coulomb effects has already been exhibited in 5(61).

Electron-Positron Scattering There are four possible normal products contributing to electron-positron, or Bhabba, scattering, namely

$$:\bar{\psi}^{(-)}\gamma_\mu\psi^{(+)}\bar{\psi}^{(+)}\gamma_\nu\psi^{(-)}: \underbrace{A_\mu A_\nu} \qquad 6(76a)$$

$$:\bar{\psi}^{(+)}\gamma_\mu\psi^{(-)}\bar{\psi}^{(-)}\gamma_\nu\psi^{(+)}: \underbrace{A_\mu A_\nu} \qquad 6(76b)$$

$$:\bar{\psi}^{(+)}\gamma_\mu\psi^{(+)}\bar{\psi}^{(-)}\gamma_\nu\psi^{(-)}: \underbrace{A_\mu A_\nu} \qquad 6(76c)$$

$$:\bar{\psi}^{(-)}\gamma_\mu\psi^{(-)}\bar{\psi}^{(+)}\gamma_\nu\psi^{(+)}: \underbrace{A_\mu A_\nu} \qquad 6(76d)$$

* Recall that $1/|\mathbf{k}|^2$ is just the momentum representation of the Coulomb potential $1/r$. Note that if $k^2 \to 0$, the Coulomb term 6(75) can be neglected relative to the transverse terms in 6(72). This means that as the exchanged photon becomes more and more 'real', only the transverse degrees of freedom associated with the pure radiation field become important.

Only 6(76a) is listed in Table 6.2. The terms 6(76a) and 6(76c) give distinctly different contributions (see below), while the remaining two terms simply contribute a factor 2 after integration over x_1 and x_2. Following the same procedure as for 6(68a) and 6(68b) we find that the S-matrix element is given as the sum of two distinct terms

$$A = e^2 \frac{1}{V^2}\left(\frac{m^4}{E_{\mathbf{p}'}E_{\mathbf{p}}E_{\mathbf{q}'}E_{\mathbf{q}}}\right)^{1/2} \int d^4k (2\pi)^4 \delta^{(4)}(p'-p-k)(2\pi)^4\delta^{(4)}(q'-q+k)$$

$$\times \bar{u}_{\mathbf{p}'\sigma'}\gamma_\mu u_{\mathbf{p}\sigma}\left[\frac{1}{(2\pi)^4}\frac{-i}{k^2-i\varepsilon}\right]\bar{v}_{\mathbf{q}\tau}\gamma_\mu v_{\mathbf{q}'\tau'} \qquad 6(77a)$$

and

$$B = -e^2 \frac{1}{V^2}\left(\frac{m^4}{E_{\mathbf{p}'}E_{\mathbf{p}}E_{\mathbf{q}'}E_{\mathbf{q}}}\right)^{1/2} \int d^4k (2\pi)^4 \delta^{(4)}(p+q-k)(2\pi)^4\delta^{(4)}(p'+q'-k)$$

$$\times \bar{u}_{\mathbf{p}'\sigma'}\gamma_\mu v_{\mathbf{q}'\tau'}\left[\frac{1}{(2\pi)^4}\frac{-i}{k^2-i\varepsilon}\right]\bar{v}_{\mathbf{q}\tau}\gamma_\mu u_{\mathbf{p}\sigma} \qquad 6(77b)$$

where we again note the overall change of sign in B relative to A. We represent the two terms 6(77a) and 6(77b) by means of the Feynman diagrams shown in Fig. 6.2. At this point it may be worthwhile to state our graphical convention for external fermion lines more systematically. This is done in Table 6.3.

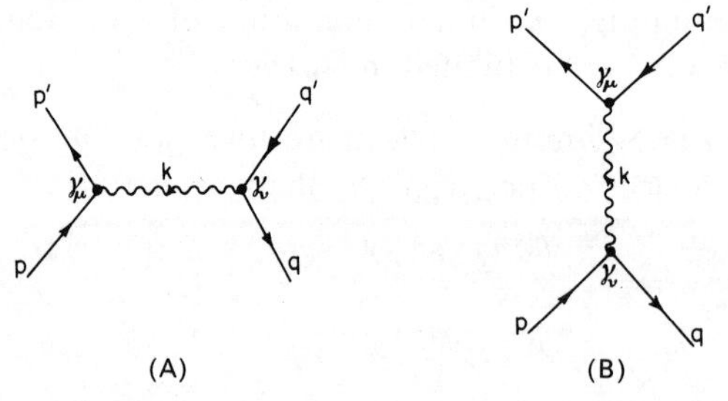

FIG. 6.2. Feynman diagrams for Bhabba scattering. Diagram A represents scattering through exchange of virtual photon. Diagram B represents virtual annihilation and recreation.

Observe that for positrons, the momentum points in the opposite direction to the arrow. This is unavoidable since, for *fermion* lines we use the arrow to distinguish electrons from positrons. For photons (and in

TABLE 6.3
CORRESPONDENCE BETWEEN SPINOR FACTORS AND EXTERNAL FERMION LINES

Factor in S-Matrix	Graphical Representation	
$\sqrt{\frac{m}{E_{\mathbf{p}'}V}}\bar{u}_{\mathbf{p}'\sigma'}\dots u_{\mathbf{p}\sigma}\sqrt{\frac{m}{E_{\mathbf{p}}V}}$		electron scattering
$\sqrt{\frac{m}{E_{\mathbf{p}}V}}\bar{v}_{\mathbf{p}\sigma}\dots v_{\mathbf{p}'\sigma'}\sqrt{\frac{m}{E_{\mathbf{p}'}V}}$		positron scattering
$\sqrt{\frac{m}{E_{\mathbf{p}'}V}}\bar{u}_{\mathbf{p}'\sigma'}\dots v_{\mathbf{q}'\tau'}\sqrt{\frac{m}{E_{\mathbf{q}'}V}}$	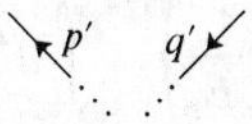	pair creation
$\sqrt{\frac{m}{E_{\mathbf{q}}V}}\bar{v}_{\mathbf{q}\tau}\dots u_{\mathbf{p}\sigma}\sqrt{\frac{m}{E_{\mathbf{p}}V}}$	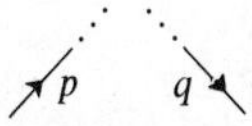	pair annihilation

general for particles which are identical to their antiparticles, e.g. π^0's) we may use the arrow to indicate the momentum direction, as in Figs. 6.1 and 6.2, but more frequently we shall simply suppress the arrow on a photon or neutral boson line.

An important point to note in connection with Bhabba scattering is that B is *not* the exchange scattering graph as in the case of Møller scattering, but instead represents virtual annihilation and recreation of the electron-positron pair. Physically, the existence of an exchange scattering matrix element in the case of electron-electron scattering is a consequence of the Pauli principle. It does not arise in the case of electron-positron scattering, since the two particles are not indistinguishable.

Compton Scattering The S-matrix element for electron-photon scattering arises from the four normal products

$$\bar{\psi}^{(-)}(x_1)\gamma_\mu\underbrace{\psi(x_1)\bar{\psi}(x_2)}\gamma_\nu\psi^{(+)}(x_2)(A_\mu^{(-)}(x_1)A_\nu^{(+)}(x_2)+A_\nu^{(-)}(x_2)A_\mu^{(+)}(x_1)) \qquad 6(78a)$$

and

$$\bar{\psi}^{(-)}(x_2)\gamma_\nu\underbrace{\psi(x_2)\bar{\psi}(x_1)}\gamma_\mu\psi^{(+)}(x_1)(A_\mu^{(-)}(x_1)A_\nu^{(+)}(x_2)+A_\nu^{(-)}(x_2)A_\mu^{(+)}(x_1)) \qquad 6(78b)$$

due to 6(61c) and 6(61d) respectively; only the first term of 6(78a) is listed in Table 6.2. The contribution of 6(78b) can be seen to be identical to that of 6(78a) after integration over x_1 and x_2 and we find, using 3(154a), 3(154b), 4(67) and 6(66),

$$(\mathbf{k}'\lambda', \mathbf{p}'\sigma'|S^{(2)}|\mathbf{k}\lambda, \mathbf{p}\sigma) = A + B$$

with

$$A = e^2 \frac{1}{V^2}\left(\frac{m^2}{4|\mathbf{k}'||\mathbf{k}|E_{\mathbf{p}'}E_{\mathbf{p}}}\right)^{1/2} \int d^4q (2\pi)^4 \delta^{(4)}(p+k-q)(2\pi)^4 \delta^{(4)}(q-p'-k')$$

$$\times\, e^{\nu}_{\mathbf{k}'\lambda'} \bar{u}_{\mathbf{p}'\sigma'} \gamma_\nu \left[\frac{1}{(2\pi)^4} \frac{-1}{\gamma . q - im}\right] \gamma_\mu u_{\mathbf{p}\sigma} e^{\mu}_{\mathbf{k}\lambda} \qquad 6(79a)$$

and

$$B = e^2 \frac{1}{V^2}\left(\frac{m^2}{4|\mathbf{k}'||\mathbf{k}|E_{\mathbf{p}'}E_{\mathbf{p}}}\right)^{1/2} \int d^4q (2\pi)^4 \delta^{(4)}(p-k'-q)(2\pi)^4 \delta^{(4)}(q-p'+k)$$

$$\times\, e^{\mu}_{\mathbf{k}\lambda} \bar{u}_{\mathbf{p}'\sigma'} \gamma_\mu \left[\frac{1}{(2\pi)^4} \frac{-1}{\gamma . q - im}\right] \gamma_\nu u_{\mathbf{p}\sigma} e^{\nu}_{\mathbf{k}'\lambda'} \qquad 6(79b)$$

where the two matrix elements A and B arise from the two terms in the normal product 6(78a). Note that the *total* matrix element is invariant under the substitution

$$k \rightleftarrows -k' \qquad e_{\mathbf{k}\lambda} \rightleftarrows e_{\mathbf{k}'\lambda'} \qquad 6(80)$$

This invariance is known as crossing symmetry and the B-contribution 6(79b) is referred to as the 'crossed' term. This terminology is suggested by the graphical representation of A and B exhibited in Fig. 6.3. The internal fermion line represents the fermion propagator function

$$\frac{1}{(2\pi)^4} \frac{-1}{\gamma . q - im} \qquad 6(81)$$

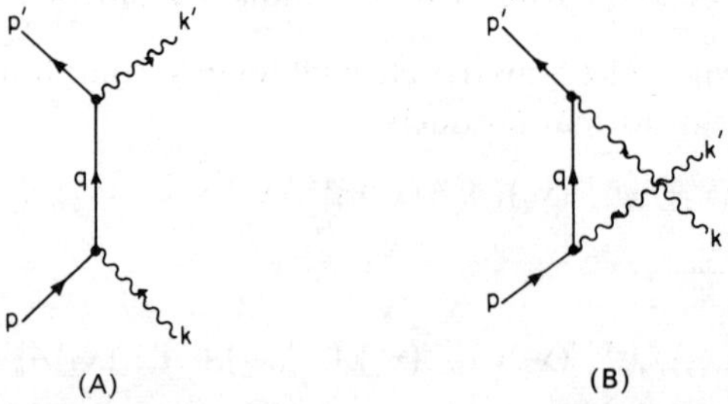

FIG. 6.3. Compton scattering graphs. Internal fermion line represents Feynman propagator 6(81). Diagram B represents 'crossed' term.

TABLE 6.4
EXTERNAL PHOTON FACTORS

Factor in S-Matrix		Graphical Representation
$\frac{1}{\sqrt{2\lvert\mathbf{k}\rvert V}}e^{\mu}_{\mathbf{k}\lambda}$	$(\lambda = 1, 2)$	k ingoing photon line
$\frac{1}{\sqrt{2\lvert\mathbf{k}'\rvert V}}e^{\mu}_{\mathbf{k}'\lambda'}$	$(\lambda = 1, 2)$	k' outgoing photon line

while the external photon lines are associated with the factors $e_{\mathbf{k}\lambda}$ and $e_{\mathbf{k}'\lambda'}$ according to Table 6.4.

In Section 6.3 we shall compute the cross section for Compton scattering based on the second-order matrix element 6(79a) and 6(79b). For this purpose it is convenient to define a *T-matrix* element from which normalization and delta-function factors have been removed. For Compton scattering, we define

$$(\mathbf{k}'\lambda', \mathbf{p}'\sigma'|S|\mathbf{k}\lambda, \mathbf{p}\sigma) = \frac{-i}{V^2}\left(\frac{m^2}{4|\mathbf{k}'||\mathbf{k}|E_{\mathbf{p}'}E_{\mathbf{p}}}\right)^{1/2}(2\pi)^4\delta^{(4)}(p'+k'-p-k)$$
$$\times(\mathbf{k}'\lambda', \mathbf{p}'\sigma'|T|\mathbf{k}\lambda, \mathbf{p}\sigma) \qquad 6(82)$$

Performing the trivial d^4q integrations in 6(79a) and 6(79b), we obtain

$$(\mathbf{k}'\lambda', \mathbf{p}'\sigma'|T^{(2)}|\mathbf{k}\lambda, \mathbf{p}\sigma) = \bar{u}_{\mathbf{p}'\sigma'}e^{\nu}_{\mathbf{k}'\lambda'}t_{\nu\mu}e^{\mu}_{\mathbf{k}\lambda}u_{\mathbf{p}\sigma} \qquad 6(83)$$

with

$$t_{\nu\mu} = ie^2\left(\gamma_\nu\frac{1}{\gamma\,.\,(p+k)-im}\gamma_\mu+\gamma_\mu\frac{1}{\gamma\,.\,(p-k')-im}\gamma_\nu\right)$$
$$= ie^2\left(\gamma_\nu\frac{\gamma\,.\,(p+k)+im}{(p+k)^2+m^2}\gamma_\mu+\gamma_\mu\frac{\gamma\,.\,(p-k')+im}{(p-k')^2+m^2}\gamma_\nu\right) \qquad 6(84)$$

as the second order T-matrix element for Compton scattering.

Electron and Photon Self-Energy Transitions The terms 6(61e)–6(61g) in the Wick decomposition of $S^{(2)}$ are self-energy terms, that is, they refer to transitions between initial and final states consisting of a single particle with the same quantum numbers. Consider first the two terms 6(61e) and 6(61f). These give rise to electron self-energy transitions, as indicated in Table 6.2. We find

$$(\mathbf{p}'\sigma'|S^{(2)}|\mathbf{p}\sigma) = e^2\frac{1}{V}\left(\frac{m^2}{E_{\mathbf{p}}E_{\mathbf{p}'}}\right)^{1/2}\int d^4q\, d^4k(2\pi)^4\delta^{(4)}(p-q-k)$$
$$\times(2\pi)^4\delta^{(4)}(q+k-p')$$
$$\times\bar{u}_{\mathbf{p}'\sigma'}\gamma_\mu\left[\frac{1}{(2\pi)^4}\frac{-1}{\gamma\,.\,q-im}\right]\gamma_\nu u_{\mathbf{p}\sigma}\left[\frac{1}{(2\pi)^4}\delta_{\mu\nu}\frac{-i}{k^2-i\varepsilon}\right] \quad 6(85)$$

where, as usual, the factor $\frac{1}{2}$ has been cancelled by the appearance of two identical terms. Diagrammatically, 6(85) is represented by the graph of Fig. 6.4. The delta functions allow one trivial integration, say d^4q, with the result

$$(\mathbf{p}'\sigma'|S^{(2)}|\mathbf{p}\sigma) = -e^2\frac{1}{V}\frac{m}{E_{\mathbf{p}}}(2\pi)^4\delta^{(4)}(p'-p)\bar{u}_{\mathbf{p}'\sigma'}\Sigma(p)u_{\mathbf{p}\sigma} \quad 6(86)$$

where

$$\Sigma(p) = \frac{-1}{(2\pi)^4}\int d^4k\gamma_\mu\frac{-1}{\gamma\,.\,(p-k)-im}\gamma_\mu\frac{-i}{k^2} \quad 6(87)$$

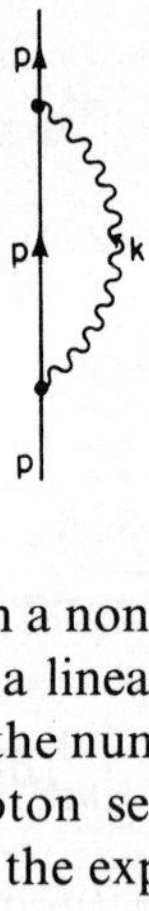

FIG. 6.4. Electron self-energy graph.

We are left with a nontrivial integration to perform. In fact the right-hand side of 6(87) is a linearly divergent integral, as is easily seen by counting powers of k in the numerator and denominator of the integrand.

For the photon self-energy transition matrix element, generated by 6(61g), we find the expression

$$(\mathbf{k}'\lambda'|S^{(2)}|\mathbf{k}\lambda) = -e^2\frac{1}{V}\left(\frac{1}{4|\mathbf{k}'|\,|\mathbf{k}|}\right)^{1/2}\int d^4q\, d^4p(2\pi)^4\delta^{(4)}(k+q-p)$$
$$\times(2\pi)^4\delta^{(4)}(k'+q-p)$$
$$\times e^\mu_{\mathbf{k}'\lambda'}\,\mathrm{Tr}\left[\gamma_\mu\frac{1}{(2\pi)^4}\frac{-1}{\gamma\,.\,q-im}\gamma_\nu\frac{1}{(2\pi)^4}\frac{-1}{\gamma\,.\,p-im}\right]e^\nu_{\mathbf{k}\lambda} \quad 6(88)$$

The amplitude 6(88) is represented graphically in Fig. 6.5. Note the overall minus sign, a characteristic of the closed fermion loop. Its appearance can be traced to the operation

$$\bar{\psi}(x_1)\gamma_\mu\psi(x_1)\bar{\psi}(x_2)\gamma_\nu\psi(x_2) = -\gamma_\mu^{\alpha\beta}\psi_\beta(x_1)\bar{\psi}_\gamma(x_2)\gamma_\nu^{\gamma\delta}\psi_\delta(x_2)\bar{\psi}_\alpha(x_1)$$
$$= -\mathrm{Tr}\,[\gamma_\mu S_F(x_1-x_2)\gamma_\nu S_F(x_2-x_1)]$$

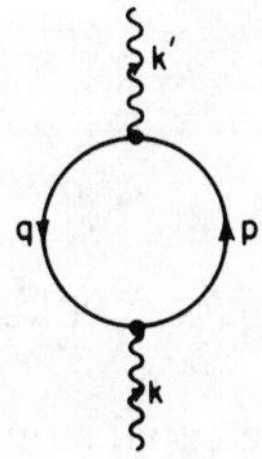

FIG. 6.5. Photon self-energy graph. Momentum follows direction of arrow in closed loop.

giving rise to the trace term in 6(88). Performing the integral over d^4q in 6(88) we find

$$(\mathbf{k}'\lambda'|S^{(2)}|\mathbf{k}\lambda) = ie^2\frac{1}{V}\frac{1}{2|\mathbf{k}|}(2\pi)^4\delta^{(4)}(k'-k)e^\mu_{\mathbf{k}'\lambda'}\Pi_{\mu\nu}(k)e^\nu_{\mathbf{k}\lambda} \qquad 6(89)$$

where

$$\Pi_{\mu\nu}(k) = \frac{i}{(2\pi)^4}\int d^4p\,\mathrm{Tr}\left[\gamma_\mu\frac{1}{\gamma\,.\,(p-k)-im}\gamma_\nu\frac{1}{\gamma\,.\,p-im}\right] \qquad 6(90)$$

is again divergent, this time quadratically. $\Pi_{\mu\nu}$ is known as the vacuum-polarization tensor.

The existence of nonzero matrix elements of the type 6(86) and 6(89) poses a problem, for it appears to imply that one-particle states are not ' steady ', that is, that

$$U(\infty, -\infty)|\mathbf{p}\sigma) \neq |\mathbf{p}\sigma)$$
$$U(\infty, -\infty)|\mathbf{k}\lambda) \neq |\mathbf{k}\lambda)$$

On the other hand, the steadiness of the one-particle states is essential for a reasonable theory of scattering. This difficulty was already alluded to in our discussion of the adiabatic hypothesis. The solution, as far as the electron self-energy is concerned, is provided by *mass renormalization**, which allows us to absorb 6(87) into the definition of the experimental electron mass and thereby ensure the steadiness of $|\mathbf{p}\sigma)$.

* See Section 6-4.

The solution of the *photon* self-energy problem is rather different. Let us write

$$\Pi_{\mu\nu}(k) = \Pi_{\mu\nu}(0) + \Pi^{(1)}_{\mu\nu}(k) \qquad 6(91)$$

where $\Pi^{(1)}_{\mu\nu}(k) = \Pi_{\mu\nu}(k) - \Pi_{\mu\nu}(0)$ tends to zero as $k \to 0$. We note that the quadratic divergence of $\Pi_{\mu\nu}(k)$ is entirely contained in the constant term $\Pi_{\mu\nu}(0)$. Indeed, applying the operator expansion*

$$\frac{1}{A+B} = \frac{1}{A} - \frac{1}{A}B\frac{1}{A} + \frac{1}{A}B\frac{1}{A}B\frac{1}{A} + \ldots \qquad 6(92)$$

to the propagator $[\gamma \,.\, (p-k) - im]^{-1}$ in 6(90) we have

$$\frac{1}{\gamma \,.\, p - \gamma \,.\, k - im} = \frac{1}{\gamma \,.\, p - im} + \frac{1}{\gamma \,.\, p - im}\gamma \,.\, k\frac{1}{\gamma \,.\, p - im} + \ldots \qquad 6(93)$$

Inserting 6(93) back into 6(90) we see that the *first* term in the expansion gives $\Pi_{\mu\nu}(0)$,

$$\Pi_{\mu\nu}(0) = \frac{i}{(2\pi)^4}\int d^4p \,\mathrm{Tr}\left[\gamma_\mu \frac{1}{\gamma \,.\, p - im}\gamma_\nu \frac{1}{\gamma \,.\, p - im}\right] \qquad 6(94)$$

while $\Pi_{\mu\nu}(k)$ is determined by the sum of the remaining (k-dependent) terms in 6(93). Since each of these terms features an additional factor of p in the denominator, it is clear that the divergence of $\Pi^{(1)}_{\mu\nu}(k)$ is at worst linear. *The entire quadratically divergent part of* $\Pi_{\mu\nu}(k)$ *is absorbed into* $\Pi_{\mu\nu}(0)$. Now $\Pi^{(1)}_{\mu\nu}(k)$ is a second-rank tensor dependent only on the 4-vector k_μ. Lorentz-covariance therefore imposes the general form

$$\Pi^{(1)}_{\mu\nu}(k) = C(k^2)\delta_{\mu\nu}k^2 + D(k^2)k_\mu k_\nu \qquad 6(95)$$

where $C(k^2)$ and $D(k^2)$ must be regular at $k^2 = 0$ to ensure that $\Pi^{(1)}_{\mu\nu}(k) \to 0$ as $k \to 0$. Similarly

$$\Pi_{\mu\nu}(0) = A\delta_{\mu\nu} \qquad 6(96)$$

where A is a constant. The key point is that, as a consequence of the conservation law 5(8) for the electromagnetic current, the *total* tensor $\Pi_{\mu\nu}(k)$ must satisfy the constraint

$$k_\mu \Pi_{\mu\nu}(k) = 0 \qquad 6(97)$$

* The expansion 6(92) is proved by iterating the operator identity

$$\frac{1}{A+B} = \frac{1}{A}(A+B-B)\frac{1}{A+B}$$
$$= \frac{1}{A} - \frac{1}{A}B\frac{1}{A+B}$$

This can be shown by applying the technique developed in the following chapter*. Applying 6(97) to

$$\Pi_{\mu\nu}(k) = A\delta_{\mu\nu} + C(k^2)\delta_{\mu\nu}k^2 + D(k^2)k_\mu k_\nu$$

we get

$$k_\nu[A + k^2C(k^2) + k^2D(k^2)] = 0 \qquad 6(98)$$

For $k^2 = 0$ with $k \neq 0$ we deduce that

$$A = 0 \qquad 6(99)$$

Hence, for arbitrary k^2

$$C(k^2) = -D(k^2)$$

and $\Pi_{\mu\nu}(k)$ must be of the form

$$\Pi_{\mu\nu}(k) = (\delta_{\mu\nu}k^2 - k_\mu k_\nu)C(k^2) \qquad 6(100)$$

Inserting the result 6(100) into 6(89), we find

$$\begin{aligned}(\mathbf{k}'\lambda'|S^{(2)}|\mathbf{k}\lambda) &\sim \delta^{(4)}(k'-k)e^{\mu}_{\mathbf{k}\lambda'}(\delta_{\mu\nu}k^2 - k_\mu k_\nu)e^{\nu}_{\mathbf{k}\lambda}C(k^2) \\ &= 0 \qquad 6(101)\end{aligned}$$

by virtue of the transversality condition

$$e_{\mathbf{k}\lambda} \,.\, k = 0 \qquad (\lambda = 1, 2) \qquad 6(102)$$

and the mass shell constraint $k^2 = 0$ for the initial and final photons. Thus current conservation ensures that one-photon states are steady.

Actually, $\Pi_{\mu\nu}(k)$ as given by 6(90) appears to violate the current conservation requirement, in that direct calculation yields the result

$$\Pi_{\mu\nu}(0) = A\delta_{\mu\nu} \neq 0 \qquad 6(103)$$

in contradiction with 6(99). The easiest way to see this is to compute

$$\begin{aligned}A = \tfrac{1}{4}\Pi_{\mu\mu}(0) &= \frac{i}{(2\pi)^4}\int d^4p \tfrac{1}{4}\,\mathrm{Tr}\,[\gamma_\mu(\gamma\,.\,p + im)\gamma_\mu(\gamma\,.\,p+im)]\frac{1}{(p^2+m^2)^2} \\ &= \frac{-i}{(2\pi)^4}\int d^4p \tfrac{1}{4}\,\mathrm{Tr}\,[4(p^2+m^2) - 2\gamma\,.\,p(\gamma\,.\,p+im)]\frac{1}{(p^2+m^2)^2} \\ &= \frac{-i}{(2\pi)^4}\int d^4p\frac{2p^2+4m^2}{(p^2+m^2)^2} \qquad 6(104)\end{aligned}$$

* See Chapter 7, Problem 5. The electromagnetic current induced in the physical vacuum state by an external electromagnetic field A^{ext}_μ is essentially $\Pi_{\mu\nu}(k)A^{\text{ext}}_\nu(k)$ in momentum space.

where we have used the property $\mathrm{Tr}\,\gamma_\mu = 0$. Thus A does not vanish. In fact it diverges quadratically! The appearance of this awkward term violates current conservation and reopens the question of the steadiness of the one-photon states, since it gives a nonvanishing contribution to the right-hand side of 6(101). Fortunately it can be disposed of by an appropriate method of manipulating the divergent integral 6(90)*.

Vacuum-Vacuum Transitions The final term 6(61h) in the Wick decomposition of the second-order S-matrix is a pure c-number and represents a vacuum-vacuum transition, as indicated in Table 6.2. We find

$$\begin{aligned}(0|S^{(2)}|0) &= \frac{-e^2}{2!}\int d^4x_1\, d^4x_2\, \mathrm{Tr}\,[\gamma_\mu S_F(x_1-x_2)\gamma_\nu S_F(x_2-x_1)]\delta_{\mu\nu}D(x_1-x_2)\\ &= -\frac{e^2}{2!}\int d^4p\, d^4q\, d^4k(2\pi)^4\delta^{(4)}(p-q+k)(2\pi)^4\delta^{(4)}(p-q+k)\\ &\quad\times \mathrm{Tr}\left[\gamma_\mu\frac{1}{(2\pi)^4}\frac{-1}{\gamma\,.\,p-im}\gamma_\nu\frac{1}{(2\pi)^4}\frac{-1}{\gamma\,.\,q-im}\right]\frac{1}{(2\pi)^4}\delta_{\mu\nu}\frac{-i}{k^2-i\varepsilon}\end{aligned} \qquad 6(105)$$

which we represent graphically by Fig. 6.6. Note that the factor $(2!)^{-1}$ has not cancelled out in this case. Performing the d^4k integration we get the divergent result

$$(0|S^{(2)}|0) = \frac{ie^2}{2!}\delta^{(4)}(0)\frac{1}{(2\pi)^4}\int d^4q\, d^4q\, \mathrm{Tr}\left[\gamma_\mu\frac{1}{\gamma\,.\,p-im}\gamma_\mu\frac{1}{\gamma\,.\,q-im}\right]\frac{1}{(q-p)^2} \qquad 6(106)$$

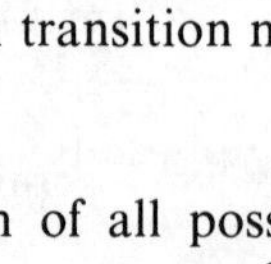

FIG. 6.6. Second-order vacuum–vacuum graph.

Actually we can simply ignore 6(106), along with all higher-order vacuum-vacuum transition matrix elements. Let us denote by

$$C = (0|S|0) \qquad 6(107)$$

the sum of all possible vacuum-vacuum graphs. Examples of fourth-order vacuum graphs are shown in Fig. 6.7 [see Problem 4]. We now

* See Schwinger (1948), Pauli (1949), Jost (1949) and also J. D. Bjorken and S. Drell, *Relativistic Quantum Mechanics*, McGraw-Hill, New York, 1964, Section 8-2. K. Johnson (1961) has shown that the automatic subtraction of $\Pi_{\mu\nu}(0)$ can be ensured by suitably redefining the fermion electromagnetic current operator, [see Chapter 7, Problem 5].

note that by virtue of energy-momentum conservation, the S-matrix can only take the vacuum state (which has $P_\mu = 0$) into itself, that is, we must have

$$S|0) = C|0) \tag{6(108)}$$

with C given by 6(107). The unitarity of S then implies that

$$|C|^2 = 1$$

so that C is just a trivial phase factor and can be disregarded.

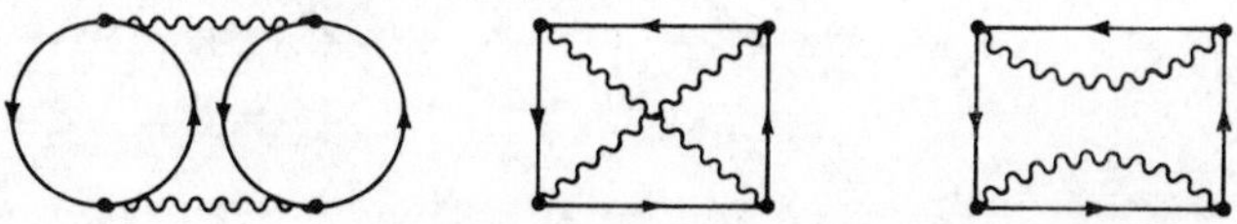

FIG. 6.7. Examples of fourth-order vacuum–vacuum graphs.

Any Feynman graph for a real physical process will be accompanied by the complete set of vacuum graphs if one calculates the perturbation series to infinite order. Thus for example a second-order Feynman graph for Compton scattering will appear in fourth order together with a disconnected second-order vacuum-vacuum graph, as shown in Fig. 6.8*.

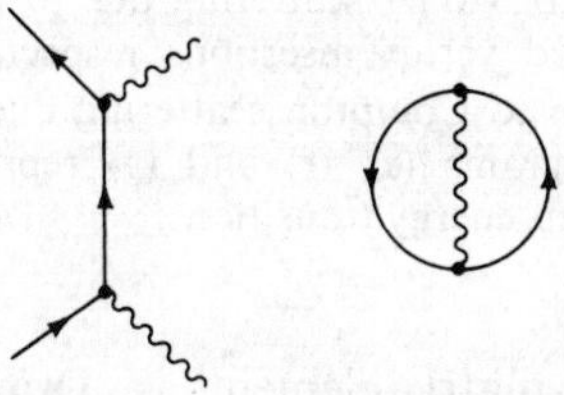

FIG. 6.8. A disconnected fourth-order graph for Compton scattering.

To sixth order it will be accompanied by disconnected fourth order graphs of the type shown in Fig. 6.7, and so on. Since in a disconnected graph, the total matrix element will simply be the product of the S-matrix factors for the disjoint portions, we conclude that the constant 6(107) appears as an *overall multiplicative factor* in the S-matrix. If S' is the S-matrix from which disconnected graphs are *omitted*, then CS' is the full S-matrix including disconnected graphs. Since, as we have seen, the

* The reader can easily check this statement by examining the Wick decomposition of the fourth-order scattering operator.

constant C is just a trivial phase factor, we shall in the future omit all disconnected graphs from consideration.

General Rules for Feynman Graphs The Wick decomposition for any order e^n in the perturbation series leads to a set of elementary processes which can be represented by means of Feynman diagrams. Some examples of fourth-order graphs are shown in Fig. 6.9. To order e^n, the rule for

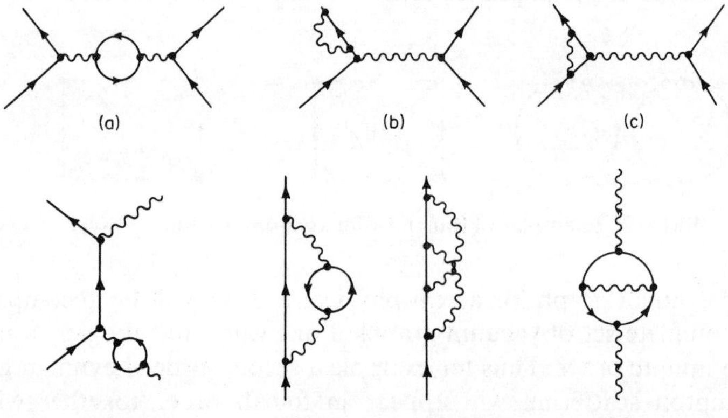

FIG. 6.9. Some fourth-order graphs. Diagrams (a), (b) and (c) represent corrections to Møller scattering due to photon self-energy, fermion self-energy and vertex insertions respectively. Diagram (d) represents a correction to Compton scattering due to a photon self-energy insertion. Diagrams (e), (f) and (g) represent fourth-order electron and photon self-energy transitions.

generating all possible S-matrix elements is: Draw all connected graphs and use the rules for associating S-matrix factors to Feynman graph components as summarized in Tables 6.3, 6.4 and 6.5. In addition, include a relative minus sign for a fermion exchange graph as in 6(68a) and 6(68b) and include a minus sign for each closed fermion loop.

Interaction with External Electromagnetic Field Let us add to the electrodynamic coupling Hamiltonian 6(57), the additional Hamiltonian

$$\mathscr{H}_I^{\text{ext}}(x) = -ie\tfrac{1}{2}[\bar{\psi}(x), \gamma_\mu \psi(x)] A_\mu^{\text{ext}}(x) \tag{6(109)}$$

for the coupling of the electron to an externally applied electromagnetic field $A_\mu^{\text{ext}}(x)$. The lowest order transition amplitude for electron scattering

TABLE 6.5
VERTICES AND PROPAGATORS IN SPINOR ELECTRODYNAMICS

Factor in S-matrix	Graphical Representation
$-e\gamma_\mu(2\pi)^4\delta^{(4)}(p'-p-k)$	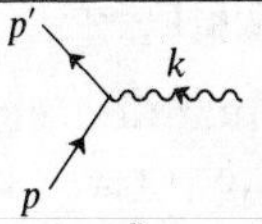 vertex
$\dfrac{1}{(2\pi)^4}\dfrac{-1}{\gamma\,.\,p-im}\left(\text{with}\int d^4p\right)$	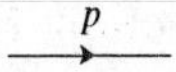 fermion propagator
$\dfrac{1}{(2\pi)^4}\delta_{\mu\nu}\dfrac{-i}{k^2}\left(\text{with}\int d^4k\right)$	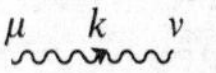 photon propagator

in the external field is then

$$(\mathbf{p}'\sigma'|S^{(1)}|\mathbf{p}\sigma) = -i\int d^4x(\mathbf{p}'\sigma'|\mathscr{H}_I^{\text{ext}}(x)|\mathbf{p}\sigma)$$

$$= -e\frac{1}{V}\left(\frac{m^2}{E_{\mathbf{p}'}E_{\mathbf{p}}}\right)^{1/2}\bar{u}_{\mathbf{p}'\sigma'}\gamma_\mu u_{\mathbf{p}\sigma}\int d^4x\,\mathrm{e}^{-ip'.x}\,\mathrm{e}^{ip.x}A_\mu^{\text{ext}}(x)$$

$$= -e\frac{1}{V}\left(\frac{m^2}{E_{\mathbf{p}'}E_{\mathbf{p}}}\right)^{1/2}\bar{u}_{\mathbf{p}'\sigma'}\gamma_\mu u_{\mathbf{p}\sigma}A_\mu^{\text{ext}}(p'-p) \qquad 6(110)$$

where $A_\mu^{\text{ext}}(k)$ is the Fourier transform of the electromagnetic field

$$A_\mu^{\text{ext}}(k) = \int A_\mu^{\text{ext}}(x)\,\mathrm{e}^{-ik.x}\,d^4x \qquad 6(111)$$

Note that the momentum k supplied at the vertex is *not* restricted to its mass shell value $k^2 = 0$ as it is for the corresponding matrix element in the case of a quantized electromagnetic field. Thus the first order scattering matrix element does not vanish in the case of an external field.

We represent the new type of vertex introduced by 6(109) by means of a $\times$ as in Fig. 6.10. To this vertex corresponds the S-matrix factor

$$-e\gamma_\mu A_\mu^{\text{ext}}(k) \qquad 6(112)$$

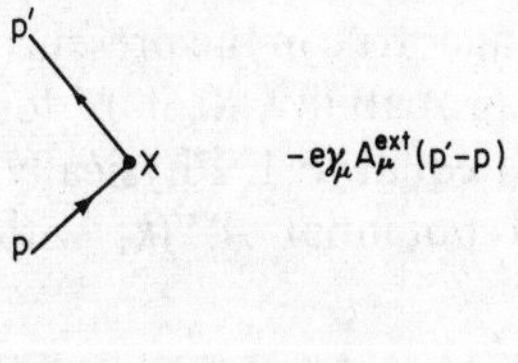

FIG. 6.10. External field vertex and associated S-matrix factor. X denotes application of external field.

where $k = p' - p$ is the momentum imparted at the vertex. This replaces the corresponding factor

$$-e\frac{1}{\sqrt{2|\mathbf{k}|V}}\gamma_\mu e^\mu_{\mathbf{k}\lambda}(2\pi)^4\delta^{(4)}(p'-p-k)$$

for a photon vertex in quantum electrodynamics. The absence of the delta function factor in the present case obviously reflects the fact that, owing to the external interaction, P_μ is not conserved between initial and final quanta.

An important special case is that of a static field $A_\mu^{\text{ext}}(\mathbf{x})$. In that case the vertex factor 6(112) becomes

$$-e\gamma_\mu A_\mu^{\text{ext}}(\mathbf{k})2\pi\delta(E_\mathbf{p}-E_{\mathbf{p}'}) \qquad 6(113)$$

with

$$A_\mu^{\text{ext}}(\mathbf{k}) = \int A_\mu^{\text{ext}}(\mathbf{x})\,\mathrm{e}^{-i\mathbf{k}.\mathbf{x}}\,d^3x \qquad 6(114)$$

Energy, though not momentum, is conserved in this case. The first order matrix element 6(110) reduces to

$$(\mathbf{p}'\sigma'|S^{(1)}|\mathbf{p}\sigma) = -2\pi\delta(E_\mathbf{p}-E_{\mathbf{p}'})\frac{m}{E_\mathbf{p}V}eA_\mu^{\text{ext}}(\mathbf{p}'-\mathbf{p})\bar{u}_{\mathbf{p}'\sigma'}\gamma_\mu u_{\mathbf{p}\sigma} \qquad 6(115)$$

The breakdown of energy-momentum conservation between initial and final quantum states has a significant effect on the vacuum-vacuum amplitude $(0|S|0)$. Examples of additional vacuum-vacuum graphs introduced by the external interaction are shown in Fig. 6.11. We have seen

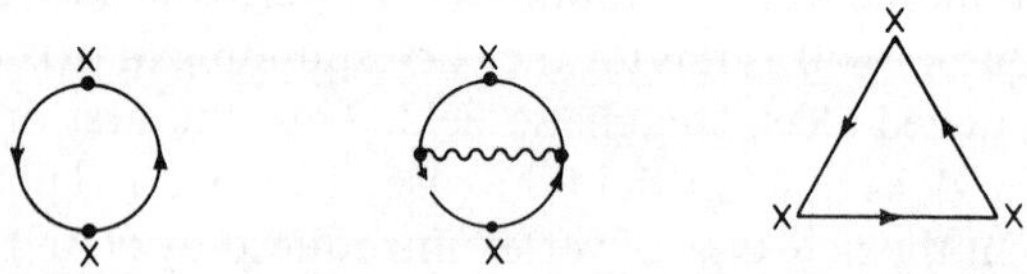

FIG. 6.11. Some additional vacuum–vacuum graphs arising from the interaction with A_μ^{ext}.

that in the absence of external fields, $(0|S|0)$ is simply a phase factor on grounds of momentum conservation. This is no longer true in the presence of the external interaction and, accordingly, the probability $|(0|S|0)|^2$ for the vacuum to remain a vacuum is no longer just equal to 1. Physically, this is due to the possibility that an external potential $A_\mu^{\text{ext}}(k)$ with

$k^2 > (2m)^2$ can create real pairs in the vacuum. Nevertheless, $(0|S|0)$ still appears as an overall multiplicative factor in the S-matrix and to calculate the probability amplitude for any given physical process relative to the amplitude for the vacuum to remain a vacuum, we simply restrict our attention to connected graphs, dividing through by $(0|S|0)$.

Neutral Pseudoscalar Meson Theory The interaction Hamiltonian 5(67) can be written in the form

$$\mathscr{H}_I(x) = -iG\tfrac{1}{2}[\bar{\psi}(x), \gamma_5\psi(x)]\phi(x) \tag{6(116)}$$

without the need for an additional antisymmetrization postulate, since, in this case*,

$$\bar{\psi}(x)\gamma_5\psi(x) = \tfrac{1}{2}[\bar{\psi}(x), \gamma_5\psi(x)]$$

The structure of 6(116) is very similar to the electrodynamic coupling 6(57) and the Wick decomposition yields an array of Feynman graphs which are topologically identical to those of spinor electrodynamics. Table 6.6 summarizes the rules for vertices and propagators and Table 6.7 lists the factors for external meson lines. The similarity of these rules to those of spinor electrodynamics is evident from inspection of the tables. We do not list the rules for external *spinor* lines, as these are identical in all respects to those given in Table 6.3. As in spinor electrodynamics, a minus sign must be included for a fermion exchange graph and for a closed fermion loop.

As an example, consider the second-order meson-nucleon scattering graphs of Fig. 6.12. These correspond to the Compton scattering graphs of Fig. 6.3. Applying the Feynman rules, we obtain the corresponding

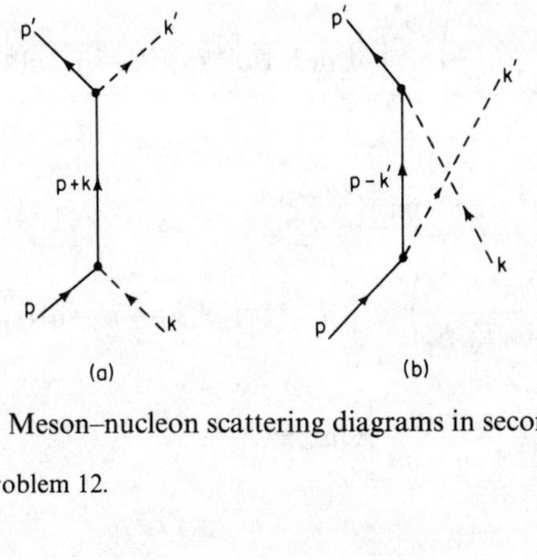

FIG. 6.12. Meson–nucleon scattering diagrams in second-order.

* See Chapter 3, Problem 12.

TABLE 6.6
VERTICES AND PROPAGATORS IN NEUTRAL PSEUDOSCALAR MESON THEORY

Factor in S-Matrix	Graphical Representation	
$-G\gamma_5(2\pi)^4\delta^{(4)}(p'-p-k)$	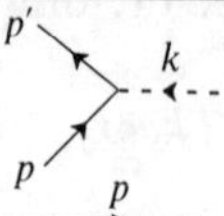	vertex
$\frac{1}{(2\pi)^4}\frac{-1}{\gamma\,.\,p-im}\left(\text{with}\int d^4p\right)$	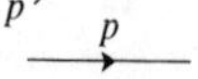	nucleon propagator
$\frac{1}{(2\pi)^4}\frac{-i}{k^2+\mu^2}\left(\text{with}\int d^4k\right)$	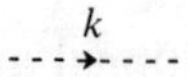	meson propagator

TABLE 6.7
EXTERNAL MESON FACTORS

Factor in S-Matrix	Graphical Representation	
$\frac{1}{\sqrt{2\omega_{\mathbf{k}}V}}$	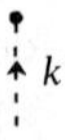	ingoing meson line
$\frac{1}{\sqrt{2\omega_{\mathbf{k}'}V}}$	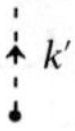	outgoing meson line

S-matrix elements, structurally similar to 6(79a) and 6(79b)

$$A = G^2\frac{1}{V^2}\left(\frac{m^2}{4\omega_{\mathbf{k}'}\omega_{\mathbf{k}}E_{\mathbf{p}'}E_{\mathbf{p}}}\right)^{1/2}\int d^4q(2\pi)^4\delta^{(4)}(p+k-q)(2\pi)^4\delta^{(4)}(q-p'-k')$$

$$\times\,\bar{u}_{\mathbf{p}'\sigma'}\gamma_5\left[\frac{1}{(2\pi)^4}\frac{-1}{\gamma\,.\,q-im}\right]\gamma_5 u_{\mathbf{p}\sigma}$$

$$B = -G^2\frac{1}{V^2}\left(\frac{m^2}{4\omega_{\mathbf{k}'}\omega_{\mathbf{k}}E_{\mathbf{p}'}E_{\mathbf{p}}}\right)^{1/2}\int d^4q(2\pi)^4\delta^{(4)}(p-k'-q)(2\pi)^4\delta^{(4)}(q-p'+k)$$

$$\times\,\bar{u}_{\mathbf{p}'\sigma'}\gamma_5\left[\frac{1}{(2\pi)^4}\frac{-1}{\gamma\,.\,q-im}\right]\gamma_5 u_{\mathbf{p}\sigma}$$

Performing the trivial integrations over d^4q we find

$$A = -G^2 \frac{1}{V^2}\left(\frac{m^2}{4\omega_{\mathbf{k}'}\omega_{\mathbf{k}}E_{\mathbf{p}}E_{\mathbf{p}'}}\right)^{1/2} (2\pi)^4\delta^{(4)}(p'+k'-p-k)\bar{u}_{\mathbf{p}'\sigma'}\gamma_5$$

$$\times \frac{\gamma.p+\gamma.k+im}{(p+k)^2+m^2}\gamma_5 u_{\mathbf{p}\sigma} \qquad 6(117a)$$

$$B = G^2 \frac{1}{V^2}\left(\frac{m^2}{4\omega_{\mathbf{k}'}\omega_{\mathbf{k}}E_{\mathbf{p}}E_{\mathbf{p}'}}\right)^{1/2} (2\pi)^4\delta^{(4)}(p'+k'-p-k)\bar{u}_{\mathbf{p}'\sigma'}\gamma_5$$

$$\times \frac{\gamma.p-\gamma.k'+im}{(p-k')^2+m^2}\gamma_5 u_{\mathbf{p}\sigma} \qquad 6(117b)$$

Note the appearance of poles at $(p+k)^2 = -m^2$ and $(p-k')^2 = -m^2$ respectively. These poles lie outside the physical region available to p and k. They correspond to placing the *intermediate* fermions in Fig. 6.12 on their mass shells, and this cannot be reconciled with the mass shell restrictions for the external lines*.

A significant difference between meson theory and quantum electrodynamics is that the meson self-energy transition matrix element (Fig. 6.13)

$$(\mathbf{k}'|S|\mathbf{k}) = iG^2 \frac{1}{V}\frac{1}{2\omega_{\mathbf{k}}}(2\pi)^4\delta^{(4)}(k'-k)\Pi(k) \qquad 6(118)$$

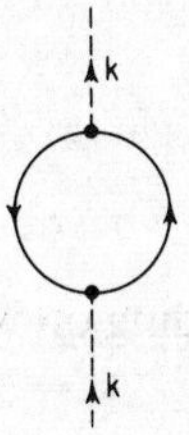

FIG. 6.13. Second-order meson self-energy transition.

with

$$\Pi(k) = \frac{i}{(2\pi)^4}\int d^4p \,\mathrm{Tr}\left[\gamma_5 \frac{1}{\gamma.(p-k)-im}\gamma_5 \frac{1}{\gamma.p-im}\right]$$

can obviously not be eliminated by appealing to current conservation. Meson mass renormalization is necessary in this case [see Section 6.4].

* We recall that the impossibility of imposing mass shell constraints on all three particles meeting at a trilinear vertex was responsible for the vanishing of the first order graphs in spinor electrodynamics. The same applies to meson theory.

SU_2-invariant π-N Coupling The extension of the rules of Table 6.6 to the SU_2-invariant coupling 5(95b) or

$$\mathscr{H}_I(x) = -iG\tfrac{1}{2}[\bar{N}(x), \gamma_5 \boldsymbol{\tau} N(x)]\boldsymbol{\phi}(x) \qquad 6(119)$$

is straightforward. We list the rules in Table 6.8. Thus, an external meson line in the isotopic spin state i ($i = 1, 2, 3$) is emitted and absorbed at a vertex τ_i *with the same S-matrix factor as in Table* 6.7. The external spinor

TABLE 6.8
VERTICES AND PROPAGATORS IN SU_2-INVARIANT PSEUDOSCALAR MESON THEORY

Factor in S-Matrix	Graphical Representation	
$-G\gamma_5\tau_i(2\pi)^4\delta^{(4)}(p'-p-k)$	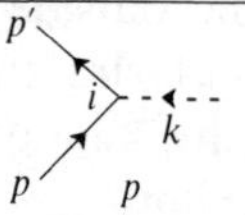	vertex
$\dfrac{1}{(2\pi)^4}\dfrac{-1}{\gamma\,.\,p-im}\ \left(\text{with} \int d^4p\right)$	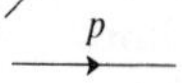	nucleon propagator
$\dfrac{1}{(2\pi)^4}\delta_{ij}\dfrac{-i}{k^2+\mu^2}\ \left(\text{with} \int d^4k\right)$	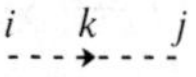	meson propagator

factors are identical to those listed in Table 6.3, except that an additional isotopic spinor factor

$$\chi_p = \begin{pmatrix}1\\0\end{pmatrix} \quad \text{or} \quad \chi_n = \begin{pmatrix}0\\1\end{pmatrix}$$

must be included for each initial and final nucleon, depending on whether the nucleon is a proton or neutron respectively*.

Other Couplings—Cancellation of Normal-Dependent Terms Sets of Feynman rules for other interaction terms can be derived in the same way as for spinor electrodynamics, but they are generally much more complicated. Fortunately, we can discount one apparent complication—the appearance of normal-dependent terms in Hamiltonian densities like 5(35), 5(73) and 5(86). Such terms are always cancelled, in the S-matrix, by other normal-dependent terms due to boson contractions. To take an

* Note that it is not necessary to include an isotopic spin factor for an external meson. External meson isotopic spin states are signalled by the index i carried by the external vertices. [See Problem 5.]

example, consider the normal-dependent term

$$\tfrac{1}{2}f^2[\bar{\psi}(x)\gamma_4\gamma_5\psi(x)]^2 \qquad 6(120)$$

in the pseudovector interaction density 5(73). To all orders of perturbation theory the contributions of 6(120) cancel with the normal-dependent terms in the meson-contraction

$$\begin{aligned}\underbrace{\partial_\mu\phi(x_1)\partial_\nu\phi(x_2)} &= (0|T\partial_\mu\phi(x_1)\partial_\nu\phi(x_2)|0)\\ &= \theta(t_1-t_2)i\partial_\mu^{(1)}\partial_\nu^{(2)}\Delta^{(+)}(x_1-x_2)\\ &\quad -\theta(t_2-t_1)i\partial_\mu^{(1)}\partial_\nu^{(2)}\Delta^{(-)}(x_1-x_2)\\ &= \partial_\mu^{(1)}\partial_\nu^{(2)}\Delta_F(x_1-x_2)+i\delta_{\mu4}\delta_{\nu4}\delta^{(4)}(x_1-x_2) \qquad 6(121)\end{aligned}$$

The first term on the right-hand side

$$-\partial_\mu\partial_\nu\Delta_F(x) = \frac{1}{(2\pi)^4}\int d^4k\, e^{ik.x}k_\mu\frac{-i}{k^2+\mu^2}k_\nu$$

yields the momentum space factor

$$k_\mu\frac{1}{(2\pi)^4}\frac{-i}{k^2+\mu^2}k_\nu$$

for an internal meson line connecting two pseudovector vertices, [see Table 6.9]. The second, normal-dependent, term on the right-hand side of 6(121) serves to cancel the effects of 6(120). For instance, the first order contribution of 6(120) to the S-matrix is given by

$$-\frac{i}{2}f^2\int d^4x\, T\bar{\psi}(x)\gamma_4\gamma_5\psi(x)\bar{\psi}(x)\gamma_4\gamma_5\psi(x) \qquad 6(122)$$

TABLE 6.9
VERTICES AND PROPAGATORS IN NEUTRAL PSEUDOVECTOR COUPLING THEORY

Factor in S-Matrix	Graphical Representation	
$-f\gamma_\mu\gamma_5 ik_\mu(2\pi)^4\delta^{(4)}(p'-p-k)$	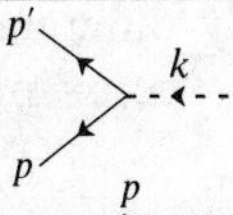	vertex
$\frac{1}{(2\pi)^4}\frac{-1}{\gamma.p-im}$ $\left(\text{with}\int d^4p\right)$		nucleon propagator
$\frac{1}{(2\pi)^4}\frac{-i}{k^2+\mu^2}$ $\left(\text{with}\int d^4k\right)$		meson propagator

Decomposing this into normal products, we get a term

$$-\frac{i}{2}f^2\int d^4x:\bar{\psi}(x)\gamma_4\gamma_5\psi(x)\bar{\psi}(x)\gamma_4\gamma_5\psi(x):$$

which is cancelled by the second-order term

$$\frac{f^2}{2!}\int d^4x_1\,d^4x_2:\bar{\psi}(x_1)\gamma_\mu\gamma_5\psi(x_1)\bar{\psi}(x_2)\gamma_\mu\gamma_5\psi(x_2):i\delta_{\mu 4}\delta_{\nu 4}\delta^{(4)}(x_1-x_2)$$

arising from the normal-dependent term in the contraction 6(121). Similarly, the term

$$-\frac{i}{2}f^2\int d^4x:\bar{\psi}(x)\gamma_4\gamma_5\underbrace{\psi(x)\bar{\psi}(x)}\gamma_4\gamma_5\psi(x):$$

in the Wick decomposition of 6(122) is cancelled by the second-order term

$$\frac{f^2}{2!}\int d^4x_1\,d^4x_2:\bar{\psi}(x_1)\gamma_\mu\gamma_5\underbrace{\psi(x_1)\bar{\psi}(x_2)}\gamma_\mu\gamma_5\psi(x_2):i\delta_{\mu 4}\delta_{\nu 4}\delta^{(4)}(x_1-x_2)$$

A general proof to all orders for pseudovector coupling theory has been sketched by Matthews (1949). The rule is: Use

$$\partial_\mu\underbrace{\phi(x_1)\partial_\nu\phi(x_2)} = \frac{1}{(2\pi)^4}\int e^{ik.(x_1-x_2)}k_\mu k_\nu\frac{-i}{k^2+\mu^2}\,d^4k$$

for the meson contraction and drop all normal-dependent terms. A similar rule holds in spin 0 electrodynamics [Rohrlich (1950)] and indeed for all cases in which the interaction Hamiltonian density exhibits a normal dependent term*. This is fortunate, since the appearance of such terms in the S-matrix would mean a violation of Lorentz invariance!

Although considerable additional work would be required to exhibit the Feynman rules for all the coupling schemes considered in Chapter 5, one particular model—the interaction 5(76) of neutral vector mesons with nucleons—can be treated as a simple modification of quantum electrodynamics. The only difference is that the photon propagator is replaced by the massive vector propagator 4(38) minus the normal-dependent term,

$$\Delta_{F\mu\nu}(k) = \left(\delta_{\mu\nu}+\frac{k_\mu k_\nu}{\mu^2}\right)\frac{-i}{k^2+\mu^2} \qquad 6(123)$$

* See Y. Takahashi, *An Introduction to Field Quantization*, Pergamon, London, 1968, for a more systematic account.

and that the external photon factors of Table 6.4 are replaced by

$$\frac{1}{\sqrt{2\omega_{\mathbf{k}}V}}\varepsilon^{\mu}_{\mathbf{k}\lambda} \qquad (\lambda = 1, 2, 3) \qquad 6(124)$$

where the $\varepsilon_{\mathbf{k}\lambda}$ are the set of three orthogonal polarization vectors 1(133).

We shall not list the explicit Feynman rules for the weak interaction coupling terms considered in Section 5-3, since, in applications to weak interaction theory, only the simple first-order term is retained. Higher order terms are not well understood, but are assumed to be unimportant, at least at low energies, due to the smallness of the weak interaction coupling constant. We shall consider higher order graphs for a four-fermion interaction only once, in connection with the self-coupled model 5(152) in Section 9-5.

We conclude this section with a summary of the Feynman rules for spin 0 electrodynamics corresponding to the Hamiltonian density 5(35) or

$$\begin{aligned}\mathscr{H}_I(x) &= ie\tfrac{1}{2}\{\phi^\dagger(x), (\partial_\mu - \overleftarrow{\partial}_\mu)\phi(x)\}A_\mu(x) \\ &\quad + e^2\tfrac{1}{2}\{\phi^\dagger(x), \phi(x)\}A_\mu(x)A_\mu(x) \qquad 6(125)\end{aligned}$$

where we have used 5(24), symmetrized the current, and dropped the normal-dependent term. Two distinct types of vertices occur in this theory, corresponding to the appearance of both trilinear and quadrilinear coupling terms in 6(125). The Feynman rules for vertices and propagators are listed in Table 6.10. External meson and photon line factors are identical to those given in Tables 6.7 and 6.4. In addition to the rules of Table 6.10, we must add the important instruction to multiply by a factor $\frac{1}{2}$ for each closed loop containing only two photon lines, as in Fig. 6.14.

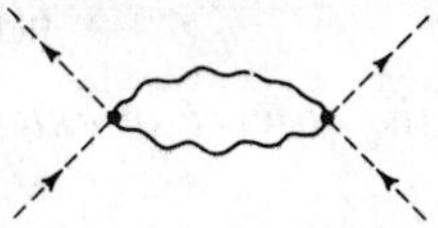

FIG. 6.14. Two-photon closed loop carrying factor $\frac{1}{2}$.

6-3 Simple applications

Introduction We now apply the results of lowest order perturbation theory to calculate cross sections and lifetimes for physical processes. We shall treat Compton scattering, Coulomb scattering, μ-decay and π-decay, where for the latter process we adopt the phenomenological coupling 5(163). We begin by considering the general relation between

TABLE 6.10

VERTICES AND PROPAGATORS FOR ELECTRODYNAMICS OF SPIN 0 BOSON

Factor in S-Matrix	Graphical Representation	
$ie(p_\mu + p'_\mu)(2\pi)^4\delta^{(4)}(p'-p-k)$	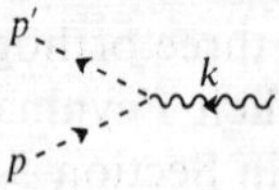	trilinear vertex
$-2ie^2\delta_{\mu\nu}(2\pi)^4\delta^{(4)}(p'-p+k'-k)$	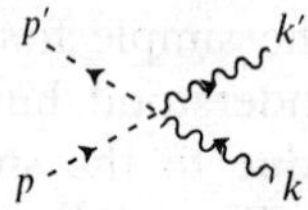	quadrilinear vertex
$\frac{1}{(2\pi)^4}\frac{-i}{p^2+\mu^2}\left(\text{with}\int d^4p\right)$		meson propagator
$\frac{1}{(2\pi)^4}\delta_{\mu\nu}\frac{-i}{k^2}\left(\text{with}\int d^4k\right)$		photon propagators

S-matrix elements, on the one hand, and physical cross sections and lifetimes on the other.

Cross Sections and Lifetimes Consider the scattering amplitude

$$S_{ba} = (b|S|a) \tag{6(126)}$$

from an initial m-particle state a to a final n-particle state b. Assuming $a \neq b$, we can replace 6(126) by the matrix element

$$R_{ba} = (b|R|a) \tag{6(127)}$$

where R is defined by

$$S = 1 + iR \tag{6(128)}$$

From the work of Section 6-2, we know that *in the absence of external fields* R_{ba} will have the general structure

$$R_{ba} = \frac{(2\pi)^4\delta^{(4)}(P_b - P_a)}{N_b N_a V^{(n+m)/2}} T_{ba} \tag{6(129)}$$

generalizing 6(82) for Compton scattering. P_a and P_b are, respectively, the initial and final 4-momenta, while N_a and N_b are products of energy-dependent normalization factors, ($\sqrt{2k_0}$ for bosons, or $\sqrt{p_0/m}$ for fermions), for initial and final particles respectively. Equation 6(129) defines the T-matrix T_{ba} for a general transition $a \to b$.

R_{ba} is the probability amplitude for a transition from the initial state a to the final state b. Accordingly, the total *transition probability* from a to b is given by the modulus squared of 6(129). We now note that the square of the factor $(2\pi)^4\delta^{(4)}(P_b - P_a)$ will give rise to the meaningless factor

$$(2\pi)^4\delta^{(4)}(0) \qquad 6(130)$$

in the transition probability. To circumvent this difficulty, we proceed in two steps and consider first the factor $\delta(0)$ arising from the energy-conservation delta-function

$$\delta(E_b - E_a) = \int_{-\infty}^{\infty} e^{i(E_b - E_a)t'}\, dt' \qquad 6(131)$$

for $E_b = E_a$. According to the formal theory of scattering, the factor $\delta(E_b - E_a)$ appearing in R_{ba} must be viewed as the limit for $\varepsilon \to 0$ of

$$\frac{1}{i}\frac{e^{i(E_b - E_a)t}\, e^{\varepsilon t}}{E_b - E_a - i\varepsilon} = \int_{-\infty}^{t} e^{i(E_b - E_a)t'}\, e^{\varepsilon t'}\, dt' \qquad 6(132)$$

with the limit $t \to \infty$ being taken at the end of the calculation*. Now in calculating physical cross sections and lifetimes we are concerned not with the *total* transition probability $|R_{ba}|^2$, but with the transition probability *per unit time*. Accordingly we must compute the *time derivative* of the modulus squared of 6(132), i.e.

$$\frac{d}{dt}\frac{e^{2\varepsilon t}}{(E_b - E_a)^2 + \varepsilon^2} = 2\varepsilon\frac{e^{2\varepsilon t}}{(E_b - E_a)^2 + \varepsilon^2} \qquad 6(133)$$

Evaluating the limit $\varepsilon \to 0$ of 6(133) with the aid of the formula

$$\delta(x) = \frac{1}{\pi}\lim_{\varepsilon \to 0}\frac{\varepsilon}{x^2 + \varepsilon^2}$$

we obtain simply

$$2\pi\delta(E_b - E_a)$$

so that Γ, the transition probability per unit time, is given by

$$\Gamma = 2\pi\delta(E_b - E_a)\left|\frac{(2\pi)^3\delta^{(3)}(\mathbf{P}_b - \mathbf{P}_a)}{N_b N_a V^{(n+m)/2}} T_{ba}\right|^2 \qquad 6(134)$$

* The explicit appearance of the factor 6(132) in the matrix element $(0|U(t, -\infty)|0)$ can be exhibited to any order of perturbation theory by a suitable manipulation of the perturbation expansion [see Section 7-1, especially the equations preceding 7(11)].

This is just $|R_{ba}|^2$ divided by the unpleasant factor $2\pi\delta(0)$. The same result is reached more directly by interpreting the factor

$$2\pi\delta(0) = \lim_{T\to\infty} \lim_{E_b\to E_a} \int_{-T/2}^{T/2} e^{i(E_b - E_a)t}\, dt$$

$$= \lim_{T\to\infty} T$$

as the (infinite) time T, during which the transition takes place. The transition probability per unit time is then obtained by dividing $|R_{ba}|^2$ by this factor.

The expression 6(134) for Γ still contains the meaningless factor $(2\pi)^3\delta^{(3)}(0)$. This is due to the fact that we have prematurely replaced the factor $V\delta_{\mathbf{P}_b\mathbf{P}_a}$ corresponding to plane waves normalized in a finite box, by the factor $(2\pi)^3\delta^{(3)}(\mathbf{P}_b - \mathbf{P}_a)$. The remedy, therefore, is to simply substitute V for $(2\pi)^3\delta^{(3)}(0)$ and write

$$\Gamma = (2\pi)^4\delta(E_b - E_a)\delta^{(3)}(\mathbf{P}_b - \mathbf{P}_a)\frac{1}{N_b^2 N_a^2 V^{n+m-1}}|T_{ba}|^2 \qquad 6(135)$$

The justification for these manipulations—passing from box normalization to continuum normalization and back again—is of course the fact that the final physical results are independent of the normalization volume.

The total cross section σ from an initial *2-particle* state a is defined as the transition probability *per unit time* and *per unit incident flux* into all accessible final states b. Thus

$$\sigma = (2\pi)^4\frac{1}{F}\frac{1}{N_a^2 V}\sum_b \frac{\delta^{(4)}(P_b - P_a)}{N_b^2 V^n}|T_{ba}|^2 \qquad 6(136)$$

where F is the incident flux factor. To determine F we proceed as follows. Let p and k be the 4-momenta of the initial particles and $p', k', q', \ldots$ denote the 4-momenta of the outgoing particles. To be definite, let us also assume that p and p' represent fermion momenta, while $k, k', q', \ldots$ represent boson momenta*. Then

$$N_a = \sqrt{2k_0}\sqrt{p_0/m}$$

$$N_b = \sqrt{2k_0'}\sqrt{2q_0'}\ldots\sqrt{p_0'/m}$$

and hence

$$\sigma = (2\pi)^4\frac{1}{F}\frac{m^2}{2k_0 p_0 V}\sum_{\mathbf{p}',\mathbf{k}',\mathbf{q}'\ldots}\frac{\delta^{(4)}(p + k - p' - k' - q' - \ldots)}{p_0' 2k_0' 2q_0' \ldots V^n}|T|^2$$

* The extension to other cases of physical interest will be evident.

Passing from box normalization to continuum normalization by means of the replacement $V^{-1}\Sigma_{\mathbf{k}} \to (2\pi)^{-3}\int d^3k$, we obtain the volume-independent result

$$\sigma = (2\pi)^4 \frac{1}{F}\frac{m^2}{2k_0 p_0 V}\frac{1}{(2\pi)^{3m}}\int \frac{d^3p'}{p'_0}\frac{d^3k'}{2k'_0}\frac{d^3q'}{2q'_0}\cdots$$
$$\times\, \delta^{(4)}(p+k-p'-k'-q'-\ldots)|T|^2 \qquad 6(137)$$

Now T is a Lorentz-invariant matrix element, since all energy-dependent normalization factors have been extracted. The integrations over the final 3-momenta are also Lorentz-invariant. Since σ itself must be Lorentz-invariant, the incident flux factor must be of the form

$$F = \frac{1}{V}\frac{1}{k_0 p_0} f \qquad 6(138)$$

where f is a Lorentz-invariant quantity. The explicit expression of f is determined by the requirement that, in the *laboratory* frame

$$\mathbf{p} = 0, \qquad p_0 = m$$

in which the initial fermion (assumed to be massive) is at rest, the incident flux reduce to

$$F = \frac{1}{V}\frac{|\mathbf{k}|}{k_0} \qquad 6(139)$$

where $|\mathbf{k}|/k_0$ and V^{-1} are respectively the velocity and density of the incident bosons, the incident wave functions having been normalized to one particle per volume V. Thus in the laboratory frame we have $f = |\mathbf{k}|p_0$ and this can be replaced by the invariant expression [Møller (1945)]

$$f = \sqrt{(p\,.\,k)^2 - p^2k^2} \qquad 6(140)$$

in an arbitrary frame. From 6(140) and 6(138) we deduce that in the *centre of momentum* frame $\mathbf{p} = -\mathbf{k}$, the incident flux is given by*

$$F = \frac{|\mathbf{k}|}{V}\left(\frac{1}{p_0}+\frac{1}{k_0}\right) \qquad 6(141)$$

Equations 6(136) and 6(138) with 6(140) provide the prescription for computing the total two-particle cross section, once the matrix element

* The result 6(141) can be written down directly, on the grounds that $\mathbf{k}(p_0^{-1}+k_0^{-1})$ is the relative velocity of the two incident particles as seen in the centre of momentum frame.

T_{ba} is known. Frequently we are also interested in computing *differential* cross sections. These are obtained by restricting the final state momentum vectors in the sum Σ_b to lie within certain differential ranges. We shall illustrate this for the case of Compton scattering presently.

Equation 6(135) can also be applied to compute the *lifetime* of a decaying particle. The latter is defined as the *inverse* of the transition probability per unit time into all accessible final states, i.e.

$$\tau^{-1} = (2\pi)^4 \frac{1}{N_a^2} \sum_b \frac{\delta^{(4)}(P_b - P_a)}{N_b^2 V^n} |T_{ba}|^2 \qquad 6(142)$$

Replacing the summation over b by continuous momentum integrals, as in 6(137), it is easy to see that τ is independent of the normalization volume V. Later in this section we shall apply 6(142) to the calculation of the μ- and π-decay lifetimes.

Unpolarized Cross Sections and Lifetimes As given by the Feynman rules, the matrix element T_{ba} always refers to a transition between specified initial and final spin states. For Compton scattering, for example, the T-matrix element is

$$(\mathbf{k}'\lambda', \mathbf{p}'\sigma'|T|\mathbf{k}\lambda, \mathbf{p}\sigma) = \bar{u}_{\mathbf{p}'\sigma'} t_{\nu\mu} u_{\mathbf{p}\sigma} e^{\nu}_{\mathbf{k}'\lambda'} e^{\mu}_{\mathbf{k}\lambda} \qquad 6(143)$$

with $t_{\nu\mu}$ given by 6(84). $e^{\mu}_{\mathbf{k}\lambda}$ and $e^{\nu}_{\mathbf{k}'\lambda'}$ are the polarization 4-vectors of the initial and final photons respectively, and σ and σ' are the spin indices of the initial and final electrons. The cross sections for given initial and final spin states, computed from 6(143), is known as the *polarized* cross section $\sigma_{\text{pol.}}$. If the electron target and incident photon beam are unpolarized and if we do not observe the polarization of the scattered photon or recoil electron, then the observed cross section is given by *averaging* $\sigma_{\text{pol.}}$ over the initial electron spin and photon polarizations and *summing* $\sigma_{\text{pol.}}$ over the final spins and polarizations. Thus for Compton scattering,

$$\sigma_{\text{unpol.}} = \tfrac{1}{2} \sum_{\lambda=1}^{2} \sum_{\lambda'=1}^{2} \tfrac{1}{2} \sum_{\sigma=1}^{2} \sum_{\sigma'=1}^{2} \sigma_{\text{pol.}} \qquad 6(144)$$

and in general

$$\sigma_{\text{unpol.}} = \tfrac{1}{2} \sum_{\text{initial spins}} \sum_{\text{final spins}} \sigma_{\text{pol.}} \qquad 6(145)$$

Similarly, to calculate the lifetime $\tau^{-1}_{\text{unpol.}}$ for the decay of an unpolarized particle into unpolarized products, we form

$$\tau^{-1}_{\text{unpol.}} = \tfrac{1}{2} \sum_{\text{initial spins}} \sum_{\text{final spins}} \tau^{-1}_{\text{pol.}} \qquad 6(146)$$

Trace Theorems For a factor $\bar{u}_{\mathbf{p}'\sigma'} t u_{\mathbf{p}\sigma}$ in a general S-matrix element, the fermion spin sums

$$\frac{1}{2}\sum_{\sigma=1}^{2}\sum_{\sigma'=1}^{2} |\bar{u}_{\mathbf{p}'\sigma'} t u_{\mathbf{p}\sigma}|^2 = \frac{1}{2}\sum_{\sigma=1}^{2}\sum_{\sigma'=1}^{2} \bar{u}_{\mathbf{p}\sigma}\gamma_4 t^\dagger \gamma_4 u_{\mathbf{p}'\sigma'}\bar{u}_{\mathbf{p}'\sigma'} t u_{\mathbf{p}\sigma} \qquad 6(147)$$

can be handled by repeated application of 1(83a), i.e.

$$\frac{1}{2}\sum_{\sigma=1}^{2}\sum_{\sigma'=1}^{2} |\bar{u}_{\mathbf{p}'\sigma'} t u_{\mathbf{p}\sigma}|^2 = \frac{1}{2}\sum_{\sigma=1}^{2} \bar{u}_{\mathbf{p}\sigma}\gamma_4 t^\dagger \gamma_4 \frac{\gamma . p' + im}{2im} t u_{\mathbf{p}\sigma}$$

$$= \frac{1}{2}\operatorname{Tr}\left[\gamma_4 t^\dagger \gamma_4 \frac{\gamma . p' + im}{2im} t \frac{\gamma . p + im}{2im}\right] \qquad 6(147a)$$

Similarly a factor $\bar{u}_{\mathbf{p}'\sigma'} t v_{\mathbf{p}\sigma}$ is handled by using 1(83a) for the sum over σ' and 1(83b) for the sum σ, i.e.

$$\frac{1}{2}\sum_{\sigma=1}^{2}\sum_{\sigma'=1}^{2} |\bar{u}_{\mathbf{p}'\sigma'} t v_{\mathbf{p}\sigma}|^2 = \frac{1}{2}\operatorname{Tr}\left[\gamma_4 t^\dagger \gamma_4 \frac{\gamma . p' + im}{2im} t \frac{\gamma . p - im}{2im}\right] \qquad 6(147b)$$

By means of 6(147a) and 6(147b) we reduce the evaluation of fermion spin sums to the calculation of a trace. The actual calculation is usually fairly lengthy, since t will normally involve multiple products of γ-matrices. As an aid to calculation we now derive a set of theorems on traces of γ-matrices, basing ourselves on the fundamental anticommutation relations 1(25).

I. The trace of an odd number of γ-matrices vanishes. That is, we have

$$\operatorname{Tr} \gamma . a_1 \gamma . a_2 \ldots \gamma . a_n = 0 \quad (n \text{ odd}) \qquad 6(148)$$

where $a_1, a_2, \ldots a_n$ are n arbitrary 4-vectors.

Proof. Writing

$$\gamma_5 \gamma . a_1 \gamma . a_2 \ldots \gamma . a_n \gamma_5 = (-1)^n \gamma . a_1 \gamma . a_2 \ldots \gamma . a_n$$

and taking the trace of both sides, we have

$$\operatorname{Tr} \gamma . a_1 \ldots \gamma . a_n = (-1)^n \operatorname{Tr} \gamma . a_1 \ldots \gamma . a_n$$

where on the left-hand side we have used the elementary property of the trace

$$\operatorname{Tr} AB = \operatorname{Tr} BA \qquad 6(149)$$

and the fact that $\gamma_5^2 = 1$ to get rid of the two γ_5 factors. The result

6(148) then follows if n is odd. In particular we have

$$\text{Tr}\,\gamma_\mu = 0 \tag{6(150)}$$

II.

$$\text{Tr}\,1 = 4 \tag{6(151)}$$

$$\text{Tr}\,\gamma\,.\,a\gamma\,.\,b = 4a\,.\,b \tag{6(152)}$$

Proof. The first statement is trivial. The second follows by writing

$$\text{Tr}\,\gamma_\rho\gamma_\sigma = \text{Tr}\,\tfrac{1}{2}(\gamma_\rho\gamma_\sigma+\gamma_\sigma\gamma_\rho) = \text{Tr}\,\delta_{\rho\sigma} = 4\delta_{\rho\sigma} \tag{6(153)}$$

III.

$$\text{Tr}\,\gamma_5 = 0 \tag{6(154)}$$

$$\text{Tr}\,\gamma_5\gamma\,.\,a\gamma\,.\,b = 0 \tag{6(155)}$$

$$\text{Tr}\,\gamma_5\gamma\,.\,a\gamma\,.\,b\gamma\,.\,c\gamma\,.\,d = 4\varepsilon_{\mu\nu\rho\sigma}a^\mu b^\nu c^\rho d^\sigma \tag{6(156)}$$

Proof. To establish 6(154), we write

$$\text{Tr}\,\gamma_5 = \text{Tr}\,\gamma_5\gamma_\mu^2 = \text{Tr}\,\gamma_\mu\gamma_5\gamma_\mu$$

with no summation over μ. Since $\gamma_5\gamma_\mu = -\gamma_\mu\gamma_5$ we get

$$\text{Tr}\,\gamma_5 = -\text{Tr}\,\gamma_5 = 0$$

For 6(155) we note that

$$\text{Tr}\,\gamma_5\gamma_\mu\gamma_\nu = 0$$

follows trivially from 6(154) in the case when $\mu = \nu$. If $\mu \neq \nu$, the trace of $\gamma_5\gamma_\mu\gamma_\nu$ reduces to a trace of $\gamma_\rho\gamma_\sigma$ with $\sigma \neq \rho$ since $\gamma_5 = \gamma_1\gamma_2\gamma_3\gamma_4$. The result then follows from 6(153). To prove the final identity 6(156) we observe that $\text{Tr}\,\gamma_5\gamma_\mu\gamma_\nu\gamma_\rho\gamma_\sigma$ vanishes unless the indices μ, ν, ρ and σ are all different. This is a consequence of 6(154) and 6(155). On the other hand, if μ, ν, ρ and σ are all different, the result is $+1$ or -1, depending on the permutation, and we find

$$\text{Tr}\,(\gamma_5\gamma_\mu\gamma_\nu\gamma_\rho\gamma_\sigma) = 4\varepsilon_{\mu\nu\rho\sigma}$$

IV.

$$\begin{aligned}\text{Tr}\,\gamma\,.\,a_1\ldots\gamma\,.\,a_n = {} & a_1\,.\,a_2\,\text{Tr}\,\gamma\,.\,a_3\ldots\gamma\,.\,a_n \\ & -a_1\,.\,a_3\,\text{Tr}\,\gamma\,.\,a_2\gamma\,.\,a_4\ldots\gamma\,.\,a_n \\ & +\ldots \\ & +a_1\,.\,a_n\,\text{Tr}\,\gamma\,.\,a_2\ldots\gamma\,.\,a_{n-1}\end{aligned} \tag{6(157)}$$

Proof. From the fundamental commutation rule $\gamma_\mu\gamma_\nu+\gamma_\nu\gamma_\mu = 2\delta_{\mu\nu}$ we deduce

$$\gamma\,.\,a_1\gamma\,.\,a_2 = -\gamma\,.\,a_2\gamma\,.\,a_1+2a_1\,.\,a_2 \tag{6(158)}$$

Equation 6(157) then follows by repeated application of 6(158) to move $\gamma . a_1$ to the extreme right, and the fact that

$$\text{Tr}\,\gamma . a_2 \ldots \gamma . a_n \gamma . a_1 = \text{Tr}\,\gamma . a_1 \gamma . a_2 \ldots \gamma . a_n$$

by virtue of 6(149).

In the particular case $n = 4$, we can use 6(157) to obtain the useful formula*

$$\begin{aligned}\text{Tr}\,\gamma . a_1 \gamma . a_2 \gamma . a_3 \gamma . a_4 = 4a_1 . a_2 a_3 . a_4 \\ -4a_1 . a_3 a_2 . a_4 \\ +4a_1 . a_4 a_2 . a_3\end{aligned} \qquad 6(159)$$

V. For n even, we have

$$\text{Tr}\,\gamma . a_1 \gamma . a_2 \ldots \gamma . a_n = \text{Tr}\,\gamma . a_n \ldots \gamma . a_1 \qquad 6(160)$$

Proof. Using the charge conjugation matrix C, we have

$$\begin{aligned}\text{Tr}\,\gamma . a_1 \gamma . a_2 \ldots \gamma . a_n &= \text{Tr}\,C\gamma . a_1 C^{-1} C\gamma . a_2 C^{-1} \ldots C\gamma . a_n C^{-1} \\ &= (-1)^n\,\text{Tr}\,\gamma^T . a_1 \gamma^T . a_2 \ldots \gamma^T . a_n \\ &= \text{Tr}\,(\gamma . a_n \ldots \gamma . a_2 \gamma . a_1)\end{aligned}$$

Further properties of the γ-matrices which are frequently useful in trace calculations are

$$\gamma_\mu \gamma . a \gamma_\mu = -2\gamma . a \qquad 6(161\text{a})$$

$$\gamma_\mu \gamma . a \gamma . b \gamma_\mu = 4a . b \qquad 6(161\text{b})$$

$$\gamma_\mu \gamma . a \gamma . b \gamma . c \gamma_\mu = -2\gamma . c \gamma . b \gamma . a \qquad 6(161\text{c})$$

$$\gamma_\mu \gamma . a \gamma . b \gamma . c \gamma . d \gamma_\mu = 2(\gamma . d \gamma . a \gamma . b \gamma . c + \gamma . c \gamma . b \gamma . a \gamma . d) \qquad 6(161\text{d})$$

These identities follow directly from the fundamental commutation relations of the Dirac matrices [see Problem 6].

Photon Polarization Sums We now turn to the question of evaluating sums over photon polarization states. A typical summation is of the form

$$\sum_{\lambda=1}^{2} |L . e_{\mathbf{k}\lambda}|^2 = \sum_{\lambda=1}^{2} |L_\mu . e^\mu_{\mathbf{k}\lambda}|^2 \qquad 6(162)$$

for a transition matrix element of the form $L_\mu e^\mu_{\mathbf{k}\lambda}$. 6(162) represents the summation over the polarization states of a single photon. L_μ can, itself,

* The formulae used to derive 1(184) are a particular consequence of this result.

be proportional to other polarization vectors $e_{\mathbf{k}'\lambda'}$, as for example in Compton scattering, where the transition matrix element is given by 6(143) and $L_\mu = e^\nu_{\mathbf{k}'\lambda'}\bar{u}_{\mathbf{p}'\sigma'}t_{\nu\mu}u_{\mathbf{p}\sigma}$.

The simplest way to evaluate a photon spin sum like 6(162) is to use the important property [Feynman (1949)]

$$L_\mu k^\mu = 0 \qquad 6(163)$$

valid for either a real or a virtual photon 4-momentum k_μ. An equivalent statement is that the transition amplitude $L_\mu e^\mu_{\mathbf{k}\lambda}$ is invariant under the substitution $e^\mu_{\mathbf{k}\lambda} \to e^\mu_{\mathbf{k}\lambda} + k^\mu$. This property is often referred to as the *gauge invariance* of the transition amplitude, since the substitution $e^\mu_{\mathbf{k}\lambda} \to e^\mu_{\mathbf{k}\lambda} + k^\mu$ represents the classical gauge transformation 1(151) in momentum space. Equation 6(163) is evident in the simplest case of a first order amplitude for emission or absorption of a single (virtual) photon of momentum $k = p' - p$. In that case we have

$$L_\mu k_\mu = \bar{u}_{\mathbf{p}'\sigma'}\gamma_\mu u_{\mathbf{p}\sigma}k_\mu = 0 \qquad 6(164)$$

by virtue of the free Dirac equations for $u_{\mathbf{p}\sigma}$ and $\bar{u}_{\mathbf{p}'\sigma'}$ as in 6(74a). A slightly more complicated case is presented by the Compton scattering matrix element 6(83). Replacing $e_{\mathbf{k}\lambda}$ by k we have

$$\begin{aligned} L_\mu k_\mu &= \bar{u}_{\mathbf{p}'\sigma'}e^\nu_{\mathbf{k}'\lambda'}t_{\nu\mu}k^\mu u_{\mathbf{p}\sigma} \\ &= ie^2\bar{u}_{\mathbf{p}'\sigma'}\left[\gamma\,.\,e_{\mathbf{k}'\lambda'}\frac{1}{\gamma\,.\,(p+k)-im}\gamma\,.\,k+\gamma\,.\,k\frac{1}{\gamma\,.\,(p-k')-im}\gamma\,.\,e_{\mathbf{k}'\lambda'}\right]u_{\mathbf{p}\sigma} \end{aligned} \qquad 6(165)$$

Using the free Dirac equations for $u_{\mathbf{p}\sigma}$ and $\bar{u}_{\mathbf{p}'\sigma'}$ we can replace $\gamma\,.\,k$ in the first term by $\gamma\,.\,(p+k)-im$ and in the second term by

$$\gamma\,.\,(k-p')+im = \gamma\,.\,(k'-p)+im$$

giving

$$\bar{u}_{\mathbf{p}'\sigma}e^\nu_{\mathbf{k}'\lambda'}t_{\nu\mu}k^\mu u_{\mathbf{p}\sigma} = ie^2\bar{u}_{\mathbf{p}'\sigma'}(\gamma\,.\,e_{\mathbf{k}'\lambda'} - \gamma\,.\,e_{\mathbf{k}'\lambda'})u_{\mathbf{p}\sigma} = 0$$

so that gauge invariance holds also in this case. From a graphical point of view we can interpret 6(165) as the insertion of a ' vertex ' $\gamma_\mu k^\mu$ into a fermion line carrying a photon-emission vertex $\gamma_\mu e^\mu_{\mathbf{k}'\lambda'}$ [see Fig. 6.15]. When the additional vertex is inserted in all possible positions (two, in this case) relative to the first photon, and the extremities of the fermion line are restricted to the mass shell, the net effect vanishes. This can be done for both real and virtual photons and irrespective of the number of original photon vertices carried by the fermion line*.

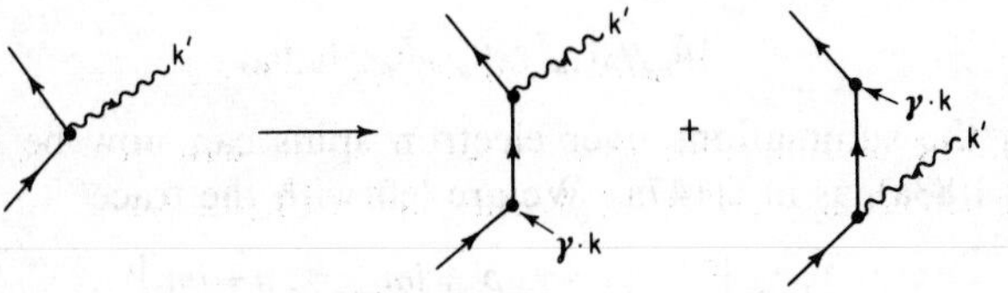

FIG. 6.15. Insertion of an additional $\gamma \, . \, k$ vertex into a fermion line, illustrating gauge invariance.

Accepting the gauge-invariance property 6(163), it is a simple exercise to show that

$$\begin{aligned}\sum_{\lambda=3}^{4} |L \, . \, e_{\mathbf{k}\lambda}|^2 &= L^* \, . \, e_{\mathbf{k}3} L \, . \, e_{\mathbf{k}3} + L^* \, . \, e_{\mathbf{k}4} L \, . \, e_{\mathbf{k}4} \\ &= 0 \end{aligned} \qquad 6(166)$$

where $e_{\mathbf{k}3}$ and $e_{\mathbf{k}4}$ are the polarization vectors 1(169) and 1(170)

$$\begin{aligned} e_{\mathbf{k}3} &= -\frac{k + \eta(k \, . \, \eta)}{k \, . \, \eta} \\ e_{\mathbf{k}4} &= i\eta \end{aligned}$$

Hence we can extend the sum over λ in 6(162) into a covariant summation over all *four* polarization vectors $e_{\mathbf{k}\lambda}$. In this way we obtain

$$\begin{aligned}\sum_{\lambda=1}^{2} |L \, . \, e_{\mathbf{k}\lambda}|^2 &= \sum_{\lambda=1}^{4} |L \, . \, e_{\mathbf{k}\lambda}|^2 \\ &= \sum_{\lambda=1}^{4} L_\rho^* e_{\mathbf{k}\lambda}^\rho L_\mu e_{\mathbf{k}\lambda}^\mu \\ &= L_\mu^* L^\mu \end{aligned} \qquad 6(167)$$

where we have used the completeness relation 1(172).

The result 6(167) yields a considerable simplification in the calculation of unpolarized cross sections in electrodynamics. For Compton scattering, for example, the summations over initial and final electron spins σ and σ' bear on the modulus squared of 6(143), i.e. on

$$\bar{u}_{\mathbf{p}\sigma} \gamma_4 t_{\sigma\rho}{}^\dagger \gamma_4 u_{\mathbf{p}'\sigma'} \bar{u}_{\mathbf{p}'\sigma'} t_{\nu\mu} u_{\mathbf{p}\sigma} e_{\mathbf{k}'\lambda'}^\sigma e_{\mathbf{k}'\lambda'}^\nu e_{\mathbf{k}\lambda}^\rho e_{\mathbf{k}\lambda}^\mu \qquad 6(168)$$

* For a general proof, see Feynman (1949). The theorem also holds if the fermion line is a closed loop, providing the latter is given by a convergent integral. See J. D. Bjorken and S. Drell, *Relativistic Quantum Fields*, McGraw-Hill, New York, 1965, Section 17-9.

The prescription is: carry out the summations over photon polarization states *first*. Applying 6(167) to the sum over final photon spins and to the average over initial photon spins, we obtain the simplified expression

$$\tfrac{1}{2}\bar{u}_{\mathbf{p}\sigma}\gamma_4 t_{\nu\mu}{}^{\dagger}\gamma_4 u_{\mathbf{p}'\sigma'}\bar{u}_{\mathbf{p}'\sigma'}t_{\nu\mu}u_{\mathbf{p}\sigma} \qquad 6(169)$$

in which the summations over electron spins can now be handled by applying 1(83a), as in 6(147a). We are left with the trace

$$\tfrac{1}{4}\,\mathrm{Tr}\left[\gamma_4 t_{\nu\mu}{}^{\dagger}\gamma_4 \frac{\gamma\,.\,p'+im}{2im} t_{\nu\mu} \frac{\gamma\,.\,p+im}{2im}\right] \qquad 6(170)$$

which is considerably easier to evaluate than the corresponding trace based on 6(168) with $\sigma \neq \nu$, $\rho \neq \mu$.*

Compton Scattering Cross Section We are now equipped to calculate the Compton scattering cross section due to the second order S-matrix element

$$(\mathbf{k}'\lambda', \mathbf{p}'\sigma'|T|\mathbf{k}\lambda, \mathbf{p}\sigma) = \bar{u}_{\mathbf{p}'\sigma'}t_{\nu\mu}u_{\mathbf{p}\sigma}e^{\nu}_{\mathbf{k}'\lambda'}e^{\mu}_{\mathbf{k}\lambda} \qquad 6(171)$$

with

$$t_{\nu\mu} = ie^2\left(\gamma_\nu \frac{\gamma\,.\,(p+k)+im}{(p+k)^2+m^2}\gamma_\mu + \gamma_\mu \frac{\gamma\,.\,(p-k')+im}{(p-k')^2+m^2}\gamma_\nu\right) \qquad 6(172)$$

Applying 6(137) with $k_0 = |\mathbf{k}|$ and $p_0 = E_{\mathbf{p}}$ we obtain the total (polarized) cross section in the form

$$\sigma = \frac{1}{(2\pi)^2}\frac{1}{F}\frac{m^2}{2|\mathbf{k}|E_{\mathbf{p}}V}\int \frac{d^3p'}{E_{\mathbf{p}'}}\frac{d^3k'}{2|\mathbf{k}'|}\delta^{(4)}(p'+k'-p-k)$$

$$\times|\bar{u}_{\mathbf{p}'\sigma'}\,t_{\nu\mu}u_{\mathbf{p}\sigma}e^{\nu}_{\mathbf{k}'\lambda'}e^{\mu}_{\mathbf{k}\lambda}|^2 \qquad 6(173)$$

where F is the incident flux factor 6(138). We shall work in the laboratory frame

$$\mathbf{p} = 0 \qquad E_{\mathbf{p}} = m \qquad 6(174)$$

in which case F is given by

$$F = \frac{1}{V}\frac{|\mathbf{k}|}{k_0} = \frac{1}{V} \qquad 6(175)$$

The *unpolarized* cross section is computed by summing 6(173) over

* In particular, the formulae 6(161a), 6(161b), 6(161c) and 6(161d) can now be applied, since the contracted indices ν and μ are always carried by γ-matrices.

initial and final spin sums, as in 6(144). As we have seen, the evaluation of

$$\sum = \tfrac{1}{4} \sum_{\sigma,\sigma'} \sum_{\lambda,\lambda'} |\bar{u}_{\mathbf{p}'\sigma'} t_{\nu\mu} u_{\mathbf{p}\sigma} e^{\nu}_{\mathbf{k}'\lambda'} e^{\mu}_{\mathbf{k}\lambda}|^2$$

reduces to the calculation of the trace 6(170) with $t_{\nu\mu}$ given by 6(172). Application of the trace theorems yields, after a fairly lengthy calculation,*

$$\sum = \frac{e^4}{2m^2}\left[\left(\frac{m^2}{p\,.\,k} - \frac{m^2}{p\,.\,k'}\right)^2 - 2\left(\frac{m^2}{p\,.\,k} - \frac{m^2}{p\,.\,k'}\right) + \frac{p\,.\,k}{p\,.\,k'} + \frac{p\,.\,k'}{p\,.\,k}\right]$$

or, in the laboratory frame with $p\,.\,k = -m|\mathbf{k}|$ and $p\,.\,k' = -m|\mathbf{k}'|$

$$\sum = \frac{e^4}{2m^2}\left[\left(\frac{m}{|\mathbf{k}|} - \frac{m}{|\mathbf{k}'|}\right)^2 + 2\left(\frac{m}{|\mathbf{k}|} - \frac{m}{|\mathbf{k}'|}\right) + \frac{|\mathbf{k}|}{|\mathbf{k}'|} + \frac{|\mathbf{k}'|}{|\mathbf{k}|}\right] \qquad 6(176)$$

To evaluate the momentum integrals in 6(173), we write

$$d^3p'\, d^3k' = d^3p' \mathbf{k}'^2\, d|\mathbf{k}'|\, d\Omega_{\mathbf{k}'}$$

where $d\Omega_{\mathbf{k}'} = d\cos\theta\, d\varphi$ is the infinitesimal solid angle element in the direction (θ, φ) of the emerging photon. The integration over $\mathbf{p}'$ is trivial, owing to the presence of the momentum-conservation delta function in 6(173): We simply set

$$\mathbf{p}' = \mathbf{k} - \mathbf{k}' \qquad 6(177)$$

since $\mathbf{p} = 0$, in the laboratory system. The $|\mathbf{k}'|$ integration can be carried out by writing

$$|\mathbf{k}'|^2\, d|\mathbf{k}'| = |\mathbf{k}'|^2 \frac{d|\mathbf{k}'|}{dE_f}\, dE_f \qquad 6(178)$$

where $E_f = |\mathbf{k}'| + E_{\mathbf{p}'}$ is the total energy of the system in the final state. The integration over E_f in 6(173) is then taken care of by the energy-conservation delta-function which instructs us to set

$$m + |\mathbf{k}| = E_{\mathbf{p}'} + |\mathbf{k}'| \qquad 6(179)$$

in the laboratory system. Since, by 6(177)

$$\begin{aligned} E_{\mathbf{p}'} &= \sqrt{(\mathbf{k}-\mathbf{k}')^2 + m^2} \\ &= \sqrt{|\mathbf{k}|^2 + |\mathbf{k}'|^2 - 2|\mathbf{k}||\mathbf{k}'|\cos\theta + m^2} \end{aligned} \qquad 6(180)$$

we easily derive the Compton condition

$$|\mathbf{k}'| = \frac{m|\mathbf{k}|}{m + |\mathbf{k}|(1-\cos\theta)} \qquad 6(181)$$

* See for example, F. Mandl, *Introduction to Quantum Field Theory*, Interscience, New York, 1960.

Summarizing the result of the above manipulations, we can write the total unpolarized cross section in the form

$$\sigma = \frac{1}{(2\pi)^2}\frac{m}{4}\frac{|\mathbf{k}'|}{|\mathbf{k}|}\int d\Omega_{\mathbf{k}}\frac{1}{E_{\mathbf{p}'}}\frac{d|\mathbf{k}'|}{dE_f}\sum \tag{6(182)}$$

with $\sum$ given by 6(176) and with all momenta and angles referring to the laboratory system. Since, from 6(181)

$$\frac{m}{|\mathbf{k}|}-\frac{m}{|\mathbf{k}'|} = \cos\theta - 1$$

we find

$$\sum = \frac{e^4}{2m^2}\left(\frac{|\mathbf{k}|}{|\mathbf{k}'|}+\frac{|\mathbf{k}'|}{|\mathbf{k}|}-\sin^2\theta\right) \tag{6(183)}$$

Moreover, by 6(180)

$$\frac{dE_f}{d|\mathbf{k}'|} = 1+\frac{dE_{\mathbf{p}'}}{d|\mathbf{k}'|} = 1+\frac{|\mathbf{k}'|-|\mathbf{k}|\cos\theta}{E_{\mathbf{p}'}}$$

so that, using 6(179) and the Compton relation 6(181) we have

$$\frac{d|\mathbf{k}'|}{dE_f} = \frac{E_{\mathbf{p}'}}{E_{\mathbf{p}'}+|\mathbf{k}'|-|\mathbf{k}|\cos\theta} = \frac{E_{\mathbf{p}'}}{m+|\mathbf{k}|(1-\cos\theta)} = \frac{E_{\mathbf{p}'}}{m}\frac{|\mathbf{k}'|}{|\mathbf{k}|}$$

and hence

$$\sigma = \frac{\alpha^2}{2m^2}\frac{|\mathbf{k}'|^2}{|\mathbf{k}|^2}\int d\Omega_{\mathbf{k}}\left(\frac{|\mathbf{k}|}{|\mathbf{k}'|}+\frac{|\mathbf{k}'|}{|\mathbf{k}|}-\sin^2\theta\right) \tag{6(184)}$$

where

$$\alpha = \frac{e^2}{4\pi} \cong \frac{1}{137} \tag{6(185)}$$

is the fine structure constant. If we are interested in the *differential* cross section $d\sigma$ for scattering into a solid angle $d\Omega_{\mathbf{k}}$ in the direction (θ, φ) then we simply omit the integration over $d\Omega_{\mathbf{k}}$ in 6(184). The result is the Klein–Nishina formula [Klein (1929)]

$$\frac{d\sigma}{d\Omega} = \frac{\alpha^2}{2m^2}\frac{|\mathbf{k}'|^2}{|\mathbf{k}|^2}\left(\frac{|\mathbf{k}|}{|\mathbf{k}'|}+\frac{|\mathbf{k}'|}{|\mathbf{k}|}-\sin^2\theta\right) \tag{6(186)}$$

for the Compton scattering of unpolarized light. Note that $|\mathbf{k}'|$ can be eliminated altogether by using 6(181).

The $\mathbf{k} = 0$ limit of Compton scattering is of particular importance. From 6(181) we have

$$\lim_{|\mathbf{k}| \to 0} \frac{|\mathbf{k}'|}{|\mathbf{k}|} = 1$$

and hence, by 6(184)

$$\sigma = \frac{\alpha^2}{2m^2} \int d\Omega_{\mathbf{k}}(2 - \sin^2 \theta) = \frac{\pi\alpha^2}{m^2} \int_{-1}^{1} d\cos\theta(1 + \cos^2 \theta)$$

$$= \frac{8\pi}{3} \frac{\alpha^2}{m^2} \qquad \text{for} \quad |\mathbf{k}| = 0 \qquad 6(187)$$

Equation 6(187) is the Thomson formula for the scattering of classical electromagnetic radiation* and is valid only at low frequencies. The Klein–Nishina formula deviates significantly from 6(187) for photon energies $|\mathbf{k}|$ of the order of the electron rest mass m. Thus quantum radiation effects become important when the wavelength $|\mathbf{k}|^{-1}$ of the radiation becomes comparable with the Compton wavelength m^{-1} of the electron. The Thomson scattering formula 6(187) is also important in that it can be shown to be an exact result (i.e. valid to all orders of perturbation theory) in the limit $\mathbf{k} \to 0$†.

Electron Coulomb Scattering As our next application we calculate the scattering amplitude for electron Coulomb scattering. The relevant S-matrix element for this process is given by 6(115). We have

$$(\mathbf{p}'\sigma'|S|\mathbf{p}\sigma) = -2\pi\delta(E_{\mathbf{p}'} - E_{\mathbf{p}})\frac{m}{E_{\mathbf{p}}V} e A_0(\mathbf{p}' - \mathbf{p})\bar{u}_{\mathbf{p}'\sigma'}\gamma_4 u_{\mathbf{p}\sigma} \qquad 6(188)$$

where $\mathbf{p}$ and $\mathbf{p}'$ are the initial and final electron momenta respectively, and where we have specialized 6(115) to the case of a Coulomb potential

$$A_0(\mathbf{x}) = \frac{Ze}{4\pi|\mathbf{x}|} \qquad 6(189)$$

with the Fourier transform

$$A_0(\mathbf{p}' - \mathbf{p}) = \int A_0(\mathbf{x})\, e^{-i(\mathbf{p}'-\mathbf{p}).\mathbf{x}}\, d^3x \qquad 6(190)$$

* See for example J. D. Jackson, *Classical Electrodynamics*, Wiley, New York, 1962, Section 14.7.

† See Thirring (1950), and also J. M. Jauch and F. Rohrlich, *Electrons and Photons*, Addison-Wesley, Reading, U.S.A., 1955, Section 11.3.

The absence of a momentum-conservation delta-function in 6(188) means that the general analysis leading to 6(135) must be amended slightly. We easily find that the transition probability per unit time is given by

$$\Gamma = 2\pi\delta(E_{\mathbf{p}'} - E_{\mathbf{p}})\frac{m^2}{E_{\mathbf{p}}^2}\frac{1}{V^2}|eA_0(\mathbf{p}'-\mathbf{p})|^2|\bar{u}_{\mathbf{p}'\sigma'}\gamma_4 u_{\mathbf{p}\sigma}|^2 \qquad 6(191)$$

The total cross section is then obtained by summing Γ over $\mathbf{p}'$ and dividing by the incident flux

$$F = \frac{1}{V}\frac{|\mathbf{p}|}{E_{\mathbf{p}}} \qquad 6(192)$$

Thus, for given initial and final spins

$$\sigma = \frac{1}{(2\pi)^2}\frac{m^2}{|\mathbf{p}|E_{\mathbf{p}}}\int d^3p'\,\delta(E_{\mathbf{p}} - E_{\mathbf{p}'})|eA_0(\mathbf{p}'-\mathbf{p})|^2|\bar{u}_{\mathbf{p}'\sigma'}\gamma_4 u_{\mathbf{p}\sigma}|^2 \qquad 6(193)$$

The *unpolarized* cross section is obtained by summing 6(193) over final spins and averaging over initial spins. Applying 6(147a) and the trace theorems, we easily find

$$\tfrac{1}{2}\sum_{\sigma=1}^{2}\sum_{\sigma'=1}^{2}|\bar{u}_{\mathbf{p}'\sigma'}\gamma_4 u_{\mathbf{p}\sigma}|^2 = \tfrac{1}{2}\,\mathrm{Tr}\left[\gamma_4\frac{\gamma\,.\,p'+im}{2im}\gamma_4\frac{\gamma\,.\,p+im}{2im}\right]$$
$$= \frac{1}{2m^2}(m^2 + E_{\mathbf{p}}^2 + \mathbf{p}\,.\,\mathbf{p}') \qquad 6(194)$$

To integrate 6(193) we write

$$\begin{aligned} d^3p' &= \mathbf{p}'^2 d|\mathbf{p}'|\,d\Omega_{\mathbf{p}'} \\ &= \mathbf{p}'^2\frac{d|\mathbf{p}'|}{dE_{\mathbf{p}'}}\,dE_{\mathbf{p}'}\,d\Omega_{\mathbf{p}'} \\ &= |\mathbf{p}'|E_{\mathbf{p}'}\,dE_{\mathbf{p}'}\,d\Omega_{\mathbf{p}'} \end{aligned} \qquad 6(195)$$

and perform the trivial integration over $E_{\mathbf{p}'}$ with the aid of the energy-conservation delta-function. Finally, we identify the Fourier transform $A_0(\mathbf{p}'-\mathbf{p})$ as

$$A_0(\mathbf{p}'-\mathbf{p}) = \frac{Ze}{|\mathbf{p}'-\mathbf{p}|^2} = \frac{Ze}{4\mathbf{p}^2\sin^2\frac{1}{2}\theta} \qquad 6(196)$$

Combining results we obtain the total unpolarized cross section in the form

$$\sigma = \frac{1}{2(2\pi)^2}\int d\Omega_{\mathbf{p}'}\left(\frac{Ze^2}{4\mathbf{p}^2\sin^2\frac{1}{2}\theta}\right)^2(m^2+E_{\mathbf{p}}^2+\mathbf{p}\cdot\mathbf{p}')$$

$$= \frac{1}{2(2\pi)^2}\int d\Omega_{\mathbf{p}'}\left(\frac{Ze^2}{4\mathbf{p}^2\sin^2\frac{1}{2}\theta}\right)^2(2E_{\mathbf{p}}^2-2\mathbf{p}^2\sin^2\tfrac{1}{2}\theta)$$

The *differential* cross section is therefore given by

$$\frac{d\sigma}{d\Omega} = \frac{Z^2\alpha^2}{4\mathbf{p}^2v^2\sin^4\frac{1}{2}\theta}(1-v^2\sin^2\tfrac{1}{2}\theta) \qquad 6(197)$$

The result 6(197) is the Mott cross section [Mott (1929)]. As $v \to 0$ it tends to the well known Rutherford formula*.

μ-Decay An important application of perturbation theory to the field of weak interactions is the calculation of the lifetime for μ-decay

$$\mu^- \to e^- + \nu_\mu + \bar{\nu}_e \qquad 6(198)$$

The interaction Lagrangian for this process is given by 5(155). The corresponding interaction Hamiltonian is

$$\mathscr{H}_I = -\frac{1}{\sqrt{2}}G\bar{\psi}_{\nu_\mu}\gamma_\lambda(1+\gamma_5)\psi_\mu\bar{\psi}_e\gamma_\lambda(1+\gamma_5)\psi_{\nu_e}+\text{herm. conj.} \qquad 6(199)$$

and the lowest order transition amplitude from an initial μ-meson state with momentum p to a final state

$$\begin{cases} e^- \text{ with momentum } q & (q^2 = -m_e^2) \\ \nu_\mu \text{ with momentum } k & (k^2 = 0) \\ \bar{\nu}_e \text{ with momentum } k' & (k'^2 = 0) \end{cases}$$

is given by

$$(\mathbf{q},\mathbf{k},\mathbf{k}'|S|\mathbf{p}) = -i\int d^4x(\mathbf{q},\mathbf{k},\mathbf{k}'|:\mathscr{H}_I(x):|\mathbf{p})$$

$$= \frac{i}{\sqrt{2}}G\left(\frac{m_\mu m_e}{E_{\mathbf{p}}E_{\mathbf{q}}}\right)^{1/2}\frac{1}{V^2}(2\pi)^4\delta^{(4)}(p-q-k-k')$$

$$\times\bar{u}_{\mathbf{k}}\gamma_\lambda(1+\gamma_5)u_{\mathbf{p}}\bar{u}_{\mathbf{q}}\gamma_\lambda(1+\gamma_5)v_{\mathbf{k}'} \qquad 6(200)$$

* Although in the nonrelativistic limit the Rutherford formula is known to be exact, it appears here as a first approximation, corresponding to the lowest order term in the perturbation expansion. For an explanation of this apparent anomaly, see Dalitz (1951).

where we have suppressed polarization indices for the sake of clarity. The absence of normalization factors $E_{\mathbf{k}}^{1/2}$ or $E_{\mathbf{k}'}^{1/2}$ is due to our choice 1(90a) and 1(90b) for the neutrino wave functions. To obtain the μ-decay lifetime τ, we apply the general formula 6(142) with the identifications $N_a = (E_{\mathbf{p}}/m_\mu)^{1/2}$, $N_b = (E_{\mathbf{q}}/m_e)^{1/2}$. Remembering to sum over final spins and average over initial spins, so as to obtain the lifetime for unpolarized particles, we obtain

$$\tau^{-1} = (2\pi)^4 \tfrac{1}{2} G^2 \frac{m_\mu m_e}{E_{\mathbf{p}}} \frac{1}{(2\pi)^9} \int \frac{d^3 q}{E_{\mathbf{q}}} d^3 k \, d^3 k' \delta^{(4)}(p-q-k-k')$$
$$\times \tfrac{1}{2} \sum_{\text{spins}} |\bar{u}_{\mathbf{k}} \gamma_\lambda (1+\gamma_5) u_{\mathbf{p}} \bar{u}_{\mathbf{q}} \gamma_\lambda (1+\gamma_5) v_{\mathbf{k}'}|^2 \qquad 6(201)$$

where we have replaced the discrete summations in 6(142) by integrations over $\mathbf{q}$, $\mathbf{k}$ and $\mathbf{k}'$. To evaluate the spin sums, we apply 1(83a) and 1(89a), 1(89b) to write

$$\tfrac{1}{2} \sum_{\text{spins}} |\bar{u}_{\mathbf{k}} \gamma_\lambda (1+\gamma_5) u_{\mathbf{p}} \bar{u}_{\mathbf{q}} \gamma_\lambda (1+\gamma_5) v_{\mathbf{k}'}|^2$$
$$= \tfrac{1}{2} \operatorname{Tr}\left[\gamma_\lambda (1+\gamma_5) \frac{\gamma \cdot p + i m_\mu}{2 i m_\mu} \gamma_4 (1+\gamma_5) \gamma_\kappa \gamma_4 \frac{\gamma \cdot k}{2i|\mathbf{k}|} \right]$$
$$\times \operatorname{Tr}\left[\gamma_\lambda (1+\gamma_5) \frac{\gamma \cdot k'}{2i|\mathbf{k}'|} \gamma_4 (1+\gamma_5) \gamma_\kappa \gamma_4 \frac{\gamma \cdot q + i m_e}{2 i m_e} \right] \qquad 6(202)$$

The right-hand side of 6(202) can be simplified with the aid of a few simple manipulations, including the fact that $(1-\gamma_5)^2 = 2(1-\gamma_5)$ and the property that the trace of a product of an odd number of γ-matrices vanishes. We thus obtain for τ^{-1}

$$\tau^{-1} = \frac{1}{2^4 (2\pi)^5} G^2 \frac{1}{E_{\mathbf{p}}} \int \frac{d^3 q}{E_{\mathbf{q}}} \frac{d^3 k}{|\mathbf{k}|} \frac{d^3 k'}{|\mathbf{k}'|} \delta^{(4)}(p-q-k-k')$$
$$\times \operatorname{Tr}[\gamma_\lambda \gamma \cdot p (1-\gamma_5) \gamma_\kappa \gamma \cdot k] \operatorname{Tr}[\gamma_\lambda \gamma \cdot k' (1-\gamma_5) \gamma_\kappa \gamma \cdot q] \qquad 6(203)$$

To evaluate the traces, we use 6(159) and 6(156) and find, after a short calculation

$$\operatorname{Tr}[\gamma_\lambda \gamma \cdot p (1-\gamma_5) \gamma_\kappa \gamma \cdot k] \operatorname{Tr}[\gamma_\lambda \gamma \cdot k' (1-\gamma_5) \gamma_\kappa \gamma \cdot q]$$
$$= 32(p \cdot q\, k \cdot k' + p \cdot k'\, k \cdot q) + \operatorname{Tr}[\gamma_5 \gamma_\lambda \gamma \cdot p \gamma_\kappa \gamma \cdot k) \operatorname{Tr}[\gamma_5 \gamma_\lambda \gamma \cdot k' \gamma_\kappa \gamma \cdot q]$$
$$= 32(p \cdot q\, k \cdot k' + p \cdot k'\, k \cdot q) + 32(p \cdot k'\, k \cdot q - p \cdot q\, k \cdot k') \qquad 6(204)$$

where, in deriving the last line, we have used the identity

$$\tfrac{1}{2} \varepsilon_{\lambda\mu\kappa\nu} \varepsilon_{\lambda\rho\kappa\sigma} = \delta_{\mu\rho}\delta_{\nu\sigma} - \delta_{\mu\sigma}\delta_{\nu\rho} \qquad 6(205)$$

Thus 6(203) becomes

$$\tau^{-1} = \frac{4}{(2\pi)^5} G^2 \frac{1}{E_{\mathbf{p}}} \int \frac{d^3q}{E_{\mathbf{q}}} \frac{d^3k}{|\mathbf{k}|} \frac{d^3k'}{|\mathbf{k}'|} \delta^{(4)}(p-q-k-k')p \,.\, k'k \,.\, q \qquad 6(206)$$

We now integrate over the neutrino momenta k and k'. Let us evaluate the integral

$$I_{\mu\nu} = \int \frac{d^3k}{|\mathbf{k}|} \frac{d^3k'}{|\mathbf{k}'|} k_\mu k'_\nu \delta^{(4)}(p-q-k-k') \qquad 6(207)$$

in the Lorentz frame in which $\mathbf{p}-\mathbf{q} = 0$. In this frame $\mathbf{k}+\mathbf{k}' = 0$ and we easily find that $I_{\mu\nu}$, viewed as a 4×4 matrix, is given by

$$-\frac{\pi}{6}(p_0-q_0)^2 \begin{pmatrix} 1 & & & \\ & 1 & & \\ & & 1 & \\ & & & 3 \end{pmatrix} \qquad 6(208)$$

Since $I_{\mu\nu}$ as given by 6(207) is a second-rank tensor, its form in an arbitrary Lorentz frame is uniquely determined from 6(208) to be

$$I_{\mu\nu} = \frac{\pi}{6}(p-q)^2 \left[\delta_{\mu\nu} + \frac{2(p-q)_\mu (p-q)_\nu}{(p-q)^2} \right] \qquad 6(209)$$

which reduces to 6(208) in the frame in which $p-q$ is purely timelike. Substituting this result into 6(206), we find

$$\tau^{-1} = \frac{G^2}{48\pi^4} \frac{1}{E_{\mathbf{p}}} \int \frac{\mathbf{q}^2\, d|\mathbf{q}|}{E_{\mathbf{q}}} d\Omega_{\mathbf{q}} [3q \cdot p(m_\mu^2 + m_e^2) - 4(q \cdot p)^2 - 2m_e^2 m_\mu^2]$$

where we now set $p = (0, 0, 0, im)$ to obtain the μ-decay lifetime at rest. Neglecting m_e/m_μ and $m_e/E_{\mathbf{q}}$ we get

$$\tau^{-1} = \frac{G^2}{48\pi^4} \frac{1}{m_\mu} \int \mathbf{q}^2\, d|\mathbf{q}|\, d\Omega_{\mathbf{q}} [3m_\mu^3 - 4m_\mu^2 |\mathbf{q}|]$$

and integrating over all $\mathbf{q}$ from 0 to the maximum momentum $\frac{1}{2}m_\mu$ we obtain the result

$$\tau^{-1} = \frac{G^2 m_\mu^5}{192\pi^3} \qquad 6(210)$$

From the observed μ-decay lifetime

$$\tau = (2{\cdot}20 \pm 0{\cdot}003) \times 10^{-6} \text{ sec.} \qquad 6(211)$$

and the experimental value $m_\mu = 206 m_e$, we obtain the value

$$G = (1{\cdot}43 \pm 0{\cdot}003) \times 10^{-49}\ \text{erg cm}^3 \qquad 6(212)$$

for the weak decay coupling constant. This is most commonly expressed in terms of inverse proton masses as

$$G = (1{\cdot}03 \pm 0{\cdot}003) \times 10^{-5}\, m_p^{-2} \qquad 6(213)$$

π-Decay The lifetime for the decay of the π^--meson into a μ^--meson and a neutrino

$$\pi^- \to \mu^- + \bar{\nu}_\mu \qquad 6(214)$$

can be computed by using the decay interaction 5(163)

$$\mathscr{L}_I = i g_\pi \bar{\psi}_\mu \gamma_\lambda (1+\gamma_5) \psi_{\nu_\mu} \partial_\lambda \phi_{\pi^-} + \text{herm. conj.}$$

and restricting ourselves to the lowest order transition matrix element

$$(\mathbf{q}_\mu, \mathbf{q}_\nu | S | \mathbf{k}) = -i \int d^4x (\mathbf{q}_\mu, \mathbf{q}_\nu | : \mathscr{H}_I(x) : | \mathbf{k})$$

$$= -g_\pi \sqrt{\frac{m_\mu}{E_{\mathbf{q}_\mu} 2\omega_{\mathbf{k}}}} \frac{1}{V^{3/2}} (2\pi)^4 \delta^{(4)}(k - q_\mu - q_\nu)$$

$$\times\, i k_\lambda \bar{u}_{\mathbf{q}_\mu} \gamma_\lambda (1+\gamma_5) v_{\mathbf{q}_\nu} \qquad 6(215)$$

where $\mathbf{k}$, $\mathbf{q}_\mu$ and $\mathbf{q}_\nu$ are the pion, μ-meson and neutrino 3-momenta respectively. This treatment of π-decay is basically phenomenological, the effects of the strong interactions being lumped into the decay constant g_π. In Section 8.2, we shall justify this procedure, using the more basic semileptonic weak interaction 5(157).

The matrix element 6(215) can be simplified by using energy-momentum conservation to write

$$\bar{u}_{\mathbf{q}_\mu} \gamma \cdot k (1+\gamma_5) v_{\mathbf{q}_\nu} = \bar{u}_{\mathbf{q}_\mu} \gamma \cdot (q_\mu + q_\nu)(1+\gamma_5) v_{\mathbf{q}_\nu}$$

$$= i m_\mu \bar{u}_{\mathbf{q}_\mu} (1+\gamma_5) v_{\mathbf{q}_\nu}$$

owing to the free Dirac equations

$$\gamma \cdot q_\nu v_{\mathbf{q}_\nu} = 0$$

$$\bar{u}_{\mathbf{q}_\mu} \gamma \cdot q_\mu = i m_\mu \bar{u}_{\mathbf{q}_\mu}$$

Applying 6(142) we obtain, for the lifetime τ,

$$\tau^{-1} = (2\pi)^4 g_\pi^2 \frac{m_\mu^3}{2\omega_\mathbf{k}} \frac{1}{(2\pi)^6} \int \frac{d^3 q_\mu}{E_{\mathbf{q}_\mu}} d^3 q_\nu \delta^{(4)}(k - q_\mu - q_\nu)$$
$$\times \sum_{\text{spins}} |\bar{u}_{\mathbf{q}_\mu}(1+\gamma_5) v_{\mathbf{q}_\nu}|^2 \qquad 6(216)$$

The summation over the muon and neutrino spins is quite simple in this case. We find, using 1(83a), 1(89b) and the trace theorems 6(152) and 6(155),

$$\sum_{\text{spins}} |\bar{u}_{\mathbf{q}_\mu}(1+\gamma_5) v_{\mathbf{q}_\nu}|^2 = \text{Tr}\left[(1+\gamma_5) \frac{\gamma \,.\, q_\nu}{2i|\mathbf{q}_\nu|} \gamma_4 (1+\gamma_5) \gamma_4 \frac{\gamma \,.\, q_\mu + i m_\mu}{2 i m_\mu} \right]$$
$$= -\frac{1}{2 m_\mu |\mathbf{q}_\nu|} \text{Tr}\,[\gamma \,.\, q_\nu (1-\gamma_5) \gamma \,.\, q_\mu]$$
$$= -\frac{2}{m_\mu |\mathbf{q}_\nu|} q_\nu \,.\, q_\mu$$
$$= \frac{1}{m_\mu |\mathbf{q}_\nu|} (m_\pi^2 - m_\mu^2) \qquad 6(217)$$

where, in deriving the last line, we have used the relation

$$-m_\pi^2 = k^2 = (q_\mu + q_\nu)^2 = -m_\mu^2 + 2 q_\mu \,.\, q_\nu$$

dictated by energy-momentum conservation. Inserting 6(217) into 6(216), we now perform the momentum integrals in the rest frame of the decaying pion

$$\mathbf{k} = 0, \qquad \omega_\mathbf{k} = m_\pi$$
$$\mathbf{q}_\nu = -\mathbf{q}_\mu$$

Thus

$$\tau^{-1} = \frac{1}{(2\pi)^2} g_\pi^2 \frac{m_\mu^2}{2 m_\pi} (m_\pi^2 - m_\mu^2) \int \frac{d^3 q_\mu}{E_{\mathbf{q}_\mu}} \frac{d^3 q_\nu}{|\mathbf{q}_\nu|} \delta^{(4)}(k - q_\mu - q_\nu)$$
$$= \frac{g_\pi^2}{(2\pi)^2} \frac{m_\mu^2}{2 m_\pi} (m_\pi^2 - m_\mu^2) \int \frac{d^3 q_\nu}{E_{\mathbf{q}_\nu} |\mathbf{q}_\nu|} \delta(m_\pi - E_{\mathbf{q}_\nu} - |\mathbf{q}_\nu|) \qquad 6(218)$$

Writing

$$d^3 q_\nu = |\mathbf{q}_\nu|^2 \, d|\mathbf{q}_\nu| \, d\Omega_{\mathbf{q}_\nu} = |\mathbf{q}_\nu|^2 \frac{d|\mathbf{q}_\nu|}{dE_f} dE_f \, d\Omega_{\mathbf{q}_\nu}$$

where E_f is the total energy of the final state

$$E_f = |\mathbf{q}_\nu| + E_{\mathbf{q}_\nu}$$

with

$$\frac{d|\mathbf{q}_\nu|}{dE_f} = \frac{1}{1+\frac{|\mathbf{q}_\nu|}{E_{\mathbf{q}_\nu}}} = \frac{E_{\mathbf{q}_\nu}}{m_\pi}$$

Since the integrand in 6(218) is angle-independent, the integration over $d\Omega_\mathbf{q}$ simply provides a factor 4π. The remaining integration over dE_f is taken care of by the energy-conservation delta-function which imposes the constraint

$$|\mathbf{q}_\nu| + \sqrt{\mathbf{q}_\nu^2 + m_\mu^2} = m_\pi$$

or

$$|\mathbf{q}_\nu| = \frac{m_\pi^2 - m_\mu^2}{2m_\pi}$$

Combining results, we obtain the formula

$$\tau^{-1} = \frac{g_\pi^2}{4\pi} m_\pi m_\mu^2 \left(1 - \frac{m_\mu^2}{m_\pi^2}\right)^2 \tag{6(219)}$$

giving the π-decay lifetime at rest in terms of the phenomenological coupling constant g_π.

6-4 Renormalization

Introduction In Section 6.2 we encountered several examples of divergent terms in the perturbation expansion. The technique for dealing with these divergences in a systematic way is known as renormalization. By means of this technique, the divergences are isolated and reinterpreted as unobservable redefinitions or *renormalizations* of the mass and coupling-constant parameters of the theory. This cannot be done for all quantum field theories, but fortunately the procedure works for the physically important case of quantum electrodynamics and for pseudoscalar meson theory. Such theories, in which all divergences can be absorbed in the coupling-constant and mass renormalizations and thereby ignored, are known as *renormalizable* field theories.

The key features of renormalization theory are most clearly exhibited in the framework of the so-called *chain approximation.* Accordingly we shall introduce the concepts of mass and coupling-constant renormalization within this approximation and then discuss briefly how these concepts are generalized to all orders of perturbation theory. In general we shall work with pseudoscalar meson theory. The treatment of quantum electrodynamics follows largely the same pattern although there are certain differences which will be pointed out.

Nucleon–Nucleon Scattering in Chain Approximation Let us consider the lowest order nucleon–nucleon scattering graph of Fig. 6.16a. To

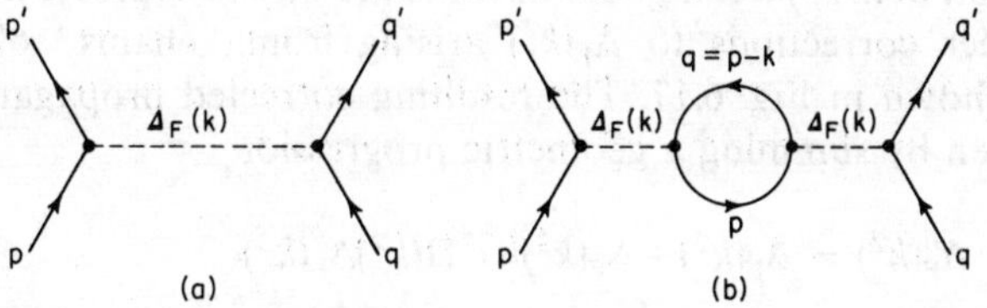

FIG. 6.16. (a) Lowest order nucleon–nucleon scattering graph. (b) Fourth-order graph giving second-order correction to $\Delta_F(k)$.

exhibit the essential structure of this graph, we write the matrix element in the form

$$M_0 = G^2\Delta_F(k^2) = G^2\frac{-i}{k^2+\mu^2} \qquad 6(220)$$

$$(k = p'-p)$$

where we have suppressed normalization factors, spinor factors like $\bar{u}_{\mathbf{p}'}\gamma_5 u_{\mathbf{p}}$, and the overall delta-function. We have retained only the coupling constants attached to the two vertices and the boson propagator $\Delta_F(k^2)$. We note that 6(220) exhibits a pole at the unphysical point $k^2 = -\mu^2$ with residue $-iG^2$.

Let us evaluate the correction to 6(220) arising from the meson self-energy insertion of Fig. 6.16b. Applying the Feynman rules, we easily find that the ' bare ' propagator $\Delta_F(k^2)$ is modified to

$$\Delta_F(k^2)+\Delta_F(k^2)\frac{-G^2}{(2\pi)^4}\int d^4q\, d^4p\, d^4k'\,\delta^{(4)}(k-p+q)\delta^{(4)}(k'-p+q)$$
$$\times \operatorname{Tr}\left[\gamma_5\frac{-1}{\gamma\,.\,q-im}\gamma_5\frac{-1}{\gamma\,.\,p-im}\right]\Delta_F(k'^2)$$

or, more concisely,

$$\Delta_F(k^2) \to \Delta_F(k^2) + \Delta_F(k^2) iG^2 \Pi(k^2) \Delta_F(k^2) \qquad 6(221)$$

where

$$\Pi(k^2) = \frac{i}{(2\pi)^4} \int d^4p \, \mathrm{Tr}\left[\gamma_5 \frac{-1}{\gamma . (p-k) - im} \gamma_5 \frac{-1}{\gamma . p - im} \right] \qquad 6(222)$$

is the analytic expression for the quadratically divergent fermion loop already encountered in 6(118). By relativistic invariance, Π can only be a function of k^2. We can go on in the same way to evaluate higher and higher order corrections to $\Delta_F(k^2)$ arising from 'chains' of fermion loops, as shown in Fig. 6.17. The resulting corrected propagator $\Delta_F'(k^2)$ is then given by summing a geometric progression

$$\begin{aligned} \Delta_F'(k^2) &= \Delta_F(k^2) + \Delta_F(k^2) iG^2 \Pi(k^2) \Delta_F(k^2) \\ &\quad + \Delta_F(k^2) iG^2 \Pi(k^2) \Delta_F(k^2) iG^2 \Pi(k^2) \Delta_F(k^2) + \ldots \\ &= \Delta_F(k^2) \frac{1}{1 - iG^2 \Pi(k^2) \Delta_F(k^2)} \\ &= \frac{-i}{k^2 + \mu^2 - G^2 \Pi(k^2)} \end{aligned} \qquad 6(223)$$

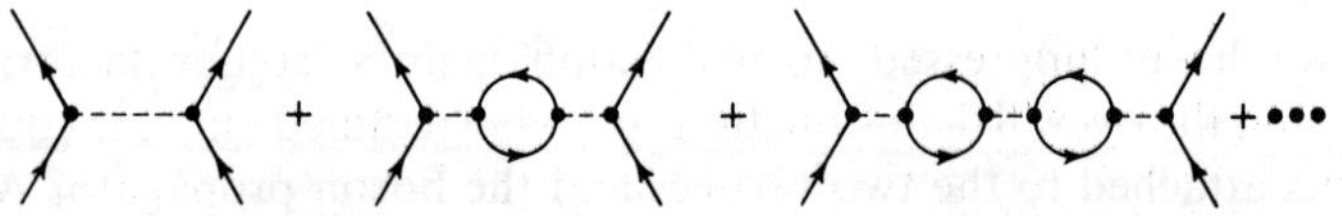

FIG. 6.17. Nucleon–nucleon scattering in chain approximation for Δ_F'.

The result 6(223) is known as the *chain approximation* to the exact meson propagator. When the bare propagator in 6(220) is replaced by its chain approximation expression 6(223), we get the nucleon–nucleon scattering matrix element

$$M = G^2 \frac{-i}{k^2 + \mu^2 - G^2 \Pi(k^2)} \qquad 6(224)$$

Invariant Integration We have asserted that

$$\Pi(k^2) = \frac{i}{(2\pi)^4}\int d^4p \,\mathrm{Tr}\left[\gamma_5 \frac{\gamma.(p-k)+im}{(p-k)^2+m^2}\gamma_5 \frac{\gamma.p+im}{p^2+m^2}\right]$$

$$= \frac{-4i}{(2\pi)^4}\int d^4p \frac{p^2+m^2-p.k}{(p^2+m^2)[(p-k)^2+m^2]} \qquad 6(225)$$

can only be a function of the Lorentz scalar k^2 on grounds of Lorentz invariance. To check this explicitly, we apply the important formula*

$$\frac{1}{ab} = \int_0^1 \frac{dz}{[az+b(1-z)]^2} \qquad 6(226)$$

due to Feynman (1949), to combine the two denominator factors in 6(225):

$$\frac{1}{p^2+m^2}\frac{1}{(p-k)^2+m^2} = \int_0^1 dz \frac{1}{[(p^2+m^2)z+[(p-k)^2+m^2](1-z)]^2}$$

$$= \int_0^1 dz \frac{1}{[p^2+m^2+(k^2-2p.k)(1-z)]^2}$$

Completing $p^2-2p.k(1-z)$ to a square, we get

$$\frac{1}{p^2+m^2}\frac{1}{(p-k)^2+m^2} = \int_0^1 dz \frac{1}{[[p-k(1-z)]^2+m^2+k^2(1-z)z]^2} \qquad 6(227)$$

We can now dispose of all k-dependent (as opposed to k^2-dependent) terms by shifting the origin of integration in 6(225)

$$p \to p+k(1-z) \qquad 6(228)$$

* This formula is proved by writing

$$\frac{1}{ab} = \frac{1}{b-a}\left(\frac{1}{a}-\frac{1}{b}\right) = \frac{1}{b-a}\int_a^b \frac{dx}{x^2}$$

and making the change of variable $x = az+b(1-z)$.

to get*

$$\Pi(k^2) = \frac{-4i}{(2\pi)^4}\int_0^1 dz \int d^4p \frac{p^2+m^2-k^2(1-z)z+p\,.\,k(1-2z)}{[p^2+m^2+k^2(1-z)z]^2} \qquad 6(229)$$

in which terms linear in p can now be dropped, on the grounds that an integral of the form $\int d^4p\, p_\mu f(p^2)$ vanishes by symmetry. The integrations over d^4p are now most easily evaluated by transforming 6(229) into an integral over a Euclidian 4-space, as follows. According to the definition of the Feynman propagator, the mass m carries a small positive imaginary part which instructs us how to perform the integration in the complex p_0-plane. Accordingly, the singularities of the integrand of 6(229) in the complex p_0-plane are located below the positive and above the negative real axes, and we can rotate the integration contour by 90°, as shown in Fig. 6.18, without crossing any singularities. The integration then ranges

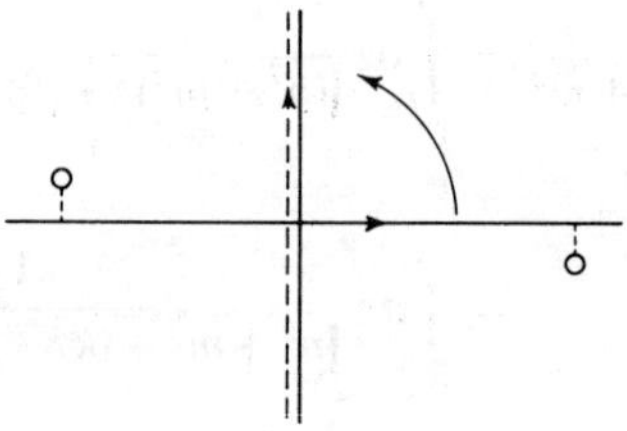

FIG. 6.18. Complex p_0-plane showing 90° rotation of integration contour.

from $p_0 = -i\infty$ to $p_0 = +i\infty$. Changing variables of integration $p_0 \to ip_0$, we substitute

$$d^4p = id^4p$$
$$\mathbf{p}^2 - p_0^2 \to \mathbf{p}^2 + p_0^2 \qquad 6(230)$$

in 6(229). The integration is now over a 4-dimensional Euclidian space and

* This procedure is not entirely correct as it stands, since a shift in origin in an integral which diverges worse than logarithmically creates additional 'surface' terms. [See J. M. Jauch and F. Rohrlich, *The Theory of Electrons and Photons*, Addison-Wesley, Reading, U.S.A., 1955, Appendix A5]. A procedure often applied is *regularization* [Pauli (1949)] in which the divergent integrals are first cut off in a Lorentz-invariant way. All manipulations then bear on finite integrals and the cutoff is set equal to infinity at the end of the calculation. [See Feynman (1949)].

we can introduce polar coordinates $p = \sqrt{p^2}$ and φ, θ, χ such that

$$\int d^4p = \int |p|^3 d|p| \int_0^{2\pi} d\varphi \int_0^{\pi} \sin\theta\, d\theta \int_0^{\pi} \sin^2\chi\, d\chi$$

$$= 2\pi^2 \int |p|^3 d|p| \qquad 6(231)$$

for an integrand depending only on p^2. The integral 6(229) is thereby reduced to a simple one-dimensional integral over $|p|$.

Mass Renormalization As the next step we expand $\Pi(k^2)$ about the point $k^2 = -\mu^2$ according to

$$\Pi(k^2) = \Pi(-\mu^2) + (k^2+\mu^2)\Pi'(-\mu^2) + \Pi_c(k^2) \qquad 6(232)$$

where

$$\Pi'(-\mu^2) = \left.\frac{d\Pi(k^2)}{dk^2}\right|_{k^2=-\mu^2} \qquad 6(233)$$

and where $\Pi_c(k^2)$, the sum of the remaining terms in the expansion, is of the form

$$\Pi_c(k^2) = (k^2+\mu^2)\pi_c(k^2) \qquad 6(234)$$

where $\pi_c(k^2)$ is zero for $k^2 = -\mu^2$. Referring back to 6(229), or rather

$$\Pi(k^2) = -\frac{4i}{(2\pi)^4} \int_0^1 dz \int d^4p \frac{p^2+m^2-k^2(1-z)z}{[p^2+m^2+k^2(1-z)z]^2} \qquad 6(235)$$

in which we have suppressed the linear term in p, we see that $\Pi(-\mu^2)$, the first term in the expansion 6(232), is quadratically divergent. On the other hand $\Pi'(-\mu^2)$, given by

$$\frac{4i}{(2\pi)^4} \int_0^1 dz \int d^4p \left\{ \frac{(1-z)z}{[p^2+m^2-\mu^2(1-z)z]^2} + \frac{2[p^2+m^2-k^2(1-z)z](1-z)z}{[p^2+m^2-\mu^2(1-z)z]^3} \right\}$$

$$6(236)$$

is only logarithmically divergent, the differentiation with respect to k^2 having lowered the power of the integration variable p^2 by one unit. Each successive differentiation with respect to k^2 lowers the power of the integration variable by one unit, so that we can conclude that $\Pi_c(k^2)$ in 6(232) is given by a *convergent* integral. Thus $\Pi_c(k^2)$ is finite and the entire quadratic and logarithmic divergence of $\Pi(k^2)$ is absorbed by the first two terms in the expansion 6(232).

Let us substitute 6(232) in the chain approximation amplitude 6(224). The important observation is that the result

$$M = G^2 \frac{-i}{k^2+\mu^2-G^2\Pi(-\mu^2)-(k^2+\mu^2)G^2\Pi'(-\mu^2)-G^2\Pi_c(k^2)} \tag{6(237)}$$

no longer displays a pole at $k^2 = -\mu^2$. This is due to the appearance of the $-G^2\Pi(-\mu^2)$ term in the denominator of 6(237). The position of the pole has been shifted from its original position at $k^2 = -\mu^2$, or, in physical terms, the *effective* meson mass has been shifted from its original value μ. This is unpleasant, as we would obviously like to continue to regard μ as the effective physical meson mass. The solution is to add to the interaction Hamiltonian a *counterterm* which cancels out the unwanted $-G^2\Pi(-\mu^2)$ term in 6(237). This is the Dyson mass renormalization prescription [Dyson (1949)]. Let us add to $\mathscr{H}_I(x)$ as given by 6(116), the counterterm

$$\mathscr{H}_I^{\delta\mu}(x) = -\tfrac{1}{2}\delta\mu^2\phi(x)\phi(x) \tag{6(238)}$$

with

$$\delta\mu^2 = -G^2\Pi(-\mu^2) \tag{6(239)}$$

and recalculate all Feynman diagrams, taking 6(238) into account. Thus the second-order self-energy loop of Fig. 6.16b will be corrected as in Fig. 6.19 by the insertion of a counterterm, which we represent by a small shaded circle. Analytically, we easily find that 6(221) is corrected by the additional term

$$\Delta_F(k^2)i\delta\mu^2\Delta_F(k^2) \tag{6(240)}$$

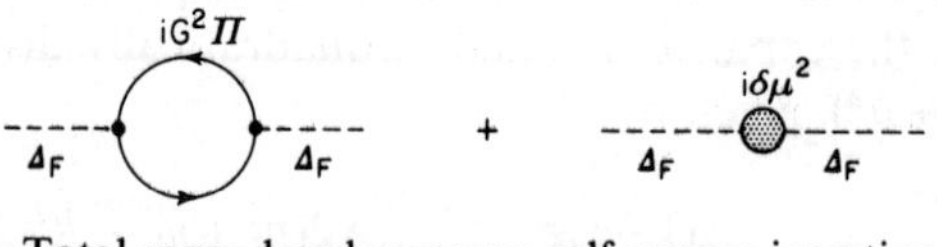

FIG. 6.19. Total second-order meson self-energy insertion in meson propagator showing $\delta\mu^2$ correction.

the Feynman rule being: Associate a factor $i\delta\mu^2$ for each mass counterterm insertion in a meson propagator $\Delta_F(k^2)$. Accordingly the term $-G^2\Pi(-\mu^2)$ in 6(237) is corrected to

$$\begin{aligned} -G^2\Pi(-\mu^2)-\delta\mu^2 &= -G^2\Pi(-\mu^2)+G^2\Pi(-\mu^2) \\ &= 0 \end{aligned}$$

which effectively disposes of the unwanted mass shift. M now becomes

$$M = G^2 \frac{-i}{(k^2+\mu^2)[1-G^2\Pi'(-\mu^2)]-G^2\Pi_c(k^2)} \qquad 6(241)$$

with the pole restored to its position at $k^2 = -\mu^2$. By the same token we have eliminated the quadratically divergent part of $\Pi(k^2)$ from 6(237) *and* disposed of the meson self-energy transition problem discussed in Section 6.2. Indeed, 6(118) is just cancelled by the additional self-energy transition amplitude due to 6(238) [see Fig. 6.20].

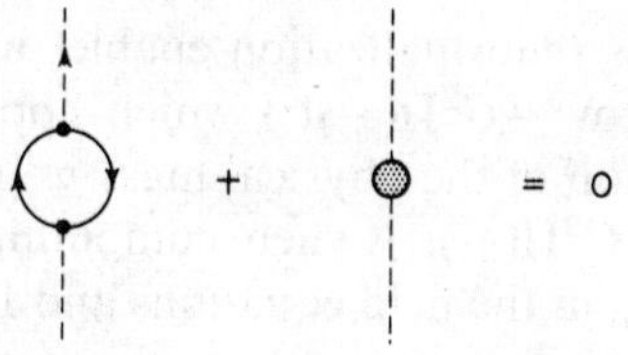

FIG. 6.20. Cancellation of second-order meson self-energy transitions.

Naturally we must attach some physical interpretation to the above subtraction procedure. Without such an interpretation the inclusion of the counterterm 6(238) in the interaction Hamiltonian would appear highly arbitrary. To see the physical meaning of the $\delta\mu^2$ term, we note that it corresponds to the addition of the term

$$\mathscr{L}_I^{\delta\mu}(x) = \tfrac{1}{2}\delta\mu^2\phi(x)\phi(x) \qquad 6(242)$$

to the interaction *Lagrangian* density. This term may be combined with the term

$$-\tfrac{1}{2}\mu^2\phi(x)\phi(x) \qquad 6(243)$$

in the *free* Lagrangian, to yield

$$-\tfrac{1}{2}\mu_0^2\phi(x)\phi(x)$$

where

$$\mu_0^2 = \mu^2 - \delta\mu^2 \qquad 6(244)$$

We see that the effect of the counterterm 6(238) is to replace μ in the *total* Lagrangian by μ_0, the so-called *bare* mass. Thus our subtraction procedure is just a mass redefinition, or 'renormalization.' The physical mass μ, which we have used in computing Feynman diagrams and whose square fixes the position of the pole in the nucleon-nucleon scattering amplitude, is *different* from the mass parameter μ_0 which appears in the Lagrangian and the Heisenberg field equations. Explicitly, the Heisenberg equation of motion 5(64) for $\phi(x)$ should really read

$$(\Box - \mu_0^2)\phi = -iG\bar{\psi}\gamma_5\psi \tag{6(245)}$$

with μ replaced by μ_0. To proceed backwards from 6(245) to our subtraction procedure, we simply rewrite 6(245) as

$$(\Box - \mu^2)\phi = -iG\bar{\psi}\gamma_5\psi - \delta\mu^2\phi \tag{6(246)}$$

regard the $\delta\mu^2$ term as part of the interaction, and transform to the interaction picture.

To summarize, mass renormalization enables us to absorb the quadratically divergent term $-G^2\Pi(-\mu^2)$ which appears in 6(237) and in 6(118) into the definition of the physical mass μ. In practice, we drop all terms of the form $-G^2\Pi(-\mu^2)$ when computing Feynman diagrams, and we replace μ by μ_0 in the field equations and Lagrangian.

Wave Function or Coupling Constant Renormalization Mass renormalization has eliminated the quadratic divergence in 6(237) but there remains the logarithmically divergent term $-G^2\Pi'(-\mu^2)$. This term alters the residue at the $k^2 = -\mu^2$ pole from its lowest order value $-iG^2$ to $-iG^2[1-G^2\Pi'(-u^2)]^{-1}$. Again, a physically equivalent statement is that the $-G^2\Pi'(-\mu^2)$ term alters the value of the *effective* coupling constant.

Let us begin by adopting the same attitude as for mass renormalization, and insist that G be the effective physical coupling constant. In that case we must add to the interaction Hamiltonian a counterterm to cancel the unwanted term

$$-G^2\Pi'(-\mu^2)(k^2+\mu^2) \tag{6(247)}$$

in the denominator of 6(237), [Takeda (1952)]. The required counterterm is given by

$$\mathscr{H}_I^C(x) = \tfrac{1}{2}C\frac{\partial\phi(x)}{\partial x_\mu}\frac{\partial\phi(x)}{\partial x_\mu} + \tfrac{1}{2}C\mu^2\phi(x)\phi(x) \tag{6(248)}$$

with

$$C = G^2\Pi'(-\mu^2) \tag{6(249)}$$

If we recalculate all Feynman graphs taking 6(247) into account, we easily find that 6(221) is corrected by the additional term

$$\Delta_F(k^2)[-iC(k^2+\mu^2)]\Delta_F(k^2) \tag{6(250)}$$

so that the logarithmically divergent term 6(247) is modified to

$$-G^2\Pi'(-\mu^2)(k^2+\mu^2)+C(k^2+\mu^2) = 0$$

by virtue of our choice 6(249) for C. Thus M becomes

$$M = G^2\frac{-i}{k^2+\mu^2-G^2\Pi_c(k^2)} \qquad 6(251)$$

with G restored to its role as the effective coupling constant, being the residue at the pole. Notice that M is now completely finite.

To give a physical interpretation for the above subtraction procedure, we observe that 6(248) corresponds to the additional term

$$\mathscr{L}_I^C = -\tfrac{1}{2}C\partial_\mu\phi\partial_\mu\phi-\tfrac{1}{2}C\mu^2\phi\phi \qquad 6(252)$$

to the interaction Lagrangian density. Combining 6(252) with the term

$$-\tfrac{1}{2}\partial_\mu\phi\partial_\mu\phi-\tfrac{1}{2}\mu^2\phi\phi$$

in the free Lagrangian density, we get

$$-\tfrac{1}{2}(1+C)\partial_\mu\phi\partial_\mu\phi-\tfrac{1}{2}(1+C)\mu^2\phi\phi \qquad 6(253)$$

so that ϕ has been replaced by

$$\phi_0 = (1+C)^{1/2}\phi \qquad 6(254)$$

the so-called ' bare ' field. As in the analogous case of mass renormalization, the ' physical ' field ϕ, which we have used to compute Feynman graphs, is now different from the field ϕ_0 which appears in the free field Lagrangian. Our subtraction procedure therefore amounts to a renormalization of the field, or equivalently, of the single particle wave functions. Note that the remaining interaction part of the Lagrangian

$$\mathscr{L}_I-\mathscr{L}_I^C = iG\bar{\psi}\gamma_5\psi\phi+\tfrac{1}{2}\delta\mu^2\phi^2 \qquad 6(255)$$

is still expressed in terms of ϕ. The crucial point is that since ϕ_0 is the field operator which appears in the free Lagrangian, it is ϕ_0 which must obey the usual equal-time commutation rules

$$[\phi_0(x), \dot{\phi}_0(x')]_{t=t'} = i\delta^{(3)}(\mathbf{x}-\mathbf{x}') \qquad 6(256)$$

Hence, the physical or *renormalized* field ϕ obeys

$$[\phi(x), \dot{\phi}(x')]_{t=t'} = iZ_3^{-1}\delta^{(3)}(\mathbf{x}-\mathbf{x}') \qquad 6(257)$$

where the (divergent) constant

$$Z_3 = 1+C = 1+G^2\Pi'(-\mu^2) \qquad 6(258)$$

is known as the meson wave function renormalization constant. For greater clarity, we now relabel

$$\phi_0 \to \phi \qquad \phi \to \phi_R \tag{6(259)}$$

so as to keep the symbol ϕ for a field which satisfies 6(256). This is no more than a change of notation. We also relabel

$$G \to G_R \tag{6(260)}$$

so that our physical coupling constant will now be called G_R. After relabelling, 6(254) becomes

$$\phi = Z_3^{1/2}\phi_R \tag{6(261)}$$

with

$$Z_3 = 1 + G_R^2\Pi'(-\mu^2) \tag{6(262)}$$

and the total Lagrangian, 6(253) plus 6(255) becomes

$$\mathscr{L} = -\tfrac{1}{2}\partial_\mu\phi\partial_\mu\phi - \tfrac{1}{2}\mu^2\phi^2 + iG_R\bar{\psi}\gamma_5\psi\phi_R + \tfrac{1}{2}\delta\mu^2\phi_R^2 \tag{6(263)}$$

This last form makes it clear that our subtraction procedure can be regarded simply as a renormalization of the *coupling constant*. Indeed, the Yukawa coupling term can be presented in the form

$$iG_R\bar{\psi}\gamma_5\psi Z_3^{-1/2}\phi = iG_0\bar{\psi}\gamma_5\psi\phi \tag{6(264)}$$

where

$$G_0 = Z_3^{-1/2}G_R \tag{6(265)}$$

is called the *bare* coupling constant*. Also, the mass counterterm can be written as

$$\tfrac{1}{2}\delta\mu^2 Z_3^{-1}\phi^2 = -\tfrac{1}{2}G_0^2\Pi(-\mu^2)\phi^2 \tag{6(266)}$$

with the aid of 6(261) and 6(265). Thus the overall effect of the subtraction is simply to replace $G(\equiv G_R)$ by G_0 in the total Lagrangian. The coupled Heisenberg field equations are now, instead of 5(63a), 5(63b) and 5(64)

$$(\gamma\,.\,\partial + m)\psi = iG_0\gamma_5\psi\phi \tag{6(267a)}$$

$$\bar{\psi}(\gamma\,.\,\overleftarrow{\partial} - m) = -iG_0\bar{\psi}\gamma_5\phi \tag{6(267b)}$$

and

$$(\Box - \mu^2)\phi = -iG_0\bar{\psi}\gamma_5\psi - \delta\mu^2\phi \tag{6(268)}$$

* By substituting 6(265) into 6(262) we get the alternative expression $Z_3^{-1} = 1 - G_0^2\Pi'(-\mu^2)$ for Z_3 in terms of the *bare* coupling constant.

with

$$\delta\mu^2 = -G_0^2\Pi(-\mu^2) \tag{6(269)}$$

As a consistency check, let us recalculate M with 6(267a), 6(267b) and 6(268) as a starting point, and show that we recover the result 6(251). Our interaction Hamiltonian in the interaction picture is now

$$\mathscr{H}_I(x) = -iG_0\bar{\psi}\gamma_5\psi\phi - \tfrac{1}{2}\delta\mu^2\phi^2$$

and we have in place of 6(224)

$$M = G_0^2 \frac{-i}{k^2+\mu^2-G_0^2\Pi(k^2)-\delta\mu^2} \tag{6(270)}$$

The counterterm 6(269) cancels the quadratically divergent term $-G_0^2\Pi(-\mu^2)$ as before and 6(270) reduces to

$$M = G_0^2 \frac{-i}{(k^2+\mu^2)[1-G_0^2\Pi'(-\mu^2)]-G_0^2\Pi_c(k^2)} \tag{6(271)}$$

To eliminate the logarithmically divergent term, we substitute

$$G_0^2 = Z_3^{-1}G_R^2 = \frac{1}{1+G_R^2\Pi'(-\mu^2)}G_R^2 \tag{6(272)}$$

into 6(271) and simplify. We find

$$\begin{aligned} M &= G_0^2[1+G_R^2\Pi'(-\mu^2)]\frac{-i}{k^2+\mu^2-G_R^2\Pi_c(k^2)} \\ &= G_0^2 Z_3 \frac{-i}{k^2+\mu^2-G_R^2\Pi_c(k^2)} \\ &= G_R^2 \frac{-i}{k^2+\mu^2-G_R^2\Pi_c(k^2)} \end{aligned} \tag{6(273)}$$

in agreement with 6(251). In some respects this is a more convenient way of performing the coupling constant renormalization. The expression 6(271) containing the logarithmically divergent term is converted to the finite expression 6(273) by absorbing the unwanted term into the definition of the coupling constant with the aid of 6(272). In this approach there is no explicit reference either to a counterterm or to the renormalization of the field.

To summarize, renormalization allows the quadratic and logarithmic divergences appearing in $\Pi(k^2)$ to be absorbed into the definition of the physical mass and coupling constant parameters. It should be clear, however, that renormalization would, in general, still be required even in the absence of divergences. If $\Pi(-\mu^2)$ and $\Pi'(-\mu^2)$ are nonzero, the interaction modifies the bare mass and coupling constant, and renormalization is needed to express M in terms of the physical, observed quantities.

In the future we shall avoid the explicit use of a counterterm in discussing wave function or coupling constant renormalization. Thus we shall consistently use the bare coupling constant G_0 in computing Feynman diagrams and renormalize as in 6(273). *This means that all Feynman rules presented in Section 6.2 are to be modified by the replacements**

$$e \to e_0 \qquad G \to G_0 \qquad f \to f_0 \tag{6(274)}$$

with the corresponding physical (renormalized) constants denoted by e_R, G_R and f_R. For mass renormalization, on the other hand, we shall adhere to the counterterm approach and continue to use μ or m to denote the physical (renormalized) mass.

Fermion Propagator in Chain Approximation We now turn to the elimination of divergences for the *fermion* self-energy graph. Let us consider the lowest order fermion propagator

$$S_F(p) = \frac{-1}{\gamma \cdot p - im} \tag{6(275)}$$

and compute the correction to $S_F(p)$ arising from the insertion of a second-order self-energy part, as shown in Fig. 6.21. Applying the Feynman rules

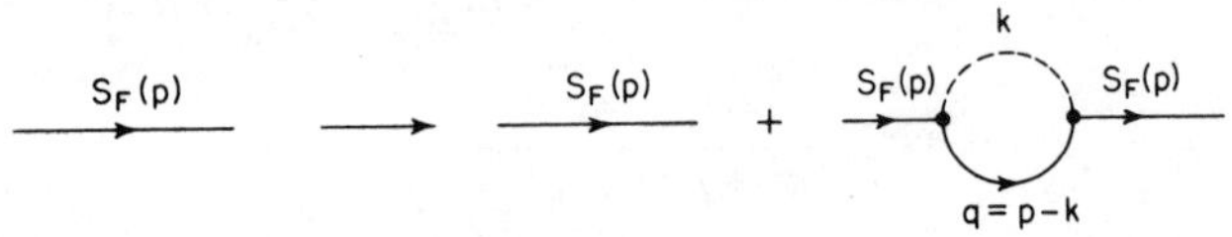

FIG. 6.21. Alteration of fermion propagator by insertion of second-order self-energy part.

with the replacement 6(274) we find that $S_F(p)$ is modified to

$$S_F(p) + S_F(p)\frac{G_0^2}{(2\pi)^4}\int d^4q\, d^4k\, d^4p'\,\delta^{(4)}(k-p+q)\delta^{(4)}(k-p'+q)$$
$$\times\left(\gamma_5\frac{-1}{\gamma \cdot q - im}\gamma_5\frac{-i}{k^2+\mu^2}\right)S_F(p')$$

* In principle this replacement also affects the lowest order graphs used in the calculations of Section 6-3. However, in lowest order $e_0 = e\ (= e_R)$, and so we can continue to regard the constant e used in Section 6-3 as the physical electrodynamic coupling constant. Note that this question does not arise if we use the counterterm approach.

or

$$S_F(p)+S_F(p)[-G^2\Sigma(p)]S_F(p) \tag{6(276)}$$

where $\Sigma(p)$ is the meson theory analog of 6(87), i.e.

$$\Sigma(p) = \frac{-1}{(2\pi)^4}\int d^4k\gamma_5\frac{-1}{\gamma\,.\,(p-k)-im}\gamma_5\frac{-i}{k^2+\mu^2} \tag{6(277)}$$

Summing all graphs shown in Fig. 6.22, we obtain the chain approximation result for the corrected propagator $S'_F(p)$

$$S'_F(p) = \frac{-1}{\gamma\,.\,p-im-G_0^2\Sigma(p)} \tag{6(278)}$$

analogous to 6(223).

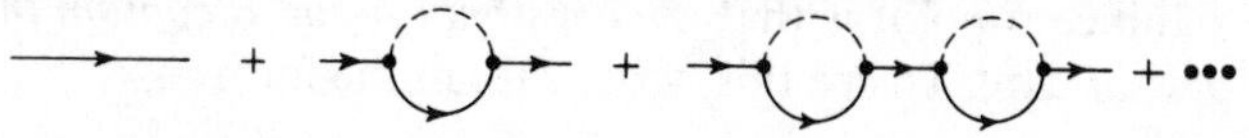

FIG. 6.22. Chain approximation for corrected fermion propagator S'_F.

The second order fermion self-energy part $\Sigma(p)$ is linearly divergent. To isolate the divergent parts, we first note that on grounds of Lorentz invariance $\Sigma(p)$ must be of the form*

$$\Sigma(p) = a(p^2)+b(p^2)\gamma\,.\,p \tag{6(279)}$$

Operating with $\Sigma(p)$ on an initial Dirac spinor $u_{\mathbf{p}\sigma}$ we have

$$\begin{aligned}\Sigma(p)u_{\mathbf{p}\sigma} &= (a(-m^2)+b(-m^2)im)u_{\mathbf{p}\sigma}\\ &= Au_{\mathbf{p}\sigma}\end{aligned}$$

where

$$A = \Sigma(p)|_{\gamma.p=im} \tag{6(280)}$$

is the ' free particle value ' of $\Sigma(p)$. If we subtract A from $\Sigma(p)$, the result must be of the form

$$\Sigma(p)-A = (\gamma\,.\,p-im)\Sigma^{(1)}(p)$$

where, like $\Sigma(p)$, $\Sigma^{(1)}$ is some uniquely defined function of p^2 and $\gamma\,.\,p$. Performing a further subtraction $\Sigma^{(1)}(p)-B$, where

$$B = \Sigma^{(1)}(p)|_{\gamma.p=im}$$

* This follows from the fact that p^2 and $\gamma\,.\,p$ are the only Lorentz scalars which can be formed with the aid of the available vectors p_μ and γ_μ; the only other possible term would be proportional to $[\gamma_\mu,\gamma_\nu]p_\mu p_\nu$ which vanishes identically.

we must again have

$$\Sigma^{(1)}(p) - B = (\gamma \,.\, p - im)\Sigma^{(2)}(p)$$

with $\Sigma^{(2)}(p)$ a uniquely defined function of p^2 and $\gamma \,.\, p$. Summarizing the result of our sequence of subtractions, we conclude that $\Sigma(p)$ must be of the form

$$\Sigma(p) = A + (\gamma \,.\, p - im)B + \Sigma_c(p) \tag{6(281)}$$

where $\Sigma_c(p)$ is of the form

$$\Sigma_c(p) = (\gamma \,.\, p - im)\sigma_c(p)$$

with $\sigma_c(p)$ vanishing for $\gamma \,.\, p = im$*. The form 6(281) is analogous to 6(232), the constants A and B producing shifts in the effective fermion mass and effective coupling constants which are to be absorbed by renormalization. The significant point is that *the constants A and B contain the entire divergent part of* $\Sigma(p)$. To see this, write $\Sigma(p)$ in the form

$$\Sigma(p) = \int d^4k R(p, k)$$

with

$$R(p, k) = \frac{-1}{(2\pi)^4}\gamma_5 \frac{-1}{\gamma \,.\, (p-k) - im}\gamma_5 \frac{-i}{k^2 + \mu^2}$$

and isolate the divergences by expanding the integrand $R(p, k)$ in powers of p_μ

$$R(p, k) = R(0, k) + p_\mu \left(\frac{\partial R(p, k)}{\partial p_\mu} \right)_{p=0} + R_c(p, k) \tag{6(282)}$$

Correspondingly, we have

$$\Sigma(p) = a + b_\mu p_\mu + c(p) \tag{6(283)}$$

where, in analogy with the meson self-energy case, $c(p)$ is given by a convergent integral, since the integrand $R(p, k)$ features two extra powers of k in the denominator. a and b_μ are given by constant divergent integrals†. Now on grounds of Lorentz invariance, b_μ must be of the form $b\gamma_\mu$ and, moreover, $c(p)$ must be a function of p^2 and $\gamma \,.\, p$ similar to 6(279). We can

* $\sigma_c(p)$ is just $(\gamma \,.\, p - im)\Sigma^{(2)}(p)$. For the explicit construction of A, B and $\Sigma_c(p)$ (in quantum electrodynamics) by means of invariant integration, see for example J. M. Jauch and F. Rohrlich, *The Theory of Electrons and Photons*, Addison-Wesley, Reading, U.S.A., 1955.

† The linear divergence contained in a can be shown to disappear upon performing invariant integration, so that a and b are both logarithmically divergent.

therefore apply to $c(p)$ the subtraction procedure previously applied to $\Sigma(p)$ and write $c(p)$ in a form similar to 6(281). Then, by redefining the coefficients a and b, we can easily recast the *entire* expression 6(283) in the (unique) form 6(281), where A and B are now guaranteed to contain the entire divergence of $\Sigma(p)$.

Having isolated the divergences of $\Sigma(p)$ in the convenient form 6(281), we now insert 6(281) into 6(278), giving

$$S_F'(p) = \frac{-1}{\gamma . p - im - G_0^2 A - G_0^2 B(\gamma . p - im) - G_0^2 \Sigma_c(p)} \tag{6(284)}$$

and perform the renormalization exactly as in the meson propagator case. To dispose of the unwanted $G_0^2 A$ term, we add to the interaction Hamiltonian density the fermion mass counterterm

$$\mathscr{H}_I^{\delta m}(x) = -\delta m \bar{\psi}(x)\psi(x) \tag{6(285)}$$

with

$$\delta m = -iG_0^2 A \tag{6(286)}$$

This amounts to a renormalization of the fermion mass: the Lagrangian density and the field equations are modified by the replacement of m by

$$m_0 = m - \delta m \tag{6(287)}$$

where m_0 is known as the bare fermion mass. To eliminate the B term in 6(284) we renormalize the coupling constant according to*

$$G_R^2 = Z_2 G_0^2 \tag{6(288)}$$

where Z_2 is a (divergent) *fermion* wave function renormalization constant given by

$$Z_2 = 1 + G_R^2 B \tag{6(289)}$$

Substituting 6(288) with 6(289) into 6(284) and suppressing the $G_0^2 A$ term, we get, just as in 6(273)

$$\begin{aligned} S_F'(p) &= \frac{-1}{(\gamma . p - im)\left[1 - \dfrac{G_R^2 B}{1 + G_R^2 B}\right] - \dfrac{G_R^2}{1 + G_R^2 B}\Sigma_c(p)} \\ &= [1 + G_R^2 B]\frac{-1}{\gamma . p - im - G_R^2 \Sigma_c(p)} \\ &= Z_2 \frac{-1}{\gamma . p - im - G_R^2 \Sigma_c(p)} \end{aligned} \tag{6(290)}$$

* There is no contradiction between 6(288) and 6(265). The point is that G_R picks up contributions from both the fermion and boson wave function renormalizations as well as from a vertex renormalization. [See the discussion of renormalization to all orders of perturbation theory, below.]

where the factor multiplying Z_2 is completely finite. The factor Z_2 is to be absorbed into the definition of the renormalized coupling constant at the two vertices joined by S'_F.

Spinor Electrodynamics In quantum electrodynamics the treatment of the electron and photon propagators in the chain approximation follows almost exactly the same pattern as in the mesodynamic case discussed above. The chief difference is that *mass* renormalization is not required for the photon propagator [see below]. Another difference is that the summation of the chain diagrams is somewhat more complicated in the case of the photon propagator, owing to the tensor indices carried by $\Pi_{\mu\nu}$. Denoting the uncorrected propagator by

$$D_{F\mu\nu}(k) = \delta_{\mu\nu}\frac{-i}{k^2} \qquad 6(291)$$

the insertion of a single fermion loop modifies 6(291) to

$$D_{F\mu\nu}(k) + D_{F\mu\lambda}(k) i e_0^2 \Pi_{\lambda\sigma}(k) D_{F\sigma\nu}(k) \qquad 6(292)$$

For the corrected propagator $D'_{F\mu\nu}$ in the chain approximation, we write the implicit equation

$$D'_{F\mu\nu}(k) = D_{F\mu\nu}(k) + D_{F\mu\lambda}(k) i e_0^2 \Pi_{\lambda\sigma}(k) D'_{F\sigma\nu}(k) \qquad 6(293)$$

which, when iterated, yields the sum of all chain graphs*. We now set, on grounds of Lorentz invariance

$$D'_{F\mu\nu}(k) = A(k^2)\delta_{\mu\nu} + B(k^2) k_\mu k_\nu \qquad 6(294)$$

and insert 6(291), 6(294) and 6(100), i.e.

$$\Pi_{\lambda\sigma}(k) = (\delta_{\lambda\sigma}k^2 - k_\lambda k_\sigma) C(k^2) \qquad 6(295)$$

into 6(293). Identifying the coefficients of $\delta_{\mu\nu}$ and $k_\mu k_\nu$ we find

$$A(k^2) = \frac{-i}{k^2 - e_0^2 k^2 C(k^2)}, \qquad B(k^2) = \frac{e_0^2}{k^2} C(k^2) A(k^2)$$

and hence

$$D'_{F\mu\nu}(k) = \delta_{\mu\nu}\frac{-i}{k^2 - e_0^2 k^2 C(k^2)} + \frac{k_\mu k_\nu}{k^2}\frac{-ie_0^2 C(k^2)}{k^2 - e_0^2 k^2 C(k^2)} \qquad 6(296)$$

* Needless to say, a corresponding equation also exists for the meson propagator Δ'_F. [See 6(392) below.]

Actually we can ignore the $k_\mu k_\nu$ term in 6(296) since, as discussed earlier in connection with the gauge invariance of the Compton scattering amplitude, the insertion of a $\gamma_\mu k_\mu$ factor into a fermion line gives a vanishing contribution to the S-matrix when the insertion is performed in all possible ways*. Effectively, we have

$$D'_{F\mu\nu}(k) = \delta_{\mu\nu}\frac{-i}{k^2 - e_0^2 k^2 C(k^2)} \qquad 6(297)$$

Since $C(k^2)$ is regular at $k^2 = 0$, the corrected propagator continues to display a pole at $k^2 = 0$ and mass renormalization is not required. This can be traced to the non-appearance in 6(295) of the quadratically divergent term $\Pi_{\mu\nu}(0) = A\delta_{\mu\nu}$. This term would contribute a mass shift proportional to A in the denominator of 6(297), but is ruled out by the current conservation constraint 6(97). Note that current conservation does not automatically guarantee that the physical photon mass is zero. The regularity of $C(k^2)$ at $k^2 = 0$ is also required. In Section 10-4 we shall exhibit an exactly soluble model, due to Schwinger, in which this second requirement is violated.

Coupling constant renormalization—known as *charge* renormalization in the case of electrodynamics—is carried out in exactly the same way as in 6(273). The quantity Π_c defined by

$$k^2 C(k^2) = k^2 C(0) + \Pi_c(k^2) \qquad 6(298)$$

is convergent, since $C(k^2)$ is logarithmically divergent. Defining the renormalized charge by

$$e_R^2 = Z_3 e_0^2 \qquad 6(299)$$

with

$$Z_3 = 1 + e_R^2 C(0) \qquad 6(300)$$

we easily find

$$D'_{F\mu\nu}(k) = \delta_{\mu\nu} Z_3 \frac{-i}{k^2 - e_R^2 \Pi_c(k^2)} \qquad 6(301)$$

This completes our treatment of renormalization theory in the chain approximation. To understand how the concepts of mass and coupling constant renormalization generalize to all orders of perturbation theory, we need a formal language with which to discuss the perturbation series as

* We stress that the $k_\mu k_\nu$ terms are 'decoupled' only insofar as the S-matrix is concerned. The insertion of $\gamma . k$ into a fermion line yields a vanishing result only if the extremities of the line are placed on the mass shell.

a whole. This is provided by Dyson's equations [Dyson (1949)]. Once more we carry out our discussion within the convenient framework of pseudoscalar meson theory. The extension to spinor electrodynamics is straightforward.

Dyson's Equations Let us generalize the lowest order fermion loop of Fig. 6.16b by introducing the concept of a *proper* meson self-energy graph. A proper self-energy graph is one which cannot be divided into two disjoint parts by cutting one meson line. Improper self-energy insertions are those which can be so divided as, for example, in Fig. 6.23. We now define

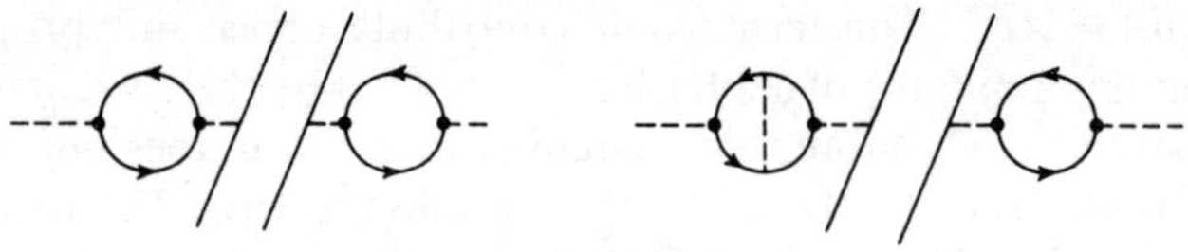

FIG. 6.23. Examples of improper meson self-energy insertions.

$\Pi^*(k^2)$ to be the *sum* of all proper meson self-energy insertions, as illustrated in Fig. 6.24. $\Pi^*(k^2)$ represents the generalization of $\Pi(k^2)$ to all orders of perturbation theory*. Inserting $\Pi^*(k^2)$ into the lowest order propagator $\Delta_F(k^2) = -i(k^2+\mu^2)^{-1}$ modifies the latter to

$$\Delta_F(k^2)+\Delta_F(k^2)iG_0^2\Pi^*(k^2)\Delta_F(k^2)$$

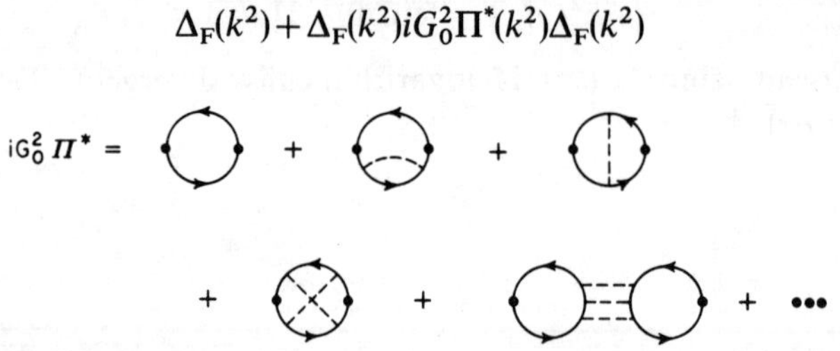

FIG. 6.24. Expansions of Π^* in terms of all proper meson self-energy graphs.

and if we insert $\Pi^*(k^2)$ an infinite number of times we clearly generate the sum of all possible graphs, both proper and improper. Summing the geometric progression as in 6(223) we obtain the exact, fully corrected, meson propagator in the form

$$\Delta_F'(k^2) = \frac{-i}{k^2+\mu^2-G_0^2\Pi^*(k^2)-\delta\mu^2} \qquad 6(302)$$

* Lorentz invariance asserts that Π^*, like Π, must be a function of the invariant k^2.

where we have taken care to include the contribution of the counterterm interaction

$$\mathscr{H}_I^{\delta\mu}(x) = -\tfrac{1}{2}\delta\mu^2\phi(x)\phi(x) \qquad 6(303)$$

where $\delta\mu^2$ will be determined shortly*.

The fermion propagator can be similarly treated. Defining $\Sigma^*(p)$ to be the sum of all proper† fermion self-energy graphs, as shown in Fig. 6.25,

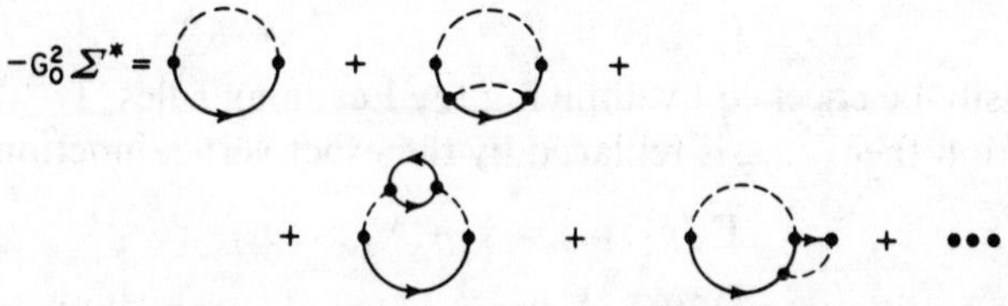

FIG. 6.25. Expansion of Σ^* in terms of all proper fermion self-energy graphs.

we can write the exact fermion propagator in the form

$$S'_F(p) = \frac{-1}{\gamma \cdot p - im - G_0^2\Sigma^*(p) + i\delta m} \qquad 6(304)$$

generalizing 6(278). Again we have included the δm counterterm due to

$$\mathscr{H}_I^{\delta m}(x) = -\delta m\bar{\psi}(x)\psi(x) \qquad 6(305)$$

where δm is to be determined below.

In addition to the exact meson and fermion propagators, we can define an exact *vertex function.* Consider the lowest order vertex graph of Fig. 6.26a. The momenta p_1, p_2 and $k = p_2 - p_1$ meeting at the vertex are *not* assumed to be restricted to their mass shell values, so that the trilinear graph of Fig. 6.26a can form part of some larger graph in which the three lines are internal. The second-order correction to the lowest order vertex graph is shown in Fig. 6.26b. It has the effect of adding to γ_5 the logarithmically divergent term

$$\frac{G_0^2}{(2\pi)^4}\int d^4k\gamma_5\frac{-1}{\gamma \cdot (p_2-k) - im}\gamma_5\frac{-1}{\gamma \cdot (p_1-k) - im}\gamma_5\frac{-i}{k^2+\mu^2} \qquad 6(306)$$

* $\delta\mu^2$ will now be given by a power series expansion, of which 6(269) represents only the first term. As in the chain approximation, $\delta\mu^2$ serves to cancel the mass shift due to $\Pi^*(-\mu^2)$ [see below].

† The definition of a proper *fermion* self-energy graph follows the same pattern as for the boson graphs, namely a proper graph is one which cannot be divided into two disjoint parts by cutting one fermion line.

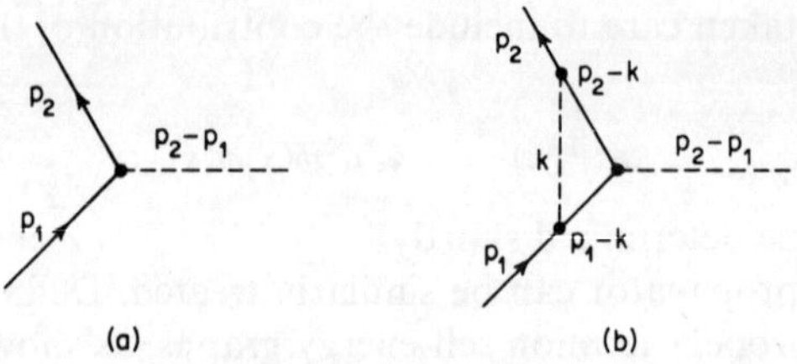

FIG. 6.26. Lowest order vertex (a) and second-order correction (b).

as can easily be checked by applying the Feynman rules. To all orders of perturbation theory, γ_5 is replaced by the exact vertex function

$$\Gamma_5(p_2, p_1) = \gamma_5 + \Lambda_5(p_2, p_1) \qquad 6(307)$$

where, in addition to 6(306), $\Lambda_5(p_2, p_1)$ includes contributions from all proper trilinear graphs, as illustrated in Fig. 6.27*. Improper graphs like those of Fig. 6.28, featuring self-energy insertions on the external lines, are excluded.

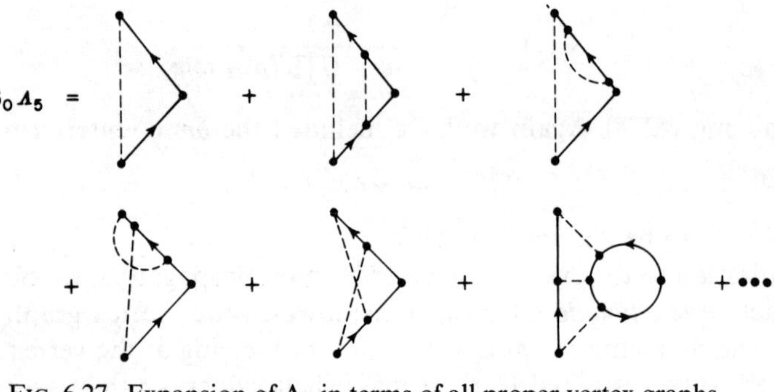

FIG. 6.27. Expansion of Λ_5 in terms of all proper vertex graphs.

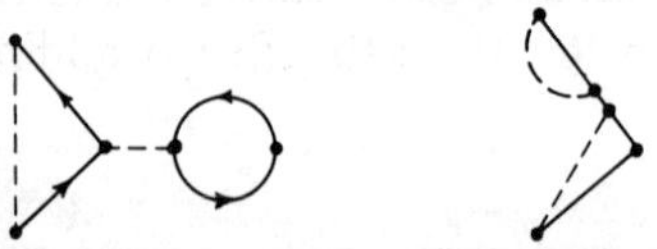

FIG. 6.28. Examples of improper vertex graphs excluded from definition of Λ_5.

* We note in passing that contributions to Λ_5 from diagrams like those of Fig. 6.29—displaying closed loops with an odd number of pseudoscalar vertices—are not a priori excluded, but vanish identically [Salam (1951)]. The corresponding property for spinor electrodynamics was first proved by Furry (1937) [see Problem 7].

FIG. 6.29. Graphs containing closed loops with odd number of vertices.

We can now convert the formal summations for Δ'_F, S'_F and Γ_5 into a set of coupled integral equations [Dyson (1949)]. We have seen that $\Delta'_F(k^2)$ has the form

$$\Delta'_F(k^2) = \frac{-i}{k^2+\mu^2-G_0^2\Pi^*(k^2)-\delta\mu^2} \qquad 6(308)$$

where $\Pi^*(k^2)$ is given by the sum of all proper self-energy insertions. From the definition of $\Gamma_5(p_2, p_1)$ as the sum of all proper vertex graphs, it is easy to see that $\Pi^*(k^2)$ is given by the expression

$$\Pi^*(k^2) = \frac{i}{(2\pi)^4}\int d^4p \operatorname{Tr}[\gamma_5 S'_F(p-k)\Gamma_5(p-k, p)S'_F(p)] \qquad 6(309)$$

obtained by replacing the lowest order fermion propagator

$$S_F(p) = \frac{-1}{\gamma\,.\,p-im}$$

in 6(222) by the exact propagator $S'_F(p)$ and by replacing γ_5 at one of the vertices by the exact vertex function $\Gamma_5(p_2, p_1)$. If we agree to represent the exact propagators and vertex function by the general blobs of Fig. 6.30,

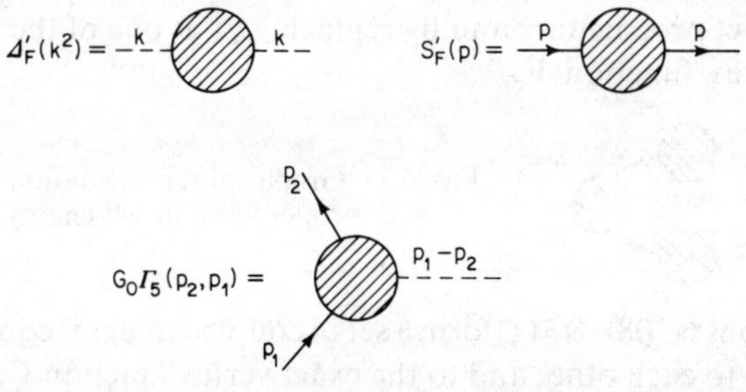

FIG. 6.30. Graphical representation of exact propagators and vertex function.

then the equality 6(309) can be represented graphically, as in Fig. 6.31. We stress that the exact vertex function can appear only at one of the corners. If we were to insert Γ_5 at both corners we would be counting

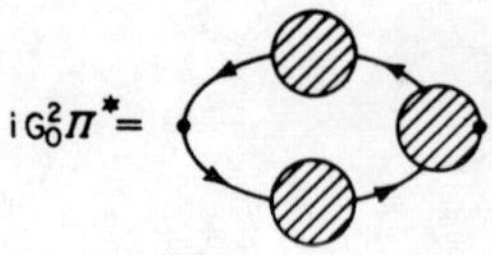

FIG. 6.31. Illustration of Eq. 6(309) for proper meson self-energy part.

each vertex correction twice. For example, the graph of Fig. 6.32 would be counted once as a correction to the vertex a and once as a correction to the vertex b, whereas in fact it only appears once in the perturbation series.

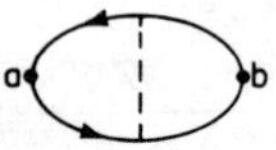

FIG. 6.32. Graph contributing to Π^*.

Turning to the exact fermion propagator

$$S'_{\mathrm{F}}(p) = \frac{-1}{\gamma \cdot p - im - G_0^2\Sigma^*(p) + i\delta m} \tag{6(310)}$$

we can write $\Sigma^*(p)$, the sum of all proper fermion self-energy graphs, in the form

$$\Sigma^*(p) = \frac{-1}{(2\pi)^4}\int d^4k\gamma_5 S'_{\mathrm{F}}(p-k)\Gamma_5(p-k, p)\Delta'_{\mathrm{F}}(k^2) \tag{6(311)}$$

as illustrated in Fig. 6.33. 6(311) is obtained from 6(277) by replacing $S_{\mathrm{F}}(p)$ and

$$\Delta_{\mathrm{F}}(k^2) = \frac{-i}{k^2+\mu^2}$$

by the exact propagators and by replacing γ_5 at one of the vertices by the exact vertex function Γ_5.

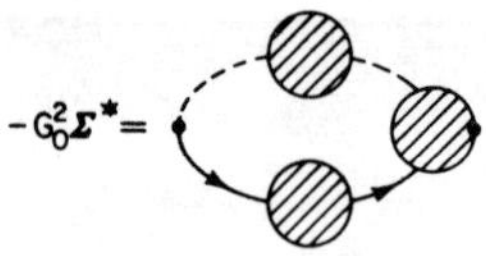

FIG. 6.33. Graphical representation of Eq. 6(311) for proper fermion self-energy part.

Equations 6(308)–6(311) form a set of coupled integral equations relating S'_{F} and Δ'_{F} to each other and to the exact vertex function Γ_5. To complete the set we need an equation for

$$\Gamma_5(p_2, p_1) = \gamma_5 + \Lambda_5(p_2, p_1) \tag{6(312)}$$

Actually, a closed expression for Λ_5 in terms of S'_{F}, Δ'_{F} and Γ_5 cannot be written down. However we can derive a *power series* integral equation. To do this we define the concept of an *irreducible* graph. If from a graph M

we omit all meson and fermion self-energy insertions and all vertex insertions, we obtain a graph M_0, known as the *skeleton* of M. This process of reducing a graph to its skeleton is illustrated in Fig. 6.34. A graph which

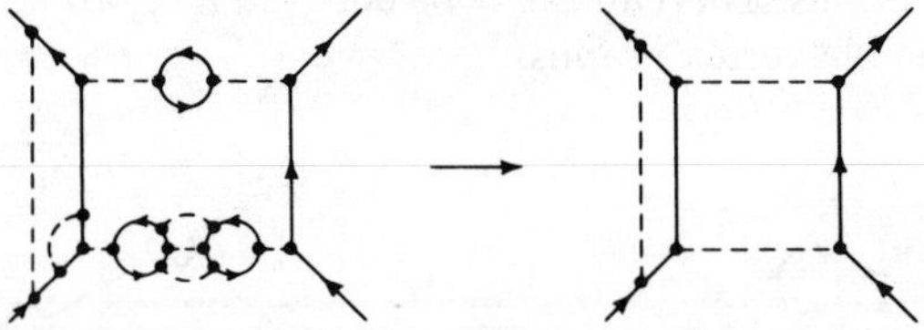

FIG. 6.34. Reduction of a Feynman graph to its (irreducible) skeleton.

is equal to its skeleton, for example the graph of Fig. 6.1, is called irreducible. We now note that there exist an infinity of irreducible vertex diagrams. The first, fifth, and sixth diagrams in the expansion of Λ_5, shown in Fig. 6.27, are irreducible. Other irreducible vertex insertions are shown in Fig. 6.35. Obviously we generate the sum of *all* vertex graphs contributing

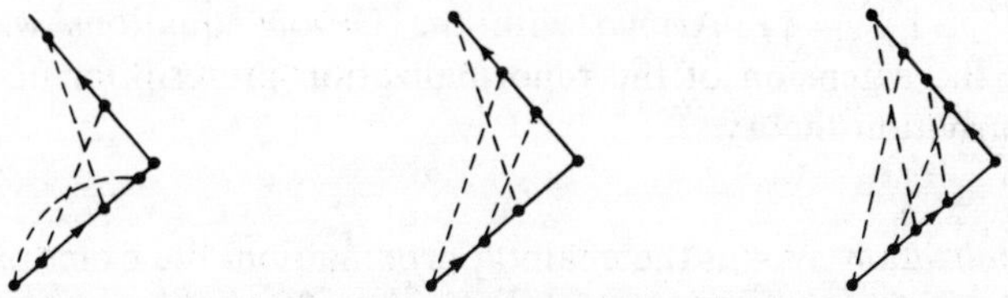

FIG. 6.35. Some higher order irreducible (skeleton) vertex insertions.

to Λ_5 by first summing only over irreducible vertex graphs and then 'dressing' each skeleton by means of the substitutions

$$S_F \to S'_F$$

$$\Delta_F \to \Delta'_F$$

$$\gamma_5 \to \Gamma_5$$

as in Fig. 6.36 for example. Thus

$$\Lambda_5(p_2, p_1) = \frac{G_0^2}{(2\pi)^4} \int d^4k \Gamma_5(p_2, p_2 - k) S'_F(p_2 - k) \Gamma_5(p_2 - k, p_1 - k) \times S'_F(p_1 - k) \Gamma_5(p_1 - k, p_1) \Delta'_F(k^2) + \dots \quad 6(313)$$

where the first term on the right-hand side represents the dressed first-order term 6(306) and the remaining terms include contributions from all irreducible skeleton graphs. We remark that each term in the series 6(313) with coefficient G_0^{2n} will contain n Δ'_F functions, $2n$ S'_F functions, and $2n+1$ Γ_5 functions. This can easily be checked by examining the structure of the irreducible vertex graphs.

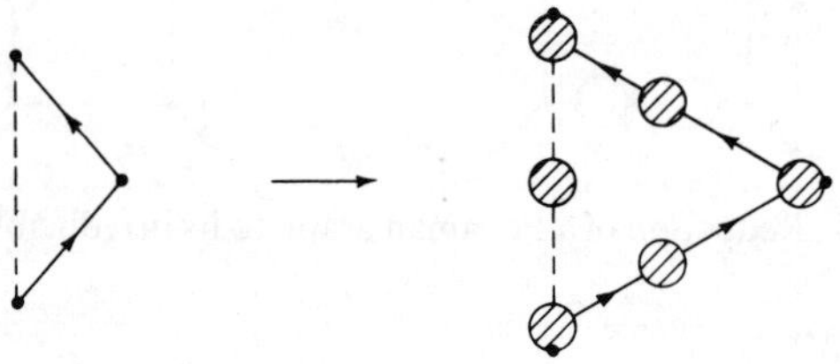

FIG. 6.36. 'Dressing' an irreducible vertex graph.

The coupled Dyson equations 6(308)–6(313) have been derived by examining the diagrammatic expansions of the exact functions Δ'_F, S'_F and Γ_5. These expansions can in turn be recovered iteratively from 6(308)–6(313) by using the zero-order approximation $\Delta_F'^{(0)} = \Delta_F$, $S_F'^{(0)} = S_F$, $\Gamma_5^{(0)} = \gamma_5$. Armed with the Dyson equations we can now discuss the extension of the renormalization prescription to all orders of perturbation theory.

Renormalization As in the chain approximation, the exact propagators Δ'_F and S'_F contain terms which shift the effective boson and fermion masses and the effective coupling constant. Thus, if we expand $\Pi^*(k^2)$ about $k^2 = -\mu^2$, as in 6(232),

$$\Pi^*(k^2) = \Pi^*(-\mu^2)+(k^2+\mu^2)\Pi^{*\prime}(-\mu^2)+\Pi_c^*(k^2) \qquad 6(314)$$

insert 6(314) into 6(308) and sandwich Δ'_F between two vertices, we see that $G_0^2\Pi^*(-\mu^2)$ produces a shift in the position of the one-particle pole—and hence in the effective meson mass—while $G_0^2\Pi^{*\prime}(-\mu^2)$ alters the effective coupling constant, as measured by the residue at the pole. Similarly, by a sequence of subtractions at $\gamma . p = im$, $\Sigma^*(p)$ can be expanded in a form similar to 6(281)

$$\Sigma^*(p) = A^*+B^*(\gamma . p-im)+\Sigma_c^*(p) \qquad 6(315)$$

with $\Sigma_c^*(p)$ of the form

$$\Sigma_c^*(p) = (\gamma . p-im)\sigma_c^*(p) \qquad 6(316)$$

where $\sigma_c^*(p)$ vanishes for $\gamma \cdot p = im$. The constant A^* produces a shift in the fermion mass in the Dyson equation for S_F', while B^* generates a further alteration in the coupling constant.

The mass shifts due to $\Pi^*(-\mu^2)$ and A^* are easily disposed of by setting

$$\delta\mu^2 = -G_0^2\Pi^*(-\mu^2) \tag{6(317)}$$

and

$$\delta m = -iG_0^2 A^* \tag{6(318)}$$

in 6(303) and 6(305) to cancel out the effects of $-G_0^2\Pi^*(-\mu^2)$ and $-G_0^2 A^*$ in the Dyson equations 6(308) and 6(310) respectively. This is the straightforward generalization of the procedure followed in the chain approximation, 6(269) and 6(286) appearing as the lowest order approximations to 6(317) and 6(318) respectively. As before, the inclusion of the counterterms 6(303) and 6(305) is interpreted as a simple renormalization of the meson and fermion masses in the original Lagrangian. We must now turn to coupling-constant renormalization. This is rather more involved than before, owing to the simultaneous appearance of Z_3 and Z_2 corrections to G_0, as well as additional corrections arising from the vertex function.

Let us consider the vertex correction $\Lambda_5(p_2, p_1)$ in 6(312). Since $\Lambda_5(p_2, p_1)$ is a Lorentz pseudoscalar, it is necessarily of the form

$$\Lambda_5(p_2, p_1) = \gamma_5 f(\gamma \cdot p_1, \gamma \cdot p_2, p_1^2, p_2^2, p_1 \cdot p_2)$$

with f some function of the five scalar quantities which can be formed with the aid of the 4-vectors p_1^μ, p_2^μ and γ^μ. If $\Lambda_5(p_2, p_1)$ is sandwiched between initial and final spinors $\bar{u}_{\mathbf{p}_2}$ and $u_{\mathbf{p}_1}$, we can move all $\gamma \cdot p_1$ factors in f to the extreme right and all $\gamma \cdot p_2$ factors to the extreme left, setting

$$\gamma \cdot p_1 = im, \qquad p_1^2 = -m^2 \tag{6(319a)}$$

and

$$\gamma \cdot p_2 = im, \qquad p_2^2 = -m^2 \tag{6(319b)}$$

The only remaining variable is $p_1 \cdot p_2$ or, equivalently, $(p_2 - p_1)^2$ and we have

$$\bar{u}_{\mathbf{p}_2}\Lambda_5(p_2, p_1)u_{\mathbf{p}_1} = \bar{u}_{\mathbf{p}_2}\gamma_5 u_{\mathbf{p}_1} g[(p_2 - p_1)^2] \tag{6(320)}$$

We now set $g(-\mu^2) = L$ and perform a subtraction at the point $(p_2 - p_1)^2 = -\mu^2$ by writing

$$\Lambda_5(p_2, p_1) = L\gamma_5 + \Lambda_{5c}(p_2, p_1) \tag{6(321)}$$

where $\Lambda_5(p_2, p_1)$ is equal to zero for 6(319a), 6(319b) and $(p_2 - p_1)^2 = -\mu^2$, that is, when all three particles meeting at the vertex are placed on their mass shells. It should be noted that this subtraction point is 'unphysical' in that the energy-momentum conservation delta-function attached to the vertex gives zero at this point. We are essentially assuming that $g[(p_2-p_1)^2]$ can be analytically continued from the physical region $(p_2-p_1)^2 \geqslant 0$ to the point $(p_2-p_1)^2 = -\mu^2$, [see Nambu (1958)]. Substituting 6(321) into 6(320), we see that the $L\gamma_5$ term in 6(321) must produce an alteration in the effective coupling constant. In the diagram of Fig. 6.37, for example, the $L\gamma_5$ term alters the residue of the scattering amplitude at the one-particle pole in Δ_F'.

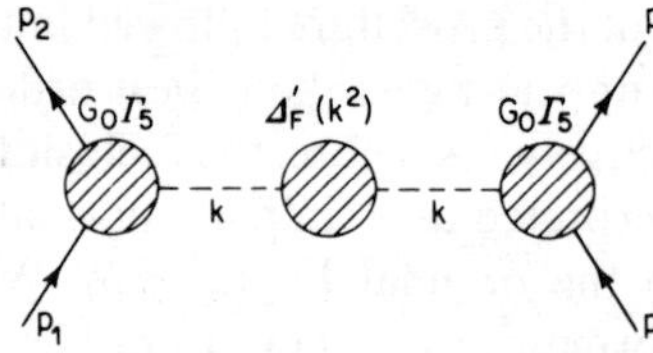

FIG. 6.37. Diagram illustrating role of $L\gamma_5$ term in vertex correction Λ_5.

Generalizing our earlier procedure for the chain approximation, we seek to eliminate the correction terms $G_0^2\Pi^{*\prime}(-\mu^2)$, $G_0^2B^*$, and L in the Dyson equations for Δ_F', S_F' and Γ_5 by absorbing them into the renormalized coupling constant G_R with the aid of renormalization constants. Let us attempt to find constants Z_3, Z_2 and Z_1 such that the functions $\mathbf{S}_F'$, $\mathbf{\Delta}_F'$ and $\mathbf{\Gamma}_5$ defined by*

$$\Delta_F'(k^2) = Z_3\mathbf{\Delta}_F'(k^2) \qquad 6(322a)$$

$$S_F'(p) = Z_2\mathbf{S}_F'(p) \qquad 6(322b)$$

$$\Gamma_5(p_2, p_1) = Z_1^{-1}\mathbf{\Gamma}_5(p_2, p_1) \qquad 6(322c)$$

satisfy the coupled Dyson equations with the $G_0^2\Pi^{*\prime}(-\mu^2)$, $G_0^2B^*$, and L terms *removed* and with G_0 replaced by a renormalized constant G_R. This is clearly the simplest generalization of the renormalization procedure followed in 6(273) or 6(290). Actually we shall find that this generalization is not quite adequate and that our requirement must be amended slightly.

Consider first the Dyson equation for $\Gamma_5(p_2, p_1)$. Each term in the expansion 6(313) for $\Lambda_5(p_2, p_1)$ with coefficient G_0^{2n} will contain $n\ \Delta_F'$ functions, $2n\ S_F'$ functions and $2n+1\ \Gamma_5$ functions, as remarked earlier. Accordingly the substitutions 6(322a), 6(322b) and 6(322c) in a term of

* The appearance of Z_1^{-1} in 6(322c) rather than Z_1 is clearly a matter of definition.

order G_0^{2n} will produce a factor

$$Z_3^n Z_2^{2n} Z_1^{-2n-1} \tag{6(323)}$$

If we now define a renormalized coupling constant

$$G_R = Z_3^{1/2} Z_2 Z_1^{-1} G_0 \tag{6(324)}$$

then all factors in 6(323) will be used up in replacing G_0^{2n} by G_R^{2n} except for one factor Z_1^{-1}. We have

$$\Lambda_5(\Delta_F', S_F', \Gamma_5, G_0) = Z_1^{-1}\mathbf{\Lambda}_5(\mathbf{\Delta}_F', \mathbf{S}_F', \mathbf{\Gamma}_5, G_R)$$

or, in an obvious notation,

$$\Lambda_5(G_0) = Z_1^{-1}\mathbf{\Lambda}_5(G_R) \tag{6(325)}$$

Writing, as in 6(321)

$$\mathbf{\Lambda}_5(G_R) = \mathbf{L}(G_R)\gamma_5 + \mathbf{\Lambda}_{5c}(G_R) \tag{6(326)}$$

we exploit the remaining Z_1^{-1} factor to get rid of the $\mathbf{L}(G_R)$ term by setting

$$Z_1 = 1 - \mathbf{L}(G_R) \tag{6(327)}$$

With this value for Z_1 we have

$$\begin{aligned}
\Gamma_5(G_0) &= \gamma_5 + \Lambda_5(G_0) \\
&= \gamma_5 + Z_1^{-1}\mathbf{L}(G_R)\gamma_5 + Z_1^{-1}\mathbf{\Lambda}_{5c}(G_R) \\
&= \frac{1}{1-\mathbf{L}(G_R)}[\gamma_5 + \mathbf{\Lambda}_{5c}(G_R)] \\
&= Z_1^{-1}\mathbf{\Gamma}_5(G_R)
\end{aligned} \tag{6(328)}$$

in agreement with our requirement 6(322c).

Next we consider the Dyson equation for $\Delta_F'(k^2)$. Substituting 6(322b) and 6(322c) for S_F' and Γ_5 in the right-hand side of 6(309) produces a factor $Z_2^2 Z_1^{-1}$ which can be partly absorbed into the G_0^2 factor multiplying $\Pi^*(k^2)$. Using 6(324) we have

$$\begin{aligned}
G_0^2\Pi^*(S_F', \Gamma_5) &= G_0^2 Z_2^2 Z_1^{-1}\mathbf{\Pi}^*(\mathbf{S}_F', \mathbf{\Gamma}_5) \\
&= Z_3^{-1} Z_1 G_R^2 \mathbf{\Pi}^*(\mathbf{S}_F', \mathbf{\Gamma}_5)
\end{aligned} \tag{6(329)}$$

or, in terms of our abbreviated notation,

$$G_0^2\Pi^* = Z_3^{-1} Z_1 G_R^2 \mathbf{\Pi}^* \tag{6(330)}$$

Like the Z_1^{-1} factor in 6(325), the Z_3^{-1} factor in 6(330) is just what is needed to dispose of the $\mathbf{\Pi}^{*\prime}(-\mu^2)$ term in the expansion of $\mathbf{\Pi}^*(k^2)$*

$$\mathbf{\Pi}^*(k^2) = (k^2+\mu^2)\mathbf{\Pi}^{*\prime}(-\mu^2)+\mathbf{\Pi}^*_c(k^2) \tag{6(331)}$$

Defining

$$Z_3 = 1+Z_1 G_R^2 \mathbf{\Pi}^{*\prime}(-\mu^2) \tag{6(332)}$$

and proceeding exactly as in the chain approximation case, we have

$$\begin{aligned}\Delta'_F(k^2) &= \frac{-i}{k^2+\mu^2-G_0^2\Pi^*(k^2)-\delta\mu^2}\\ &= \frac{-i}{(k^2+\mu^2)[1-Z_3^{-1}Z_1G_R^2\mathbf{\Pi}^{*\prime}(-\mu^2)]-Z_3^{-1}Z_1G_R^2\,\mathbf{\Pi}^*_c(k^2)}\\ &= [1+Z_1G_R^2\mathbf{\Pi}^{*\prime}(-\mu^2)]\frac{-i}{k^2+\mu^2-Z_1G_R^2\,\mathbf{\Pi}^*_c(k^2)}\\ &= Z_3\frac{-i}{k^2+\mu^2-Z_1G_R^2\,\mathbf{\Pi}^*_c(k^2)}\end{aligned} \tag{6(333)}$$

with the $\mathbf{\Pi}^{*\prime}(-\mu^2)$ term removed. Equation 6(333) is of the form 6(322a) but it does not quite fulfil our requirement regarding $\mathbf{\Delta}_F(k^2)$, since $\mathbf{\Pi}^*_c(k^2)$ appears multiplied by a factor Z_1. Actually the Z_1 factor is essential and we must conclude that our requirement for $\mathbf{\Delta}'_F$ was inadequately formulated†. We shall return to this point later.

The treatment of the Dyson equation for $S'_F(p)$ follows the same pattern. Substituting 6(322a), 6(322b) and 6(322c) into the expression 6(311) for Σ^* we have

$$G_0^2\Sigma^*(\Delta'_F, S'_F, \Gamma_5) = G_0^2Z_3Z_2Z_1^{-1}\Sigma^*(\mathbf{\Delta}'_F, \mathbf{S}'_F, \mathbf{\Gamma}_5)$$

or, using 6(324)

$$G_0^2\Sigma^* = Z_2^{-1}Z_1G_R^2\mathbf{\Sigma}^* \tag{6(334)}$$

which, like 6(330), displays the additional Z_1 factor. Writing

$$\mathbf{\Sigma}^*(p) = \mathbf{B}^*(\gamma\,.\,p-im)+\mathbf{\Sigma}^*_c(p) \tag{6(335)}$$

as in 6(315) with the A^* term suppressed, we dispose of the B^* term by defining

$$Z_2 = 1+Z_1G_R^2\mathbf{B}^* \tag{6(336)}$$

* We drop the $\mathbf{\Pi}^*(-\mu^2)$ term, since it is effectively cancelled by the mass counterterm.
† The Z_1 factor is needed to remove so-called *overlap divergences*, [see below].

A now familiar manipulation yields

$$S'_F(p) = \frac{-1}{\gamma \, . \, p - im - G_0^2\Sigma^*(p) + i\delta m}$$

$$= Z_2 \frac{-1}{\gamma \, . \, p - im - Z_1 G_R^2 \Sigma_c^*(p)} \qquad 6(337)$$

Again, 6(337) is of the form 6(322b) but the appearance of the Z_1 factor indicates that our requirement for S'_F was too restrictive.

External Lines The final step in the renormalization procedure is the demonstration that—as in the simple chain approximation calculation 6(273)—the overall factors Z_3, Z_2 and Z_1^{-1} in 6(322a), 6(322b) and 6(322c) can be absorbed into the coupling constants attached to the external lines *consistently with* 6(324). Consider for example the diagram of Fig. 6.37. It corresponds to a matrix element of the general form

$$M = \bar{u}_{\mathbf{p}_2}\Gamma_5(p_2, p_1)u_{\mathbf{p}_1}\, G_0^2\Delta'_F(k^2)\bar{u}_{\mathbf{p}'_2}\Gamma_5(p'_2, p'_1)u_{\mathbf{p}'_1} \qquad 6(338)$$

Substituting 6(322a) and 6(322c), we get

$$M = \bar{u}_{\mathbf{p}_2}\boldsymbol{\Gamma}_5 u_{\mathbf{p}_1} Z_1^{-2} Z_3 G_0^2 \boldsymbol{\Delta}'_F(k^2)\bar{u}_{\mathbf{p}'_2}\boldsymbol{\Gamma}_5(p'_2, p'_1)u_{\mathbf{p}'_1} \qquad 6(339)$$

To convert the factor G_0^2 to G_R^2, we need an additional factor Z_2^2. This is provided by additional self-energy insertions on the external fermion lines, as shown in Fig. 6.38. We have so far failed to take these corrections into account. When we do, the net effect is to replace $u_{\mathbf{p}}$ and $\bar{u}_{\mathbf{p}}$ in matrix elements like 6(338) by

$$\bar{u}'_{\mathbf{p}} = Z_2^{1/2}\bar{u}_{\mathbf{p}} \qquad \text{and} \qquad u'_{\mathbf{p}} = Z_2^{1/2}u_{\mathbf{p}} \qquad 6(340)$$

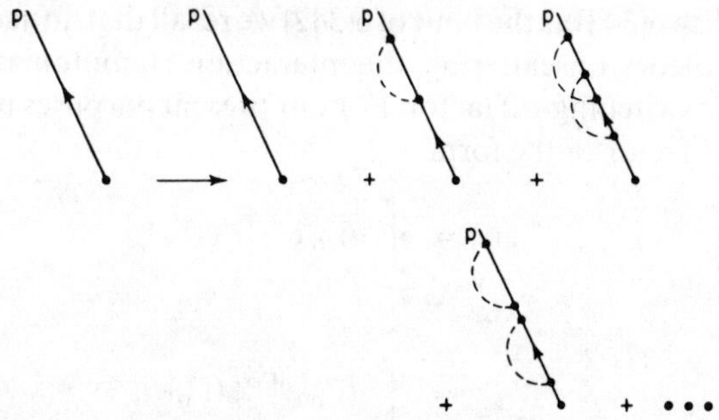

FIG. 6.38. Corrections to spinor factor $\bar{u}_{\mathbf{p}}$ due to self-energy insertions on external line.

respectively [see below]. This provides the required $Z_2^{\frac{1}{2}}$ factor in 6(339). Similarly, the inclusion of boson self-energy parts on external boson lines has the effect of multiplying each external boson factor, say $(2\omega_{\mathbf{k}})^{-1/2}$, by $Z_3^{1/2}$. We leave it to the reader to prove that—as in 6(339)—the inclusion of the $Z_3^{1/2}$ and $Z_2^{1/2}$ factors for the external lines is always sufficient to absorb all Z factors arising from the renormalizations 6(322a), 6(322b) and 6(322c) [see Problem 8].

The proof of 6(340) to all orders of perturbation theory is rather involved [see Dyson (1949), (1951)] and is given here only in lowest order. Consider the second-order correction to $\bar{u}_{\mathbf{p}}$, represented by the second graph in the expansion of Fig. 6.38, (together with the associated δm insertion, not shown explicitly). In this approximation

$$\bar{u}'_{\mathbf{p}} = \bar{u}_{\mathbf{p}} + \bar{u}_{\mathbf{p}}[-G_0^2\Sigma(p) + i\delta m]S_F(p) \qquad 6(341)$$

Using 6(281) and the expression $S_F(p) = -(\gamma \,.\, p - im)^{-1}$ for the bare fermion propagator, we get

$$\bar{u}'_{\mathbf{p}} = \bar{u}_{\mathbf{p}} + \bar{u}_{\mathbf{p}} G_0^2 B(\gamma \,.\, p - im)\frac{1}{\gamma \,.\, p - im} \qquad 6(342)$$

All other terms in 6(281) drop out when operating on $\bar{u}_{\mathbf{p}}$ by virtue of the mass shell constraint $\gamma \,.\, p = im$. Now the expression 6(342) is ambiguous, being of the form 0/0 when $\gamma \,.\, p$ is put equal to im. In fact, the correct value for 6(342) is, as we shall show,

$$\bar{u}'_{\mathbf{p}} = \bar{u}_{\mathbf{p}} + \tfrac{1}{2}\bar{u}_{\mathbf{p}} G_0^2 B \qquad 6(343)$$

or, using 6(289) in the form* $Z_2 = (1 - G_0^2 B)^{-1}$,

$$\bar{u}'_{\mathbf{p}} = \bar{u}_{\mathbf{p}}(1 + \tfrac{1}{2}G_0^2 B) \sim \bar{u}_{\mathbf{p}} Z_2^{1/2} \qquad 6(344)$$

Equation 6(344) is just the result 6(340) to second order in G_0^2.

To establish 6(343) as the limit of 6(342) we recall that, in accordance with the formal theory of scattering, the interaction Hamiltonian $H_I(t)$ carries an adiabatic switching-off factor. For our present purposes it is convenient to write this factor in the form

$$g(t) = \int_{-\infty}^{\infty} d\Gamma_0 \, e^{-i\Gamma_0 t} g(\Gamma_0)$$

$$= \int_{-\infty}^{\infty} d\Gamma_0 \, e^{i\Gamma . x} g(\Gamma_0) \qquad 6(345)$$

* Combining 6(288) with 6(289) we have $Z_2 = 1 + Z_2 G_0^2 B$ or $Z_2^{-1} = 1 - G_0^2 B$.

where $\Gamma_\mu = (0, 0, 0, i\Gamma_0)$ and where we suppose that $g(\Gamma_0)$ is strongly peaked at $\Gamma_0 = 0$*. From the normalization $g(t = 0) = 1$ we have the condition

$$\int g(\Gamma_0)\, d\Gamma_0 = 1$$

If we now recalculate the correction 6(341) to the external fermion line, using the interaction

$$\mathscr{H}_I(x) = -iGg(t)\tfrac{1}{2}[\bar{\psi}(x), \gamma_5\psi(x)]\phi(x) - \delta m[g(t)]^2\bar{\psi}(x)\psi(x)$$

we find that $g(t)$ acts in much the same way as an external field, providing an additional momentum 'kick' Γ_μ at each vertex, with the instruction to integrate over all additional momenta. In place of 6(341) we find [see Problem 9],

$$\bar{u}'_{\mathbf{p}} = \bar{u}_{\mathbf{p}} + \bar{u}_{\mathbf{p}} \int d\Gamma_0\, d\Gamma'_0 g(\Gamma_0) g(\Gamma'_0)[-G_0^2\Sigma(p-\Gamma) + i\delta m]S_F(p-\Gamma-\Gamma') \quad 6(347)$$

Thus 6(342) is replaced by

$$\bar{u}'_{\mathbf{p}} = \bar{u}_{\mathbf{p}} + \bar{u}_{\mathbf{p}} \int d\Gamma_0\, d\Gamma'_0 g(\Gamma_0) g(\Gamma'_0) G_0^2 B[\gamma\,.\,(p-\Gamma) - im] \frac{1}{\gamma\,.\,(p-\Gamma-\Gamma') - im} \quad 6(348)$$

in which the ambiguity is no longer present. We can now replace the factor $\gamma\,.\,(p-\Gamma) - im$ by

$$\gamma\,.\left(p - \frac{\Gamma - \Gamma'}{2}\right) - im \quad 6(349)$$

since the integrand in 6(348) is otherwise symmetrical in Γ and Γ', and add to 6(349) the term $\frac{1}{2}(\gamma\,.\,p - im)$, which is effectively zero when operating on $\bar{u}_{\mathbf{p}}$. The numerator and denominator in 6(348) cancel, leaving a factor $\frac{1}{2}$, and we recover the result 6(343) upon taking the limit $g(\Gamma_0) \to \delta(\Gamma_0)$.

For details regarding the extension of the above procedure to higher orders of perturbation, we refer the reader to Dyson's 1951 paper. Here we simply note that to all orders of perturbation theory 6(341) is replaced by

$$\bar{u}'_{\mathbf{p}} = \bar{u}_{\mathbf{p}} + \bar{u}_{\mathbf{p}}[-G_0^2\Sigma^*(p) + i\delta m]S'_F(p) \quad 6(350)$$

as illustrated in Fig. 6.39. Using 6(310) we have

$$[-G_0^2\Sigma^*(p) + i\delta m]S'_F(p) = [-S_F'^{-1}(p) - \gamma\,.\,p + im]S'_F(p)$$

* The precise form of $g(t)$ is not important [see Dyson (1951)].

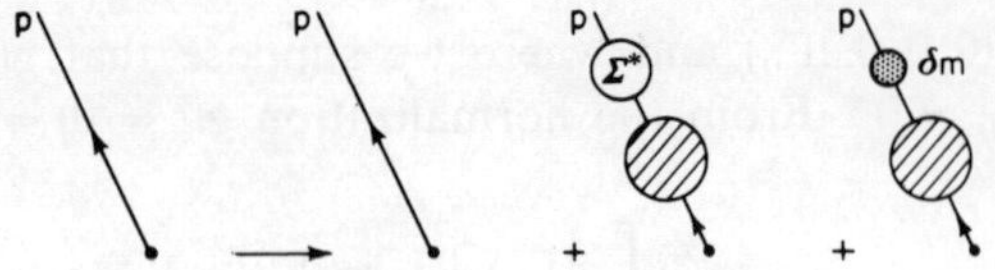

FIG. 6.39. Diagram illustrating Eq. 6(350).

so that 6(350) reduces to

$$\bar{u}'_{\mathbf{p}} = -\bar{u}_{\mathbf{p}}(\gamma \cdot p - im)S'_{\mathrm{F}}(p)$$

Dyson's result 6(340) therefore indicates that

$$(\gamma \cdot p - im)S'_{\mathrm{F}}(p)|_{\gamma . p = im} \xrightarrow[g(t)\to 1]{} -Z_2^{1/2} \tag{6(351)}$$

with the adiabatic cut-off serving to define the otherwise ambiguous left-hand side*. Similarly, the appearance of the $Z_3^{1/2}$ factor for external boson lines indicates that

$$(k^2+\mu^2)\Delta'_{\mathrm{F}}(k^2)|_{k^2=-\mu^2} \xrightarrow[g(t)\to 1]{} -iZ_3^{1/2} \tag{6(352)}$$

Spinor Electrodynamics—The Ward Identity The renormalization of spinor electrodynamics follows the same pattern as that of neutral pseudoscalar meson theory. There is one important simplification, however, due to the existence of an identity linking the electron propagator S'_{F} to the electron-photon vertex function Γ_μ. S'_{F} is now given by

$$S'_{\mathrm{F}}(p) = \frac{-1}{\gamma \cdot p - im - e_0^2\Sigma^*(p) + i\delta m} \tag{6(353)}$$

with

$$\Sigma^*(p) = \frac{-1}{(2\pi)^4}\int d^4k\gamma_\mu S'_{\mathrm{F}}(p-k)\Gamma_\mu(p-k, p)D'_{\mathrm{F}}(k^2) \tag{6(354)}$$

Here $D'_{\mathrm{F}}(k^2)$ is the exact photon propagator and Γ_μ is the exact vertex function

$$\Gamma_\mu(p_2, p_1) = \gamma_\mu + \Lambda_\mu(p_2, p_1) \tag{6(355)}$$

with

$$\Lambda_\mu(p_2, p_1) = \frac{e_0^2}{(2\pi)^4}\int d^4k\Gamma_\nu(p_2, p_2-k)S'_{\mathrm{F}}(p_2-k)\Gamma_\mu(p_2-k, p_1-k)$$
$$\times S'_{\mathrm{F}}(p_1-k)\Gamma_\nu(p_1-k, p_1)D'_{\mathrm{F}}(k^2) + \ldots \tag{6(356)}$$

* Note that a 'straightforward' evaluation (without g(t)) of the left-hand side, based on the formula 6(337) for $S'_{\mathrm{F}}(p)$, would yield Z_2 rather than $Z_2^{1/2}$! [See also Section 7-5.]

The relation linking S'_F to Γ_μ is then

$$\Gamma_\mu(p, p) = -\frac{\partial}{\partial p_\mu} S_F'^{-1}(p) \qquad 6(357)$$

and is known as Ward's identity [Ward (1950)]*. Equation 6(357) is trivially satisfied by the bare vertex γ_μ and free propagator

$$S_F(p) = -(\gamma . p - im)^{-1}$$

Using 6(353) and 6(355), we can write Ward's identity in the form of a relation between $\Sigma^*(p)$ and the vertex correction $\Lambda_\mu(p, p)$:

$$\Lambda_\mu(p, p) = -\frac{\partial}{\partial p_\mu} e_0^2 \Sigma^*(p) \qquad 6(358)$$

The Ward identity is essentially a consequence of the fact that the electromagnetic field is coupled to a conserved current. The simplest proof of 6(358) exploits the fact that in the presence of a constant external electromagnetic field a_μ, the proper self-energy part $e_0^2\Sigma^*(p)$ is modified to

$$e_0^2 \Sigma_a{}^*(p) = e_0^2 \Sigma^*(p) + e_0 a_\mu \Lambda_\mu(p, p) + \ldots \qquad 6(359)$$

where the right-hand side represents a power series in a_μ. The first order term in a_μ represents the effect of a single interaction with the external field. The point is that when γ_μ is inserted in all possible ways into $\Sigma^*(p)$ the effect is to replace $\Sigma^*(p)$ by the vertex correction $\Lambda_\mu(p, p)$, as illustrated in Fig. 6.40†. Now gauge invariance requires that

$$\Sigma_a{}^*(p) = \Sigma^*(p - e_0 a) = \Sigma^*(p) - e_0 a_\mu \left(\frac{\partial \Sigma^*(p)}{\partial p_\mu} \right)_{a=0} + \ldots \qquad 6(360)$$

Comparing 6(360) with 6(359) we recover Ward's identity 6(358).

A significant consequence of Ward's identity is that the vertex and fermion wave function renormalization constants are equal in quantum electrodynamics, i.e. that

$$Z_1 = Z_2 \qquad 6(361)$$

To show this we sandwich 6(357) between Dirac spinors

$$\bar{u}_{\mathbf{p}} \Gamma_\mu(p, p) u_{\mathbf{p}} = -\bar{u}_{\mathbf{p}} \frac{\partial}{\partial p_\mu} S_F'^{-1}(p) u_{\mathbf{p}} \qquad 6(362)$$

* A generalized form of this identity, namely

$$(p'_\mu - p_\mu)\Gamma_\mu(p', p) = S_F'^{-1}(p) - S_F'^{-1}(p')$$

has been established by Takahashi (1957).

† Note that a constant external field transfers zero momentum.

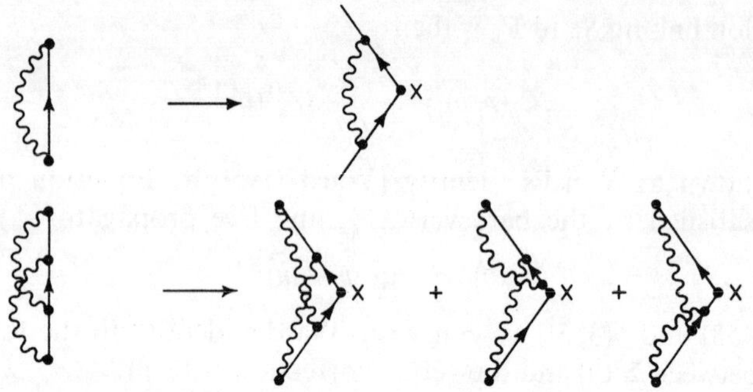

FIG. 6.40. Insertion of electromagnetic vertex into self-energy graphs illustrating Eq. 6(359).

and use the relations

$$S'_F(p) = Z_2 \mathbf{S}'_F(p) \qquad 6(363a)$$

$$\Gamma_\mu(p_2, p_1) = Z_1^{-1} \mathbf{\Gamma}_\mu(p_2, p_1) \qquad 6(363b)$$

analogous to 6(322b) and 6(322c). As in meson theory, $\mathbf{S}_F$ is characterized by the fact that

$$\mathbf{S}_F(p) \simeq \frac{-1}{\gamma \cdot p -- im} \quad \text{near the mass shell}$$

but $\mathbf{\Gamma}_\mu$ differs from its meson theoretic analog in that we now have*

$$\bar{u}_{\mathbf{p}} \mathbf{\Gamma}_\mu(p, p) u_{\mathbf{p}} = \bar{u}_{\mathbf{p}} \gamma_\mu u_{\mathbf{p}} \qquad 6(364)$$

Thus 6(362) becomes

$$Z_1^{-1} \bar{u}_{\mathbf{p}} \gamma_\mu u_{\mathbf{p}} = Z_2^{-1} \bar{u}_{\mathbf{p}} \gamma_\mu u_{\mathbf{p}}$$

or $Z_1 = Z_2$. The equality of Z_1 and Z_2 implies that the renormalized charge

$$e_R = Z_3^{1/2} Z_2 Z_1^{-1} e_0$$

is simply

$$e_R = Z_3^{1/2} e_0 \qquad 6(365)$$

the effect of the vertex function renormalization cancelling with that of the fermion wave function renormalization. This result is of fundamental

* In the meson theoretic case, $\mathbf{\Gamma}_5(p_2, p_1)$ is normalized to γ_5 at $(p_2 - p_1)^2 = -\mu^2$ (with $\gamma \cdot p_1 = \gamma \cdot p_2 = im$). This difference is dictated by the fact that in electrodynamics we perform the subtraction in $\Gamma_\mu(p_2, p_1)$ at the point $p_1 = p_2$ (with $\gamma \cdot p_2 = \gamma \cdot p_1 = im$).

importance. Generalized to the electromagnetic interaction of other charged particles, it implies that the electromagnetic coupling is *universal*, i.e. of the same strength for all particles, [see Section 8.1].

External Field Interactions We have seen that external fermion and boson lines carry factors $Z_2^{1/2}$ and $Z_3^{1/2}$ respectively, arising from the insertion of self-energy parts on the external lines. This situation is altered, however, when the external line is replaced by an interaction with an external field. Consider, for example, the interaction with an external electromagnetic field $A_\mu^{\text{ext}}(x)$. To first order in A_μ^{ext} the electron scattering amplitude is given by 6(110). If we continue to work to first order in $A_\mu^{\text{ext}}(x)$ but take into account the electron-photon interaction to *all* orders of perturbation theory, 6(110) will be modified by replacing γ_μ by $\Gamma_\mu(p', p)$, $\bar{u}_{\mathbf{p}'}$ by $Z_2^{1/2}\bar{u}_{\mathbf{p}'}$, $u_{\mathbf{p}}$ by $Z_2^{1/2}u_{\mathbf{p}}$ and $A_\mu^{\text{ext}}(k)$ by

$$A_\mu^{\text{ext}}(k) + D'_{F\mu\lambda}(k)ie_0^2\Pi^*_{\lambda\nu}(k)A_\nu(k) \qquad 6(366)$$

where $\Pi^*_{\lambda\nu}$ is the exact proper photon self-energy insertion generalizing 6(90). The replacement of $A_\mu^{\text{ext}}(k)$ by 6(366) is illustrated in Fig. 6.41. Setting

$$\Pi^*_{\lambda\nu}(k) = (\delta_{\lambda\nu}k^2 - k_\lambda k_\nu)C^*(k^2)$$

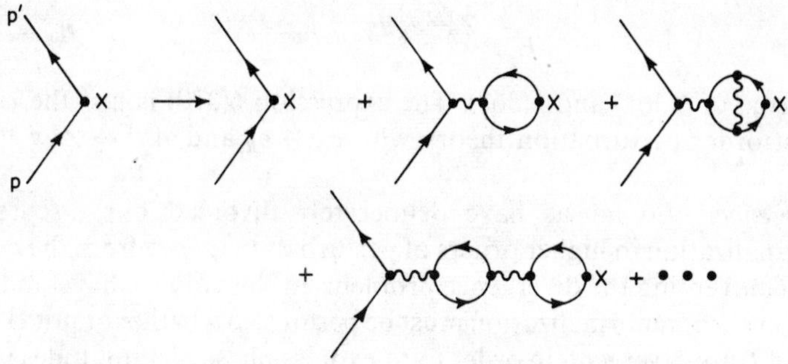

FIG. 6.41. Vacuum polarization corrections to scattering in an external field, illustrating Eq. 6(366).

on grounds of current conservation, as in 6(100), and noting that

$$D'_{F\mu\lambda}(k) = \delta_{\mu\lambda}\frac{-i}{k^2 - e_0^2k^2C^*(k^2)} \qquad 6(367)$$

where we have dropped $k_\mu k_\lambda$ terms which are effectively zero, we find that 6(366) reduces to

$$\left(1+\frac{e_0^2k^2C^*(k^2)}{k^2-e_0^2k^2C^*(k^2)}\right)A_\mu^{\text{ext}}(k) = k^2\frac{1}{k^2-e_0^2k^2C^*(k^2)}A_\mu^{\text{ext}}(k)$$
$$= ik^2D'_{F\mu\nu}(k)A_\nu^{\text{ext}}(k)$$
$$= ik^2Z_3\mathbf{D}'_{F\mu\nu}(k)A_\nu^{\text{ext}}(k)$$

where $\mathbf{D}'_{F\mu\nu}$—the renormalized photon propagator analogous to $\mathbf{\Delta}'_F$—is characterized by the fact that

$$\mathbf{D}'_{F\mu\nu}(k) \cong \delta_{\mu\nu}\frac{-i}{k^2} \qquad \text{near} \quad k^2 = 0 \qquad 6(368)$$

We conclude that to first order in A_μ^{ext} and to *all* orders in the electron-photon interaction, the electron scattering amplitude is given by

$$-e_0\frac{1}{V}\left(\frac{m^2}{E_{\mathbf{p}'}E_{\mathbf{p}}}\right)^{1/2}Z_2Z_1^{-1}\bar{u}_{\mathbf{p}'\sigma'}\mathbf{\Gamma}_\mu(p',p)u_{\mathbf{p}\sigma}ik^2Z_3\mathbf{D}'_{F\mu\nu}(k)A_\nu^{\text{ext}}(k) \quad 6(369)$$

where we have substituted 6(363b) for $\Gamma_\mu(p',p)$. Absorbing a factor $Z_2Z_1^{-1}Z_3^{1/2}$ to replace e_0 by e_R, we are left with an additional factor $Z_3^{1/2}$ which can be absorbed into a renormalization of the external field*. Note that in the limit $p' \to p$, 6(369) reduces to

$$-\frac{m}{E_{\mathbf{p}}V}Z_3^{1/2}e_R\bar{u}_{\mathbf{p}'\sigma'}\gamma_\mu u_{\mathbf{p}\sigma}A_\mu^{\text{ext}}(0) \qquad 6(370)$$

by virtue of 6(364) and 6(368). The expression 6(370) is just the result of lowest order perturbation theory with $e_0 \to e_R$ and $A_\mu^{\text{ext}} \to Z_3^{1/2}A_\mu^{\text{ext}}$.

Divergences So far as have deliberately divorced our discussion of renormalization in higher orders of perturbation theory from the question of circumventing the divergence problem. In this way we have underlined the point that renormalization must be performed whether or not $\Pi^*(-\mu^2)$, B^* and L are divergent, in order to re-express physical amplitudes in terms of *measured* masses and coupling constants. However, the most significant aspect of renormalization theory is its practical value as a means of circumventing the divergence difficulty. In our discussion of the chain approximation for Δ'_F and S'_F in neutral pseudoscalar meson theory,

* This can be justified by regarding the external field as a phenomenological representation of a quantized field. We thereby enlarge our physical system to include the source of the external field.

we have seen that renormalization eliminates all divergences contained in Π and Σ. The more general statement—that our renormalization procedure based on 6(322a), 6(322b) and 6(322c) eliminates all the divergences contained in Π^*, Σ^* and Λ_5 to all orders of perturbation theory—has been shown by Dyson (1949) and Salam (1951). The rigorous proof of this important statement (together with the analogous statement for spinor electrodynamics) depends on a detailed analysis of the degree of divergence of higher order graphs, [Weinberg (1960)] and lies outside the scope of this book*.

In this connection, the appearance of the Z_1 factor in the Dyson equations in 6(333) and 6(337) seems to indicate that physical amplitudes are still dependent on the (divergent) constant Z_1. In fact however, the Dyson-Salam analysis indicates that the Z_1 factors serve to cancel *additional* divergences, still present both in $\mathbf{\Pi}_c^*(k^2)$ and $\mathbf{\Sigma}_c^*(p)$, which are known as *overlap divergences*. In renormalizing Π^* we made the replacements $\Gamma_5 = Z_1^{-1}\mathbf{\Gamma}_5$, $S_F' = Z_2\mathbf{S}_F'$ in the integral $\int \mathrm{Tr}\,[\gamma_5 S_F' \Gamma_5 S_F']$ for Π^* and then isolated the coupling constant correction term by means of the expansion 6(331). In this process, we have treated the two 'external' vertices in an unsymmetric way. Graphs such as those of Fig. 6.32 can be viewed as the insertion of a second-order vertex correction at either *a* or *b*. The two insertions 'overlap' but in writing down the Dyson equation 6(309) for Π^* we were obliged to make a choice and replace γ_5 by Γ_5 at only *one* of the vertices. This asymmetry in treatment is unavoidable, but it has the effect of masking the fact that the overlap graphs contain vertex divergences at *each* of the two vertices. By replacing Γ_5 by $Z_1^{-1}\mathbf{\Gamma}_5$ at only one of the two vertices, we have removed only a part of the overlap divergence. We may expect, and detailed analysis [Salam (1951)] supports this conclusion, that the removal of the divergence at the second vertex produces an additional factor Z_1^{-1} which is then cancelled by the Z_1 factor in 6(333). Defining $\mathbf{\Pi}_1^*(k^2)$ by

$$\mathbf{\Pi}^*(k^2) = Z_1^{-1}\mathbf{\Pi}_1^*(k^2) \tag{6(371)}$$

we have, from 6(333) and 6(322a)

$$\mathbf{\Delta}_F'(k^2) = \frac{-i}{k^2+\mu^2-G_R^2\mathbf{\Pi}_{1c}^*(k^2)} \tag{6(372a)}$$

where $\mathbf{\Pi}_{1c}^*$, and hence $\mathbf{\Delta}_F'$, is *finite* according to Salam's analysis. The

* For a thorough account of Weinberg's analysis and its application to renormalization theory, see J. D. Bjorken and S. Drell, *Relativistic Quantum Fields*, McGraw-Hill, New York, 1965, Chapter 19.

expression 6(332) for Z_3 can now be written as

$$Z_3 = 1 + G_R^2 \mathbf{\Pi}_1^{*\prime}(-\mu^2) \qquad 6(372b)$$

where $\mathbf{\Pi}_1^{*\prime}(-\mu^2)$, like $\Pi'(-\mu^2)$ in the chain approximation, is given by a logarithmically divergent integral.

We can characterize the overlap problem by saying that the integral $\int \mathrm{Tr}\,[\gamma_5 S_F' \Gamma_5 S_F']$ effectively behaves like $\int \mathrm{Tr}\,[\Gamma_5 S_F' \Gamma_5 S_F']$ from the point of view of renormalization. In fact, if we use 6(312) and 6(313) to replace γ_5 by the infinite series $\Gamma_5 - \int \Gamma_5 S_F' \Gamma_5 S_F' \Gamma_5 \Delta_F' - \ldots$ we will automatically generate the additional factor Z_1^{-1} upon performing the renormalization, since we now have the requisite number of factors in each term. Salam's 1951 analysis is essentially a rigorous justification of this expansion procedure*.

In a similar way, the integral $\int \gamma_5 S_F' \Gamma_5 \Delta_F'$ for Σ^* contains overlap divergences which are only partly eliminated by substituting $\Gamma_5 = Z_1^{-1}\mathbf{\Gamma}_5$ at one of the vertices. An additional factor Z_1^{-1} must be extracted in order to remove the remaining divergence at the γ_5-vertex. This factor is then cancelled by the Z_1 factor in 6(337), yielding a finite result for $\mathbf{S}_F'$.

The overlap divergences do not appear in the vertex correction $\Lambda_5(p_2, p_1)$. In this case, the replacements 6(322a), 6(322b) and 6(322c) remove all divergences arising from self-energy and vertex insertions in the irreducible vertex graphs, leaving an integral $\mathbf{\Lambda}_5$ which can be shown to be logarithmically divergent. This remaining divergence, known as the *skeleton divergence*, is removed by the subtraction procedure 6(326), leaving a remainder $\mathbf{\Lambda}_{5c}$ which is finite†.

Actually we must qualify our statement that mass and coupling constant renormalizations are sufficient to absorb all the divergences of pseudoscalar meson theory. There is an extra source of divergence, due to meson-meson scattering graphs of the type shown in Fig. 6.42, which cannot be

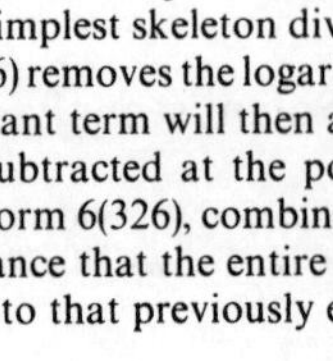

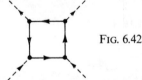

FIG. 6.42. Meson–meson scattering graph.

* See also J. D. Bjorken and S. Drell, op. cit.

† The simplest skeleton divergence is, of course, 6(306). To see that the subtraction procedure 6(326) removes the logarithmic divergence in this case, expand 6(306) about $p_1 = p_2 = 0$. The constant term will then absorb the entire logarithmic infinity. The finite remainder can now be subtracted at the point $(p_2 - p_1)^2 = -\mu^2$, $\gamma \cdot p_2 = im$, $\gamma \cdot p_1 = im$. To recover the (unique) form 6(326), combine the two constant subtraction terms to give $L\gamma_5$. This provides the assurance that the entire logarithmic divergence has been absorbed in L. The argument is similar to that previously employed for $\Sigma(p)$.

so absorbed. An additional counterterm of the form

$$\mathscr{H}_I(x) = \lambda\phi^4(x) \tag{6(373)}$$

is required to compensate this divergence. This leads to an additional term of the form $\lambda\phi^3$ in the Heisenberg equation of motion for ϕ*.

Pseudoscalar meson theory, spinor electrodynamics and also the electrodynamics of spin 0 bosons [Salam (1952)] are *renormalizable* field theories, in that mass and coupling constant renormalization, (and, eventually, the inclusion of a $\lambda\phi^4$ term in the case of pseudoscalar meson theory), are sufficient to absorb all the divergences†. The same applies to the interaction of neutral vector bosons with spin $\frac{1}{2}$ fermions, *provided* that the vector boson is coupled to a conserved current, as in 5(76), [Salam (1960), Kamefuchi (1960)]. However, the Yang–Mills coupling of an isotopic triplet of vector bosons is not renormalizable, except in the case of zero vector mass, [Komar (1960), Umezawa (1960)]. Also the *electromagnetic* interaction of charged vector bosons is nonrenormalizable [see, however, Lee (1962)]. Derivative coupling theories are nonrenormalizable, (except for spin 0 electrodynamics), the reason being, essentially, that each additional vertex in a higher order graph introduces an additional power of the integration variable k [see Table 6.9]. A derivative coupling theory can, in general, be ' renormalized ' only by introducing an *infinite* number of counterterms. Finally, we note that four-fermion coupling theories are also nonrenormalizable‡.

Application—The Anomalous Moment and the Lamb Shift The most important application of renormalization theory is to the calculation of the anomalous magnetic moment and the Lamb shift in quantum electrodynamics. Consider the second order corrections to the lowest order scattering amplitude 6(110) in the external field A_μ^{ext}. These corrections are shown in Fig. 6.43. We have seen that after mass renormalization the net effect of diagrams (*a*)–(*d*) is to contribute a factor $(Z_2^{1/2})^2 = Z_2$ to the

*The corresponding 4-photon graph in spinor electrodynamics can be shown to be convergent on grounds of gauge invariance, so that a counterterm is not required in that case [Feynman (1949), Karplus (1950b)]. A further simplification in the case of electrodynamics is that the overlap problem can be bypassed completely with the aid of Ward's identity [see for example J. M. Jauch and F. Rohrlich, *The Theory of Electrons and Photons*, Addison-Wesley, Reading, U.S.A., (1955), Chapter 10].

†Naturally, renormalizability does not guarantee that the renormalized perturbation series converges. In fact, it does not converge for pseudoscalar meson theory, as the coupling constant is too strong. Whether the renormalized series converges in electrodynamics remains an unanswered question. [See, in this connection, Dyson (1952).]

‡For a general discussion, see H. Umezawa, *Quantum Field Theory*, North-Holland, Amsterdam, 1956, Chapter XV.

renormalized charge. A further contribution Z_1^{-1} comes from the logarithmically divergent part of the vertex correction of Fig. 6.43e*.

$$\Lambda_\mu(p', p) = \frac{e_0^2}{(2\pi)^4} \int d^4k \gamma_\nu \frac{-1}{\gamma . (p'-k) - im} \gamma_\mu \frac{-1}{\gamma . (p-k) - im} \gamma_\nu \frac{-i}{k^2} \quad 6(374)$$

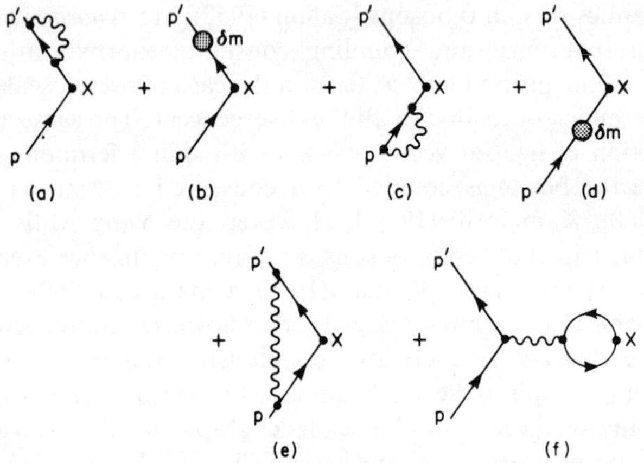

FIG. 6.43. Second-order corrections to electron scattering in an external electromagnetism field.

where $p'^2 = p^2 = -m^2$. We have seen that renormalization theory gives a unique prescription for separating out the convergent part $\Lambda_{\mu c}(p', p)$: write $\Lambda_\mu(p', p)$ in the form

$$\Lambda_\mu(p', p) = L\gamma_\mu + \Lambda_{\mu c}(p', p)$$

where $\Lambda_{\mu c}(p', p)$ vanishes for $p' = p$ with $\gamma . p' = \gamma . p = im$. The constant L can then be absorbed into the definition of the renormalized charge via the renormalization constant $Z_1 = 1 - L$. Moreover, L contains the entire logarithmic divergence. The finite remainder $\Lambda_{\mu c}(p', p)$ must then represent the observable effect of Fig. 6.43e on the scattering amplitude. When this calculation is carried out explicitly, the result is [Feynman (1949)]

$$\Lambda_{\mu c}(q) = \frac{e_R^2}{8\pi^2}\left[-\frac{1}{2m}\Sigma_{\mu\nu} q_\nu + \gamma_\mu \frac{2q^2}{3m^2}\left(\log\frac{m}{\lambda_{\min}} - \frac{3}{8}\right)\right] \quad 6(375)$$

$$(q = p' - p)$$

* In fact the Z_1^{-1} and Z_2 corrections cancel by virtue of Ward's identity. The only effective charge renormalization to this order comes from the vacuum polarization graph Fig. 6.43f.

for small q and for $p^2 = p'^2 = -m^{2*}$. $\Sigma_{\mu\nu}$ is the antisymmetric spin tensor 1(42). The constant $\lambda_{\min}$ represents a fictitious photon mass introduced in the photon propagator to avoid a divergence in the integral 6(374) for small k, known as the *infrared divergence.* Thus $-i/k^2$ in 6(374) has been replaced by $-i(k^2+\lambda_{\min}^2)^{-1}$. The origin of the infrared divergence is well understood and has nothing to do with the divergences for large k which are absorbed by renormalization. Its origin is the fact that in any scattering experiment the electrons can radiate photons whose energy and momentum is sufficiently small to be undetected by the apparatus. If our apparatus has an energy resolution ΔE, then photons of energy less than ΔE will remain undetected. When the production amplitude for these ' soft ' photons is combined with the infrared divergent scattering amplitude, the infrared divergence always disappears†.

The finite vertex part 6(375) alters the first order scattering amplitude 6(110) to

$$-e_R\frac{1}{V}\left(\frac{m^2}{E_{\mathbf{p}'}E_{\mathbf{p}}}\right)^{1/2}\bar{u}_{\mathbf{p}'\sigma'}[\gamma_\mu+\Lambda_{\mu c}(p',p)]u_{\mathbf{p}\sigma}A_\mu^{\text{ext}}(q) \tag{6(376)}$$

The additional term is called a ' radiative correction '. A further radiative correction arises from the vacuum polarization graph of Fig. 6.43f. In accordance with 6(366), this graph modifies $A_\mu^{\text{ext}}(q)$ in 6(376) to

$$A_\mu^{\text{ext}}(q)+D_{F\mu\lambda}(q)ie_0^2\Pi_{\lambda\nu}(q)A_\nu(q) \tag{6(377)}$$

where

$$\Pi_{\lambda\nu}(q) = (\delta_{\lambda\nu}q^2-q_\lambda q_\nu)C(q^2) \tag{6(378)}$$

is the lowest-order photon self-energy loop and where $D_{F\mu\nu}$ is the bare photon propagator 6(291). Again, renormalization theory prescribes a unique procedure for separating out the finite part of $\Pi_{\lambda\nu}(q)$. As only the $q^2C(q^2)$ term in 6(378) contributes effectively, the correction 6(377) reduces to

$$[1+e_0^2C(q^2)]A_\mu^{\text{ext}}(q) \tag{6(379)}$$

and the finite part of $C(q^2)$ is isolated by expanding $C(q^2)$ as in 6(298)

$$q^2C(q^2) = q^2C(0)+\Pi_c(q^2)$$

* For an evaluation of $\Lambda_{\mu c}$ when the external electron lines are off the mass shell, see Karplus (1950a). See also J. M. Jauch and F. Rohrlich, op. cit.

† This was first shown by Bloch and Nordsieck [Bloch (1937)]. For a detailed account, see Yennie (1961) and Yennie's review article in *Lectures on Strong and Electromagnetic Interactions*, Brandeis Summer Inst. 1963.

The logarithmically divergent $C(0)$ term is then absorbed into the definition of the renormalized charge, with e_0 in 6(379) being replaced by e_R. The explicit computation of $\Pi_c(q^2)$ yields the result*

$$\Pi_c(q^2) = -\frac{1}{4\pi^2}\frac{1}{15}\frac{q^4}{m^2} \qquad 6(380)$$

for small q^2. Accordingly, the effect of the vacuum polarization is to replace $A_\mu^{\text{ext}}(q)$ in 6(376) by

$$\left(1-\frac{e_R^2}{4\pi^2}\frac{1}{15}\frac{q^2}{m^2}\right)A_\mu^{\text{ext}}(q) \qquad 6(381)$$

Thus the effect is just to add a constant $-\frac{1}{5}$ to the $-\frac{3}{8}$ in 6(375).

Physically, the first term in 6(375) represents the effect of an anomalous electron magnetic moment on the scattering amplitude. To see this we first exhibit the effect of the ordinary Dirac moment

$$\mu = -\frac{e_R}{2m} \qquad 6(382)$$

in the scattering amplitude 6(376), by applying the Gordon decomposition [Gordon (1928)] of the Dirac current into an orbital and a spin part†

$$\bar{u}_{\mathbf{p}'\sigma'}\gamma_\mu u_{\mathbf{p}\sigma} = \bar{u}_{\mathbf{p}'\sigma'}\left[\frac{p'_\mu+p_\mu}{2im}-\frac{1}{2m}\Sigma_{\mu\nu}(p'_\nu-p_\nu)\right]u_{\mathbf{p}\sigma} \qquad 6(383)$$

The second term in 6(383) represents the contribution of the Dirac moment 6(382). When 6(383) is inserted back into 6(376), the result is a term proportional to

$$\left(-\frac{e_R}{2m}\right)\bar{u}_{\mathbf{p}'\sigma'}\tfrac{1}{2}\Sigma_{\mu\nu}u_{\mathbf{p}\sigma}(q_\mu A_\nu^{\text{ext}}(q)-q_\nu A_\mu^{\text{ext}}(q))$$

representing the transition amplitude due to the coupling of the magnetic moment tensor to the field strength tensor $F_{\mu\nu} = \partial_\mu A_\nu - \partial_\nu A_\mu$. If we now examine the first term in 6(375), we see that it has the same form as the

* See Feynman (1949) and also J. M. Jauch and F. Rohrlich, op. cit.

† To establish 6(383), use the Dirac equations $\gamma\,.\,pu_{\mathbf{p}\sigma} = imu_{\mathbf{p}\sigma}$ and $\bar{u}_{\mathbf{p}'\sigma'}\gamma\,.\,p' = im\bar{u}_{\mathbf{p}'\sigma'}$ to write

$$\bar{u}_{\mathbf{p}'\sigma'}\gamma_\mu u_{\mathbf{p}\sigma} = \frac{1}{2im}[\bar{u}_{\mathbf{p}'\sigma'}\gamma_\mu\gamma_\nu p_\nu u_{\mathbf{p}\sigma}+\bar{u}_{\mathbf{p}'\sigma'}\gamma_\nu p'_\nu\gamma_\mu u_{\mathbf{p}\sigma}]$$

and apply the identity $\gamma_\mu\gamma_\nu = \delta_{\mu\nu}+i\Sigma_{\mu\nu}$.

Dirac moment term in 6(383). We conclude, therefore, that the electron develops an additional 'anomalous' magnetic moment

$$\Delta\mu = \frac{e_R^2}{8\pi^2}\left(-\frac{e_R}{2m}\right) = \frac{\alpha}{2\pi}\mu \qquad 6(384)$$

by virtue of its interaction with the quantized electromagnetic field. This result, first derived by Schwinger (1949) is in very good agreement with experiment. The formula 6(384) gives

$$\Delta\mu = 0{\cdot}0011614\mu \qquad 6(385)$$

as against an experimentally measured value [Franken (1956)]

$$\Delta\mu = (0{\cdot}001165 \pm 0{\cdot}000011)\mu \qquad 6(386)$$

The inclusion of fourth-order radiative corrections to electron scattering modifies 6(385) to [Sommerfield (1957)]

$$\Delta\mu = 0{\cdot}0011596\mu \qquad 6(387)$$

We stress that 6(375) is only an approximation for small q. For arbitrary q, the 'static' moment 6(384) is replaced by a function of q^2 known as the anomalous magnetic moment form factor*.

In addition to predicting the correct value for the anomalous magnetic moment, quantum electrodynamics is also successful in accounting for the splitting of the $2s_{1/2}$ and $2p_{1/2}$ levels in hydrogen. These two levels are degenerate in the Dirac theory. The splitting was first measured by Lamb and Retherford [Lamb (1947)] and is known as the *Lamb shift*. One of the earliest relativistic calculations of the Lamb shift is based directly on the result 6(375).† One can regard 6(375) as a modification of the external field A_μ^{ext} seen by the electron. If the external field is taken to be the Coulomb field of a hydrogen atom, then the modification represented by $\Lambda_{\mu c}$ will give rise to a shift in the atomic energy levels and in particular to the $2s_{1/2}$–$2p_{1/2}$ splitting. One must take care, however, to combine the level shift due to 6(375) with an earlier nonrelativistic calculation by Bethe (1947), to take into account the contribution from emission and reabsorption of virtual photons with momenta below the $\lambda_{\min}$ cutoff in 6(375). When this is done, the dependence on $\lambda_{\min}$ cancels out. We do not give the details of these calculations, but simply quote

* See Section 8-1.

† See Feynman (1948), (1949). The calculation is complicated by the necessity for dealing with longitudinal and timelike photons [French (1949)] owing to the use of a cutoff at the finite 'photon mass' $\lambda_{\min}$ in 6(375). The calculation can also be performed by using a (non covariant) cutoff at $k_{\min}$, but the noncovariance of this procedure makes the renormalization more delicate. [See J. D. Bjorken and S. Drell, *Relativistic Quantum Mechanics*, McGraw-Hill, New York, 1964, Section 8.6.]

the resulting value of 1052 Mc/sec. for the level shift [French (1949), Kroll (1949)] as compared with a measured value of (1057·77±0·10) Mc/sec. [Triebwasser (1953)]. Later calculations, taking into account fourth-order effects, raised the theoretical value to bring it into much better agreement with experiment. There remains a small discrepancy, however. The latest theoretical calculation [Soto (1966)] yields the result 1057·64 Mc/sec., as against an experimental value which now stands at (1058·05±0·10) Mc/sec. [Robiscoe (1966)].

Green's Functions—Dyson's Equations in Configuration Space We conclude this section on renormalization theory by rewriting Dyson's equation for neutral pseudoscalar meson theory in configuration space. This will be of particular importance for the work of Chapter 7.

Let us define the Fourier transforms

$$\Delta'_F(x_1-x_2) = \frac{1}{(2\pi)^4}\int d^4k\, e^{ik.(x_1-x_2)}\Delta'_F(k^2) \qquad 6(388a)$$

$$S'_F(x_1-x_2) = \frac{1}{(2\pi)^4}\int d^4p\, e^{ip.(x_1-x_2)}S'_F(p) \qquad 6(388b)$$

and

$$\Gamma_5(x_1-y, y-x_2) = \frac{1}{(2\pi)^8}\int d^4p\, d^4q\, e^{ip.(x_1-y)}\, e^{-iq.(y-x_2)}\Gamma_5(p, q) \qquad 6(388c)$$

of the exact boson and fermion propagators and vertex function. Recalling that the lowest order functions

$$\Delta_F(x) = \frac{1}{(2\pi)^4}\int d^4k\, e^{ik.x}\Delta_F(k^2) \qquad 6(389a)$$

and

$$S_F(x) = \frac{1}{(2\pi)^4}\int d^4p\, e^{ip.x}S_F(p) \qquad 6(389b)$$

are respectively the Green's functions for the free Klein–Gordon and Dirac theories, we shall refer to $\Delta'_F(x)$ and $S'_F(x)$, as well as $\Gamma_5(x, y)$, as *exact* Green's functions for the coupled field system. Using 6(388a), 6(388b) and 6(389a), 6(389b), it is a simple matter to transform the perturbation expansions for $\Delta'_F(k^2)$ and $S'_F(p)$ into configuration space. We find

$$\Delta'_F(x_1-x_2) = \Delta_F(x_1-x_2) - G_0^2\int d^4y_1\int d^4y_2\Delta_F(x_1-y_1)$$
$$\times \mathrm{Tr}\,[\gamma_5 S_F(y_1-y_2)\gamma_5 S_F(y_2-y_1)]\Delta_F(y_2-x_2) + \ldots \qquad 6(390)$$

and

$$S'_F(x_1-x_2) = S_F(x_1-x_2)+G_0^2 \int d^4y_1\, d^4y_2 S_F(x_1-y_1)$$
$$\times \gamma_5 S_F(y_1-y_2)\gamma_5 \Delta_F(y_2-y_1) S_F(y_2-x_2)+\ldots \qquad 6(391)$$

These expansions are conveniently represented by means of configuration space Feynman diagrams. We introduce the convention that $\Delta_F(x_1-x_2)$ or $S_F(x_1-x_2)$ is to be represented by a line from x_2 to x_1, as in Fig. 6.44.

FIG. 6.44. Configuration space graphs for $\Delta_F(x_1-x_2)$ and $S_F(x_1-x_2)$.

Adopting the further convention that a vertex at, say, y_1 carries the instruction to integrate over y_1, we illustrate the expansions 6(390) and 6(391) in Fig. 6.45. Our rules for configuration space graphs are summarized in

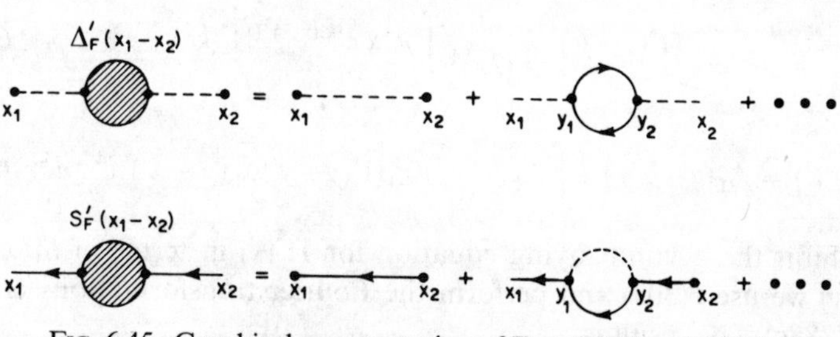

FIG. 6.45. Graphical representation of Eqs. 6(390) and 6(391).

Table 6.11. As in the case of ordinary momentum space graphs, we must supplement these rules by the instruction to include a minus sign for each closed fermion loop.

TABLE 6.11

RULES FOR FEYNMAN GRAPHS IN CONFIGURATION SPACE

Factor	Graphical Representation
$-G_0\gamma_5$ $\left(\text{with} \int d^4y\right)$	y
$S_F(x_1-x_2)$	x_1 ——— x_2
$\Delta_F(x_1-x_2)$	x_1 - - - - - - x_2

Let us write the Dyson equation 6(308) for $\Delta'_F(k^2)$ in the form

$$\Delta'_F(k^2) = \frac{1}{\Delta_F^{-1}(k^2) - iG_0^2\Pi^*(k^2)} \qquad 6(390)$$

where for the sake of convenience we have incorporated the $\delta\mu^2$ term into the mass, setting

$$\Delta_F(k^2) = \frac{-i}{k^2+\mu_0^2} \qquad 6(391)$$

This avoids the necessity for carrying along the explicit mass counterterm in the calculation*. A simple manipulation now allows us to rewrite 6(390) as

$$\Delta'_F(k^2) = \Delta_F(k^2) + \Delta_F(k^2) iG^2\Pi^*(k^2)\Delta'_F(k^2) \qquad 6(392)$$

This form of the equation is particularly easy to transform to configuration space. Using 6(388a) and 6(389a) and defining

$$\Pi^*(y-y') = \frac{1}{(2\pi)^4}\int d^4k\, e^{ik.(y-y')}\Pi^*(k^2) \qquad 6(393)$$

we get

$$\Delta'_F(x-x') = \Delta_F(x-x') + i\int \Delta_F(x-y)G_0^2\Pi^*(y-y')\Delta'_F(y'-x')\, d^4y\, d^4y' \quad 6(394)$$

To exhibit the accompanying equation for $\Pi^*(x)$ in terms of $S'_F(x)$ and $\Gamma_5(x, y)$ we use 6(309) and perform the Fourier transformations 6(388b) and 6(388c). The result is

$$\Pi^*(x-y) = i\int d^4x'\, d^4y'\, \mathrm{Tr}\,[\gamma_5 S'_F(x-x')\Gamma_5(x'-y, y-y')S'_F(y'-x)] \quad 6(395)$$

as represented graphically in Fig. 6.46.

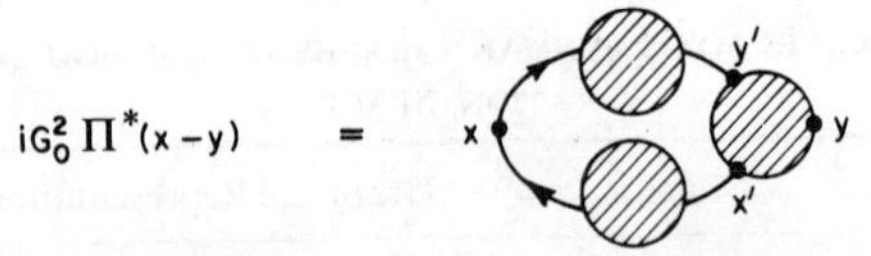

FIG. 6.46. Illustration of Eq. 6(395).

* In subsequent work we shall use the same symbol $\Delta_F(k)$ (or $\Delta_F(x)$ in configuration space), to denote the propagators $-i(k^2+\mu_0^2)^{-1}$ and $-i(k^2+\mu^2)$. This should not result in any confusion. To determine whether μ or μ_0 is being used, simply check whether the mass counterterm appears in the perturbation expansion. If it does, then $\Delta_F(k^2) = -i(k^2+\mu^2)^{-1}$. If not, then it has been incorporated into the mass term, as in 6(391). The same will apply to the bare fermion propagator $S_F(p)$ [see below].

For the fermion propagator S'_F we follow a parallel procedure, writing 6(310) in the form

$$S'_F(p) = S_F(p) + S_F(p)[-G_0^2\Sigma^*(p)]S'_F(p) \qquad 6(396)$$

where we have incorporated the δm counterterm into the mass defining

$$S_F(p) = \frac{-1}{\gamma \,.\, p - im_0} \qquad 6(397)$$

Using 6(388b) and 6(389b) and defining

$$\Sigma^*(y-y') = \frac{1}{(2\pi)^4}\int d^4p\, e^{ip(y-y')}\Sigma^*(p) \qquad 6(398)$$

we find

$$S'_F(x-x') = S_F(x-x') - \int S_F(x-y)G_0^2\Sigma^*(y-y')S'_F(y'-x')\, d^4y\, d^4y' \qquad 6(399)$$

and from 6(311) we get the accompanying equation for $\Sigma^*(x-y)$

$$\Sigma^*(x-y) = -\int d^4x'\, d^4y'\gamma_5 S'_F(x-x')\Gamma_5(x'-y', y'-y)\Delta_F(y'-x) \qquad 6(400)$$

represented graphically in Fig. 6.47. The details of these derivations are left to the reader [see Problem 10].

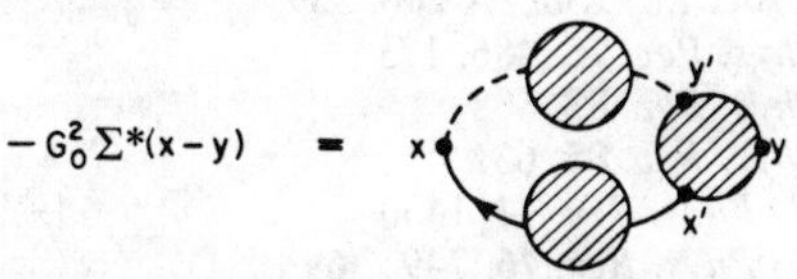

FIG. 6.47. Illustration of Eq. 6(400).

PROBLEMS

1. Check that the Wick decomposition of a fourfold time-ordered product is correctly given by 6(55).
2. Write down all normal products contained in 6(61a), 6(61b) and 6(61c). Which correspond to 4-momentum-conserving S-matrix elements?
3. Compute the second-order S-matrix elements for photon-positron scattering and for electron-positron annihilation into two photons.
4. Exhibit those terms in the S-matrix expansion which correspond to the fourth-order vacuum graphs of Fig. 6.7.
5. Compute the matrix element for $\pi - N$ scattering to second order in G, assuming the SU_2-invariant coupling 6(119). Extract the scattering amplitudes for

$\pi^+p \to \pi^+p$, $\pi^-p \to \pi^-p$ and $\pi^-p \to \pi^0 n$. Show that if R_{ij} is the scattering amplitude from an initial meson isospin state $i = 1, 2, 3$ to a final meson isospin state $j = 1, 2, 3$ then

$$R_{ij} = R^{\frac{1}{2}}\tfrac{1}{3}(\delta_{ij}+\tfrac{1}{2}[\tau_i, \tau_j]) + R^{\frac{3}{2}}\tfrac{1}{3}(2\delta_{ij}-\tfrac{1}{2}[\tau_i, \tau_j])$$

where $R^{\frac{1}{2}}$ and $R^{\frac{3}{2}}$ are the scattering amplitudes for *total* isospin $\frac{1}{2}$ and $\frac{3}{2}$ respectively. Exhibit $R^{\frac{1}{2}}$ and $R^{\frac{3}{2}}$ to second order in G.

6. Establish the identities 6(161a), 6(161b) and 6(161c).
7. Show that a closed loop with an odd number of emerging photon lines vanishes identically. [Hint: Use the properties of the charge conjugation matrix C.]
8. Consider an arbitrary irreducible Feynman graph with f external fermion lines, b external boson lines and n vertices. Perform the 'dressing' substitutions $\gamma_5 \to \Gamma_5$, $S_F \to S'_F$, $\Delta_F \to \Delta'_F$ for the internal lines and the subsequent renormalizations 6(322a), 6(322b) and 6(322c). Prove that all Z factors are absorbed into G_R after inclusion of $fZ_2^{1/2}$ factors and $bZ_3^{1/2}$ factors for the external fermion and boson lines respectively.
9. Establish the result 6(347) for the second order correction to $\bar{u}_{\mathbf{p}}$ under the effect of the adiabatic switching factor $g(t)$.
10. Derive Eqs. 6(395) and 6(400). Rewrite 6(313) as an equation in configuration space and represent the result graphically.

REFERENCES

Bethe, H. A. (1947) *Phys. Rev.* **72**, 339
Bloch, F. (1937) (with A. Nordsieck) *Phys. Rev.* **52**, 54
Dalitz, R. H. (1951) *Proc. Roy. Soc.* **A 206**, 509
Dyson, F. J. (1949) *Phys. Rev.* **75**, 486, 1736
Dyson, F. J. (1951) *Phys. Rev.* **83**, 608
Dyson, F. J. (1952) *Phys. Rev.* **85**, 631
Feynman, R. P. (1948) *Phys. Rev.* **74**, 1430
Feynman, R. P. (1949) *Phys. Rev.* **76**, 749; 769
Franken, P. (1956) (with S. Liebes) *Phys. Rev.* **104**, 1197
French, J. B. (1949) (with V. Weisskopf) *Phys. Rev.* **75**, 1240
Furry, W. H. (1937) *Phys. Rev.* **51**, 125
Gordon, W. (1928) *Z. Physik* **50**, 630
Johnson, K. (1961) *Nucl. Phys.* **25**, 431
Jost, R. (1949) (with J. Rayski) *Helv. Phys. Acta* **22**, 457
Kamefuchi, S. (1960) *Nucl. Phys.* **18**, 691
Karplus, R. (1950a) (with N. M. Kroll) *Phys. Rev.* **77**, 536
Karplus, R. (1950b) (with M. Neuman) *Phys. Rev.* **80**, 380
Klein, O. (1929) (with Y. Nishina) *Z. Physik* **52**, 853
Komar, A. (1960) (with A. Salam) *Nucl. Phys.* **21**, 624
Kroll, N. H. (1949) (with W. E. Lamb) *Phys. Rev.* **75**, 388
Lamb, W. E. (1947) (with R. C. Retherford) *Phys. Rev.* **72**, 241
Lee, T. D. (1962) (with C. N. Yang) *Phys. Rev.* **128**, 885
Matthews, P. T. (1949) *Phys. Rev.* **76**, 684 L.; Erratum **76**, 1489 L

Møller, C. (1945) *Kgl. Danske Vidensk. Selsk.* **23** No. 1, especially Section 2
Mott, N. F. (1929) *Proc. Roy. Soc.* **A 124**, 425
Nambu, Y. (1958) *Nuovo Cimento* **9**, 610
Pauli, W. (1949) (with F. Villars) *Rev. Mod. Phys.* **21**, 434
Robiscoe, R. T. (1966) (with B. L. Cosens) *Phys. Rev. Letters* **17**, 69
Rohrlich, F. (1950) *Phys. Rev.* **77**, 357; **80**, 666
Salam, A. (1951) *Phys. Rev.* **82**, 217; **84**, 426
Salam, A. (1952) *Phys. Rev.* **86**, 731
Salam, A. (1960) *Nucl. Phys.* **18**, 681
Schwinger, J. (1948) *Phys. Rev.* **74**, 1439
Schwinger, J. (1949) *Phys. Rev.* **76**, 790
Sommerfield, C. M. (1957) *Phys. Rev.* **107**, 328
Soto, M. F. Jr. (1966) *Phys. Rev. Letters* **17**, 1153
Takahashi, Y. (1957) *Nuovo Cimento* **6**, 370
Takeda, G. (1952) *Prog. Theoret. Phys.* **7**, 359
Thirring, W. (1950) *Phil. Mag.* **41**, 1193
Triebwasser, S. (1953) (with E. S. Dayhoff and W. E. Lamb) *Phys. Rev.* **89**, 98
Umezawa, H. (1961) (with S. Kamefuchi) *Nucl. Phys.* **23**, 399
Ward, J. C. (1950) *Phys. Rev.* **78**, 1824
Weinberg, S. (1960) *Phys. Rev.* **118**, 838
Wick, G. C. (1950) *Phys. Rev.* **80**, 268
Yennie, D. R. (1961) (with S. C. Frautschi and H. Suura) *Ann. Phys.* **13**, 379

7

Vacuum Expectation Values and the S-Matrix

7-1 Introduction

We have remarked earlier that as interacting relativistic quantum field theories are insoluble, one must either employ perturbation theory or else make certain assumptions on the nature of the solutions which one expects on physical grounds. For the latter approach, the Heisenberg picture provides the most convenient framework.

In this chapter we shall first re-express perturbation theory in the context of the Heisenberg picture. This will be done in Sections 7-1, 7-2 and 7-3. Then, in Sections 7-4, 7-5 and 7-6 we shall introduce the more modern nonperturbative approach.

7-2 In and out states and fields

In and Out States Let $|a)$ be an eigenstate of the *free field* Hamiltonian and momentum operators in the interaction picture

$$H_0^{ip}|a) = E|a) \qquad 7(1)$$

$$\mathbf{P}_0^{ip}|a) = \mathbf{k}|a) \qquad 7(2)$$

where we remark that $\mathbf{P}_0^{ip}$ is just $\mathbf{P}^{ip}$; in contrast to the Hamiltonian, the expression for the field momentum in terms of the field operators,

$$\mathbf{P} = -\sum_A \int d^3x \pi_A \nabla \phi_A \qquad 7(3)$$

is formally the same as for free fields:

$$\mathbf{P} = \mathbf{P}_0 \qquad 7(4)$$

We now prove the following important theorem: the structures

$$|a\,\text{in}\rangle = U(0, -\infty)|a) \qquad 7(5a)$$

$$|a\,\text{out}\rangle = U(0, +\infty)|a) \qquad 7(5b)$$

are both eigenstates of the total energy and momentum operators with eigenvalues E and **k**:

$$H|a^{\text{in}}_{\text{out}}\rangle = E|a^{\text{in}}_{\text{out}}\rangle \qquad 7(6)$$

$$\mathbf{P}|a^{\text{in}}_{\text{out}}\rangle = \mathbf{k}|a^{\text{in}}_{\text{out}}\rangle \qquad 7(7)$$

The distinction between the in and out states will be clarified later. $U(t_1, t_2)$ is the unitary operator introduced in Section 6-1 which moves the interaction picture state vector forward in time. $U(0, -\infty)$, for example, is given by

$$U(0, -\infty) = \sum_{n=0}^{\infty} \frac{(-i)^n}{n!} \int_{-\infty}^{0} dt_1 \ldots dt_n T[H_I^{ip}(t_1) \ldots H_I^{ip}(t_n)] \qquad 7(8)$$

where, in line with the remarks following 6(40), an adiabatic factor $e^{\varepsilon t}$ must be associated to each factor $H_I^{ip}(t)$.

The proof of 7(7) is immediate. Writing

$$\begin{aligned}\mathbf{P}^{ip}|a\,\text{in}\rangle &= \mathbf{P}^{ip}U(0, -\infty)|a) \\ &= [\mathbf{P}^{ip}, U(0, -\infty)]|a) + \mathbf{k}|a\,\text{in}\rangle \end{aligned} \qquad 7(9)$$

for $|a\,\text{in}\rangle$, for example, we evaluate the commutators $[\mathbf{P}^{ip}, U(0, -\infty)]$ by observing that the interaction picture field operators ϕ^{ip} which feature in H_I^{ip} satisfy

$$[\mathbf{P}^{ip}, \phi_A^{ip}] = i\nabla\phi_A^{ip}$$

Hence, we have

$$[\mathbf{P}^{ip}, H_I^{ip}] = \nabla H_I^{ip} = 0$$

and the commutator in 7(9) vanishes. Equation 7(9) is then equivalent to 7(7), since the interaction and Heisenberg pictures coincide at $t = 0$:

$$\mathbf{P}^{ip} = \mathbf{P}^{ip}(0) = \mathbf{P}$$

To prove 7(6), one could proceed in a similar fashion and write

$$\begin{aligned}H_0^{ip}|a\,\text{in}\rangle &= H_0^{ip}U(0, -\infty)|a) \\ &= [H_0^{ip}, U(0, -\infty)]|a) + E|a\,\text{in}\rangle\end{aligned}$$

One must then proceed to compute the commutator $[H_0^{ip}, U(0, -\infty)]$ which will now be non-zero [Gell-Mann (1951)]. Here we shall follow an alternative approach and evaluate the time integrals in 7(8) explicitly. In the process we shall make direct contact with the basic equations of the formal theory of scattering*.

To evaluate the time integrals occurring in $U(t_1, t_2)$ it is more convenient to work with the perturbation expansion in the form 6(29). We must then calculate the quantity

$$U(t, -\infty)|a) = |a) - i \int_{-\infty}^{t} dt_1\, e^{\varepsilon t_1} H_I^{ip}(t_1)|a)$$

$$+(-i)^2 \int_{-\infty}^{t} dt_1 \int_{-\infty}^{t_1} dt_2\, e^{\varepsilon(t_1+t_2)} H_I^{ip}(t_1) H_I^{ip}(t_2)|a) + \dots \qquad 7(10)$$

which for $t = 0$ is equal to $|a\text{ in}\rangle$. We begin by evaluating the second term on the right-hand side. Writing H_0^S and H_I^S for the (time-independent) free and interaction Hamiltonians in the Schrödinger picture, we have

$$\int_{-\infty}^{t} dt_1\, e^{\varepsilon t_1} H_I^{ip}(t_1)|a) = \int_{-\infty}^{t} dt_1\, e^{\varepsilon t_1}\, e^{iH_0^S t_1} H_I^S\, e^{-iH_0^S t_1}|a)$$

$$= \int_{-\infty}^{t} dt_1\, e^{-i(E - H_0^S + i\varepsilon)t_1} H_I^S|a)$$

$$= e^{-i(E - H_0^S + i\varepsilon)t} \frac{i}{E - H_0^S + i\varepsilon} H_I^S|a)$$

where, in deriving the second line, we have used the fact that $H_0^S|a) = E|a)$. Turning to the third term on the right-hand side of 7(10), we have

$$\int_{-\infty}^{t} dt_1 \int_{-\infty}^{t_1} dt_2\, e^{\varepsilon(t_1+t_2)} H_I^{ip}(t_1) H_I^{ip}(t_2)|a)$$

$$= \int_{-\infty}^{t} dt_1\, e^{\varepsilon t_1} H_I^{ip}(t_1)\, e^{-i(E - H_0^S + i\varepsilon)t_1} \frac{i}{E - H_0^S + i\varepsilon} H_I^S|a)$$

$$= \int_{-\infty}^{t} dt_1\, e^{\varepsilon t_1}\, e^{iH_0^S t_1} H_I^S\, e^{-i(E + i\varepsilon)t_1} \frac{i}{E - H_0^S + i\varepsilon} H_I^S|a)$$

$$= i^2\, e^{-i(E - H_0^S + 2i\varepsilon)t} \frac{1}{E - H_0^S + 2i\varepsilon} H_I^S \frac{1}{E - H_0^S + i\varepsilon} H_I^S|a)$$

* See for example, E. Merzbacher, *Quantum Mechanics*, Wiley, New York, 1961, Chapter 21.

Proceeding in the same way, we get for the $(n+1)$th term in the expansion 7(10):

$$\int_{-\infty}^{t} dt_1 \int_{-\infty}^{t_1} dt_2 \ldots \int_{-\infty}^{t_{n-1}} dt_n\, e^{\varepsilon(t_1+\ldots+t_n)} H_I^{ip}(t_1)\ldots H_I^{ip}(t_n)|a)$$

$$= i^n\, e^{-i(E-H_0^S+ni\varepsilon)t} \frac{1}{E-H_0^S+ni\varepsilon} H_I^S \frac{1}{E-H_0^S+(n-1)i\varepsilon} H_I^S \ldots \frac{1}{E-H_0^S+i\varepsilon} H_I^S|a)$$

Therefore, setting $t = 0$ and replacing all infinitesimals $\varepsilon, 2\varepsilon, \ldots, n\varepsilon, \ldots$ by ε in the limit $\varepsilon \to 0$,

$$|a \text{ in}\rangle = U(0, -\infty)|a)$$

$$= |a) + \frac{1}{E-H_0^S+i\varepsilon} H_I^S|a) + \frac{1}{E-H_0^S+i\varepsilon} H_I^S \frac{1}{E-H_0^S+i\varepsilon} H_I^S|a) + \ldots \quad 7(11)$$

As the next step, let us convert the expansion 7(11) to an implicit equation for $|a \text{ in}\rangle$ by writing it in the form

$$|a \text{ in}\rangle = |a) + \frac{1}{E-H_0^S+i\varepsilon} H_I^S\left(|a) + \frac{1}{E-H_0^S+i\varepsilon} H_I^S|a) + \ldots\right)$$

or

$$|a \text{ in}\rangle = |a) + \frac{1}{E-H_0^S+i\varepsilon} H_I^S|a \text{ in}\rangle \quad 7(12)$$

The equation 7(12), valid in the limit $\varepsilon \to 0$, is known as the Lippmann–Schwinger equation [Lippmann (1950)]. It leads directly to the desired result 7(6). We have

$$(E-H_0^S+i\varepsilon)|a \text{ in}\rangle = (E-H_0^S+i\varepsilon)|a) + H_I^S|a \text{ in}\rangle$$

$$= i\varepsilon|a) + H_I^S|a \text{ in}\rangle \quad 7(13)$$

since $H_0^S|a) = E|a)$. We now pass to the limit $\varepsilon \to 0$. Provided that $|a \text{ in}\rangle$ remains well defined in the limit $\varepsilon \to 0$, we get

$$(E-H_0^S-H_I^S)|a \text{ in}\rangle = 0$$

or, since $H^S = H$

$$H|a \text{ in}\rangle = E|a \text{ in}\rangle \quad 7(14)$$

As the final step we must ascertain whether $|a \text{ in}\rangle = U(0, -\infty)|a)$ remains finite in the limit $\varepsilon \to 0$. Now, in fact, $U(0, -\infty)$ does exhibit a singularity at $\varepsilon \to 0$ owing to the presence of the disconnected closed loops. For pseudoscalar $\pi^0 - N$ coupling, for instance, the simplest such term is

$$-\tfrac{1}{2}\int_{-\infty}^{0} dt_1 \int_{-\infty}^{0} dt_2 \int d^3x_1\, d^3x_2\, \mathrm{Tr}\,[\gamma_5 S_F(x_1-x_2)\gamma_5 S_F(x_2-x_1)]$$

$$\times \Delta_F(x_1-x_2)\, e^{\varepsilon t_1}\, e^{\varepsilon t_2}$$

or, upon expressing $S_F(x)$ and $\Delta_F(x)$ in terms of their Fourier transforms,

$$\int_{-\infty}^{0} dt_1\, dt_2 \int dp_0\, dq_0\, dk_0\, e^{-ip_0(t_1-t_2)}\, e^{-iq_0(t_2-t_1)}\, e^{-ik_0(t_1-t_2)}$$

$$\times e^{\varepsilon t_1}\, e^{\varepsilon t_2} f(p_0, q_0, k_0)$$

Integrating over t_1 we get

$$\int_{-\infty}^{0} dt_2 \int dE_1 \frac{i}{E_1+i\varepsilon}\, e^{iE_1 t_2}\, e^{\varepsilon t_2} F(E_1)$$

where we have changed integration variables to $E_1 = p_0 - q_0 + k_0$, q_0 and k_0 and defined $F(E_1) = \int dq_0\, dk_0 f(E_1, q_0, k_0)$. The integration over E_1 may be performed by closing the contour at infinity in the lower complex E_1-plane, since $t_2 < 0$. Evaluation of the residue at the pole $E_1 = -i\varepsilon$ yields the result

$$2\pi \int_{-\infty}^{0} dt_2\, e^{2\varepsilon t_2} F(-i\varepsilon) = \frac{\pi}{\varepsilon} F(-i\varepsilon)$$

which exhibits a singularity at $\varepsilon = 0$. To eliminate disconnected closed loop singularities at $\varepsilon = 0$, we must therefore divide 7(13) through by the c-number factor $(0|U(0, -\infty)|0)$ and redefine $|a\ \mathrm{in}\rangle$ as the limit of the quotient

$$\frac{U(0, -\infty)|a)}{(0|U(0, -\infty)|0)}$$

for $\varepsilon \to 0$. In practice, the factor $(0|U(0, -\infty)|0)$ always cancels out in applications and we suppress it for the sake of convenience; its necessity should not be overlooked, however.

We conclude the proof of 7(6) and 7(7) with the obvious remark that the out-states can be similarly treated. Corresponding to 7(12) we have the Lippmann–Schwinger equation for $|a \text{ out}\rangle$

$$|a \text{ out}\rangle = |a) + \frac{1}{E - H_0^S - i\varepsilon} H_I^S |a \text{ out}\rangle \qquad 7(15)$$

which differs from 7(12) only in the sign of the $i\varepsilon$ term in the denominator.

In non-relativistic potential scattering theory, the in-solution to the Lippmann–Schwinger equation represents an exact solution to the Schrödinger equation which reduces asymptotically to a superposition of a plane wave and an outgoing spherical wave*. The states $|a \text{ in}\rangle$ which we have encountered here are the generalization of this construction to relativistic quantum field theory. For example the state $|\mathbf{kq}; \text{in}\rangle = U(0, -\infty)|\mathbf{kq})$ is an eigenstate of the Hamiltonian representing two incoming free particles of momentum $\mathbf{k}$ and $\mathbf{q}$, together with outgoing waves consisting of all possible collision products including bosons, fermion pairs, etc.

S-Matrix S-matrix elements are easily expressed in terms of the in and out states as follows. We have

$$S_{ba} = \frac{(b|S|a)}{(0|S|0)} = \frac{(b|U(\infty, -\infty)|a)}{(0|U(\infty, -\infty)|0)} \qquad 7(16)$$

where the denominator factor is necessary to cancel the overall factor arising from the sum of all disconnected closed loop diagrams. Using 7(5a) and 7(5b), we have

$$(b|U(\infty, -\infty)|a) = (b|U(\infty, 0)U(0, -\infty)|a) = \langle b \text{ out}|a \text{ in}\rangle$$

$$(0|U(\infty, -\infty)|0) = (0|U(\infty, 0)U(0, -\infty)|0) = \langle 0 \text{ out}|0 \text{ in}\rangle$$

and therefore

$$S_{ba} = \frac{\langle b \text{ out}|a \text{ in}\rangle}{\langle 0 \text{ out}|0 \text{ in}\rangle} \qquad 7(17)$$

We have seen in Section 6-2 that in the absence of external fields the vacuum–vacuum matrix element

$$\langle 0 \text{ out}|0 \text{in}\rangle = (0|U(\infty, -\infty)|0) \qquad 7(18)$$

is just a phase factor. If we ignore this phase factor, then for both the vacuum and the one-particle states we obviously have

$$\begin{aligned} |0 \text{ out}\rangle &= |0 \text{ in}\rangle \\ |\mathbf{k} \text{ out}\rangle &= |\mathbf{k} \text{ in}\rangle \end{aligned} \qquad 7(19)$$

* See E. Merzbacher, *Quantum Mechanics*, Wiley, New York, 1961, Chapters 12 and 21.

since the 0 and 1 particle states are unaffected by the scattering operator:

$$U(\infty, -\infty)|0) = |0)$$

$$U(\infty, -\infty)|\mathbf{k}) = |\mathbf{k})$$

Strictly speaking, however, $|0\ \text{out}\rangle$ and $|0\ \text{in}\rangle$ differ by the phase factor 7(18), and so do $|\mathbf{k}\ \text{out}\rangle$ and $|\mathbf{k}\ \text{in}\rangle$. Also note that when external fields are present the vacuum–vacuum transition amplitude $\langle 0\ \text{out}|0\ \text{in}\rangle$ is no longer simply a phase factor.

Møller Wave Matrices The operators

$$\Omega^{(\pm)} = U(0, \mp\infty)$$

which transform eigenstates of H_0^{ip} into eigenstates of H are known as the Møller wave matrices [Møller (1945)]. We may write 7(5a) and 7(5b) as

$$|a\ \text{in}\rangle = \Omega^{(+)}|a) \qquad 7(20a)$$

$$|a\ \text{out}\rangle = \Omega^{(-)}|a) \qquad 7(20b)$$

The Møller wave matrices satisfy

$$\Omega^{(\pm)}H_0^{ip} = H\Omega^{(\pm)} \qquad 7(21)$$

and are related to the S-operator by

$$\begin{aligned} S &= U(\infty, -\infty) \\ &= U(\infty, 0)U(0, -\infty) \\ &= \Omega^{(-)-1}\Omega^{(+)} \end{aligned} \qquad 7(22)$$

Bound States It was shown in Section 6-1 that the time development operator $U(t_1, t_2)$ satisfies

$$U(t_1, t_2)^\dagger U(t_1, t_2) = 1$$

and the group property

$$U(t_1, t_2)U(t_2, t_3) = U(t_1, t_3)$$

The latter property implies the existence of the inverse

$$U(t_1, t_2)^{-1} = U(t_2, t_1)$$

which, in turn, guarantees the unitarity of $U(t_1, t_2)$:

$$U(t_1, t_2)^\dagger = U(t_1, t_2)^{-1}$$

It follows that the Møller wave matrices must satisfy

$$\Omega^{(\pm)\dagger}\Omega^{(\pm)} = 1 \qquad 7(23a)$$

and

$$\Omega^{(\pm)}\Omega^{(\pm)\dagger} = 1 \qquad 7(23b)$$

It can easily be seen that 7(23b) is incompatible with the existence of bound states. Bound states are discrete one-particle eigenstates of H to which there does not correspond any state in the spectrum of the free Hamiltonian H_0^{ip}. Denoting such states by $|\beta\rangle$ we must have the orthogonality and completeness relations

$$\langle a\,\text{in}|\beta\rangle = \langle a\,\text{out}|\beta\rangle = 0 \qquad 7(24)$$

$$\sum_a |a^{\text{in}}_{\text{out}}\rangle\langle a^{\text{in}}_{\text{out}}| + \sum_\beta |\beta\rangle\langle\beta| = 1 \qquad 7(25)$$

On the other hand, from 7(23b) and the completeness of the eigenstates $|a)$ of H_0^{ip} follows

$$\sum_a \Omega^{(\pm)}|a)(a|\Omega^{(\pm)\dagger} = 1$$

or

$$\sum_a |a^{\text{in}}_{\text{out}}\rangle\langle a^{\text{in}}_{\text{out}}| = 1 \qquad 7(26)$$

implying that the $|a\,\text{in}\rangle$ or $|a\,\text{out}\rangle$ form a complete set, in contradiction with 7(25).

Since 7(23b) is firmly grounded in perturbation theory, we must conclude that a thoroughgoing perturbative approach destroys the bound states. This is understandable, since the perturbative technique relies on the decomposition of the Hamiltonian into free and interacting parts and the assumption that the interaction Hamiltonian can be switched off when the particles are far apart in the asymptotic region*. When composite particles are present, this is clearly inadequate. When the interaction is turned off, all composite particles are decomposed into their elementary constituents, whereas we know that from a physical standpoint the forces responsible for the formation of the composite particle should be retained in the asymptotic region. An analogous problem was encountered with the self-interactions of elementary particles, where it was necessary to ensure that the interacting particles retain their physical masses from $t = -\infty$ to $t = +\infty$. This was handled by subtracting from the interaction Lagrangian the part responsible for the mass shifts by means of the

* See in this connection K. Nishijima (1958) (1965).

Dyson mass renormalization prescription, but an analogous approach for the much more complex problem of composite particles does not exist.

This failure of perturbation theory is one of the motivations for the more abstract reformulation of field theory—to be taken up later in this chapter—in which the nature of the spectrum of H is the subject of explicit *assumptions*. For the present, however, we suppose that bound states may be ignored and rely on the perturbative constructions 7(5a) and 7(5b).

In and Out Fields We now construct the operators

$$U^{(\pm)}(t) = e^{iHt}\Omega^{(\pm)}\, e^{-iHt} \tag{7(27)}$$

and define for each Heisenberg field ϕ^A and its canonically conjugate field π^A

$$\begin{aligned} \phi_{\text{in}}^A(\mathbf{x}, t) &= U^{(+)}(t)\phi^A(\mathbf{x}, t)U^{(+)}(t)^{-1} \\ \pi_{\text{in}}^A(\mathbf{x}, t) &= U^{(+)}(t)\pi^A(\mathbf{x}, t)U^{(+)}(t)^{-1} \end{aligned} \tag{7(28a)}$$

$$\begin{aligned} \phi_{\text{out}}^A(\mathbf{x}, t) &= U^{(-)}(t)\phi^A(\mathbf{x}, t)U^{(-)}(t)^{-1} \\ \pi_{\text{out}}^A(\mathbf{x}, t) &= U^{(-)}(t)\pi^A(\mathbf{x}, t)U^{(-)}(t)^{-1} \end{aligned} \tag{7(28b)}$$

We have seen that the Heisenberg fields ϕ^A and π^A satisfy the canonical equal-time commutation relations; it follows from 7(28a) and 7(28b) that $(\phi_{\text{in}}^A, \pi_{\text{in}}^A)$ and $(\phi_{\text{out}}^A, \pi_{\text{out}}^A)$ satisfy the same rules. Moreover, ϕ_{in}^A, π_{in}^A, ϕ_{out}^A, and π_{out}^A are Heisenberg fields. To see this, let us exhibit their time behaviour explicitly. For ϕ_{in}^A, for example, we have, using 7(27)

$$\begin{aligned} \phi_{\text{in}}^A(\mathbf{x}, t) &= e^{iHt}\Omega^{(+)}\, e^{-iHt}\phi^A(\mathbf{x}, t)\, e^{iHt}\Omega^{(+)-1}\, e^{-iHt} \\ &= e^{iHt}\Omega^{(+)}\phi^A(\mathbf{x}, 0)\Omega^{(+)-1}\, e^{-iHt} \end{aligned} \tag{7(29)}$$

where we have used the integrated form of the Heisenberg equation of motion for $\phi^A(x)$, i.e.

$$\phi^A(\mathbf{x}, t) = e^{iHt}\phi^A(\mathbf{x}, 0)\, e^{-iHt}$$

From 7(29) we deduce that

$$\phi_{\text{in}}^A(\mathbf{x}, t) = e^{iHt}\phi_{\text{in}}^A(\mathbf{x}, 0)\, e^{-iHt}$$

and similarly

$$\pi_{\text{in}}^A(\mathbf{x}, t) = e^{iHt}\pi_{\text{in}}^A(\mathbf{x}, 0)\, e^{-iHt}$$

Hence, $\phi_{\text{in}}^A(\mathbf{x}, t)$ and $\pi_{\text{in}}^A(\mathbf{x}, t)$ obey Heisenberg equations of motion:

$$\dot{\phi}_{\text{in}}^A(\mathbf{x}, t) = \frac{1}{i}[\phi_{\text{in}}^A(\mathbf{x}, t), H] \tag{7(30a)}$$

$$\dot{\pi}_{\text{in}}^A(\mathbf{x}, t) = \frac{1}{i}[\pi_{\text{in}}^A(\mathbf{x}, t), H] \tag{7(30b)}$$

Although ϕ_{in}^A, π_{in}^A, ϕ_{out}^A, and π_{out}^A are Heisenberg picture fields, they have the remarkable property that *they satisfy free field equations of motion.* If we differentiate 7(29) with respect to t and use 7(21), we get

$$\begin{aligned} -i\dot{\phi}_{\text{in}}^A(\mathbf{x}, t) &= \mathrm{e}^{iHt}(H\Omega^{(+)}\phi^A(\mathbf{x}, 0)\Omega^{(+)-1} - \Omega^{(+)}\phi^A(\mathbf{x}, 0)\Omega^{(+)-1}H)\,\mathrm{e}^{-iHt} \\ &= \mathrm{e}^{iHt}\Omega^{(+)}[H_0^{ip}, \phi^A(\mathbf{x}, 0)]\Omega^{(+)-1}\,\mathrm{e}^{-iHt} \end{aligned}$$

Recalling that $H_0^{ip}(t) = H_0^{ip}(0) = H_0(0)$, we now obtain by a simple manipulation the result

$$\begin{aligned} -i\dot{\phi}_{\text{in}}^A(\mathbf{x}, t) &= U^{(+)}(t)[H_0(t), \phi^A(\mathbf{x}, t)]U^{(+)}(t)^{-1} \\ &= [H_{0\,\text{in}}(t), \phi_{\text{in}}^A(\mathbf{x}, t)] \end{aligned} \tag{7(31a)}$$

where $H_{0\,\text{in}} \equiv H_0[\phi_{\text{in}}^A, \pi_{\text{in}}^A]$ is the free field Hamiltonian with all field operators replaced by in-fields. Similarly, we find

$$-i\dot{\pi}_{\text{in}}^A(\mathbf{x}, t) = [H_{0\,\text{in}}(t), \pi_{\text{in}}^A(\mathbf{x}, t)] \tag{7(31b)}$$

Equations 7(31a) and 7(31b) are just the free field equations of motion. ϕ_{out}^A and π_{out}^A can be shown to obey free field equations of motion in exactly the same way.

The key to understanding this result lies in recognizing that, by the definition 7(28a), 7(28b) and 7(21),

$$\begin{aligned} H_{0\,\text{in}}(t) &= \mathrm{e}^{iHt}\Omega^{(+)}\,\mathrm{e}^{-iHt}H_0(t)\,\mathrm{e}^{iHt}\Omega^{(+)-1}\,\mathrm{e}^{-iHt} \\ &= \mathrm{e}^{iHt}\Omega^{(+)}H_0(0)\Omega^{(+)-1}\,\mathrm{e}^{-iHt} \\ &= \mathrm{e}^{iHt}\Omega^{(+)}H_0^{ip}\Omega^{(+)-1}\,\mathrm{e}^{-iHt} \\ &= H \end{aligned} \tag{7(32)}$$

In other words, if the total Hamiltonian $H[\phi^A, \pi^A]$ is expressed in terms of the ϕ_{in}^A and π_{in}^A by inverting 7(28a), the result is just the free field Hamiltonian,

$$H[\phi^A, \pi^A] = H_0[\phi_{\text{in}}^A, \pi_{\text{in}}^A]$$

Thus 7(30a), 7(30b) and 7(31a), 7(31b) are immediately seen to be equivalent.

We may say that the in (or out) fields *diagonalize* the interacting field Hamiltonian.

Since the in and out fields satisfy canonical equal-time commutation relations and free field equations of motion, their Fourier decompositions will be formally identical to the corresponding free field decompositions. Thus, for a neutral Klein–Gordon field for example we have, for all times

$$\phi^{\text{in}}(x) = \frac{1}{\sqrt{V}} \sum_{\mathbf{k}, k_0 = \omega_\mathbf{k}} \frac{1}{\sqrt{2\omega_\mathbf{k}}} (a_\mathbf{k}^{\text{in}} \,\mathrm{e}^{ik.x} + a_\mathbf{k}^{\text{in}\dagger} \,\mathrm{e}^{-ik.x}) \tag{7(33)}$$

where the creation and destruction operators satisfy the usual commutation relations, and similarly for $\phi_{\text{out}}(x)$. For future reference we recall the inverse relations

$$a_\mathbf{k}^{\text{in}} = \frac{i}{\sqrt{V}} \frac{1}{\sqrt{2\omega_\mathbf{k}}} \int d^3x \,\mathrm{e}^{-i\mathbf{k}.\mathbf{x} + i\omega_\mathbf{k} t} \overleftrightarrow{\frac{\partial}{\partial t}} \phi^{\text{in}}(\mathbf{x}, t) \tag{7(34a)}$$

and

$$a_\mathbf{k}^{\text{in}\dagger} = \frac{i}{\sqrt{V}} \frac{1}{\sqrt{2\omega_\mathbf{k}}} \int d^3x \phi^{\text{in}}(\mathbf{x}, t) \overleftrightarrow{\frac{\partial}{\partial t}} \,\mathrm{e}^{i\mathbf{k}.\mathbf{x} - i\omega_\mathbf{k} t} \tag{7(34b)}$$

formally identical to 3(10a) and 3(10b). For a spin $\frac{1}{2}$ in-field, on the other hand, we have the expansions

$$\psi^{\text{in}}(x) = \frac{1}{\sqrt{V}} \sum_{\mathbf{k}, k_0 = E_\mathbf{k}} \sum_{\sigma=1}^{2} \sqrt{\frac{m}{E_\mathbf{k}}} (u_{\mathbf{k}\sigma} c_{\mathbf{k}\sigma}^{\text{in}} \,\mathrm{e}^{ik.x} + v_{\mathbf{k}\sigma} \, d_{\mathbf{k}\sigma}^{\text{in}\dagger} \,\mathrm{e}^{-ik.x}) \tag{7(35a)}$$

$$\bar{\psi}^{\text{in}}(x) = \frac{1}{\sqrt{V}} \sum_{\mathbf{k}, k_0 = E_\mathbf{k}} \sum_{\sigma=1}^{2} \sqrt{\frac{m}{E_\mathbf{k}}} (\bar{u}_{\mathbf{k}\sigma} c_{\mathbf{k}\sigma}^{\text{in}\dagger} \,\mathrm{e}^{-ik.x} + \bar{v}_{\mathbf{k}\sigma} \, \mathrm{d}_{\mathbf{k}\sigma}^{\text{in}} \,\mathrm{e}^{ik.x}) \tag{7(35b)}$$

formally identical to 3(154a) and 3(154b).

We now prove an important property of the in-fields which serves to relate them to the in-states 7(5a): *the creation operators $a_\mathbf{k}^{\text{in}\dagger}$, $c_{\mathbf{k}\sigma}^{\text{in}\dagger}$, $d_{\mathbf{k}\sigma}^{\text{in}\dagger}$ etc. when successively applied to the physical vacuum state $|0\,\text{in}\rangle$, generate the set of all in-states.* For example, the 1-particle state $|\mathbf{k}\text{ in}\rangle$ consisting of a single physical neutral scalar boson, is given by

$$|\mathbf{k}\text{ in}\rangle = a_\mathbf{k}^{\text{in}\dagger}|0\rangle$$

while

$$\frac{1}{\sqrt{2!}} a_\mathbf{k}^{\text{in}\dagger} a_{\mathbf{k}'}^{\text{in}\dagger}|0\rangle$$

is the scattering state $|\mathbf{kk}'\text{ in}\rangle$. To show this, we must show that the in-fields ϕ^A_{in} have the same properties with respect to the $|a\text{ in}\rangle$ as do the interaction picture field operators ϕ^{ip}_A with respect to the unperturbed states $|a)$. A general matrix element

$$\langle a_{\text{in}}|\phi^A_{\text{in}}(x_1)\phi^B_{\text{in}}(x_2)\ldots\phi^N_{\text{in}}(x_n)|b_{\text{in}}\rangle \tag{7(36)}$$

of a product of n in-field operators between arbitrary in-states $|a\text{ in}\rangle$ and $|b\text{ in}\rangle$ must be equal to

$$(a|\phi^{ip}_A(x_1)\phi^{ip}_B(x_2)\ldots\phi^{ip}_N(x_n)|b) \tag{7(37)}$$

Now by the definition 7(28a) and the relation $\phi_A(\mathbf{x}, 0) = \phi^{ip}_A(\mathbf{x}, 0)$, we see that the matrix element 7(36) is equal to

$$\begin{aligned}
&\langle a\text{ in}|U^{(+)}(t_1)\phi_A(x_1)U^{(+)}(t_1)^{-1}\ldots U^{(+)}(t_n)\phi_N(x_n)U^{(+)}(t_n)^{-1}|b\text{ in}\rangle\\
&\quad= \langle a\text{ in}|e^{iHt_1}\Omega^{(+)}\phi^{ip}_A(\mathbf{x}_1, 0)\Omega^{(+)-1}e^{-iHt_1}\ldots e^{iHt_n}\Omega^{(+)}\phi^{ip}_N(\mathbf{x}_n, 0)\Omega^{(+)-1}\\
&\qquad\qquad\times e^{-iHt_n}|b\text{ in}\rangle\\
&\quad= \langle a\text{ in}|\Omega^{(+)}e^{iH^{ip}_0t_1}\phi^{ip}_A(\mathbf{x}_1, 0)e^{-iH^{ip}_0t_1}\ldots e^{iH^{ip}_0t_n}\phi^{ip}_N(\mathbf{x}_n, 0)\\
&\qquad\qquad\times e^{-iH^{ip}_0t_n}\Omega^{(+)-1}|b\text{ in}\rangle\\
&\quad= \langle a\text{ in}|\Omega^{(+)}\phi^{ip}_A(x_1)\ldots\phi^{ip}_N(x_n)\Omega^{(+)-1}|b\text{ in}\rangle
\end{aligned}$$

To complete the proof of the equality of 7(36) and 7(37), we note that by 7(20a) we have $|b\text{ in}\rangle = \Omega^{(+)}|b)$ and $\langle a\text{ in}|\Omega^{(+)} = (a|$. This yields the desired result

$$\langle a\text{ in}|\phi^A_{\text{in}}(x_1)\ldots\phi^N_{\text{in}}(x_n)|b\text{ in}\rangle = (a|\phi^{ip}_A(x_1)\ldots\phi^{ip}_N(x_n)|b) \tag{7(38)}$$

An analogous statement holds for the matrix elements of products of out-fields taken between out-states.

Next we derive a relation between in and out fields. From 7(28a), 7(28b) and 7(27) we have, at time $t = 0$,

$$\begin{aligned}
\phi_{\text{out}}(\mathbf{x}, 0) &= \Omega^{(-)}\phi(\mathbf{x}, 0)\Omega^{(-)-1}\\
&= \Omega^{(-)}\Omega^{(+)-1}\phi_{\text{in}}(\mathbf{x}, 0)\Omega^{(+)}\Omega^{(-)-1}
\end{aligned}$$

We now translate this relation forward in time, multiplying on the left by e^{iHt} and on the right by e^{-iHt}. Since

$$H\Omega^{(-)}\Omega^{(+)-1} = \Omega^{(-)}H^{ip}_0\Omega^{(+)-1} = \Omega^{(-)}\Omega^{(+)-1}H$$

we get the relation

$$\begin{aligned}
\phi_{\text{out}}(x) = e^{iHt}\phi_{\text{out}}(\mathbf{x}, 0)e^{-iHt} &= \Omega^{(-)}\Omega^{(+)-1}e^{iHt}\phi_{\text{in}}(\mathbf{x}, 0)e^{-iHt}\Omega^{(+)}\Omega^{(-)-1}\\
&= \Omega^{(-)}\Omega^{(+)-1}\phi_{\text{in}}(x)\Omega^{(+)}\Omega^{(-)-1}
\end{aligned} \tag{7(39)}$$

valid for arbitrary times. To write this relation in a more appropriate notation, let us introduce the scattering operator S_{in} satisfying

$$\langle b\ \text{in}|S_{\text{in}}|a\ \text{in}\rangle = (b|S|a) = S_{ba} \tag{7(40)}$$

S_{in} is the scattering operator in the basis spanned by the in-states. Since $|a\ \text{in}\rangle = \Omega^{(+)}|a)$ and $\langle b\ \text{in}|\Omega^{(+)} = (b|$, the above definition is equivalent to

$$(b|\Omega^{(+)-1}S_{\text{in}}\Omega^{(+)}|a) = (b|S|a)$$

or

$$S_{\text{in}} = \Omega^{(+)}S\Omega^{(+)-1} \tag{7(41)}$$

Using 7(22), i.e.

$$S = \Omega^{(-)-1}\Omega^{(+)}$$

we can express S_{in} solely in terms of Møller wave matrices:

$$\begin{aligned} S_{\text{in}} &= \Omega^{(+)}\Omega^{(-)-1}\Omega^{(+)}\Omega^{(+)-1} \\ &= \Omega^{(+)}\Omega^{(-)-1} \end{aligned} \tag{7(42)}$$

Comparing with 7(39) we therefore find

$$\phi_{\text{out}} = S_{\text{in}}^{-1}\phi_{\text{in}}S_{\text{in}} \tag{7(43)}$$

Thus, the scattering operator S_{in} transforms in-fields into out-fields. Its effect on state vectors is exhibited by

$$\begin{aligned} S_{\text{in}}|a\ \text{out}\rangle &= \Omega^{(+)}\Omega^{(-)-1}\Omega^{(-)}|a) = \Omega^{(+)}|a) \\ &= |a\ \text{in}\rangle \end{aligned} \tag{7(44)}$$

Yang–Feldman Equations It is possible to relate ϕ_{in} directly to ϕ without explicit reference to the Møller wave matrix. This will now be done.

We shall illustrate the method for the simplest case of a neutral scalar field ϕ satisfying the equation of motion

$$(\Box - \mu^2)\phi = j \tag{7(45)}$$

Here μ is the *physical* boson mass and j is given as the sum of a term independent of ϕ and of a term $-\delta\mu^2\phi$ coming from the Dyson mass-renormalization counterterm. For example, for pseudoscalar coupling to spin $\frac{1}{2}$ fermions

$$\mathscr{L}_I = iG_0\bar{\psi}\gamma_5\psi\phi + \tfrac{1}{2}\delta\mu^2\phi^2 + \delta m\bar{\psi}\psi \tag{7(46)}$$

we have

$$j = -iG_0\bar{\psi}\gamma_5\psi - \delta\mu^2\phi \tag{7(47)}$$

Note that for a wide class of Lagrangians of physical interest we will have

$$j = -\frac{\partial \mathscr{L}_I}{\partial \phi} \qquad 7(48)$$

Writing now

$$\begin{aligned}\phi_{\text{in}}(\mathbf{x}, 0) &= \Omega^{(+)}\phi(\mathbf{x}, 0)\Omega^{(+)-1} \\ &= \phi(\mathbf{x}, 0)+[\Omega^{(+)}, \phi(\mathbf{x}, 0)]\Omega^{(+)-1} \\ &= \phi(\mathbf{x}, 0)+[U(0, -\infty), \phi^{ip}(\mathbf{x}, 0)]U(0, -\infty)^{-1} \qquad 7(49)\end{aligned}$$

where we have used the fact that at $t = 0$, $\phi(\mathbf{x}, 0) = \phi^{ip}(\mathbf{x}, 0)$, we compute the commutator $[U(0, -\infty), \phi^{ip}(\mathbf{x}, 0)]$. Using 7(8) we find

$$[U(0, -\infty), \phi^{ip}(\mathbf{x}, 0)] = \sum_{n=1}^{\infty} \frac{(-i)^n n}{n!} \int_{-\infty}^{0} d^4x_1 \ldots \int_{-\infty}^{0} d^4x_n$$
$$\times T([\mathscr{H}_I^{ip}(x_1), \phi^{ip}(\mathbf{x}, 0)]\mathscr{H}_I^{ip}(x_2) \ldots \mathscr{H}_I^{ip}(x_n)) \qquad 7(50)$$

where $\mathscr{H}_I^{ip} = -\mathscr{L}_I^{ip}$ is the interaction Hamiltonian density in the interaction picture. The commutator in 7(50) is easily evaluated. We have

$$\begin{aligned}[\mathscr{H}_I^{ip}(x_1), \phi^{ip}(\mathbf{x}, 0)] &= \frac{\partial \mathscr{H}_I^{ip}(x_1)}{\partial \phi^{ip}(x_1)}[\phi^{ip}(x_1), \phi^{ip}(\mathbf{x}, 0)] \\ &= j^{ip}(x_1)[\phi^{ip}(x_1), \phi^{ip}(\mathbf{x}, 0)] \\ &= -ij^{ip}(x_1)\Delta(\mathbf{x}-\mathbf{x}_1, -t_1)\end{aligned}$$

where we have used the free field commutation relations 3(89). Hence

$$[U(0, -\infty), \phi^{ip}(\mathbf{x}, 0)] = \sum_{n=1} \frac{(-i)^{n-1}}{(n-1)!} \int_{-\infty}^{0} d^4x_1 \ldots \int_{-\infty}^{0} d^4x_n$$
$$\times \Delta(\mathbf{x}-\mathbf{x}_1, -t_1)T(j^{ip}(x_1)\mathscr{H}_I^{ip}(x_2) \ldots \mathscr{H}_I^{ip}(x_n))$$

If we now relabel $x_1 \to y, x_2 \to x_1, \ldots x_n \to x_{n-1}$ we get

$$\begin{aligned}[U(0, -\infty), \phi^{ip}(\mathbf{x}, 0)] &= -\int_{-\infty}^{0} d^4y\Delta(\mathbf{x}-\mathbf{y}, -y_0) \sum_{n=0} \frac{(-i)^n}{n!} \int_{-\infty}^{0} d^4x_1 \ldots \\ &\quad \times \int_{-\infty}^{0} d^4x_n T(j^{ip}(y)\mathscr{H}_I^{ip}(x_1) \ldots \mathscr{H}_I^{ip}(x_n)) \\ &= -\int_{-\infty}^{0} d^4y\Delta(\mathbf{x}-\mathbf{y}, -y_0)T(j^{ip}(y)U(0, -\infty)) \qquad 7(51)\end{aligned}$$

We now use the fact that $y_0 < 0$ to manipulate the factor $T(j^{ip}(y)U(0, -\infty))$ as follows:

$$\begin{aligned} T(j^{ip}(y)U(0, -\infty)) &= U(0, y_0)j^{ip}(y)U(y_0, -\infty) \\ &= U(y_0, 0)^{-1}j^{ip}(y)U(y_0, 0)U(0, y_0)U(y_0, -\infty) \\ &= U(y_0, 0)^{-1}j^{ip}(y)U(y_0, 0)U(0, -\infty) \\ &= j(y)U(0, -\infty) \end{aligned} \qquad 7(52)$$

where, in deriving the last line, we have applied the equation 6(27) which relates operators in the interaction and Heisenberg pictures. Combining 7(49), 7(51) and 7(52), we get

$$\phi_{\text{in}}(\mathbf{x}, 0) = \phi(\mathbf{x}, 0) - \int_{-\infty}^{0} d^4y\Delta(\mathbf{x}-\mathbf{y}, -y_0)j(y) \qquad 7(53)$$

expressing ϕ_{in} at $t = 0$ solely in terms of the Heisenberg operators ϕ and j. We now subject 7(53) to a translation in time. Since ϕ_{in}, ϕ and j are all Heisenberg picture operators, we have at time $x_0 = t$

$$\begin{aligned} \phi_{\text{in}}(\mathbf{x}, x_0) &= e^{iHt}\phi_{\text{in}}(\mathbf{x}, 0)\, e^{-iHt} \\ &= e^{iHt}\phi(\mathbf{x}, 0)\, e^{-iHt} - \int_{-\infty}^{0} d^4y\Delta(\mathbf{x}-\mathbf{y}, -y_0)\, e^{iHt}j(y)\, e^{-iHt} \\ &= \phi(\mathbf{x}, x_0) - \int_{-\infty}^{0} d^4y\Delta(\mathbf{x}-\mathbf{y}, -y_0)j(\mathbf{y}, y_0+x_0) \end{aligned} \qquad 7(54)$$

Finally, making the change of integration variable $y_0' = y_0 + x_0$ we get the Yang–Feldman equation [Yang (1950)]

$$\phi(\mathbf{x}, x_0) = \phi_{\text{in}}(\mathbf{x}, x_0) + \int_{-\infty}^{x_0} d^4y\Delta(x-y, x_0-y_0)j(y, y_0)$$

or

$$\phi(x) = \phi_{\text{in}}(x) - \int_{-\infty}^{\infty} d^4y\Delta_R(x-y)j(y) \qquad 7(55a)$$

where we have used the retarded function

$$\Delta_R(x) = -\theta(x_0)\Delta(x) \qquad 7(56a)$$

defined by 3(101a).

In a similar fashion one can derive the Yang–Feldman equation

$$\phi(x) = \phi_{out}(x) - \int_{-\infty}^{\infty} d^4y \Delta_A(x-y) j(y) \qquad 7(55b)$$

relating ϕ to ϕ_{out}; Δ_A is the advanced function

$$\Delta_A(x) = \theta(-x_0)\Delta(x) \qquad 7(56b)$$

Combining 7(55a) and 7(55b) we get a direct relation between ϕ_{in} and ϕ_{out}

$$\phi_{out} = \phi_{in} + \int_{-\infty}^{\infty} d^4y [\Delta_A(x-y) - \Delta_R(x-y)] j(y)$$

$$= \phi_{in} + \int_{-\infty}^{\infty} d^4y \Delta(x-y) j(y) \qquad 7(57)$$

where we have used the relation

$$\Delta_A(x) - \Delta_R(x) = [\theta(-x_0) + \theta(x_0)]\Delta(x) = \Delta(x)$$

We can easily check that the Yang–Feldman equations 7(55a) and 7(55b) are consistent with the equation of motion 7(45). From 7(55a) for instance we get, using the inhomogeneous Klein–Gordon equation for Δ_R,

$$(\Box - \mu^2)\phi(x) = -\int_{-\infty}^{\infty} d^4y (\Box_x - \mu^2)\Delta_R(x-y) j(y)$$

$$= \int_{-\infty}^{\infty} d^4y \delta^{(4)}(x-y) j(y)$$

$$= j(x)$$

in agreement with 7(45).

7-3 Green's functions and reduction formulae

Green's Functions We have seen in Chapter 6 that for a renormalizable theory—pseudoscalar meson theory for example—the three Green's functions $S'_F(x)$, $\Delta'_F(x)$ and $\Gamma_5(x, y)$ (or $S'_F(x)$, $\Delta'_{F\mu\nu}(x)$ and $\Gamma_\mu(x, y)$ in the case of quantum electrodynamics), play a crucial role in the perturbation expansion. We now ask how these Green's functions may be expressed in terms of the Heisenberg picture.

Let us consider for example the exact Feynman propagator

$$\Delta'_F(x_1 - x_2) = \frac{1}{(2\pi)^4} \int d^4k \, e^{ik.(x_1 - x_2)} \Delta'_F(k) \qquad 7(58)$$

in pseudoscalar meson theory. We shall show that Δ'_F is just the vacuum expectation value of the time-ordered product $T\phi(x_1)\phi(x_2)$:

$$\Delta'_F(x_1 - x_2) = \frac{\langle 0 \text{ out}| T\phi(x_1)\phi(x_2)|0 \text{ in}\rangle}{\langle 0 \text{ out}|0 \text{ in}\rangle} \qquad 7(59)$$

where $|0 \text{ in}\rangle$ and $|0 \text{ out}\rangle$ are the *physical* vacuum states

$$|0 \text{ in}\rangle = U(0, -\infty)|0) \qquad 7(60)$$

and

$$|0 \text{ out}\rangle = U(0, +\infty)|0) \qquad 7(61)$$

which, in the absence of external fields, differ only by a phase factor. For the sake of comparison, we recall that the *free* boson propagator

$$\Delta_F(x_1 - x_2) = \frac{1}{(2\pi)^4} \int d^4k \, e^{ik.(x_1 - x_2)} \Delta_F(k) \qquad 7(62)$$

with

$$\Delta_F(k) = \frac{-i}{k^2 + \mu^2 - i\varepsilon}$$

is given by

$$\Delta_F(x_1 - x_2) = (0|T\phi^{ip}(x_1)\phi^{ip}(x_2)|0) \qquad 7(63)$$

where $|0)$ is the free field vacuum.

To proceed with the proof, we recall the relation (6.27)

$$\phi(\mathbf{x}, t) = U(t, 0)^{-1}\phi^{ip}(\mathbf{x}, t)U(t, 0) \qquad 7(64)$$

between Heisenberg and interaction picture fields. Then by 7(60), 7(61) and 7(64) we have, for $t_1 > t_2$

$$\begin{aligned}\langle 0 \text{ out}| T\phi(x_1)\phi(x_2)|0 \text{ in}\rangle &= (0|U(0, \infty)^\dagger U^{-1}(t_1, 0)\phi^{ip}(x_1)U(t_1, 0) \\ &\quad \times U^{-1}(t_2, 0)\phi^{ip}(x_2)U(t_2, 0)U(0, -\infty)|0) \qquad (t_1 > t_2)\end{aligned}$$

For the time being we ignore the factor $\langle 0 \text{ out}|0 \text{ in}\rangle$ in the denominator of 7(59). We now recall that $U(t_1, t_2)$ satisfies

$$U(t_1, t_2)U(t_2, t_3) = U(t_1, t_3)$$
$$U(t_1, t_2) = U(t_2, t_1)^{-1}$$
$$U(t_1, t_2)^\dagger = U(t_1, t_2)^{-1} = U(t_2, t_1)$$

as shown in Section 6-1. Hence, for $t_1 > t_2$ we can write

$$\begin{aligned}\langle 0\,\text{out}|T\phi(x_1\phi(x_2)|0\,\text{in}\rangle &= (0|U(\infty, 0)U(0, t_1)\phi^{ip}(x_1)U(t_1, 0)\\ &\quad\times U(0, t_2)\phi^{ip}(x_2)U(t_2, 0)U(0, -\infty)|0)\\ &= (0|U(\infty, t_1)\phi^{ip}(x_1)U(t_1, t_2)\phi^{ip}(x_2)U(t_2, -\infty)|0)\\ &= (0|TU(\infty, t_1)\phi^{ip}(x_1)U(t_1, t_2)\phi^{ip}(x_2)\\ &\quad\times U(t_2, -\infty)|0) \qquad (t_1 > t_2) \qquad 7(65)\end{aligned}$$

where in the last line we have inserted the time-ordering symbol since all factors are already in chronological order. We can also write 7(65) in the compact form

$$\langle 0\,\text{out}|T\phi(x_1)\phi(x_2)|0\,\text{in}\rangle = (0|T\phi^{ip}(x_1)\phi^{ip}(x_2)S|0) \qquad 7(66)$$

with $S = U(\infty, -\infty)$. This form is easily seen to be valid also for $t_2 > t_1$. Hence for all time orderings

$$\langle 0\,\text{out}|T\phi(x_1)\phi(x_2)|0\,\text{in}\rangle$$
$$= \sum_{n=0}^{\infty}\frac{(-i)^n}{n!}\int_{-\infty}^{\infty} d^4y_1 \ldots \int_{-\infty}^{\infty} d^4y_n (0|T(\phi^{ip}(x_1)\phi^{ip}(x_2)\mathscr{H}_I^{ip}(y_1)\ldots\mathscr{H}_I^{ip}(y_n))|0) \qquad 7(67)$$

where $\mathscr{H}_I^{ip}(y)$ is the interaction Hamiltonian density

$$\mathscr{H}_I^{ip}(y) = -iG_0[\bar{\psi}^{ip}(y), \gamma_5\psi^{ip}(y)]\phi^{ip}(y) - \tfrac{1}{2}\delta\mu^2(\phi^{ip}(y))^2 - \delta m[\bar{\psi}^{ip}(y), \psi^{ip}(y)]$$

Let us apply Wick's theorem to the T-product on the left-hand side of 7(67). The only term which survives when the vacuum expectation value is taken is the one in which all pairs of fermion and boson fields are contracted. The expansion 7(67) thus takes the form

$$\langle 0\,\text{out}|T\phi(x_1)\phi(x_2)|0\,\text{in}\rangle = \Delta_F(x_1 - x_2)$$
$$-G_0^2\int d^4y_1\int d^4y_2\Delta_F(x_1 - y_1)\,\text{Tr}\,[\gamma_5 S_F(y_1 - y_2)\gamma_5 S_F(y_2 - y_1)]\Delta_F(y_2 - x_2)$$
$$+ \ldots$$

The terms on the right-hand side are just those which appear in the expansion 6(390) of the dressed boson propagator $\Delta_F'(x_1-x_2)$. The first two terms correspond to the configuration space diagrams shown in Fig. 7.1a and 7.1b respectively, while other terms correspond to the diagrams of Fig. 7.1c, 7.1d, etc*. In addition we get disconnected diagrams

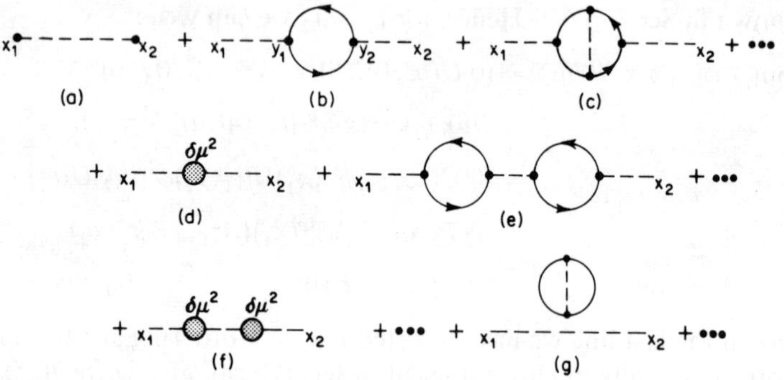

FIG. 7.1. Diagrammatic expansion of 7(67).

of the type shown in Fig. 7.1g. The effect of the latter is simply to multiply $\Delta_F'(x_1-x_2)$ by an infinite phase factor, as discussed in Section 6-2. This phase factor cancels with the denominator in 7(59) since, by the same argument which gave 7(65), we have

$$\langle 0\ \text{out}|0\ \text{in}\rangle = (0|U(\infty,-\infty)|0) = (0|S|0) \tag{7(68)}$$

The same procedure can be applied to represent the fermion Green's function $S_F'(x-y)$ in the form

$$S_F'(x-y) = \frac{\langle 0\ \text{out}|\,T\psi(x)\bar{\psi}(y)|0\ \text{in}\rangle}{\langle 0\ \text{out}|0\ \text{in}\rangle} \tag{7(69)}$$

Again the crucial relation is

$$\langle 0\ \text{out}|\,T\psi(x)\bar{\psi}(y)|0\ \text{in}\rangle = (0|\,T\psi^{ip}(x)\bar{\psi}^{ip}(y)S|0)$$

from which we recover 6(391) by expanding the right-hand side. For the vertex function Γ_5 we find that

$$\frac{\langle 0\ \text{out}|\,T\psi(x)\bar{\psi}(y)\phi(z)|0\ \text{in}\rangle}{\langle 0\ \text{out}|0\ \text{in}\rangle} = -G_0\int d^4x'\,d^4y'\,d^4z'$$

$$\times\, S_F'(x-x')\Gamma_5(x'-z', z'-y')S_F'(y'-y)\Delta_F'(z'-z) \tag{7(70)}$$

* The reader is urged to check these assertions by carrying out the expansion of 7(67) to terms of order G_0^4 explicitly.

where the right-hand side is represented graphically in Fig. 7.2, the blobs on the external legs in Fig. 7.2 representing the fermion and boson propagators S_F' and Δ_F'. The central blob represents the vertex function Γ_5. The verification of 7(69) and 7(70) is left as an exercise for the reader. [Problem 2.]

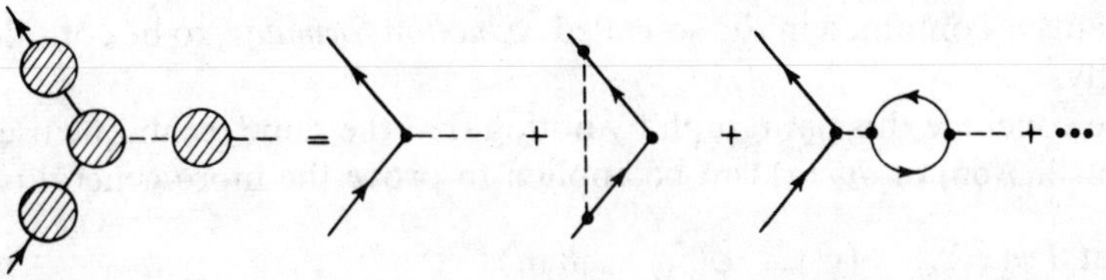

FIG. 7.2. Diagrammatic expansion of 7(70).

The extension of the above techniques to a general higher order Green's function

$$\langle 0\ \text{out}|\,T\psi(x_1)\ldots\bar{\psi}(y_1)\ldots\phi(z_1)\ldots|0\ \text{in}\rangle$$

is straightforward. We have, in general

$$\langle 0\ \text{out}|\,T\underbrace{\psi(x_1)\ldots}_{n}\underbrace{\bar{\psi}(y_1)\ldots}_{n}\underbrace{\phi(z_1)\ldots}_{m}|0\ \text{in}\rangle$$

$$= (0|\,T\underbrace{\psi^{ip}(x_1)\ldots}_{n}\underbrace{\bar{\psi}^{ip}(y_1)\ldots}_{n}\underbrace{\phi^{ip}(z_1)\ldots}_{m}S|0) \qquad 7(71a)$$

where the right-hand side may be evaluated by means of the perturbation expansion for $S = U(\infty, -\infty)$. In this way we can express 7(71a) in terms of configuration space Feynman graphs with $2n+m$ external legs. For example, the four-point Green's function*

$$\frac{\langle 0\ \text{out}|\,T\phi(z_1)\phi(z_2)\phi(z_3)\phi(z_4)|0\ \text{in}\rangle}{\langle 0\ \text{out}|0\ \text{in}\rangle}$$

is represented diagrammatically in Fig. 7.3 where, as in Fig. 7.2, the blobs on the external legs represent boson self-energy diagrams which

FIG. 7.3. Exact 4-point Green's function.

* Four-point Green's functions are discussed in detail in Section 9.1.

have been separated from the remainder of the interaction represented by the central blob. The basic difference between a Green's function and the corresponding *S-matrix element* lies in the fact that the former is defined for all values of the momenta of the external lines, whereas the latter is defined only on the mass shells of the external momenta. The precise statement of the relation between Green's functions and S-matrix elements is contained in the so-called *reduction formulae*, to be considered shortly.

We conclude this paragraph by noting that the same technique used to establish 7(66) or 7(71a) can be applied to prove the more general result

$$\langle a \text{ out}| T\psi(x_1) \ldots \bar{\psi}(y_1) \ldots \phi(z_1) \ldots |b \text{ in}\rangle$$
$$= (a| T\psi^{ip}(x_1) \ldots \bar{\psi}^{ip}(y_1) \ldots \phi^{ip}(z_1) \ldots S|b) \qquad 7(71b)$$

where $|a)$ and $|b)$ are any two unperturbed states. Thus, the most general matrix element of a time-ordered product of Heisenberg field operators can be analyzed in terms of perturbation theory.

Dyson Equations The representation of Green's functions as vacuum expectation values of Heisenberg fields can be used to provide an alternative derivation of Dyson's equations. For example we can derive the Dyson equation 6(399) for S'_F by applying $\gamma\,.\,\partial + m$ to both sides of 7(69) and using the equation of motion

$$(\gamma\,.\,\partial + m_0)\psi(x) = iG_0\gamma_5\psi(x)\phi(x)$$

where m_0 is the ' bare ' mass $m - \delta m$. Writing the T-product in the form

$$T\psi(x)\bar{\psi}(y) = \tfrac{1}{2}[\psi(x), \bar{\psi}(y)] + \tfrac{1}{2}\epsilon(x_0 - y_0)\{\psi(x), \bar{\psi}(y)\}$$

where

$$\epsilon(x_0) = 2\theta(x_0) - 1 = \begin{cases} +1 & x_0 > 0 \\ -1 & x_0 < 0 \end{cases}$$

and noting that $\partial_0\epsilon(x_0) = 2\delta(x_0)$, we easily obtain

$$\left(\gamma\,.\,\frac{\partial}{\partial x} + m_0\right) S'_F(x-y) = \left(\gamma\,.\,\frac{\partial}{\partial x} + m_0\right)\frac{\langle 0 \text{ out}| T\psi(x)\bar{\psi}(y)|0 \text{ in}\rangle}{\langle 0 \text{ out}|0 \text{ in}\rangle}$$

$$= -i\delta^{(4)}(x-y) + \frac{iG_0\langle 0 \text{ out}| T\gamma_5\psi(x)\bar{\psi}(y)\phi(x)|0 \text{ in}\rangle}{\langle 0 \text{ out}|0 \text{ in}\rangle}$$

Applying 7(70) to the matrix element on the right-hand side, we find

$$\left(\gamma \cdot \frac{\partial}{\partial x} + m_0\right) S'_F(x-y) = -i\delta^{(4)}(x-y) + iG_0^2 \int \Sigma^*(x-z) S'_F(z-y)\, d^4z \qquad 7(72)$$

where Σ^* is given by

$$\Sigma^*(x-y) = -\int d^4x'\, d^4y'\, \gamma_5 S'_F(x-x')\Gamma_5(x'-y', y'-y)\Delta'_F(y'-x) \qquad 7(73)$$

Comparison with 6(400) shows that $\Sigma^*(x-y)$ is just the sum of all proper fermion self-energy graphs in configuration space. Equation 7(72) is almost the Dyson equation 6(399); to exhibit the equivalence more explicitly, we integrate 7(72) with the aid of the free fermion Green's function S_F satisfying

$$\left(\gamma \cdot \frac{\partial}{\partial x} + m_0\right) S_F(x-y) = -i\delta^{(4)}(x-y)$$

This gives

$$S'_F(x-x') = S_F(x-x') - \int S_F(x-y) G_0^2 \Sigma^*(y-y') S'_F(y'-x')\, d^4y\, d^4y' \qquad 7(74)$$

which is identical to 6(399). By the same procedure one can derive the Dyson equation for the exact boson propagator, namely

$$\Delta'_F(x-x') = \Delta_F(x-x') + i \int \Delta_F(x-y) G_0^2 \Pi^*(y-y') \Delta'_F(y'-x')\, d^4y\, d^4y' \qquad 7(75)$$

with Π^* given by 6(395).

Green's Function in Terms of In-fields The formula

$$\Delta'_F(x_1-x_2) = \frac{(0|T\phi^{ip}(x_1)\phi^{ip}(x_2)S|0)}{(0|S|0)}$$

which serves to provide the perturbation expansion of the boson Green's function, can also be expressed as

$$\Delta'_F(x_1-x_2) = \frac{\langle 0\,\text{in}|T\phi^{\text{in}}(x_1)\phi^{\text{in}}(x_2)S[\phi^{\text{in}}]|0\,\text{in}\rangle}{\langle 0\,\text{in}|S[\phi^{\text{in}}]|0\,\text{in}\rangle} \qquad 7(76)$$

where $S[\phi^{\text{in}}]$ is the S operator in which all interaction picture fields have been replaced by the corresponding in-fields. This follows directly from the relation 7(38) which expresses the fact that the ϕ^{in} bear the same relation to the in-states $|a\,\text{in}\rangle$ as the ϕ^{ip} to the unperturbed states $|a)$.

We then have the perturbation expansion

$$\Delta_F'(x_1-x_2) = \sum_{n=0} \frac{(-i)^n}{n!} \int_{-\infty}^{\infty} d^4y_1 \ldots \int_{-\infty}^{\infty} d^4y_n$$
$$\times \langle 0\ \text{in}|T(\mathscr{H}_I^{\text{in}}(y_1)\ldots\mathscr{H}_I^{\text{in}}(y_n)\phi^{\text{in}}(x_1)\phi^{\text{in}}(x_2))|0\ \text{in}\rangle_C \quad 7(77)$$

where

$$\mathscr{H}_I^{\text{in}}(y) = -iG_0[\bar{\psi}^{\text{in}}(y), \gamma_5\psi^{\text{in}}(y)]\phi^{\text{in}}(y) - \tfrac{1}{2}\delta\mu^2(\phi^{\text{in}}(y))^2 - \delta m[\bar{\psi}^{\text{in}}(y), \psi^{\text{in}}(y)] \quad 7(78)$$

and where the subscript C indicates that only connected diagrams are to be retained. From a practical point of view there is no difference in content between 7(67) and 7(77), but the second form has the attractive feature that one works with Heisenberg picture fields and states throughout.

Similar remarks apply to the fermion propagator and to the ' vertex ' matrix element 7(70). Instead of in-fields, one can of course also use out-fields.

Reduction Formulae In the preceding paragraphs we have seen how to express Green's functions as vacuum expectation values of time-ordered products of Heisenberg fields. We now wish to express scattering matrix elements in terms of Green's function. This is accomplished by means of so-called *reduction formulae.*

For simplicity we consider the scattering of two neutral spin 0 bosons. Our starting point is the expression for the S-matrix element

$$(\mathbf{k}'\mathbf{q}'|S|\mathbf{kq})$$

for the transition $(\mathbf{k}, \mathbf{q}) \to (\mathbf{k}', \mathbf{q}')$, as given by perturbation theory. Expanding the scattering operator $S = U(\infty, -\infty)$ as in 6(39), let us apply Wick's theorem to pick up the term featuring the normal product $a_{\mathbf{k}'}{}^{\dagger}a_{\mathbf{q}'}{}^{\dagger}a_{\mathbf{k}}a_{\mathbf{q}}$. We have

$$(\mathbf{k}'\mathbf{q}'|S|\mathbf{kq}) = (\mathbf{k}'\mathbf{q}'|\mathbf{kq})$$
$$+ \int d^4x'\, d^4y'\, d^4x\, d^4y (\mathbf{k}'|\phi^{ip}(x')|0)(\mathbf{q}'|\phi^{ip}(y')|0)$$
$$\times (0|\frac{\delta}{\delta\phi^{ip}(x')}\frac{\delta}{\delta\phi^{ip}(y')}\frac{\delta}{\delta\phi^{ip}(x)}\frac{\delta}{\delta\phi^{ip}(y)}U(\infty, -\infty)|0)$$
$$\times (0|\phi^{ip}(x)|\mathbf{k})(0|\phi^{ip}(y)|\mathbf{q})\frac{1}{(0|U(\infty, -\infty)|0)} \quad 7(79)$$

To evaluate the functional derivatives we shall use the following formula [Nishijima (1965)]

$$(\Box_y - \mu^2)T(\phi^{ip}(y)a^{ip}(x_1)b^{ip}(x_2)\ldots) = i\frac{\delta}{\delta\phi^{ip}(y)}T(a^{ip}(x_1)b^{ip}(x_2)\ldots) \quad 7(80)$$

where a, b are any set of field operators, which may include ϕ. The above relation arises as follows. Expanding the T-products by means of Wick's theorem, we have for the left-hand side

$$\begin{aligned}T(\phi^{ip}(y)a^{ip}(x_1)b^{ip}(x_2)\ldots) &= :\phi^{ip}(y)a^{ip}(x_1)b^{ip}(x_2)\ldots: \\ &+ :\underbrace{\phi^{ip}(y)a^{ip}(x_1)}b^{ip}(x_2)\ldots: + :\phi^{ip}(y)a^{ip}(x_1)b^{ip}(x_2)\ldots: + \ldots \\ &+ :\phi^{ip}(y)a^{ip}(x_1)b^{ip}(x_2)\ldots: + :\phi^{ip}(y)a^{ip}(x_1)b^{ip}(x_2)\ldots: + \ldots \\ &+ :\phi^{ip}(y)a^{ip}(x_1)b^{ip}(x_2)\ldots: + \ldots\end{aligned}$$

The only terms in the Wick expansion which will give a non-zero contribution to the left-hand side of 7(80) are those which feature a contraction of ϕ^{ip} with one of the field operators in the set $a^{ip}, b^{ip}\ldots$. All other terms will give a vanishing contribution, since $(\Box_y - \mu^2)\phi^{ip}(y) = 0$. Moreover, the only nonvanishing contractions of ϕ^{ip} with members of the set $a^{ip}, b^{ip}\ldots$ will occur for those members of the set which coincide with ϕ^{ip}. We thus obtain, for the left-hand side of 7(80),

$$\begin{aligned}&(\Box_y - \mu^2)T(\phi^{ip}(y)a^{ip}(x_1)b^{ip}(x_2)\ldots) \\ &\quad = i\delta^{(4)}(x_1 - y)\delta_{\phi a}:b^{ip}(x_2)\ldots: + i\delta^{(4)}(x_2 - y)\delta_{\phi b}:a^{ip}(x_1)\ldots: + \ldots \\ &\quad + i\delta^{(4)}(x_1 - y)\delta_{\phi a}:b^{ip}(x_2)\ldots: + i\delta^{(4)}(x_2 - y)\delta_{\phi b}:a^{ip}(x_1)\ldots: + \ldots \\ &\quad + i\delta^{(4)}(x_1 - y)\delta_{\phi a}:b^{ip}(x_2)\ldots: + \ldots\end{aligned}$$

where the symbol $\delta_{\phi a}$, for example, is equal to 1 for $a \equiv \phi$ and 0 for $a \not\equiv \phi$. To show that this is identical to the right-hand side of 7(80) it suffices to write the latter in the form

$$\begin{aligned}&i\frac{\delta}{\delta\phi(y)}Ta^{ip}(x_1)b^{ip}(x_2)\ldots \\ &\qquad = i\delta(x_1 - y)\delta_{\phi a}Tb^{ip}(x_2)\ldots + i\delta(x_2 - y)\delta_{\phi b}Ta^{ip}(x_1)\ldots + \ldots\end{aligned}$$

and expand the T-products by means of Wick's theorem.

We now apply 7(80) to 7(79) to get

$$(\mathbf{k}'\mathbf{q}'|S|\mathbf{kq}) = (\mathbf{k}'\mathbf{q}'|\mathbf{kq}) + (-i)^4 \int d^4x'\, d^4y'\, d^4x\, d^4y$$

$$\times \frac{1}{\sqrt{2\omega_{\mathbf{k}'}V}}\,\mathrm{e}^{-ik'.x'}\frac{1}{\sqrt{2\omega_{\mathbf{q}'}V}}\,\mathrm{e}^{-iq'.y'}\frac{1}{\sqrt{2\omega_{\mathbf{k}}V}}\,\mathrm{e}^{ik.x}\frac{1}{\sqrt{2\omega_{\mathbf{q}}V}}\,\mathrm{e}^{iq.y}$$

$$\times(\Box_{x'}-\mu^2)(\Box_{y'}-\mu^2)(\Box_x-\mu^2)(\Box_y-\mu^2)$$

$$\times(0|T\phi^{ip}(x')\phi^{ip}(y')\phi^{ip}(x)\phi^{ip}(y)U(\infty,-\infty)|0)$$

$$\times\frac{1}{(0|U(\infty,-\infty)|0)}$$

where we have noted that

$$(0|\phi^{ip}(x)|\mathbf{k}) = \frac{1}{\sqrt{2\omega_{\mathbf{k}}V}}\,\mathrm{e}^{ik.x}$$

$$(\mathbf{k}|\phi^{ip}(x)|0) = \frac{1}{\sqrt{2\omega_{\mathbf{k}}V}}\,\mathrm{e}^{-ik.x}$$

As the final step we use the relation

$$\frac{(0|T\phi^{ip}(x')\phi^{ip}(y')\phi^{ip}(x)\phi^{ip}(y)U(\infty,-\infty)|0)}{(0|U(\infty,-\infty)|0)} = \frac{\langle 0\text{ out}|T\phi(x')\phi(y')\phi(x)\phi(y)|0\text{ in}\rangle}{\langle 0\text{ out}|0\text{ in}\rangle} \qquad 7(81)$$

which follows from 7(71a) and 7(68), to obtain the *reduction formula*

$$(\mathbf{k}'\mathbf{q}'|S|\mathbf{kq}) = (\mathbf{k}'\mathbf{q}'|\mathbf{kq})$$

$$+(-i)^4 \int d^4x'\, d^4y'\, d^4x\, d^4y$$

$$\times \frac{1}{\sqrt{2\omega_{\mathbf{k}'}V}}\,\mathrm{e}^{-ik'.x'}\frac{1}{\sqrt{2\omega_{\mathbf{q}'}V}}\,\mathrm{e}^{-iq'.y'}\frac{1}{\sqrt{2\omega_{\mathbf{k}}V}}\,\mathrm{e}^{ik.x}\frac{1}{\sqrt{2\omega_{\mathbf{q}}V}}\,\mathrm{e}^{iq.y}$$

$$\times(\Box_{x'}-\mu^2)(\Box_{y'}-\mu^2)(\Box_x-\mu^2)(\Box_y-\mu^2)$$

$$\times\frac{\langle 0\text{ out}|T\phi(x')\phi(y')\phi(x)\phi(y)|0\text{ in}\rangle}{\langle 0\text{ out}|0\text{ in}\rangle} \qquad 7(82)$$

expressing the 2 boson → 2 boson scattering amplitude in terms of the four-point Green's function

$$\frac{\langle 0 \text{ out}| T\phi(x')\phi(y')\phi(x)\phi(y)|0 \text{ in}\rangle}{\langle 0 \text{ out}|0 \text{ in}\rangle}$$

The reduction formula 7(82) has a simple graphical interpretation. Let us represent the four-point Green's function as in Fig. 7.3. Then for each external line the reduction formula features a factor $-i(\Box - \mu^2)$, or in momentum space $i(k^2 + \mu^2)$. This is to be taken on the mass shell where it cancels the propagator factor $-i(k^2 + \mu^2)^{-1}$ for the external leg. Thus 7(82) is just the statement that the S-matrix element is equal to the corresponding Green's function with the external propagators removed and the external 4-momenta restricted to their mass shell values.

Reduction formulae for general n-particle → m-particle S-matrix elements can be derived by the same procedure as above. The result will express the S-matrix element in terms of the corresponding Green's function—the vacuum expectation value of the T-product of $n + m$ fields. In Section 7-5 we shall give an alternative derivation of the reduction formulae which does not rely on perturbation theory.

7-4 Nonperturbative reformulation

The foregoing developments have all been based on perturbation theory and in particular on the statement that $|a \text{ in}\rangle$ and $|a \text{ out}\rangle$ as given by 7(5a) and 7(5b) are eigenstates of the total energy and momentum operators of the field. There are, however, certain basic drawbacks to the perturbative approach. First one has no guarantee that the perturbation series converges; in fact for strong interactions it does not. Secondly, there is considerable doubt as to whether the Møller wave matrices $\Omega^{(+)}$ and $\Omega^{(-)}$—which transform the unperturbed states $|a)$ into $|a \text{ in}\rangle$ and $|a \text{ out}\rangle$—exist in any formal mathematical sense. Van Hove (1951, 1952) has in fact shown for a model field theory with trilinear coupling that eigenstates of H_0^{ip} and H cannot lie in the same Hilbert space. Finally, we have seen earlier that the use of 7(5a) for all states $|a \text{ in}\rangle$ excludes treatment of bound states.

We now turn to a more modern—and also more 'abstract'—approach to field theory. Instead of relying on the perturbation expansion of the S-matrix, we start from a set of physically plausible assumptions about the nature of the solution to the set of coupled field equations characterizing the system.

We begin by summarizing some basic properties of interacting quantum fields.

(a) The fields ϕ_A and π_A obey canonical equal-time commutation rules

$$[\phi_A(x), \pi_{A'}(x')]_{t=t'} = i\delta_{AA'}\delta^{(3)}(\mathbf{x}-\mathbf{x}')$$

$$[\phi_A(x), \phi_{A'}(x')]_{t=t'} = [\pi_A(x), \pi_{A'}(x')]_{t=t'} = 0$$

or corresponding anticommutation rules in the case of Fermi–Dirac theory. The microcausality requirement—that local observables commute for spacelike separations—follows on grounds of Lorentz invariance.

(b) In the absence of external c-number fields, the total Hamiltonian is invariant under infinitesimal space-time translations and rotations. The generators of these transformations, P_μ and $M_{\mu\nu}$, are therefore conserved. In particular, for the translation generators we have, by virtue of the equal-time commutation relations,

$$\partial_\mu \phi_A(x) = i[\phi_A(x), P_\mu] \qquad 7(83a)$$

$$\partial_\mu \pi_A(x) = i[\pi_A(x), P_\mu] \qquad 7(83b)$$

or equivalently, since P_μ is conserved

$$\phi_A(x) = e^{-iP.x}\phi_A(0)\, e^{iP.x} \qquad 7(84a)$$

$$\pi_A(x) = e^{-iP.x}\pi_A(0)\, e^{iP.x} \qquad 7(84b)$$

(c) From the invariance of P_μ under space–time displacements, we deduce that

$$[P_\mu, P_\nu] = 0 \qquad 7(85a)$$

It follows that all four components of P_μ are simultaneously diagonalizable.

(d) Lorentz invariance requires P_μ to transform as a four-vector. For pure Lorentz transformations generated by $-\omega_{i4}M_{i4}$ this requirement reduces to

$$[H, M_{i4}] = P_i$$
$$[P_j, M_{i4}] = \delta_{ij}H \qquad 7(85b)$$

Let us now assume that we are ignorant of the results of perturbation theory and cannot rely on 7(5a) and 7(5b) for the construction of the eigenstates of H. What can we say about these eigenstates on general grounds? First we note that since all four components of P_μ are simultaneously diagonalizable, we may take our eigenstates of H to be eigenstates of P_μ. Furthermore, we shall assume that [Källén (1952)]

(1) All eigenstates of P_μ have timelike or lightlike four-momenta and non-negative energies, that is

$$k^2 = k_\mu k_\mu \leqslant 0 \qquad k_0 \geqslant 0$$

(2) The lowest eigenstate of P_μ, called the vacuum state $|0\rangle$*, is unique and Lorentz-invariant. The Lorentz-invariance of $|0\rangle$ implies that

$$M_{\mu\nu}|0\rangle = 0 \tag{7(86a)}$$

so that by 7(85b)

$$H|0\rangle = \mathbf{P}|0\rangle = 0 \tag{7(86b)}$$

(3) There exist discrete 1-particle states $|\mathbf{k}\rangle, |\mathbf{p}\rangle \ldots$ with $k^2 = -\mu^2$, $p^2 = -m^2, \ldots$ corresponding to single particles of physical mass $\mu, m, \ldots$. These states may be either *elementary* particle states, which correspond to fields featured in the Lagrangian, or *bound* states, which do not. We further assume that the spectrum contains *continuum* states with $2, 3, \ldots n \ldots$ incoming particles (elementary or bound), as well as the corresponding out-states.

(4) The set of all in-states (or out-states) forms a complete set of states spanning the Hilbert space. Here it is assumed that if the system supports additional conserved quantum numbers (charge or baryon number, for example), these quantum numbers are included in the specification of the in- and out-states.

The existence of physically reasonable in- and out-states is thus postulated. The S-matrix element for a transition from an initial state $|k_1 \ldots k_n \text{ in}\rangle$ characterized by ingoing momenta $k_1 \ldots k_n$ to a final state of outgoing momenta $k_1' \ldots k_m'$ is then given by the probability amplitude

$$\langle k_1' \ldots k_m' \text{ out}|k_1 \ldots k_n \text{ in}\rangle$$

as in 7(17), with the phase factor $\langle 0 \text{ out}|0 \text{ in}\rangle$ omitted.

It is now obvious how to define in and out fields. First, for each particle type and each momentum value $\mathbf{k}$ we define in and out creation operators with the usual properties relative to the in and out states respectively. For neutral spinless particles of mass μ, for example, we consider the states $|\mathbf{k} \text{ in}\rangle, |\mathbf{k}, \mathbf{q} \text{ in}\rangle \ldots$ etc. with $k^2 = q^2 = -\mu^2$ and define operators $a_\mathbf{k}^{\text{in}}$ and $a_\mathbf{k}^{\text{in}\dagger}$ with the properties

$$a_\mathbf{k}^{\text{in}\dagger}|0\rangle = |\mathbf{k} \text{ in}\rangle \qquad a_\mathbf{k}^{\text{in}}|\mathbf{k} \text{ in}\rangle = |0\rangle$$

$$a_\mathbf{k}^{\text{in}\dagger}|\mathbf{q} \text{ in}\rangle = |\mathbf{k}, \mathbf{q} \text{ in}\rangle \qquad a_\mathbf{k}^{\text{in}}|\mathbf{k}, \mathbf{q} \text{ in}\rangle = |\mathbf{q} \text{ in}\rangle$$

etc.

* For the remainder of this chapter we shall neglect the phase factor $\langle 0 \text{ out}|0 \text{ in}\rangle$. We shall not distinguish, therefore, between $|0 \text{ in}\rangle$ and $|0 \text{ out}\rangle$ or between the 1-particle states $|\mathbf{k} \text{ in}\rangle$ and $|\mathbf{k} \text{ out}\rangle$.

or, more generally, in the notation of Section 3-2,

$$a_{\mathbf{k}_i}|n_{\mathbf{k}_1}, n_{\mathbf{k}_2}, \ldots n_{\mathbf{k}_i}, \ldots; \text{in}\rangle = \sqrt{n_{\mathbf{k}_i}}|n_{\mathbf{k}_1}, n_{\mathbf{k}_2}, \ldots n_{\mathbf{k}_i}-1, \ldots; \text{in}\rangle \qquad 7(87a)$$

$$a_{\mathbf{k}_i}^{\dagger}|n_{\mathbf{k}_1}, n_{\mathbf{k}_2}, \ldots n_{\mathbf{k}_i}, \ldots; \text{in}\rangle = \sqrt{n_{\mathbf{k}_i}+1}|n_{\mathbf{k}_1}, n_{\mathbf{k}_2}, \ldots n_{\mathbf{k}_i}+1, \ldots; \text{in}\rangle \qquad 7(87b)$$

where $n_{\mathbf{k}}$ represents the number of incoming particles of momentum $\mathbf{k}_i$. Using the creation and destruction operators, we then build up the in and out fields in the usual way to satisfy the free field equations and commutation relations. Thus, for the neutral scalar case, we define

$$\phi_{\text{in}}(\mathbf{x}, t) = \frac{1}{\sqrt{V}} \sum_{\mathbf{k}k_0=\omega_{\mathbf{k}}} \frac{1}{\sqrt{2\omega_{\mathbf{k}}}}(a_{\mathbf{k}}^{\text{in}}\,\mathrm{e}^{ik.x} + a_{\mathbf{k}}^{\text{in}\dagger}\,\mathrm{e}^{-ik.x}) \qquad 7(88)$$

with $\omega_{\mathbf{k}} = (\mathbf{k}^2+\mu^2)^{1/2}$, and similarly for the out field.

From the assumption that the in and out states are eigenstates of H, it follows that the in and out fields, as defined above, are Heisenberg picture fields—that is, they satisfy the Heisenberg equations of motion. The argument is essentially elementary and we illustrate it for the neutral scalar case. The only non-zero matrix elements $\langle a\ \text{in}|\phi\ \text{in}|b\ \text{in}\rangle$ are those for which $|a\ \text{in}\rangle$ and $|b\ \text{in}\rangle$ differ by one incoming neutral scalar boson of mass μ. Let us denote the energy-momentum eigenvalues of $|a\ \text{in}\rangle$ and $|b\ \text{in}\rangle$ by p_a and p_b respectively:

$$\begin{aligned} P^{\mu}|a\ \text{in}\rangle &= p_a^{\mu}|a\ \text{in}\rangle \\ P^{\mu}|b\ \text{in}\rangle &= p_b^{\mu}|b\ \text{in}\rangle \end{aligned} \qquad 7(89)$$

If $p_b^0 > p_a^0$, the matrix element $\langle a\ \text{in}|\phi_{\text{in}}|b\ \text{in}\rangle$ singles out the *destruction* operator for momentum $\mathbf{k} = \mathbf{p}_b - \mathbf{p}_a$ in the Fourier expansion 7(88); if $p_b^0 < p_a^0$, it singles out the *creation* operator for $\mathbf{k} = \mathbf{p}_a - \mathbf{p}_b$. In either case the matrix element will feature a factor $\exp i(p_b - p_a) . x$ from the expansion 7(88), i.e.

$$\langle a\ \text{in}|\phi_{\text{in}}(x)|b\ \text{in}\rangle = \mathrm{e}^{i(p_b-p_a).x}\langle a\ \text{in}|\phi_{\text{in}}(0)|b\ \text{in}\rangle \qquad 7(90)$$

On the other hand, using 7(89) we obtain

$$\begin{aligned} \langle a\ \text{in}|\phi_{\text{in}}(x)|b\ \text{in}\rangle &= \langle a\ \text{in}|\mathrm{e}^{-iP.x}\,\mathrm{e}^{iP.x}\phi_{\text{in}}(x)\,\mathrm{e}^{-iP.x}\,\mathrm{e}^{iP.x}|b\ \text{in}\rangle \\ &= \mathrm{e}^{i(p_b-p_a)x}\langle a\ \text{in}|\mathrm{e}^{iP.x}\phi_{\text{in}}(x)\,\mathrm{e}^{-iP.x}|b\ \text{in}\rangle \end{aligned}$$

so that, comparing with 7(90), we have

$$\langle a\ \text{in}|\phi_{\text{in}}(0)|b\ \text{in}\rangle = \langle a\ \text{in}|\mathrm{e}^{iP.x}\phi_{\text{in}}(x)\,\mathrm{e}^{-iP.x}|b\ \text{in}\rangle$$

Since this holds for all non-zero matrix elements of ϕ, we infer that

$$\phi_{\text{in}}(x) = \mathrm{e}^{-iP.x}\phi_{\text{in}}(0)\,\mathrm{e}^{iP.x} \qquad 7(91)$$

from which it follows that $\phi_{\text{in}}(x)$ obeys the Heisenberg equation of motion.

Finally, we note that from 7(88) we have

$$\langle 0|\phi_{\text{in}}(x)|\mathbf{k}\rangle = \frac{1}{\sqrt{V}}\frac{1}{\sqrt{2\omega_{\mathbf{k}}}}\mathrm{e}^{ik.x} \qquad 7(92a)$$

$$\langle \mathbf{k}|\phi_{\text{in}}(x)|0\rangle = \frac{1}{\sqrt{V}}\frac{1}{\sqrt{2\omega_{\mathbf{k}}}}\mathrm{e}^{-ik.x} \qquad 7(92b)$$

for the matrix elements of ϕ_{in} between the vacuum and one-particle states.

To summarize, the abstract approach to field theory *postulates* the existence of the in and out eigenstates of the Hamiltonian. The free fields $\phi^A_{\text{in}}(x)$ and $\phi^A_{\text{out}}(x)$ are then *constructed* by defining their matrix elements within the in and out manifolds respectively. To proceed further with this reformulation we require a connection between the free fields $\phi^A_{\substack{\text{in}\\ \text{out}}}(x)$ and the interacting fields $\phi^A(x)$. This connection is provided by the *asymptotic condition* which will be discussed in the following section.

7-5 The asymptotic condition

One-Particle Wave Functions For the present we restrict our attention to those $\phi^A_{\substack{\text{in}\\ \text{out}}}(x)$ which refer to *elementary* particle states in the spectrum of H*. To each elementary particle there corresponds a field in the Lagrangian. Let us consider neutral spin 0 particles for example, and ask for the matrix elements of the corresponding field operator $\phi(x)$ between the vacuum and one-particle states. By 7(84a) we have

$$\begin{aligned}\langle 0|\phi(x)|\mathbf{k}\rangle &= \langle 0|\mathrm{e}^{-iP.x}\phi(0)\,\mathrm{e}^{iP.x}|\mathbf{k}\rangle \\ &= \mathrm{e}^{ik\cdot x}\langle 0|\phi(0)|\mathbf{k}\rangle \end{aligned} \qquad 7(93a)$$

where we have used the fact that $|0\rangle$ and $|\mathbf{k}\rangle$ are eigenstates of P_μ with eigenvalues 0 and k_μ respectively. Similarly

$$\langle \mathbf{k}|\phi(x)|0\rangle = \mathrm{e}^{-ik.x}\langle \mathbf{k}|\phi(0)|0\rangle \qquad 7(93b)$$

Thus the matrix elements 7(93a) and 7(93b) have the same x dependence as 7(92a) and 7(92b) respectively.

Now on covariance grounds we have

$$\langle 0|\phi(0)|\mathbf{k}\rangle = C\frac{1}{\sqrt{V}}\frac{1}{\sqrt{2\omega_{\mathbf{k}}}} \qquad 7(94a)$$

$$\langle \mathbf{k}|\phi(0)|0\rangle = C^*\frac{1}{\sqrt{V}}\frac{1}{\sqrt{2\omega_{\mathbf{k}}}} \qquad 7(94b)$$

* For the treatment of composite particles, see Section 9-4.

where C is a constant; the factor $(2\omega_{\mathbf{k}})^{-1/2}$ is necessary for consistency with the normalization convention $\langle \mathbf{k}|\mathbf{k}'\rangle = \delta_{\mathbf{kk}'}$.* For convenience we shall take C to be real. For further characterization of C, we *define* the boson Green's function

$$\Delta_F'(x-y) = \langle 0|T\phi(x)\phi(y)|0\rangle \qquad 7(95)$$

The fact that Δ_F' depends only on the difference $x-y$ is a consequence of the invariance of the vacuum under space–time displacements, since for $x_0 > y_0$

$$\begin{aligned}\Delta_F'(x, y) &= \langle 0|e^{-iP.x}\phi(0)\,e^{iP.x}\,e^{-iP.y}\phi(0)\,e^{iP.y}|0\rangle \\ &= \langle 0|\phi(0)\,e^{iP.(x-y)}\phi(0)|0\rangle = \Delta_F'(x-y)\end{aligned}$$

and similarly for $x_0 < y_0$. We now take $x_0 > y_0$, insert a sum over a complete set of states between $\phi(x)$ and $\phi(y)$, and isolate the contribution of the one-particle states $|\mathbf{k}\rangle$. Since the latter satisfy the mass shell restriction $k_0 = \omega_{\mathbf{k}}$ and have positive energies, we get

$$\begin{aligned}\underset{(x_0 > y_0)}{\Delta_F'(x-y)} &= \int \langle 0|\phi(x)|\mathbf{k}\rangle\langle \mathbf{k}|\phi(y)|0\rangle\delta(k_0-\omega_{\mathbf{k}})\,d^4k + \ldots \\ &= C^2\frac{1}{(2\pi)^3}\int \frac{1}{2\omega_{\mathbf{k}}}\,e^{ik.(x-y)}\delta(k_0-\omega_{\mathbf{k}})\,d^4k + \ldots \\ &= C^2\frac{1}{(2\pi)^4}\int \frac{1}{2\omega_{\mathbf{k}}}\,e^{ik.(x-y)}\frac{i}{k_0-\omega_{\mathbf{k}}+i\varepsilon}\,d^4k + \ldots \qquad 7(96a)\end{aligned}$$

In writing down the second line we have gone over from discrete to continuum normalization, replacing $V^{-1/2}$ in 7(94a) and 7(94b) by $(2\pi)^{-3/2}$. In deriving the third line we have, (a) written the integral over k_0 in the form of an integral over the infinite contour shown in Fig. 7.4, and (b) shifted the pole at $k_0 = \omega_{\mathbf{k}}$ by an amount $i\varepsilon$ and modified the contour to lie along the real axis. In a similar way we obtain the contribution

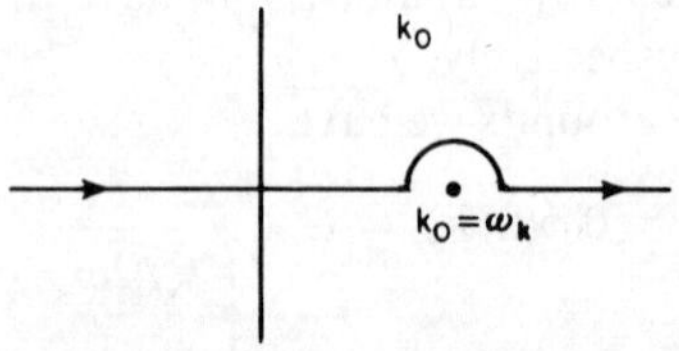

FIG. 7.4. Integration contour in complex k_0-plane for 7(96a).

* See the discussion following Eq. 3(25).

of the one-particle states for the opposite time ordering $x_0 < y_0$:

$$\Delta'_F(x-y) = C^2 \frac{1}{(2\pi)^4} \int \frac{1}{2\omega_{\mathbf{k}}} e^{-ik(x-y)} \frac{i}{k_0-\omega_{\mathbf{k}}+i\varepsilon} d^4k + \ldots$$

$(x_0 < y_0)$

$$= C^2 \frac{1}{(2\pi)^4} \int \frac{1}{2\omega_{\mathbf{k}}} e^{ik(x-y)} \frac{-i}{k_0+\omega_{\mathbf{k}}-i\varepsilon} d^4k + \ldots \qquad 7(96b)$$

Hence, defining the Fourier transform $\Delta'_F(k)$ by

$$\Delta'_F(x-y) = \frac{1}{(2\pi)^4} \int e^{ik(x-y)} \Delta'_F(k)\, d^4k$$

for all x_0 and y_0, we can identify the contribution of the one-particle states to $\Delta'_F(k)$ as

$$C^2 \frac{-i}{k^2+\mu^2-i\varepsilon}$$

by the same argument which gave 3(95). Thus we have shown that C^2 determines the residue of $\Delta'_F(k)$ at the pole $k^2 = -\mu^2$:

$$\Delta'_F(k) = C^2 \frac{-i}{k^2+\mu^2-i\varepsilon} + \text{terms regular at } k^2 = -\mu^2 \qquad 7(96c)$$

At the present stage we can go further only if we are willing to rely on perturbation theory. In that case, we may identify $\Delta'_F(x_1-x_2)$ with the function given by 6(388a) and the residue of the pole of $\Delta'_F(k)$ at $k^2 = -\mu^2$ with the wave function renormalization constant Z_3 given by

$$Z_3 = 1 + G_R^2 \mathbf{\Pi}_1^{*\prime}(-\mu^2) \qquad 7(97a)$$

or, equivalently*

$$Z_3^{-1} = 1 - G_0^2 \Pi^{*\prime}(-\mu^2) \qquad 7(97b)$$

Thus, perturbation theory indicates that

$$C = Z_3^{1/2} = (1 + G_R^2 \mathbf{\Pi}_1^{*\prime}(-\mu^2))^{1/2} \qquad 7(98)$$

to within a phase factor. In the following we shall simply write $C = Z_3^{1/2}$ as a matter of notation, setting

$$\langle 0|\phi(x)|\mathbf{k}\rangle = Z_3^{1/2} \frac{1}{\sqrt{V}} \frac{1}{\sqrt{2\omega_{\mathbf{k}}}} e^{ik.x} \qquad 7(99a)$$

$$\langle \mathbf{k}|\phi(x)|0\rangle = Z_3^{1/2} \frac{1}{\sqrt{V}} \frac{1}{\sqrt{2\omega_{\mathbf{k}}}} e^{-ik.x} \qquad 7(99b)$$

* To derive 7(97b) from 7(97a), combine 6(371) with 6(330).

without necessarily committing ourselves to the perturbative identifications 7(97a) or 7(97b); in Section 8-4 we shall give a nonperturbative characterization of Z_3.

Analogous considerations apply to other fields. For a charged scalar field we have, for example

$$\langle 0|\phi(x)|\mathbf{k}, Q = 1\rangle = Z_3^{1/2}\frac{1}{\sqrt{V}}\frac{1}{\sqrt{2\omega_{\mathbf{k}}}}\,\mathrm{e}^{ik.x}$$

but

$$\langle 0|\phi(x)|\mathbf{k}, Q = -1\rangle = 0$$

since $\phi(x)$ destroys one unit of charge; this follows from the formula expressing Q as the generator of infinitesimal phase transformations*.

$$\delta\phi = i\alpha\phi = \frac{1}{i}[\phi, -\alpha Q]$$

Similarly for Dirac fields we have, for a one-particle state $|\mathbf{k}\sigma\rangle$ [see Problem 3],

$$\langle 0|\psi(x)|\mathbf{k}\sigma\rangle = Z_2^{1/2}\frac{1}{\sqrt{V}}\sqrt{\frac{m}{E_{\mathbf{k}}}}\,u_{\mathbf{k}\sigma}\,\mathrm{e}^{ik.x} \qquad 7(100)$$

Asymptotic Condition We now turn to the connection between the interacting fields $\phi^A(x)$ and the free fields $\phi^A_{\substack{\text{in}\\\text{out}}}(x)$. On physical grounds the simplest assumption one can make is that interacting fields go over into free fields as $t \to \pm\infty$:

$$\lim_{x_0\to-\infty}\langle a|\phi^A(x)|b\rangle = \lim_{x_0\to-\infty}\langle a|\phi^A_{\text{in}}(x)|b\rangle$$

However, this statement is unsatisfactory, as it stands, since the matrix elements on the right-hand side involve oscillating exponentials for which the limit $x_0 \to -\infty$ is undefined. The precise statement has been given by Lehmann, Symanzik and Zimmermann, [Lehmann (1955)] as follows:

Setting

$$f_{\mathbf{k}}(\mathbf{x}, t) = \frac{1}{\sqrt{V}}\frac{1}{\sqrt{2\omega_{\mathbf{k}}}}\,\mathrm{e}^{i\mathbf{k}.\mathbf{x}-i\omega_{\mathbf{k}}t} \qquad 7(101)$$

we write the expansion of a neutral scalar field $\phi(x)$ in the form

$$\phi(\mathbf{x}, t) = \sum_{\mathbf{k}}(a_{\mathbf{k}}(t)f_{\mathbf{k}}(\mathbf{x}, t) + a_{\mathbf{k}}^{\dagger}(t)f_{\mathbf{k}}^{*}(\mathbf{x}, t)) \qquad 7(102)$$

* See the corresponding discussion for charged Dirac fields following 5(119).

with the inversion

$$a_{\mathbf{k}}(t) = i\int d^3x f_{\mathbf{k}}^*(\mathbf{x},t)\overleftrightarrow{\partial}_t\phi(\mathbf{x},t) \tag{7(103)}$$

We also expand $\phi^{\text{in}}(x)$ according to

$$\phi^{\text{in}}(\mathbf{x},t) = \sum_{\mathbf{k}}(a_{\mathbf{k}}^{\text{in}}f_{\mathbf{k}}(\mathbf{x},t)+a_{\mathbf{k}}^{\text{in}\dagger}f_{\mathbf{k}}^*(\mathbf{x},t)) \tag{7(104)}$$

with the inverse relation

$$a_{\mathbf{k}}^{\text{in}} = i\int d^3x f_{\mathbf{k}}^*(\mathbf{x},t)\overleftrightarrow{\partial}_t\phi^{\text{in}}(\mathbf{x},t) \tag{7(105)}$$

and similar formulae for the out-operators. The coefficients $a_{\mathbf{k}}(t)$ must of course depend on time, since $\phi(x)$ does not obey the Klein–Gordon equation; however, the $a_{\mathbf{k}}^{\text{in}}$ are independent of time. Lehmann, Symanzik and Zimmermann now set

$$\lim_{t\to-\infty}\langle a|a_{\mathbf{k}}(t)|b\rangle = Z_3^{1/2}\langle a|a_{\mathbf{k}}^{\text{in}}|b\rangle \tag{7(106a)}$$

and

$$\lim_{t\to+\infty}\langle a|a_{\mathbf{k}}(t)|b\rangle = Z_3^{1/2}\langle a|a_{\mathbf{k}}^{\text{out}}|b\rangle \tag{7(106b)}$$

for any two normalizable state vectors $|a\rangle$ and $|b\rangle$. Equations 7(106a) and 7(106b) are known as the asymptotic condition for a neutral scalar field*. For the time being we shall simply *postulate* the asymptotic condition and explore its consequences. The latter include the Yang–Feldman equations and the reduction formulae. Afterwards we shall return to the question of proving the asymptotic condition from first principles.

Note the presence of the $Z_3^{1/2}$ factor in 7(106a) and 7(106b). Its necessity is apparent if we consider the matrix element

$$\langle 0|a_{\mathbf{k}}(t)|\mathbf{q}\rangle$$

for which we have a degenerate case of the asymptotic condition, namely

$$\begin{aligned}\langle 0|a_{\mathbf{k}}(t)|\mathbf{q}\rangle &= i\int d^3x f_{\mathbf{k}}^*(x)\overleftrightarrow{\partial}_t\langle 0|\phi(x)|\mathbf{q}\rangle\\ &= Z_3^{1/2}\, i\int d^3x f_{\mathbf{k}}^*(x)\overleftrightarrow{\partial}_t f_{\mathbf{q}}(x)\\ &= Z_3^{1/2}\delta_{\mathbf{kq}}\\ &= Z_3^{1/2}\langle 0|a_{\mathbf{k}}^{\text{in}}|\mathbf{q}\rangle\end{aligned}$$

* Strictly speaking, the asymptotic condition as formulated by Lehmann, Symanzik and Zimmermann applies to the operators 7(105), with the additional modification $f_{\mathbf{k}}\to f_\alpha$ where f_α is a *normalizable* solution of the Klein–Gordon equation, i.e. a *wave-packet* solution. In practice, however, we may approximate these wave packets by plane waves.

for *all* t, owing to 7(99a) and 7(99b). Frequently the asymptotic condition is formulated in terms of the renormalized field operator $\phi_R = Z_3^{-1/2}\phi$ introduced in Section 6-4. In that case, of course, the factor $Z_3^{1/2}$ will not appear explicitly: we have in place of 7(106a)

$$\lim_{t\to-\infty} \langle a|a_{\mathbf{k}}^R(t)|b\rangle = \langle a|a_{\mathbf{k}}^{\text{in}}|b\rangle$$

To recapitulate, the asymptotic condition must be applied to the expansion coefficients $a_{\mathbf{k}}(t)$ rather than the field operator $\phi(x)$ themselves, to avoid taking undefined limits of oscillating exponentials. Furthermore, the condition requires a proportionality factor $Z_3^{1/2}$ for consistency with 7(99a) and 7(99b). Last but not least, it can only be applied to the *matrix elements* of ϕ as in 7(106a) and 7(106b); this is known as *weak operator convergence.* The stronger statement

$$a_{\mathbf{k}}(t) \xrightarrow[t\to-\infty]{} Z_3^{1/2} a_{\mathbf{k}}^{\text{in}}$$

known as *strong operator convergence*, immediately runs into contradictions. It would mean that, for example

$$[a_{\mathbf{k}}(t), a_{\mathbf{k}'}^{\dagger}(t)] \xrightarrow[t\to-\infty]{} Z_3[a_{\mathbf{k}}^{\text{in}}, a_{\mathbf{k}'}^{\text{in}\dagger}]$$

On the other hand, the commutator $[a_{\mathbf{k}}(t), a_{\mathbf{k}'}^{\dagger}(t)]$ is given by 7(103) in terms of the *equal-time* commutation relations for $\phi(x)$. As the latter are the same as for free fields we have

$$[a_{\mathbf{k}}(t), a_{\mathbf{k}'}^{\dagger}(t)] = \delta_{\mathbf{k}\mathbf{k}'} = [a_{\mathbf{k}}^{\text{in}}, a_{\mathbf{k}'}^{\text{in}\dagger}]$$

for all times. To have consistency we would therefore have to set $Z_3 = 1$, and we shall see in Section 8-2 that this is impossible for interacting fields.

The formulation of the asymptotic condition for a Dirac field follows the same pattern as for the scalar field. Defining

$$\begin{aligned} u_{\mathbf{k}\sigma}(x) &= \frac{1}{\sqrt{V}}\sqrt{\frac{m}{E_{\mathbf{k}}}}\, u_{\mathbf{k}\sigma}\, e^{ik.x} \\ v_{\mathbf{k}\sigma}(x) &= \frac{1}{\sqrt{V}}\sqrt{\frac{m}{E_{\mathbf{k}}}}\, v_{\mathbf{k}\sigma}\, e^{-ik.x} \end{aligned} \qquad 7(107)$$

we write the expansions 7(35a) and 7(35b) in the form

$$\psi_{\text{in}}(x) = \sum_{\mathbf{k}} (c_{\mathbf{k}\sigma}^{\text{in}} u_{\mathbf{k}\sigma}(x) + d_{\mathbf{k}\sigma}^{\text{in}\,\dagger} v_{\mathbf{k}\sigma}(x)) \qquad 7(108a)$$

$$\psi_{\text{in}}^{\dagger}(x) = \sum_{\mathbf{k}} (c_{\mathbf{k}\sigma}^{\text{in}\,\dagger} u_{\mathbf{k}\sigma}^{\dagger}(x) + d_{\mathbf{k}\sigma}^{\text{in}} v_{\mathbf{k}\sigma}^{\dagger}(x)) \qquad 7(108b)$$

with the inversions

$$c_{\mathbf{k}\sigma}^{\text{in}} = \int d^3x u_{\mathbf{k}\sigma}{}^{\dagger}(x)\psi_{\text{in}}(x) \qquad 7(109a)$$

$$c_{\mathbf{k}\sigma}^{\text{in}\,\dagger} = \int d^3x \psi_{\text{in}}{}^{\dagger}(x)u_{\mathbf{k}\sigma}(x) \qquad 7(109b)$$

$$d_{\mathbf{k}\sigma}^{\text{in}} = \int d^3x \psi_{\text{in}}{}^{\dagger}(x)v_{\mathbf{k}\sigma}(x) \qquad 7(109c)$$

$$d_{\mathbf{k}\sigma}^{\text{in}\,\dagger} = \int d^3x v_{\mathbf{k}\sigma}{}^{\dagger}(x)\psi_{\text{in}}(x) \qquad 7(109d)$$

formally identical to 3(155a), 3(155b), 3(155c) and 3(155d). The statement of the asymptotic condition is then

$$\lim_{t\to-\infty} \langle a|c_{\mathbf{k}\sigma}(t)|b\rangle = Z_2^{1/2}\langle a|c_{\mathbf{k}\sigma}^{\text{in}}|b\rangle \qquad 7(110a)$$

$$\lim_{t\to+\infty} \langle a|c_{\mathbf{k}\sigma}(t)|b\rangle = Z_2^{1/2}\langle a|c_{\mathbf{k}\sigma}^{\text{out}}|b\rangle \qquad 7(110b)$$

for arbitrary normalizable states $|a\rangle$ and $|b\rangle$, together with corresponding relations for the antiparticle operators $d_{\mathbf{k}\sigma}$. The operators $c_{\mathbf{k}\sigma}(t)$ are defined by 7(109a) with $\psi_{\text{in}}(x)$ replaced by $\psi(x)$.

We now turn to some direct applications of the asymptotic condition.

Yang–Feldman Equations As a first application, let us recover the Yang–Feldman equations for the scalar field. From the asymptotic condition 7(106a) we have

$$Z_3^{1/2}\langle a|a_{\mathbf{k}}^{\text{in}}|b\rangle = \lim_{t\to-\infty} \langle a|a_{\mathbf{k}}(t)|b\rangle$$

$$= \lim_{t\to-\infty} i\int d^3x f_{\mathbf{k}}^*(x)\overleftrightarrow{\partial}_t\langle a|\phi(x)|b\rangle \qquad 7(111)$$

Next we use the identity

$$\int d^3x \int_{-\infty}^{t_1} dt\partial_t F(x) = \int_{t=t_1} d^3x F(x) - \int_{t=-\infty} d^3x F(x) \qquad 7(112)$$

to write 7(111) in the form

$$Z_3^{1/2}\langle a|a_{\mathbf{k}}^{\text{in}}|b\rangle = i\int_{t=t_1} d^3x f_{\mathbf{k}}^*(x)\overleftrightarrow{\partial}_t\langle a|\phi(x)|b\rangle$$

$$-i\int d^3x\int_{-\infty}^{t_1} dt\,\partial_t\{f_{\mathbf{k}}^*\overleftrightarrow{\partial}_t\langle a|\phi(x)|b\rangle\}$$

$$= \langle a|a_{\mathbf{k}}(t_1)|b\rangle + i\int d^3x\int_{-\infty}^{t_1} dt\,\partial_t^2 f_{\mathbf{k}}^*\langle a|\phi(x)|b\rangle$$

$$-i\int d^3x\int_{-\infty}^{t_1} dt f_{\mathbf{k}}^*\langle a|\partial_t^2\phi(x)|b\rangle$$

Substituting

$$\partial_t^2 f_{\mathbf{k}}^* = (\boldsymbol{\nabla}^2 - \mu^2)f_{\mathbf{k}}^*$$

we perform two successive integrations by parts over the spatial variables and drop surface terms. This yields

$$Z_3^{1/2}\langle a|a_{\mathbf{k}}^{\text{in}}|b\rangle = \langle a|a_{\mathbf{k}}(t_1)|b\rangle + i\int d^3x\int_{-\infty}^{t_1} dt f_{\mathbf{k}}^*(x)(\Box - \mu^2)\langle a|\phi(x)|b\rangle \quad 7(113a)$$

In a similar way we can obtain

$$Z_3^{1/2}\langle a|a_{\mathbf{k}}^{\text{in}\dagger}|b\rangle = \langle a|a_{\mathbf{k}}^{\dagger}(t_1)|b\rangle - i\int d^3x\int_{-\infty}^{t_1} dt f_{\mathbf{k}}(x)(\Box - \mu^2)\langle a|\phi(x)|b\rangle \quad 7(113b)$$

We now multiply 7(113a) by $f_{\mathbf{k}}(x_1)$, 7(113b) by $f_{\mathbf{k}}^*(x_1)$, sum over **k** and add the two resulting expressions, with the result

$$Z_3^{1/2}\langle a|\phi^{\text{in}}(x_1)|b\rangle = \langle a|\phi(x_1)|b\rangle$$

$$+i\int d^3x\int_{-\infty}^{t_1} dt\sum_{\mathbf{k}}(f_{\mathbf{k}}(x_1)f_{\mathbf{k}}^*(x) - f_{\mathbf{k}}(x)f_{\mathbf{k}}^*(x_1))(\Box - \mu^2)\langle a|\phi(x)|b\rangle \quad 7(114)$$

Now by comparison with 3(82a) and 3(82b) we have

$$\sum_{\mathbf{k}}(f_{\mathbf{k}}(x_1)f_{\mathbf{k}}^*(x) - f_{\mathbf{k}}(x)f_{\mathbf{k}}^*(x_1)) = \frac{1}{V}\sum_{\mathbf{k}}\frac{1}{2\omega_{\mathbf{k}}}(e^{ik(x_1-x)} - e^{-ik(x_1-x)})$$

$$= i\Delta^{(+)}(x_1 - x) + i\Delta^{(-)}(x_1 - x)$$

$$= i\Delta(x_1 - x)$$

so that 7(114) may be written

$$Z_3^{1/2}\langle a|\phi^{\text{in}}(x)|b\rangle = \langle a|\phi(x)|b\rangle + \int d^4y\Delta_R(x-y)\langle a|j(y)|b\rangle \quad 7(115)$$

where we have relabelled $x_1 \to x$ and $x \to y$, recalled the definition 7(56) and used the equation of motion

$$(\Box - \mu^2)\phi = j \quad 7(116)$$

Equation 7(115) is just the Yang–Feldman equation

$$\phi(x) = Z_3^{1/2}\phi^{\text{in}}(x) - \int d^4y\Delta_R(x-y)j(y) \quad 7(117\text{a})$$

which, although it is written as an operator equation, is to be understood in the sense of weak operator convergence—that is, as no more than the statement 7(115)*. An analogous procedure relates $\phi(x)$ to $\phi^{\text{out}}(x)$ by

$$\phi(x) = Z_3^{1/2}\phi^{\text{out}}(x) - \int d^4y\Delta_A(x-y)j(y) \quad 7(117\text{b})$$

from which we derive the equation relating ϕ^{in} directly to ϕ^{out};

$$\phi^{\text{out}}(x) = \phi^{\text{in}}(x) - Z_3^{-1/2}\int d^4y\Delta(x-y)j(y) \quad 7(117\text{c})$$

An important point to note is that we can recover the asymptotic conditions 7(106a) and 7(106b) from the Yang–Feldman equations 7(117a) and 7(117b). To do this we simply form the integral expressions 7(103) and 7(105) for both sides of 7(117a), take the matrix element between arbitrary states $|a\rangle$ and $|b\rangle$, and take the limit $t \to -\infty$. The interaction term will vanish, the region of integration shrinking to zero by virtue of the retarded character of Δ_R and we recover 7(106a). Similarly, 7(106b) follows from 7(117b). We shall see later how this point can be put to use to actually prove the asymptotic condition from first principles.

The attentive reader will have noted that the above Yang–Feldman equations differ from the corresponding equations derived in Section 7-3 by the appearance of extra $Z_3^{1/2}$ factors multiplying ϕ^{in} and ϕ^{out}. This apparent discrepancy requires an explanation; we shall take up this point after discussing the reduction formulae in which a similar discrepancy occurs. Of course, the factor $Z_3^{1/2}$ will not appear when one works with the *renormalized* field operator $\phi_R(x) = Z_3^{-1/2}\phi(x)$. In terms of ϕ_R, the

* The same applies to the Yang–Feldman equations 7(55a) and 7(55b), derived by the perturbation approach, although it is less obvious.

Yang–Feldman equation 7(117a), for example, reads

$$\phi_R(x) = \phi^{\text{in}}(x) - \int d^4y \Delta_R(x-y) j_R(y)$$

with $j_R = (\Box - \mu^2)\phi_R$.

Reduction Formulae We now turn to the derivation of the reduction formulae from the asymptotic condition [Lehmann (1955)]. We treat the same example as in Section 7-3, namely the scattering of two neutral spinless particles. The *S*-matrix element to be considered is

$$\langle \mathbf{k}'\mathbf{q}' \text{ out} | \mathbf{kq} \text{ in} \rangle$$

Writing $|\mathbf{kq} \text{ in}\rangle = a_{\mathbf{k}}^{\text{in}\dagger}|\mathbf{q} \text{ in}\rangle$ we get, using the asymptotic condition 7(106a) and the adjoint of 7(103),

$$\begin{aligned}\langle \mathbf{k}'\mathbf{q}' \text{ out}|\mathbf{kq} \text{ in}\rangle &= \langle \mathbf{k}'\mathbf{q}' \text{ out}|a_{\mathbf{k}}^{\text{in}\dagger}|\mathbf{q} \text{ in}\rangle \\ &= \lim_{t\to-\infty} Z_3^{-1/2}\langle \mathbf{k}'\mathbf{q}' \text{ out}|a_{\mathbf{k}}^{\dagger}(t)|\mathbf{q} \text{ in}\rangle \\ &= \lim_{t\to-\infty} Z_3^{-1/2} i \int d^3x \langle \mathbf{k}'\mathbf{q}' \text{ out}|\phi(x)|\mathbf{q} \text{ in}\rangle \overleftrightarrow{\partial}_t f_{\mathbf{k}}(x)\end{aligned}$$

where we have assumed that $Z_3 \neq 0$. We now appeal to the identity 7(112) for the special case $t_1 = +\infty$ to write

$$\begin{aligned}\langle \mathbf{k}'\mathbf{q}' \text{ out}|\mathbf{kq} \text{ in}\rangle = \lim_{t\to+\infty} Z_3^{-1/2} i \int d^3x \langle \mathbf{k}'\mathbf{q}' \text{ out}|\phi(x)|\mathbf{q} \text{ in}\rangle \overleftrightarrow{\partial}_t f_{\mathbf{k}}(x) \\ - Z_3^{-1/2} i \int d^3x \int_{-\infty}^{\infty} dt \partial_t \{\langle \mathbf{k}'\mathbf{q}' \text{ out}|\phi(x)|\mathbf{q} \text{ in}\rangle \overleftrightarrow{\partial}_t f_{\mathbf{k}}(x)\}\end{aligned} \quad 7(118)$$

The second term on the right-hand side can now be handled in the same way as in the derivation of the Yang–Feldman equation; it becomes simply

$$-Z_3^{-1/2} i \int d^4x f_{\mathbf{k}}(x)(\Box - \mu^2)\langle \mathbf{k}'\mathbf{q}' \text{ out}|\phi(x)|\mathbf{q} \text{ in}\rangle$$

On the other hand, the first term on the right-hand side of 7(118) is, by 7(106b), just equal to

$$\begin{aligned}\langle \mathbf{k}'\mathbf{q}' \text{ out}|a_{\mathbf{k}}^{\text{out}\dagger}|\mathbf{q} \text{ in}\rangle &= \langle \mathbf{k}'\mathbf{q}' \text{ out}|a_{\mathbf{k}}^{\text{out}\dagger}|\mathbf{q} \text{ out}\rangle \\ &= \langle \mathbf{k}'\mathbf{q}' \text{ out}|\mathbf{kq} \text{ out}\rangle \\ &= \delta_{\mathbf{kk}'}\delta_{\mathbf{qq}'} + \delta_{\mathbf{kq}'}\delta_{\mathbf{qk}'}\end{aligned}$$

where we have used the fact that for one-particle states $|\mathbf{q}\,\text{in}\rangle = |\mathbf{q}\,\text{out}\rangle$. Combining results, we have

$$\langle \mathbf{k}'\mathbf{q}'\,\text{out}|\mathbf{kq}\,\text{in}\rangle = \langle \mathbf{k}'\mathbf{q}'\,\text{out}|\mathbf{kq}\,\text{out}\rangle$$
$$-Z_3^{-1/2} i \int d^4x f_{\mathbf{k}}(x)(\Box - \mu^2)\langle \mathbf{k}'\mathbf{q}'\,\text{out}|\phi(x)|\mathbf{q}\rangle \quad 7(119)$$

We have thus 'converted' or *reduced* the particle of momentum $\mathbf{k}$ from the state vector $|\mathbf{kq}\rangle$ into the field operator $\phi(x)$. Proceeding in the same fashion, we can reduce the remaining particles from the state vectors so as to obtain the complete reduction formula expressing the S-matrix element in terms of time-ordered products of field operators. To reduce the $\mathbf{k}'$ particle, for example, we write

$$\langle \mathbf{k}'\mathbf{q}'\,\text{out}|\phi(x)|\mathbf{q}\rangle = \langle \mathbf{q}'|a_{\mathbf{k}'}^{\text{out}}\phi(x)|\mathbf{q}\rangle$$
$$= \lim_{t' \to +\infty} Z_3^{-1/2}\langle \mathbf{q}'|a_{\mathbf{k}'}(t')\phi(x)|\mathbf{q}\rangle$$
$$= \lim_{t' \to +\infty} Z_3^{-1/2} i \int d^3x' f_{\mathbf{k}'}^*(x')\overleftrightarrow{\partial}_{t'}\langle \mathbf{q}'|\phi(x')\phi(x)|\mathbf{q}\rangle$$

Since t' is equal to $+\infty$ in this last expression, we can replace $\phi(x')\phi(x)$ by $T\phi(x')\phi(x)$ in the matrix element and get

$$\langle \mathbf{k}'\mathbf{q}'\,\text{out}|\phi(x)|\mathbf{q}\rangle = \lim_{t' \to -\infty} Z_3^{-1/2} i \int d^3x' f_{\mathbf{k}'}^*(x')\overleftrightarrow{\partial}_{t'}\langle \mathbf{q}'|T\phi(x')\phi(x)|\mathbf{q}\rangle$$
$$+Z_3^{-1/2} i \int d^3x' \int_{-\infty}^{\infty} dt'\partial_{t'}\{f_{\mathbf{k}'}^*(x')\overleftrightarrow{\partial}_{t'}\langle \mathbf{q}'|T\phi(x')\phi(x)|\mathbf{q}\rangle\} \quad 7(120)$$

We now observe that the first term on the right-hand side will give a vanishing contribution when substituted into 7(119). To see this, we note that it features the time-ordered product $T\phi(x')\phi(x)$ for $x_0' = -\infty$; the latter may therefore be replaced by $\phi(x)\phi(x')$, giving

$$\lim_{t' \to -\infty} Z_3^{-1/2} i \int d^3x' f_{\mathbf{k}'}^*(x')\overleftrightarrow{\partial}_{t'}\langle \mathbf{q}'|\phi(x)\phi(x')|\mathbf{q}\rangle = \langle \mathbf{q}'|\phi(x)a_{\mathbf{k}'}|\mathbf{q}\rangle$$
$$= \delta_{\mathbf{k}'\mathbf{q}}\langle \mathbf{q}'|\phi(x)|0\rangle$$

Since by 7(99b) this is just proportional to $f_{\mathbf{q}'}^*(x)$, it will give zero when operated upon by the Klein–Gordon operator in 7(119). As for the second term on the right-hand side of 7(120), it can be manipulated into the form

$$-Z_3^{-1/2}i\int d^4x' f^*_{\mathbf{k}'}(x')(\Box'-\mu^2)\langle\mathbf{q}'|T\phi(x')\phi(x)|\mathbf{q}\rangle$$

by the same procedure as before, so that substituting into 7(119) we get

$$\begin{aligned}\langle\mathbf{k}'\mathbf{q}'\text{ out}|\mathbf{kq}\text{ in}\rangle &= \langle\mathbf{k}'\mathbf{q}'\text{ out}|\mathbf{kq}\text{ out}\rangle\\ &\quad +(-Z_3^{-1/2}i)^2\int d^4x'\,d^4x\,f^*_{\mathbf{k}'}(x')f_{\mathbf{k}}(x)(\Box'-\mu^2)(\Box-\mu^2)\\ &\quad\times\langle\mathbf{q}'|T\phi(x')\phi(x)|\mathbf{q}\rangle \qquad 7(121)\end{aligned}$$

This process can evidently be continued; the final result is

$$\begin{aligned}\langle\mathbf{k}'\mathbf{q}'\text{ out}|\mathbf{kq}\text{ in}\rangle &= \langle\mathbf{k}'\mathbf{q}'\text{ out}|\mathbf{kq}\text{ out}\rangle\\ &\quad +(-Z_3^{-1/2}i)^4\int d^4x'\,d^4y'\,d^4x\,d^4y\,f^*_{\mathbf{k}'}(x')f^*_{\mathbf{q}'}(y')f_{\mathbf{k}}(x)f_{\mathbf{q}}(y)\\ &\quad\times(\Box_{x'}-\mu^2)(\Box_{y'}-\mu^2)(\Box_x-\mu^2)(\Box_y-\mu^2)\\ &\quad\times\langle 0|T\phi(x')\phi(y')\phi(x)\phi(y)|0\rangle \qquad 7(122)\end{aligned}$$

which again differs from 7(82) in the appearance of extra $Z_3^{-1/2}$ factors. (The phase factor $\langle 0\text{ out}|0\text{ in}\rangle$ which appears in 7(82) has been consistently neglected in the present calculation.) Also note that 7(122) may be written in terms of renormalized field operators ϕ_R, in which case the $Z_3^{-1/2}$ factors no longer appear.

It is worth stressing at this point the importance of *microscopic causality*—the commutativity of the field operator $\phi(x)$ at space-like separations. Microscopic causality ensures the Lorentz invariance of the time-ordered product $T\phi(x')\phi(y')\phi(x)\phi(y)$ and this in turn guarantees the Lorentz invariance of the reduction formula 7(122).

The extension of the reduction technique to more general scattering processes is straightforward. For example, the S-matrix element $\langle\mathbf{k}', \mathbf{p}'\sigma'\text{ out}|\mathbf{k}, \mathbf{p}\sigma\text{ in}\rangle$ for scattering of neutral spin 0 bosons ($\mathbf{k}$) on spin $\frac{1}{2}$ fermions ($\mathbf{p}$, σ) is given by

$$\begin{aligned}\langle\mathbf{k}', \mathbf{p}'\sigma'\text{ out}|\mathbf{k}, \mathbf{p}\sigma\text{ in}\rangle &= \langle\mathbf{k}', \mathbf{p}'\sigma'\text{ out}|\mathbf{k}, \mathbf{p}\sigma\text{ out}\rangle\\ &\quad +(Z_3Z_2)^{-1}\int d^4x'\,d^4y'\,d^4x\,d^4y\,f^*_{\mathbf{k}'}(x')f_{\mathbf{k}}(x)\\ &\quad\times(\Box_{x'}-\mu^2)(\Box_x-\mu^2)\bar{u}_{\mathbf{p}'\sigma'}(y')(\gamma\,.\,\partial_{y'}+m)\\ &\quad\times\langle 0|T\psi(y')\bar{\psi}(y)\phi(x')\phi(x)|0\rangle(\gamma\,.\,\overleftarrow{\partial}_y-m)u_{\mathbf{p}\sigma}(y) \qquad 7(123)\end{aligned}$$

The proof is left as an exercise [see Problem 4].

The basic importance of the reduction formulae, and hence of the asymptotic condition, is that they are entirely independent of perturbation theory. They can therefore be applied to strong interactions where perturbation theory breaks down. Moreover, they are completely independent of the detailed form of the Lagrangian. This is certainly a valuable feature, since for strong interactions we have as yet no inkling of what the correct Lagrangian is (if indeed it exists!). The price we pay for this generality is that we have no means of evaluating completely the vacuum expectation values which enter into the reduction formulae. Nevertheless, the few general properties which we do know—in particular Lorentz invariance and microscopic causality—are sufficient to derive dispersion relations [Bogoliubov (1958); Bremermann (1958); Dyson (1958); Lehmann (1958)], and these have had considerable success in dealing with strong interactions. These developments lie outside the scope of this book, as do the powerful results of axiomatic field theory*.

$Z_3^{1/2}$ Factors and the Adiabatic Hypothesis We now take up the question of the apparent discrepancy between the Yang–Feldman equations and reduction formulae derived on the basis of the asymptotic condition, and those derived from perturbation theory in Section 7-3. For example, the reduction formula 7(122) features an additional factor $(Z_3^{-1/2})^4$ not present in 7(82).

The resolution of this difficulty lies in recognizing that 7(55a), 7(55b) and 7(82) were derived on the basis of interaction picture perturbation theory, and the latter relies implicitly on the use of the adiabatic hypothesis. Thus the perturbative expansions of the right-hand side of 7(55a), 7(55b) and 7(82) will always feature a factor $e^{\varepsilon t}$ multiplying the interaction Hamiltonian $H_I^{ip}(t)$ wherever it appears. When this factor is properly taken into account, results in the limit $\varepsilon \to 0$ will always agree with the results of the non-perturbative formulae 7(117a), 7(117b) or 7(122). In fact the problem encountered here is essentially identical to the problem of correctly computing the wave function renormalization factor for external lines discussed in Section 6-4.

Consider for example the reduction formula 7(82). If we recall its diagrammatic interpretation, we observe that each external boson propagator in Fig. 7.3 has a residue Z_3 at the one-particle pole $k^2 = -\mu^2$. Hence a naive application of 7(82)—neglecting the adiabatic switching-off

* See for example L. Klein (ed.), *Dispersion Relations and the Abstract Approach to Field Theory*, Gordon & Breach, New York, 1961; R. F. Streater and A. S. Wightman, *PCT, Spin Statistics and All That*, Benjamin, New York, (1964); G. Barton, *Introduction to Advanced Field Theory*, Wiley, New York, (1963).

factor $e^{\varepsilon t}$ that is concealed in the perturbative expansion of $\langle 0|T\phi\phi\phi\phi|0\rangle$—would yield a wave function renormalization factor Z_3 for each external line. The correct factor $Z_3^{1/2}$ is obtained by taking into account the adiabatic switching-off factor; then, as indicated in Section 6-4,

$$(k^2+\mu^2)\Delta'_F(k^2,\varepsilon)|_{k^2=-\mu^2} \xrightarrow[\varepsilon\to 0]{} -iZ_3^{1/2} \qquad 7(124)$$

On the other hand, the non-perturbative formula 7(122) gives the correct factor $Z_3^{1/2}$ directly, owing to the presence of the $Z_3^{-1/2}$ factors.

This difference between results based on perturbation theory on the one hand, and the asymptotic condition on the other, is best illustrated by applying the two methods to the one-particle wave function $\langle 0|\phi(x)|\mathbf{k}\rangle$. If we apply the asymptotic condition we get the reduction formula

$$\langle 0|\phi(x)|\mathbf{k}\rangle = -iZ_3^{-1/2}\int d^4y \frac{1}{\sqrt{2\omega_{\mathbf{k}}V}} e^{ik.y}(\Box_y-\mu^2)\langle 0|T\phi(x)\phi(y)|0\rangle \qquad 7(125)$$

for $k^2 = -\mu^2$, which relates the one-particle wave function to the boson propagator function

$$\Delta'_F(x-y) = \langle 0|T\phi(x)\phi(y)|0\rangle$$

Introducing the Fourier transform $\Delta'_F(k)$ by

$$\Delta'_F(x-y) = \frac{1}{(2\pi)^4}\int e^{ik.(x-y)}\Delta'_F(k)\, d^4k$$

gives

$$\langle 0|\phi(x)|\mathbf{k}\rangle = iZ_3^{-1/2}\frac{1}{\sqrt{2\omega_{\mathbf{k}}V}} e^{ik.x}(k^2+\mu^2)\Delta'_F(k)\Big|_{k^2=-\mu^2} \qquad 7(126)$$

and, since by 7(96c)

$$\Delta'_F(k) = \frac{-iZ_3}{k^2+\mu^2} + \text{terms regular at } k^2 = -\mu^2$$

we recover the result

$$\langle 0|\phi(x)|\mathbf{k}\rangle = Z_3^{1/2}\frac{1}{\sqrt{2\omega_{\mathbf{k}}V}} e^{ik.x}$$

Alternatively we can proceed as in Section 7-3, basing the reduction formula on perturbation theory. Writing

$$\langle 0|\phi(x)|\mathbf{k}\rangle = (0|T\phi^{ip}(x)S|\mathbf{k})$$

$$= \int d^4y (0|T\phi^{ip}(x)\frac{\delta}{\delta\phi^{ip}(y)}S|0)(0|\phi^{ip}(y)|\mathbf{k}) \qquad 7(127)$$

we use 7(80) to get

$$\langle 0|\phi(x)|\mathbf{k}\rangle = -i\int d^4y(\Box_y-\mu^2)(0|T\phi^{ip}(x)\phi^{ip}(y)S|0)(0|\phi^{ip}(y)|\mathbf{k})$$

$$= -i\int d^4y\frac{1}{\sqrt{2\omega_{\mathbf{k}}V}}\,e^{ik.y}(\Box_y-\mu^2)\langle 0|T\phi(x)\phi(y)|0\rangle \quad 7(128)$$

in which the explicit $Z_3^{-1/2}$ factor is missing. A naive application of 7(128) would therefore yield

$$\langle 0|\phi(x)|\mathbf{k}\rangle = i\frac{1}{\sqrt{2\omega_{\mathbf{k}}V}}\,e^{ik.x}(k^2+\mu^2)\Delta_F'(k)\Big|_{k^2=-\mu^2}$$

$$= Z_3\frac{1}{\sqrt{2\omega_{\mathbf{k}}V}}\,e^{ik.x}$$

instead of 7(99a). However, we have not taken into account the adiabatic switching-off factor concealed in the perturbation expansion of the right-hand side of 7(128). When this is done we get, by virtue of 7(124).

$$\langle 0|\phi(x)|\mathbf{k}\rangle = i\frac{1}{\sqrt{2\omega_{\mathbf{k}}V}}\,e^{ik.x}\lim_{\varepsilon\to 0}(k^2+\mu^2)\Delta_F'(k,\varepsilon)\Big|_{k^2=-\mu^2}$$

$$= Z_3^{1/2}\frac{1}{\sqrt{2\omega_{\mathbf{k}}V}}\,e^{ik.x}$$

in agreement with 7(99a).

Proof of the Asymptotic Condition We now turn to the important question of deriving the asymptotic conditions 7(106a) and 7(106b) from first principles. We have seen earlier that the asymptotic condition can be recovered from Yang–Feldman equations*. The latter were of course derived from the asymptotic condition itself, but if we can construct an independent nonperturbative proof of the Yang–Feldman equations we can evidently use the above fact to deduce the asymptotic condition. This is the procedure followed by Zimmermann (1958).

To proceed with the proof, let us *define* Heisenberg fields ϕ^{in} and ϕ^{out} in terms of the field ϕ by the equations

$$Z_3^{1/2}\phi^{\text{in}}(x) = \phi(x)+\int \Delta_R(x-y)(\Box_y-\mu^2)\phi(y)\,d^4y \qquad 7(129a)$$

$$Z_3^{1/2}\phi^{\text{out}}(x) = \phi(x)+\int \Delta_A(x-y)(\Box_y-\mu^2)\phi(y)\,d^4y \qquad 7(129b)$$

* See the paragraph following 7(117c).

where Z_3 is assumed to be nonzero and where 7(129a) and 7(129b) are to be understood in the sense of weak operator convergence—that is, they have meaning only when sandwiched between two normalizable state vectors. It follows from 7(129a) and 7(129b) that

$$\lim_{t\to-\infty} \langle a|a_{\mathbf{k}}(t)|b\rangle = Z_3^{1/2}\langle a|a_{\mathbf{k}}^{\text{in}}|b\rangle \qquad 7(130\text{a})$$

$$\lim_{t\to+\infty} \langle a|a_{\mathbf{k}}(t)|b\rangle = Z_3^{1/2}\langle a|a_{\mathbf{k}}^{\text{out}}|b\rangle \qquad 7(130\text{b})$$

with the usual definitions

$$a_{\mathbf{k}}(t) = i\int d^3x f_{\mathbf{k}}^*(\mathbf{x}, t)\overleftrightarrow{\partial}_t\phi(\mathbf{x}, t)$$

$$a_{\mathbf{k}}^{\substack{\text{in}\\\text{out}}} = i\int d^3x f_{\mathbf{k}}^*(\mathbf{x}, t)\overleftrightarrow{\partial}_t\phi^{\substack{\text{in}\\\text{out}}}(\mathbf{x}, t)$$

Now to show that 7(129a) and 7(129b) really are the Yang–Feldman equations and that 7(130a) and 7(130b) are the asymptotic conditions we want, we must prove that ϕ^{in} and ϕ^{out} as defined by 7(129a) and 7(129b) are just the usual in and out fields. That is, we must show that ϕ^{in} and ϕ^{out} obey the correct free field equations of motion and commutation relations.

First we note that the free field equations

$$(\Box - \mu^2)\phi^{\text{in}} = (\Box - \mu^2)\phi^{\text{out}} = 0 \qquad 7(131)$$

are an immediate consequence of the constructions 7(129a) and 7(129b). Next, using 7(131), we easily determine the matrix elements of ϕ^{in} and ϕ^{out} between the vacuum state and an arbitrary physical state $|a\rangle$. Denoting the energy momentum eigenvalue of $|a\rangle$ by k^a, i.e.

$$P_\mu|a\rangle = k_\mu^a|a\rangle$$

we have, since ϕ^{in} is a Heisenberg field

$$\langle a|\phi^{\text{in}}(x)|0\rangle = \mathrm{e}^{-ik^a.x}\langle a|\phi^{\text{in}}(0)|0\rangle$$

Hence

$$i\partial_\mu\langle a|\phi^{\text{in}}(x)|0\rangle = k_\mu^a\langle a|\phi^{\text{in}}(x)|0\rangle$$

and

$$(-\partial_\mu\partial_\mu + \mu^2)\langle a|\phi^{\text{in}}(x)|0\rangle = (k_\mu^a k_\mu^a + \mu^2)\langle a|\phi^{\text{in}}(x)|0\rangle$$

Since the left-hand side is zero by 7(131), the only nonzero matrix elements $\langle a|\phi^{\text{in}}(x)|0\rangle$ are those for which $(k^a)^2 = -\mu^2$, that is, for which $|a\rangle$ is the one-particle state of mass μ. For these non zero matrix elements we have

$$\langle \mathbf{k}|\phi^{\text{in}}(x)|0\rangle = \frac{1}{\sqrt{V}}\frac{1}{\sqrt{2\omega_{\mathbf{k}}}}\,\mathrm{e}^{-ik.x} \qquad 7(132)$$

with $k_0 = \omega_{\mathbf{k}} = (\mathbf{k}^2+\mu^2)^{1/2}$, since by the definition 7(129a) and the normalization 7(99a) and 7(99b) we have

$$\begin{aligned} Z_3^{1/2}\langle \mathbf{k}|\phi^{\text{in}}(x)|0\rangle &= \langle \mathbf{k}|\phi(x)|0\rangle + \int \Delta_R(x-y)(\Box_y-\mu^2)\langle \mathbf{k}|\phi(y)|0\rangle\, d^4y \\ &= Z_3^{1/2}\frac{1}{\sqrt{V}}\frac{1}{\sqrt{2\omega_{\mathbf{k}}}}\,\mathrm{e}^{-ik.x} \end{aligned}$$

The integral term vanishes, since

$$\begin{aligned} (\Box_y-\mu^2)\langle \mathbf{k}|\phi(y)|0\rangle &= (\Box_y-\mu^2)\,\mathrm{e}^{-ik.x}\langle \mathbf{k}|\phi(0)|0\rangle \\ &= -(k^2+\mu^2)\langle \mathbf{k}|\phi(y)|0\rangle = 0 \end{aligned}$$

Thus ϕ^{in} has the correct property that, operating on the vacuum, it creates properly normalized one-particle states only. The same also applies to the out-field defined by 7(129b).

The above results are necessary for the identification of ϕ^{in} and ϕ^{out} as the in and out fields, but they are not sufficient. We must still show that ϕ^{in} and ϕ^{out} satisfy the free field commutation relations

$$[\phi^{\text{in}}(x), \phi^{\text{in}}(y)] = [\phi^{\text{out}}(x), \phi^{\text{out}}(y)] = i\Delta(x-y) \qquad 7(133)$$

As a preliminary, let us calculate the vacuum expectation value of the commutators, which we are equipped to do by 7(132). We find

$$\begin{aligned} \langle 0|\phi^{\text{in}}(x)\phi^{\text{in}}(y)|0\rangle &= \sum_{\mathbf{k}}\langle 0|\phi^{\text{in}}(x)|\mathbf{k}\rangle\langle \mathbf{k}|\phi^{\text{in}}(y)|0\rangle \\ &= \frac{1}{V}\sum_{\mathbf{k}}\frac{1}{2\omega_{\mathbf{k}}}\,\mathrm{e}^{ik.(x-y)} \\ &= \frac{1}{(2\pi)^3}\int\frac{d^3k}{2\omega_{\mathbf{k}}}\,\mathrm{e}^{ik.(x-y)} \\ &= i\Delta^{(+)}(x-y) \end{aligned}$$

and

$$\langle 0|\phi^{\text{in}}(y)\phi^{\text{in}}(x)|0\rangle = -i\Delta^{(-)}(x-y)$$

and, hence

$$\langle 0|[\phi^{\text{in}}(x), \phi^{\text{in}}(y)]|0\rangle = i\Delta(x-y) \qquad 7(134a)$$

Similarly

$$\langle 0|[\phi^{\text{out}}(x), \phi^{\text{out}}(y)]|0\rangle = i\Delta(x-y) \qquad 7(134b)$$

We are now in a position to show that ϕ^{in} and ϕ^{out} satisfy the free field commutation relations 7(133). We proceed in two steps. First we show that

$$[\phi^{\text{in}}(x), \phi^{\text{in}}(y)] = [\phi^{\text{out}}(x), \phi^{\text{out}}(y)]$$

and, secondly, that the commutator of the in-fields is a *c*-number, so that

$$[\phi^{\text{in}}(x), \phi^{\text{in}}(y)] = \langle 0|[\phi^{\text{in}}(x), \phi^{\text{in}}(y)]|0\rangle$$

To prove the first statement we start with the identity

$$\int d^4x \int d^4y\, f_{\mathbf{q}}^*(x) f_{\mathbf{k}}(y)(\Box_x-\mu^2)(\Box_y-\mu^2)T\phi(x)\phi(y)$$

$$= \int d^4y \int d^4x\, f_{\mathbf{q}}^*(x) f_{\mathbf{k}}(y)(\Box_x-\mu^2)(\Box_y-\mu^2)T\phi(x)\phi(y) \qquad 7(135)$$

This simple interchange in the order of integration actually requires justification. In particular, the microcausality requirement that $\phi(x)$ and $\phi(y)$ commute for spacelike separations must be invoked [Zimmermann (1958)]. This will be assumed. Equation 7(135) and all equations following below will be understood in the sense of weak operator convergence, that is, as valid only when sandwiched between two normalizable state vectors. We now manipulate the left-hand side of 7(135). By the reverse procedure to that which gave the reduction formulae, we have

$$-i \int d^4y\, f_{\mathbf{k}}(y)(\Box_y-\mu^2)T\phi(x)\phi(y)$$

$$= -i \int d^3y \int_{-\infty}^{\infty} dy_0 \frac{\partial}{\partial y_0}\left\{T\phi(x)\phi(y)\frac{\overleftrightarrow{\partial}}{\partial y_0} f_{\mathbf{k}}(y)\right\}$$

$$= -i \int d^3y\, T\phi(x)\phi(y)\frac{\overleftrightarrow{\partial}}{\partial y_0} f_{\mathbf{k}}(y)\Big|_{y_0=-\infty}^{y_0=+\infty}$$

$$= -\ T\phi(x)a_{\mathbf{k}}{}^{\dagger}(y_0)\Big|_{y_0=-\infty}^{y_0=+\infty}$$

$$= \phi(x)a_{\mathbf{k}}^{\text{in}\dagger} - a_{\mathbf{k}}^{\text{out}\dagger}\phi(x) \qquad 7(136)$$

Dealing with the integral over x in the same way, we get

$$\int d^4x \int d^4y\, f_{\mathbf{q}}^*(x) f_{\mathbf{k}}(y)(\Box_x - \mu^2)(\Box_y - \mu^2) T\phi(x)\phi(y)$$
$$= a_{\mathbf{k}}^{\text{out}\dagger} a_{\mathbf{q}}^{\text{out}} - a_{\mathbf{k}}^{\text{out}\dagger} a_{\mathbf{q}}^{\text{in}} - a_{\mathbf{q}}^{\text{out}} a_{\mathbf{k}}^{\text{in}\dagger} + a_{\mathbf{q}}^{\text{in}} a_{\mathbf{k}}^{\text{in}\dagger}$$

Correspondingly, we find for the integral on the right-hand side

$$\int d^4y \int d^4x\, f_{\mathbf{q}}^*(x) f_{\mathbf{k}}(y)(\Box_x - \mu^2)(\Box_y - \mu^2) T\phi(x)\phi(y)$$
$$= a_{\mathbf{q}}^{\text{out}} a_{\mathbf{k}}^{\text{out}\dagger} - a_{\mathbf{q}}^{\text{out}} a_{\mathbf{k}}^{\text{in}\dagger} - a_{\mathbf{k}}^{\text{out}\dagger} a_{\mathbf{q}}^{\text{in}} + a_{\mathbf{k}}^{\text{in}\dagger} a_{\mathbf{q}}^{\text{in}}$$

so that by 7(135) we have

$$[a_{\mathbf{q}}^{\text{out}}, a_{\mathbf{k}}^{\text{out}\dagger}] = [a_{\mathbf{q}}^{\text{in}}, a_{\mathbf{k}}^{\text{in}\dagger}]$$

We can get the corresponding relations

$$[a_{\mathbf{q}}^{\text{out}\dagger}, a_{\mathbf{k}}^{\text{out}\dagger}] = [a_{\mathbf{q}}^{\text{in}\dagger}, a_{\mathbf{k}}^{\text{in}\dagger}]$$
$$[a_{\mathbf{q}}^{\text{out}}, a_{\mathbf{k}}^{\text{out}}] = [a_{\mathbf{q}}^{\text{in}}, a_{\mathbf{k}}^{\text{in}}]$$

in the same way by replacing $f_{\mathbf{q}}^* f_{\mathbf{k}}$ in 7(135) by $f_{\mathbf{q}} f_{\mathbf{k}}$ and $f_{\mathbf{q}}^* f_{\mathbf{k}}^*$ respectively. This completes the proof of the first statement:

$$[\phi^{\text{in}}(x), \phi^{\text{in}}(y)] = [\phi^{\text{out}}(x), \phi^{\text{out}}(y)]$$

To prove that the commutator is a c-number, we start with the identity

$$\int d^4x \int d^4y\, f_{\mathbf{q}}^*(x) f_{\mathbf{k}}(y)(\Box_x - \mu^2)(\Box_y - \mu^2) T\phi(x)\phi(y)\phi(z)$$
$$= \int d^4y \int d^4x\, f_{\mathbf{q}}^*(x) f_{\mathbf{k}}(y)(\Box_x - \mu^2)(\Box_y - \mu^2) T\phi(x)\phi(y)\phi(z)$$

and manipulate the integrals in exactly the same way as before. The result is

$$\phi(z)[a_{\mathbf{q}}^{\text{in}}, a_{\mathbf{k}}^{\text{in}\dagger}] = [a_{\mathbf{q}}^{\text{out}}, a_{\mathbf{k}}^{\text{out}\dagger}]\phi(z)$$

or, using our previous result,

$$[[a_{\mathbf{q}}^{\text{in}}, a_{\mathbf{k}}^{\text{in}\dagger}], \phi(z)] = 0$$

Thus the commutator $[a_{\mathbf{q}}^{\text{in}}, a_{\mathbf{k}}^{\text{in}\dagger}]$ commutes with the basic field $\phi(z)$; the same argument can be used to show that it commutes with any other basic field in the Lagrangian—i.e. $\psi(x)$ for the case of interacting mesons and nucleons. Thus $[a_{\mathbf{q}}^{\text{in}}, a_{\mathbf{k}}^{\text{in}\dagger}]$ must be a c-number*. The corresponding results for $[a_{\mathbf{q}}^{\text{in}\dagger}, a_{\mathbf{k}}^{\text{in}\dagger}]$ and $[a_{\mathbf{q}}^{\text{in}}, a_{\mathbf{k}}^{\text{in}}]$ complete the proof.

* This is often expressed by the statement that the basic fields, in terms of which the theory is formulated, form an *irreducible ring* of operators.

PROBLEMS

1. Derive the Yang–Feldman equations for $\psi(x)$ and $\bar{\psi}(x)$ in pseudoscalar coupling theory, using the perturbative approach of Section 7-2.
2. Check Eqs. 7(69) and 7(70). Exhibit the photon propagator and vertex function in quantum electrodynamics in terms of vacuum expectation values.
3. Derive all non zero matrix elements of $\psi(x)$ and $\bar{\psi}(x)$ between one-particle states and the vacuum, and between one-*anti*particle states and the vacuum.
4. Derive the reduction formula 7(123).
5. Assuming the interaction Lagrangian

$$\mathscr{L}_I(x) = ie_0\tfrac{1}{2}[\bar{\psi}(x), \gamma_\mu\psi(x)]B_\mu^{\text{ext}}(x)$$

for the coupling of a Dirac field to an external electromagnetic field $B_\mu^{\text{ext}}(x)$, calculate the vacuum expectation value of the electromagnetic current operator $j_\mu(x) = i\frac{1}{2}[\bar{\psi}(x), \gamma_\mu\psi(x)]$ to first order in B_μ^{ext}. Show that current conservation imposes the requirement

$$k_\mu \Pi_{\mu\nu}(k) = 0$$

where $\Pi_{\mu\nu}(k)$ is the vacuum polarization tensor 6(90). Show that the quadratically divergent photon self-energy term 6(103)—which violates the above requirement—disappears if one adopts the limit prescription

$$\langle 0|\tfrac{1}{2}[\bar{\psi}(x), \gamma_\mu\psi(x)]|0\rangle$$

$$= \lim_{x'\to x} \operatorname{Tr} \gamma_\mu \langle 0|T\psi(x)\bar{\psi}(x')|0\rangle \exp\left[-ie_0 \int_{x'}^{x} d\xi_\mu B_\mu(\xi)\right]$$

as the definition of the vacuum current [Schwinger (1959), Johnson (1961a, b)].
6. Show that current conservation implies $k_\mu \Pi^*_{\mu\nu}(k) = 0$ where $\Pi^*_{\mu\nu}(k)$ is the sum of all proper photon self-energy loops.

REFERENCES

Bogoliubov, N. (1958) (with B. Medvedev and M. Polivanov) *Fortschr. Physik* **6**, 169

Bremermann, H. J. (1958) (with R. Oehme and J. G. Taylor) *Phys. Rev.* **109**, 2178

Dyson, F. J. (1958) *Phys. Rev.* **110**, 1460

Gell-Mann, M. (1951) (with F. E. Low) *Phys. Rev.* **84**, 350

Johnson, K. (1961a) *Nucl. Phys.* **25**, 431

Johnson, K. (1961b) *Nuovo Cimento* **20**, 773

Källén, G. (1952) *Helv. Phys. Acta* **25**, 417

Lehmann, H. (1955) (with K. Symanzik and W. Zimmermann) *Nuovo Cimento* **1**, 205

Lehman, H. (1958) *Nuovo Cimento* **10**, 579

Lippmann, B. A. (1950) (with J. Schwinger) *Phys. Rev.* **79**, 469

Møller, C. (1945) *Kgl. Danske Vidensk. Selsk., Matt.-Fys. Medd.* **23**, No. 1

Nishijima, K. (1958) *Phys. Rev.* **111**, 995

Nishijima, K. (1965) In *Proc. Trieste Seminar, 1965* (International Atomic Energy Agency)

Schwinger, J. (1959) *Phys. Rev. Letters* **3**, 296
Van Hove, L. (1951) *Bulletin Acad. Roy. Belgique* **37**, 1055
Van Hove, L. (1952) *Physica* **18**, 145
Yang, C. N. (1950) (with D. Feldman) *Phys. Rev.* **79**, 972
Zimmermann, W. (1958) *Nuovo Cimento* **10**, 597

8

Currents, Coupling Constants and Sum Rules

8-1 Currents and renormalized coupling constants

Scattering in an External Electromagnetic Field In perturbation theory the true physical coupling constant, say e_R for spinor electrodynamics, is given in terms of the corresponding bare constant e_0 by the relation

$$e_R = Z_3^{1/2} Z_2 Z_1^{-1} e_0 \tag{8(1)}$$

where $Z_3^{1/2}$, Z_2, and Z_1^{-1} are divergent constants as defined in Section 6.4. We now propose to relate the above renormalized coupling constant to the one-particle expectation value of the electromagnetic current operator

$$-e_0 j_\mu(x) = \Box A_\mu(x) \tag{8(2)}$$

We consider the problem of an electron interacting both with photons and with an external electromagnetic field A_μ^{ext}. The total Lagrangian for this system is

$$\mathscr{L} = \mathscr{L}_0 + \mathscr{L}_I^{\text{photon}} + \mathscr{L}_I^{\text{ext}} \tag{8(3)}$$

Here $\mathscr{L}_0$ is the free Lagrangian, $\mathscr{L}_I^{\text{photon}}$ represents the interaction with the quantized electromagnetic field and $\mathscr{L}_I^{\text{ext}}$ the interaction with the external source:

$$\mathscr{L}_I^{\text{ext}}(x) = e_0 j_\mu(x) A_\mu^{\text{ext}}(x) \tag{8(4)}$$

We wish to calculate the transition amplitude for Coulomb scattering to first order in the external field, but to *all* orders in the photon-electron interaction.

Denoting the quantum fields (i.e. ψ and A_μ) appearing in j_μ by $\phi_A(A = 1 \ldots N)$, and their canonically conjugate fields by π_A, we apply a time-dependent unitary transformation

$$|a, t\rangle' = V(t)|a\rangle \tag{8(5a)}$$

$$\phi'_A(\mathbf{x}, t) = V(t)\phi_A(\mathbf{x}, t)V(t)^{-1} \qquad \pi'_A(\mathbf{x}, t) = V(t)\pi_A(\mathbf{x}, t)V(t)^{-1} \tag{8(5b)}$$

to the state vectors and field operators of the Heisenberg picture. We shall choose the new picture defined by 8(5a) and 8(5b) to be such that the time evolution of the state vectors $|a, t\rangle'$ is determined only by the external source A_μ^{ext}:

$$i\partial_t |a, t\rangle' = H_I'^{\text{ext}}(t)|a, t\rangle' \tag{8(6)}$$

whereas the ϕ'_A and π'_A obey *source-free* equations of motion

$$\dot{\phi}'_A(\mathbf{x}, t) = \frac{1}{i}[\phi'_A(\mathbf{x}, t), H'(t) - H_I'^{\text{ext}}(t)] \tag{8(7a)}$$

$$\dot{\pi}'_A(\mathbf{x}, t) = \frac{1}{i}[\pi'_A(\mathbf{x}, t), H'(t) - H_I'^{\text{ext}}(t)] \tag{8(7b)}$$

where in 8(6) and 8(7a, b) $H_I'^{\text{ext}} = VH_I^{\text{ext}}V^{-1}$ is the external source interaction Hamiltonian in the transformed picture, i.e.

$$H_I'^{\text{ext}}(t) = \int \mathscr{H}_I'^{\text{ext}}(x)\, d^3x \tag{8(8)}$$

with

$$\begin{aligned} \mathscr{H}_I'^{\text{ext}}(x) &= -\mathscr{L}_I'^{\text{ext}}(x) \\ &= -e_0 j'_\mu(x) A_\mu^{\text{ext}}(x) \end{aligned} \tag{8(9)}$$

The new picture characterized by 8(6), 8(7a) and 8(7b) can be defined by suitably choosing the unitary operator $V(t)$. We leave it as an exercise to the reader to show that if $V(t)$ satisfies

$$iV^{-1}\dot{V} = H_I^{\text{ext}} = -e_0 \int j_\mu(x) A_\mu^{\text{ext}}(x)\, d^3x \tag{8(10)}$$

then 8(6) and 8(7a, b) will be ensured*. [See Problem 1.]

Perturbation theory in the new picture can be set up by repeating the same steps as in the interaction picture. Asymptotic states at $t = \pm\infty$ are now eigenstates of the source-free Hamiltonian

$$H' - H_I'^{\text{ext}} \tag{8(11)}$$

this being a time-independent operator by virtue of 8(7a) and 8(7b):

$$\dot{H}' - \dot{H}_I'^{\text{ext}} = \frac{1}{i}[H' - H_I'^{\text{ext}}, H' - H_I'^{\text{ext}}] = 0$$

* The new picture introduced here is to be sharply distinguished from the picture introduced by Furry [Furry (1951)] in which the *state vectors* obey the source-free Schrodinger equation while the field operators interact only with the external source.

We can also require the asymptotic states to be eigenstates of the canonical momentum operator

$$\mathbf{P}' = -\sum_A \int \pi'_A \nabla \phi'_A \, d^3x \qquad 8(12)$$

The latter commutes with $H' - H_I'^{\text{ext}}$ by virtue of the space translation invariance of the source-free equations of motion. Choosing our asymptotic states to be the one-electron states

$$|\mathbf{p}\sigma\rangle \qquad 8(13a)$$

and

$$|\mathbf{q}\tau\rangle \qquad 8(13b)$$

with $p^2 = q^2 = -m^2$, we see that the probability amplitude for a transition from 8(13a) to 8(13b) under the effect of the interaction with the external source is

$$-i \int d^4x \langle \mathbf{q}\tau | H_I'^{\text{ext}}(x) | \mathbf{p}\sigma \rangle = i \int d^4x \langle \mathbf{q}\tau | e_0 j'_\mu(x) | \mathbf{p}\sigma \rangle A_\mu^{\text{ext}}(x) \qquad 8(14)$$

to first order in the external field, but to all orders in the *quantum* electromagnetic interaction. The effects of the latter, represented by 8(11), are implicitly contained in the asymptotic states $|\mathbf{p}\sigma\rangle$ and $|\mathbf{q}\tau\rangle$ and in the current operator $j'_\mu(x)$.

Writing

$$j'_\mu(x) = e^{[-i\mathbf{P}' \cdot \mathbf{x} + i(H' - H_I'^{\text{ext}})x_0]} j'_\mu(0) \, e^{[i\mathbf{P}' \cdot \mathbf{x} - i(H' - H_I'^{\text{ext}})x_0]} \qquad 8(15)$$

we can factor out the space-time dependence of the matrix element $\langle \mathbf{q}\tau | j'_\mu(x) | \mathbf{p}\sigma \rangle$ in the usual way and express the transition amplitude 8(14) in the form

$$i \langle \mathbf{q}\tau | e_0 j_\mu(0) | \mathbf{p}\sigma \rangle A_\mu^{\text{ext}}(p-q) \qquad 8(16)$$

where we have suppressed the prime on j_μ and set

$$A_\mu^{\text{ext}}(p-q) = \int d^4x \, e^{i(p-q).x} A_\mu^{\text{ext}}(x) \qquad 8(17)$$

Now the matrix element $\langle \mathbf{q}\tau | e_0 j_\mu(0) | \mathbf{p}\sigma \rangle$ must be of the general form

$$\langle \mathbf{q}\tau | e_0 j_\mu(0) | \mathbf{p}\sigma \rangle = Z_3^{1/2} \sqrt{\frac{m}{E_\mathbf{q} V}} \, \bar{u}_{\mathbf{q}\tau} \, O_\mu(\mathbf{q}, \mathbf{p}) u_{\mathbf{p}\sigma} \sqrt{\frac{m}{E_\mathbf{p} V}} \qquad 8(18)$$

where $O_\mu(\mathbf{q}, \mathbf{p})$ is vector quantity like j_μ and a 4×4 matrix in the spinor space of the electron; the factor $Z_3^{1/2}$ has been extracted for later convenience and the factor $(m^2/E_\mathbf{q} E_\mathbf{p} V^2)^{1/2}$ is included for normalization purposes, as in 7(94a) and 7(94b). Since $O_\mu(\mathbf{q}, \mathbf{p})$ is a vector constructed from the Dirac matrices and the two 4-vectors p and q, it must be of the general form

$$O_\mu(\mathbf{q}, \mathbf{p}) = i\gamma_\mu O_1 + i\Sigma_{\mu\nu} k_\nu O_2 + k_\mu O_3 + P_\mu O_4 + i\Sigma_{\mu\nu} P_\nu O_5 \qquad 8(19)$$

where we have set $\Sigma_{\mu\nu} = [\gamma_\mu, \gamma_\nu]/2i$ and

$$k = q - p$$

$$P = q + p \qquad 8(20)$$

and where $O_i (i = 1 \ldots 5)$ are functions of the invariants $p^2, q^2, p \cdot q, \gamma \cdot p$, and $\gamma \cdot q$.

$$O_i = O_i(p^2, q^2, p \cdot q, \gamma \cdot p, \gamma \cdot q) \qquad 8(21)$$

The invariants p^2 and q^2 can be eliminated, since they are restricted to their mass shell values $p^2 = q^2 = -m^2$. Moreover, by using the commutation rules of the Dirac matrices, $\gamma \cdot p$ and $\gamma \cdot q$ can be shifted to the extreme right and left, so as to operate on $u_{\mathbf{p}\sigma}$ and $\bar{u}_{\mathbf{q}\tau}$ respectively. Using the Dirac equations

$$\gamma \cdot p u_\mathbf{p} = imu_\mathbf{p} \qquad 8(22a)$$

$$\bar{u}_\mathbf{q} \gamma \cdot q = im\bar{u}_\mathbf{q} \qquad 8(22b)$$

we can eliminate both $\gamma \cdot p$ and $\gamma \cdot q$, leaving only $p \cdot q$ which can be replaced by

$$(q-p)^2 = -2m^2 - 2p \cdot q$$

Thus we may set

$$O_i = O_i(k^2)$$

Moreover, one can easily check that

$$\bar{u}_{\mathbf{q}\tau} \Sigma_{\mu\nu} P_\nu u_{\mathbf{p}\sigma} = -ik_\mu \bar{u}_{\mathbf{q}\tau} u_{\mathbf{p}\sigma} \qquad 8(23)$$

and*

$$\bar{u}_{\mathbf{q}\tau} \Sigma_{\mu\nu} k_\nu u_{\mathbf{p}\sigma} = -2m\bar{u}_{\mathbf{q}\tau} \gamma_\mu u_{\mathbf{p}\sigma} - iP_\mu \bar{u}_{\mathbf{q}\tau} u_{\mathbf{p}\sigma} \qquad 8(24)$$

by using the Dirac equations 8(22a) and 8(22b). Thus the $i\Sigma_{\mu\nu}P_\nu$ term in 8(19) can be amalgamated with the k_μ term, while the P_μ term can be

* See Eq. (6.383).

absorbed into the $i\gamma_\mu$ and $i\Sigma_{\mu\nu}k_\nu$ terms. Thus we obtain

$$O_\mu(\mathbf{q}, \mathbf{p}) = F_1(k^2)i\gamma_\mu + F_2(k^2)i\Sigma_{\mu\nu}k_\nu + F_3(k^2)k_\mu \tag{8(25)}$$

Finally, the conservation of the electromagnetic current

$$\partial_\mu j_\mu(x) = 0 \tag{8(26)}$$

imposes the condition

$$k_\mu\langle \mathbf{q}\tau|ej_\mu(0)|\mathbf{p}\sigma\rangle = \bar{u}_{\mathbf{q}\tau}k_\mu O_\mu u_{\mathbf{p}\sigma} = 0$$

or

$$\bar{u}_{\mathbf{q}\tau}(F_1 i\gamma\,.\,k + F_2 i\Sigma_{\mu\nu}k_\mu k_\nu + F_3 k^2)u_{\mathbf{p}\sigma} = 0 \tag{8(27)}$$

The first term on the left-hand side of 8(27) vanishes by virtue of the Dirac equations 8(22a) and 8(22b), while the second term is identically zero by virtue of the antisymmetric character of $\Sigma_{\mu\nu}$. Hence, by current conservation*

$$F_3 = 0 \tag{8(28)}$$

and the most general form for the matrix element $\langle \mathbf{q}\tau|e_0 j_\mu(0)|\mathbf{p}\sigma\rangle$ is

$$\langle \mathbf{q}\tau|e_0 j_\mu(0)|\mathbf{p}\sigma\rangle = \left(\frac{m^2}{E_\mathbf{q}E_\mathbf{p}V^2}\right)^{1/2} Z_3^{1/2} \times \bar{u}_{\mathbf{q}\tau}[F_1(k^2)i\gamma_\mu + F_2(k^2)i\Sigma_{\mu\nu}k_\nu]u_{\mathbf{p}\sigma} \tag{8(29)}$$

We now insert 8(29) into 8(16). Taking the limit $q = p$, we obtain the result

$$-\frac{m}{E_\mathbf{p}V}Z_3^{1/2}F_1(0)\bar{u}_{\mathbf{p}\tau}\gamma_\mu u_{\mathbf{p}\sigma}A_\mu^{\text{ext}}(0) \tag{8(30)}$$

which, when compared with the perturbation theory result 6(370) yields the identification

$$F_1(0) = e_R \tag{8(31)}$$

where e_R is the true physical (renormalized) coupling constant. For finite $q-p$ the discussion given in Section 6.4† shows that

$$F_2(0) = \Delta\mu \tag{8(32)}$$

where $\Delta\mu$ is the anomalous magnetic moment of the electron. The functions

* The term $F_3 k_\mu$ can also be ruled out by time-reflection invariance [Ernst (1960)].
† See the discussion relative to Eq. (6.375).

$F_1(k^2)$ and $F_2(k^2)$ are known respectively as the charge and (anomalous) magnetic moment *form factors* of the electron.

We can combine 8(29) and 8(31) to get the important equality

$$\lim_{\mathbf{q}\to\mathbf{p}} \langle \mathbf{q}\tau | e_0 j_\mu(0) | \mathbf{p}\sigma \rangle = \frac{m}{E_{\mathbf{p}} V} Z_3^{1/2} e_R i \bar{u}_{\mathbf{p}\tau} \gamma_\mu u_{\mathbf{p}\sigma} \qquad 8(33)$$

which gives the expectation value of the charge operator

$$e_0 Q = -i \int e_0 j_4(x)\, d^3x \qquad 8(34)$$

for a one-electron state in the form

$$\begin{aligned}\langle \mathbf{p}\tau | e_0 Q | \mathbf{p}\sigma \rangle &= \frac{m}{E_{\mathbf{p}}} Z_3^{1/2} e_R \bar{u}_{\mathbf{p}\tau} \gamma_4 u_{\mathbf{p}\sigma} \\ &= Z_3^{1/2} e_R \delta_{\sigma\tau} \end{aligned} \qquad 8(35)$$

Perturbation Theory An alternative approach to the matrix element $\langle \mathbf{q}\tau | e_0 j_\mu(0) | \mathbf{p}\sigma \rangle$ is to apply perturbation theory based on 7(71b). Setting

$$j_\mu(0) = i\bar{\psi}(0)\gamma_\mu \psi(0)$$

we have

$$\langle \mathbf{q}\tau | i e_0 \bar{\psi}(0) \gamma_\mu \psi(0) | \mathbf{p}\sigma \rangle = (\mathbf{q}\tau | i e_0 \bar{\psi}^{ip}(0) \gamma_\mu \psi^{ip}(0) S | \mathbf{p}\sigma)$$

where the right-hand side may be expanded by means of 7(71b) and Wick's theorem; the first few terms in the series are shown in Fig. 8.1. We obtain the result

$$\begin{aligned}\langle \mathbf{q}\tau | e_0 j_\mu(0) | \mathbf{p}\sigma \rangle &= -e_0 Z_2 k^2 D_F'(k^2) \left(\frac{m^2}{E_{\mathbf{q}} E_{\mathbf{p}} V^2} \right)^{1/2} \bar{u}_{\mathbf{q}\tau} \Gamma_\mu(q, p) u_{\mathbf{p}\sigma} \\ &= -e_0 k^2 Z_3 \mathbf{D}_F'(k^2) \left(\frac{m^2}{E_{\mathbf{q}} E_{\mathbf{p}} V^2} \right)^{1/2} \bar{u}_{\mathbf{q}\tau} \mathbf{\Gamma}_\mu(q, p) u_{\mathbf{p}\sigma} \end{aligned} \qquad 8(36)$$

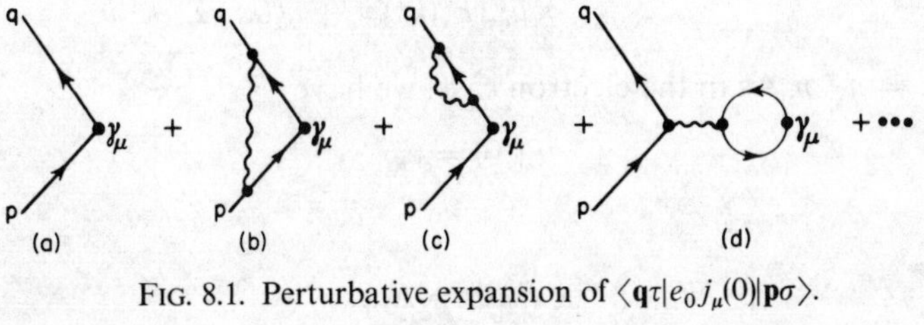

FIG. 8.1. Perturbative expansion of $\langle \mathbf{q}\tau | e_0 j_\mu(0) | \mathbf{p}\sigma \rangle$.

[See Problem 2.] The appearance of the factor Z_2 can be traced to the self energy diagrams on the external lines as, for example, in Fig. 8.1c; to get Z_2 and not Z_2^2, one must apply the relation 6(351) based on the use of the adiabatic switching-off factor $e^{\varepsilon t}$. Finally, in deriving 8(36) we have set $\Gamma_\mu = Z_1^{-1}\mathbf{\Gamma}_\mu, D'_F = Z_3\mathbf{D}_F$, and used the Ward identity $Z_2 Z_1^{-1} = 1$. Comparing 8(36) with 8(29) we see that the electromagnetic form factors are closely related to the renormalized photon vertex function $\mathbf{\Gamma}_\mu(p, q)$, taken on the electron mass shell $\gamma \,.\, p = \gamma \,.\, q = im$. From 8(36), 6(364) and 6(368), we obtain, in the limit $\mathbf{q} \to \mathbf{p}$,

$$\lim_{\mathbf{q}\to\mathbf{p}} \langle \mathbf{q}\tau | e_0 j_\mu(0) | \mathbf{p}\sigma \rangle = \frac{m}{E_\mathbf{p} V} Z_3 e_0 i \bar{u}_{\mathbf{p}\tau} \gamma_\mu u_{\mathbf{p}\sigma}$$

$$= \frac{m}{E_\mathbf{p} V} Z_3^{1/2} e_R i \bar{u}_{\mathbf{p}\tau} \gamma_\mu u_{\mathbf{p}\sigma} \qquad 8(37)$$

which agrees with 8(33).

Nucleon Electromagnetic Form Factors The results 8(29), 8(31) and 8(32) can be extended to the one-nucleon matrix elements of the electromagnetic current of the hadrons. In this case we must formally include in the total Lagrangian the (unknown) strong coupling Lagrangian $\mathscr{L}^{\text{strong}}$

$$\mathscr{L} = \mathscr{L}_0 + \mathscr{L}_I^{\text{strong}} + \mathscr{L}_I^{\text{photon}} + \mathscr{L}_I^{\text{ext}} \qquad 8(38)$$

but this does not affect the remainder of the procedure. The separation of the effects of the external source

$$\mathscr{L}_I^{\text{ext}} = e_0 J_\mu A_\mu^{\text{ext}} \qquad 8(39)$$

—where J_μ is now the electromagnetic current of the hadrons—goes through unchanged and we find, for a one-proton state for example,

$$^p\langle \mathbf{q}\tau | e_0 J_\mu(0) | \mathbf{p}\sigma \rangle^p = \left(\frac{m^2}{E_\mathbf{q} E_\mathbf{p} V^2} \right)^{1/2} Z_3^{1/2}$$

$$\times \bar{u}_{\mathbf{q}\tau} [F_1^p(k^2) i\gamma_\mu + F_2^p(k^2) i\Sigma_{\mu\nu} k_\nu] u_{\mathbf{p}\sigma} \qquad 8(40)$$

with $k = q - p$. As in the electron case, we have

$$F_1^p(0) = e_R \qquad 8(41a)$$

and

$$F_2^p(0) = \Delta\mu_p \qquad 8(41b)$$

where $\Delta\mu_p$ is the anomalous magnetic moment of the proton; experimentally,

$$\Delta\mu_p = -1{\cdot}79\frac{e_R}{2m_p} \qquad 8(42)$$

For a one neutron state, we have instead, setting $m_p = m_n = m$

$$^n\langle \mathbf{q}\tau|e_0 J_\mu(0)|\mathbf{p}\sigma\rangle^n = \left(\frac{m^2}{E_\mathbf{q}E_\mathbf{p}V^2}\right)^{1/2} Z_3^{1/2}$$

$$\times\, \bar{u}_{\mathbf{q}\tau}[F_1^n(k^2)i\gamma_\mu + F_2^n(k^2)i\Sigma_{\mu\nu}k_\nu]u_{\mathbf{p}\sigma} \qquad 8(43)$$

with

$$F_1^n(0) = 0 \qquad 8(44a)$$

and

$$F_2^n(0) = \Delta\mu_n \qquad 8(44b)$$

where, experimentally

$$\Delta\mu_n = 1{\cdot}91\frac{e_R}{2m_n} \qquad 8(45)$$

We can apply the isotopic spin formalism and regard $|\mathbf{p}\sigma\rangle^p$ and $|\mathbf{p}\sigma\rangle^n$ as two states of the nucleon doublet

$$|\mathbf{p}\sigma\rangle^N = \begin{pmatrix} |\mathbf{p}\sigma\rangle^p \\ |\mathbf{p}\sigma\rangle^n \end{pmatrix} \qquad 8(46)$$

Equations 8(40) and 8(43) can then be combined into the single equation

$$^N\langle \mathbf{q}\tau|e_0 J_\mu(0)|\mathbf{p}\sigma\rangle^N = \left(\frac{m^2}{E_\mathbf{q}E_\mathbf{p}V^2}\right)^{1/2} Z_3^{1/2}$$

$$\times\, \chi^\dagger \bar{u}_{\mathbf{q}\tau}\{[F_1^S(k^2) + \tau_3 F_1^V(k^2)]i\gamma_\mu + [F_2^S(k^2) + \tau_3 F_2^V(k^2)]i\Sigma_{\mu\nu}k_\nu\}u_{\mathbf{p}\sigma}\chi \qquad 8(47)$$

where

$$F_i^S = \tfrac{1}{2}(F_i^p + F_i^n) \qquad 8(48a)$$

$$F_i^V = \tfrac{1}{2}(F_i^p - F_i^n) \qquad 8(48b)$$

and where

$$\chi^\dagger\tau_3\chi = \begin{cases} +1 \text{ for a proton} \\ -1 \text{ for a neutron} \end{cases} \qquad 8(49)$$

Equation 8(47) exhibits the one-nucleon expectation value of j_μ as the sum of an isoscalar contribution

$$^{N}\langle \mathbf{q}\tau | e_0 J_\mu^S(0) | \mathbf{p}\sigma \rangle^{N} = \left(\frac{m^2}{E_\mathbf{q} E_\mathbf{p} V^2} \right)^{1/2} Z_3^{1/2} \chi^\dagger \chi \times \bar{u}_{\mathbf{q}\tau} [F_1^S(k^2) i\gamma_\mu + F_2^S(k^2) i\Sigma_{\mu\nu} k_\nu] u_{\mathbf{p}\sigma} \qquad 8(50a)$$

and an isovector contribution

$$^{N}\langle \mathbf{q}\tau | e_0 J_\mu^V(0) | \mathbf{p}\sigma \rangle^{N} = \left(\frac{m^2}{E_\mathbf{q} E_\mathbf{p} V^2} \right)^{1/2} Z_3^{1/2} \chi^\dagger \tau_3 \chi \times \bar{u}_{\mathbf{q}\tau} [F_1^V(k^2) i\gamma_\mu + F_2^V(k^2) i\Sigma_{\mu\nu} k_\nu] u_{\mathbf{p}\sigma} \qquad 8(50b)$$

The functions 8(48a) and 8(48b) are accordingly known as the isoscalar and isovector form factors respectively. From 8(41a), 8(41b), 8(44a) and 8(44b) we have, at $k^2 = 0$,

$$F_1^S(0) = \tfrac{1}{2} e_R \qquad F_2^S(0) = \tfrac{1}{2}(\Delta\mu_p + \Delta\mu_n) \qquad 8(51a)$$

and

$$F_1^V(0) = \tfrac{1}{2} e_R \qquad F_2^V(0) = \tfrac{1}{2}(\Delta\mu_p - \Delta\mu_n) \qquad 8(51b)$$

Current Conservation and Universality of e_R In Section 6.4 it was shown that for spinor electrodynamics, vertex function and electron wave function effects cancel each other by virtue of Ward's identity. This cancellation is expressed by the equality

$$Z_1 = Z_2 \qquad 8(52)$$

and is a consequence of the fact that the electromagnetic field is coupled to a conserved current.

We now extend the result 8(52) to the interaction of strongly interacting particles with the electromagnetic field. Let us consider the case of a proton interacting with an external electromagnetic field A_μ^{ext} with the interaction Lagrangian density

$$\mathscr{L}_I^{\text{ext}} = e_0 i \bar{\psi} \gamma_\mu \psi A_\mu^{\text{ext}} \qquad 8(53)$$

For simplicity we shall neglect the interaction of the proton with the quantized electromagnetic field, so that the total Lagrangian for the system is simply

$$\mathscr{L} = \mathscr{L}_0 + \mathscr{L}_I^{\text{strong}} + \mathscr{L}_I^{\text{ext}} \qquad 8(54)$$

and the relation between the renormalized and unrenormalized coupling constants is

$$e_R = Z_2^{st}(Z_1^{st})^{-1}e_0 \tag{8(55)}$$

where Z_2^{st} and Z_1^{st} are the wave function and vertex function renormalization constants due to the effects of the strong interactions. We wish to show that the strong interactions do not renormalize e_0, that is, that

$$e_R = e_0 \tag{8(56)}$$

The equal-time commutation relations give the commutator of $\psi(x)$ with the total charge operator

$$e_0 Q = \int e_0 \bar{\psi}(x)\gamma_4\psi(x)\, d^3x \tag{8(57)}$$

in the form

$$[\psi(x), e_0 Q] = e_0\psi(x) \tag{8(58)}$$

as in 5(119). Let us take the matrix element of the left-hand side between the vacuum and a one-proton state, and insert a sum over a complete set of states

$$\langle 0|[\psi, e_0 Q]|\mathbf{q}\tau\rangle^p = \sum_n \langle 0|\psi|n\rangle\langle n|e_0 Q|\mathbf{q}\tau\rangle^p - \sum_m \langle 0|e_0 Q|m\rangle\langle m|\psi|\mathbf{q}\tau\rangle^p \tag{8(59)}$$

Since the electromagnetic current is conserved, we have

$$[Q, H] = 0 \tag{8(60)}$$

so that Q can connect only states with the same energy eigenvalue. Thus

$$\begin{aligned}\sum_m \langle 0|e_0 Q|m\rangle\langle m|\psi|\mathbf{q}\tau\rangle^p &= \langle 0|e_0 Q|0\rangle\langle 0|\psi|\mathbf{q}\tau\rangle^p \\ &= 0\end{aligned} \tag{8(61)}$$

since $\langle 0|e_0 Q|0\rangle = 0$. Applying the same argument to the first term on the right-hand side of 8(59), we have

$$\sum_n \langle 0|\psi|n\rangle\langle n|e_0 Q|\mathbf{q}\tau\rangle^p = \sum_\sigma \langle 0|\psi|\mathbf{q}\sigma\rangle^{pp}\langle \mathbf{q}\sigma|e_0 Q|\mathbf{q}\tau\rangle^p \tag{8(62)}$$

Now from 8(40) and 8(41a) we have, as in 8(35)

$$^p\langle \mathbf{q}\sigma|e_0 Q|\mathbf{q}\tau\rangle^p = e_R\delta_{\sigma\tau} \tag{8(63)}$$

where we have set $Z_3 = 1$, in accordance with the fact that we are neglecting the proton-photon interaction. Combining 8(59), 8(61), 8(62) and

8(63), we find

$$\langle 0|[\psi, e_0 Q]|\mathbf{q}\tau\rangle^p = e_R \langle 0|\psi|\mathbf{q}\tau\rangle^p \qquad 8(64)$$

and by comparison with 8(58) we obtain the result

$$e_R = e_0$$

The above argument, due to Fubini and Furlan [Fubini (1965)] shows that in the absence of interactions with the quantized electromagnetic field, the strong interactions do not renormalize the electromagnetic coupling constant. An equivalent statement is that $Z_2^{st} = Z_1^{st}$ or that the proton wave function renormalization of e_0 cancels with the vertex renormalization. This cancellation continues to hold when we take into account the coupling with the quantized electromagnetic field although the Fubini–Furlan argument can no longer be so directly applied*. When the interaction with the quantized electromagnetic field is included, the result $e_R = e_0$ for the proton charge is replaced by

$$e_R = Z_3^{1/2} e_0$$

where Z_3 is given by a sum of contributions from both electron and proton closed loops, as illustrated in Fig. 8.2. There *is* therefore a renormalization of e_0 by the strong interactions when photon interactions are included *but this renormalization is the same for protons and electrons.* The result is that if the proton and electron have the same *bare* charge, their *physical* charges will also be equal.

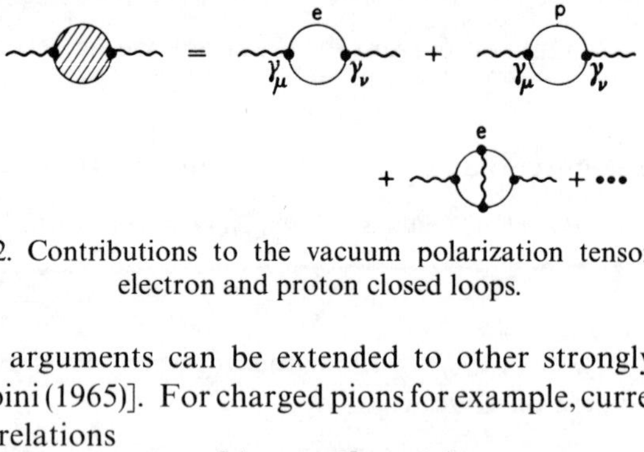

FIG. 8.2. Contributions to the vacuum polarization tensor from electron and proton closed loops.

The above arguments can be extended to other strongly interacting particles [Fubini (1965)]. For charged pions for example, current conservation and the relations

$$[\phi_{\pi^+}, e_0 Q] = e_0 \phi_{\pi^+} \qquad 8(65)$$

$$^{\pi^+}\langle \mathbf{k}|e_0 Q|\mathbf{k}\rangle^{\pi^+} = e_R \qquad 8(66)$$

* If we were to take into account the proton-photon interaction, the right-hand side of 8(62) would pick up additional contributions from intermediate states featuring one proton together with any number of zero energy photons. For a proof that the proton wave function renormalization of the electric charge cancels with the vertex renormalization in the case of interacting nucleons, pions and photons see Ward (1950).

analogous to 8(58) and 8(63) combine to give $e_R = e_0$ in the absence of photon coupling.

The Pion-Nucleon Vertex In analogy with 8(29) and 8(31) there exists a relation between the renormalized pion-nucleon coupling constant, and the one-nucleon matrix elements of the pion ' current '

$$\mathbf{j}(x) = (\Box - \mu^2)\boldsymbol{\phi}(x) \tag{8(67)}$$

where $\boldsymbol{\phi}$ represents the π-meson triplet 5(92a), 5(92b) and 5(92c). The matrix element ${}^N\langle \mathbf{q}\tau|\mathbf{j}(0)|\mathbf{p}\sigma\rangle^N$ must be of the general form

$${}^N\langle \mathbf{q}\tau|\mathbf{j}(0)|\mathbf{p}\sigma\rangle^N = -Z_3^{1/2}\sqrt{\frac{m}{E_\mathbf{q}V}}\chi^\dagger \bar{u}_{\mathbf{q}\tau}\mathbf{0}(\mathbf{q},\mathbf{p})u_{\mathbf{p}\sigma}\chi\sqrt{\frac{m}{E_\mathbf{p}V}} \tag{8(68)}$$

where $\mathbf{0}(\mathbf{q}, \mathbf{p})$ is a 4×4 matrix in spinor space and a 2×2 matrix in isotopic spin space. Since the left-hand side of 8(68) transforms as a Lorentz pseudoscalar and isotopic vector, $\mathbf{0}(\mathbf{q}, \mathbf{p})$ must be of the form

$$\mathbf{0}(\mathbf{q},\mathbf{p}) = i\gamma_5 \boldsymbol{\tau} K(k^2) \qquad (k = q-p)$$

where $\boldsymbol{\tau} = (\tau_1, \tau_2, \tau_3)$ is the usual set of three Pauli matrices; the argument presented in the paragraph following 8(21) can be applied here to show that K can only be a function of k^2.

As in the electrodynamic case, the matrix element

$${}^N\langle \mathbf{q}\tau|\mathbf{j}(0)|\mathbf{p}\sigma\rangle^N = -\left(\frac{m^2}{E_\mathbf{q}E_\mathbf{p}V^2}\right)^{1/2} Z_3^{1/2}\chi^\dagger\boldsymbol{\tau}\chi$$

$$\times \bar{u}_{\mathbf{q}\tau} i\gamma_5 u_{\mathbf{p}\sigma} K(k^2) \qquad (k = q-p) \tag{8(69)}$$

appears in the transition amplitude for scattering of a nucleon in an external π-mesonic field, and in analogy with 8(41a), we have, for $(q-p)^2 = -\mu^2$*

$$K(-\mu^2) = G_R \tag{8(70)}$$

where G_R is the renormalized $\pi - N$ coupling constant. Note that here (as in the discussion relative to 6(321)), we assume that the value of the form factor at $(q-p)^2 = -\mu^2$ can be obtained by analytic continuation from the physical region $(q-p)^2 \geqslant 0$.

* See for example Federbush (1958). Actually the scattering of a nucleon in an external mesonic field is more a mathematical concept than a physically realizable process. Nevertheless it occurs as an intermediate stage in real processes like nucleon-nucleon scattering.

We note for future reference the following alternative form of 8(69),

$$ {}^{N}\langle \mathbf{q}\tau|\boldsymbol{\phi}(0)|\mathbf{p}\sigma\rangle^{N} = \left(\frac{m^2}{E_{\mathbf{q}}E_{\mathbf{p}}V^2}\right)^{1/2} Z_3^{1/2}\chi^{\dagger}\boldsymbol{\tau}\chi $$

$$ \times \bar{u}_{\mathbf{q}\tau} i\gamma_5 u_{\mathbf{p}\sigma} \frac{K(k^2)}{k^2+\mu^2} \qquad (k = q-p) \qquad 8(71) $$

obtained by using 8(67).

π − N Scattering To exhibit the role played by the matrix element 8(69) in real physical processes, we consider meson-nucleon scattering. Applying the reduction technique developed in Section 7.5 to the $\pi - N$ scattering amplitude we get

$$ \langle \mathbf{k}', \mathbf{p}'\sigma' \text{ out}|\mathbf{k}, \mathbf{p}\sigma \text{ in}\rangle = \delta_{\mathbf{k}'\mathbf{k}}\delta_{\mathbf{p}'\mathbf{p}}\delta_{\sigma'\sigma} - Z_3^{-1}\int d^4x\, d^4x' f^*_{\mathbf{k}'}(x') f_{\mathbf{k}}(x) $$

$$ \times (\Box_{x'} - \mu^2)(\Box_x - \mu^2)\langle \mathbf{p}'\sigma'|T\phi(x')\phi(x)|\mathbf{p}\sigma\rangle \quad 8(72) $$

where $f_{\mathbf{k}}(x)$ is given by 7(101). Here, for simplicity, we have restricted our attention to the $\pi^0 - p$ scattering amplitude in order to suppress charge indices, but the generalization to other charge states is immediate. Note that 8(72) differs from 7(123) by the fact that the nucleons are now retained in the state vectors. Let us now bring the factor $(\Box_{x'} - \mu^2)$ $(\Box_x - \mu^2)$ into the T-product. Writing

$$ T\phi(x')\phi(x) = \tfrac{1}{2}\epsilon(x_0' - x_0)[\phi(x'), \phi(x)] + \tfrac{1}{2}\{\phi(x'), \phi(x)\} $$

we have

$$ \frac{\partial}{\partial x_0} T\phi(x')\phi(x) = \tfrac{1}{2}\epsilon(x_0' - x_0)[\phi(x'), \dot{\phi}(x)] + \tfrac{1}{2}\{\phi(x'), \dot{\phi}(x)\} $$

$$ - \delta(x_0' - x_0)[\phi(x'), \phi(x)] $$

$$ = T\phi(x')\dot{\phi}(x) - \delta(x_0 - x_0')[\phi(x'), \phi(x)] $$

$$ = T\phi(x')\dot{\phi}(x) $$

since the equal-time commutator of $\phi(x)$ and $\phi(y)$ vanishes, and similarly

$$ \frac{\partial^2}{\partial x_0^2} T\phi(x')\phi(x) = T\phi(x')\ddot{\phi}(x) - \delta(x_0' - x_0)[\phi(x'), \dot{\phi}(x)] $$

$$ = T\phi(x')\ddot{\phi}(x) - i\delta^{(4)}(x' - x) $$

Hence, setting

$$ j(x) = (\Box - \mu^2)\phi(x) $$

we have

$$(\Box_x - \mu^2) T\phi(x')\phi(x) = T\phi(x')j(x) + i\delta^{(4)}(x'-x)$$

Proceeding in the same way we find

$$(\Box_{x'} - \mu^2)(\Box_x - \mu^2) T\phi(x')\phi(x) = (\Box_{x'} - \mu^2) T\phi(x')j(x) + i(\Box_{x'} - \mu^2)\delta^{(4)}(x-x') \qquad 8(73)$$

with

$$(\Box_{x'} - \mu^2) T\phi(x')j(x) = Tj(x')j(x) - \delta(x_0' - x_0)[\dot{\phi}(x'), j(x)] - \frac{\partial}{\partial x_0'}(\delta(x_0' - x_0)[\phi(x'), j(x)]) \qquad 8(74)$$

The equal-time commutators appearing in the last two terms can be evaluated once we assume a definite form for the interaction Lagrangian. For a simple trilinear Yukawa coupling for example, we have

$$[\phi(x'), j(x)] = [\dot{\phi}(x'), j(x)] = 0 \qquad \text{for} \quad x_0' = x_0$$

by virtue of the equal-time commutation rules. In neutral pseudoscalar meson theory, however, j contains terms proportional to ϕ^3, as indicated in Section 6.4*. We have, in that case

$$[\phi(x'), j(x)] = 0, \qquad [\dot{\phi}(x'), j(x)] \sim \phi^2(x)\delta^{(3)}(x-x') \qquad 8(75)$$

for $x_0 = x_0'$. Thus the resulting term in 8(74) is proportional to $\delta^{(4)}(x-x')$ and will give rise in 8(76) below to a term depending only on the momentum transfer $(p'-p)^2$. For our purposes such terms can be ignored. Omitting also the second term on the right-hand side of 8(73), we are left with

$$-Z_3^{-1}\frac{1}{\sqrt{2\omega_{\mathbf{k}}V}}\cdot\frac{1}{\sqrt{2\omega_{\mathbf{k}'}V}}\int d^4x'\, d^4x\, e^{-ik'.x'}\, e^{ik.x}\langle\mathbf{p}'\sigma'|Tj(x')j(x)|\mathbf{p}\sigma\rangle \qquad 8(76)$$

as the relevant term in the $\pi - N$ scattering amplitude 8(72)†. We now apply the relation

$$j(x+a) = e^{-iP.a} j(x)\, e^{iP.a} \qquad 8(77)$$

* See the remarks relative to (6.373).

† A modified form of 8(76) having suitable analytic properties in the complex k_0-plane serves as the springboard for the axiomatic derivation of dispersion relations for $\pi - N$ scattering. See for example Gasiorowicz (1960), or Bjorken and Drell, op. cit., Section 18.11. In the language of dispersion theory, the equal-time commutator terms neglected in 8(76) can be absorbed into so-called *subtraction terms.*

valid for an arbitrary space-time displacement a to write

$$\langle \mathbf{p}'\sigma'| Tj(x')j(x)|\mathbf{p}\sigma\rangle = \langle \mathbf{p}'\sigma'|\, e^{iP.a}\, e^{-iP.a}(Tj(x')j(x))\, e^{iP.a}\, e^{-iP.a}|\mathbf{p}\sigma\rangle$$
$$= e^{i(p'-p)a}\langle \mathbf{p}'\sigma'| Tj(x'+a)j(x+a)|\mathbf{p}\sigma\rangle$$

or, choosing $a = -x$

$$\langle \mathbf{p}'\sigma'| Tj(x')j(x)|\mathbf{p}\sigma\rangle = e^{-i(p'-p)x}\langle \mathbf{p}'\sigma'| Tj(x'-x)j(0)|\mathbf{p}\sigma\rangle \qquad 8(78)$$

Thus, 8(76) takes the form

$$-Z_3^{-1}\frac{1}{\sqrt{2\omega_{\mathbf{k}}V}}\frac{1}{\sqrt{2\omega_{\mathbf{k}'}V}}\int d^4x'\, d^4x\, e^{i(p+k-p'-k')x}\, e^{-ik'.(x'-x)}$$
$$\times \langle \mathbf{p}'\sigma'| Tj(x'-x)j(0)|\mathbf{p}\sigma\rangle$$
$$= -Z_3^{-1}\frac{1}{\sqrt{2\omega_{\mathbf{k}}V}}\frac{1}{\sqrt{2\omega_{\mathbf{k}'}V}}(2\pi)^4\delta^{(4)}(p+k-p'-k')\int d^4y$$
$$\times e^{-ik'y}\langle \mathbf{p}'\sigma'| Tj(y)j(0)|\mathbf{p}\sigma\rangle$$

or, introducing a sum over a complete set of physical states $|q\alpha\rangle$

$$-Z_3^{-1}\frac{1}{\sqrt{2\omega_{\mathbf{k}}V}}\frac{1}{\sqrt{2\omega_{\mathbf{k}'}V}}(2\pi)^4\delta^{(4)}(p+k-p'-k')$$
$$\times\left\{\int_0^\infty dy_0\int d^3y\, e^{-ik'.y}\sum_{q,\alpha}\langle \mathbf{p}'\sigma'| j(y)|q\alpha\rangle\langle q\alpha| j(0)|\mathbf{p}\sigma\rangle\right\}$$
$$+\int_{-\infty}^0 dy_0\int d^3y\, e^{-ik'.y}\sum_{q,\alpha}\langle \mathbf{p}'\sigma'| j(0)|q\alpha\rangle\langle q\alpha| j(y)|\mathbf{p}\sigma\rangle \qquad 8(79)$$

Here q denotes the eigenvalues of the total energy-momentum operator, while α stands for whatever other quantum numbers are required to complete the characterization of the physical states. Setting

$$\langle \mathbf{p}'\sigma'| j(y)|q\alpha\rangle = e^{i(q-p').y}\langle \mathbf{p}'\sigma'| j(0)|q\alpha\rangle \qquad 8(80a)$$
$$\langle q\alpha| j(y)|\mathbf{p}\sigma\rangle = e^{i(p-q).y}\langle q\alpha\ \ | j(0)|\mathbf{p}\sigma\rangle \qquad 8(80b)$$

we perform the integrals over $\mathbf{y}$ and y_0 in 8(79) to obtain

$$+iZ_3^{-1}\frac{1}{\sqrt{2\omega_{\mathbf{k}}V}}\frac{1}{\sqrt{2\omega_{\mathbf{k}'}V}}(2\pi)^4\delta^{(4)}(p+k-p'-k')\sum_{q,\alpha}\langle \mathbf{p}'\sigma'| j(0)|q\alpha\rangle$$
$$\times\langle q\alpha| j(0)|\mathbf{p}\sigma\rangle(2\pi)^3\left[\delta^{(3)}(\mathbf{q}-\mathbf{p}-\mathbf{k})\frac{1}{q_0-E_{\mathbf{p}}-\omega_{\mathbf{k}}-i\varepsilon}\right.$$
$$\left.+\delta^{(3)}(\mathbf{q}-\mathbf{p}+\mathbf{k}')\frac{1}{q_0-E_{\mathbf{p}}+\omega_{\mathbf{k}'}-i\varepsilon}\right] \qquad 8(81)$$

as the expression for the relevant part of the $\pi^0 - p$ scattering amplitude.

The role played by the one-nucleon matrix elements of j in $\pi - N$ scattering is now apparent. Let us isolate the contribution of the one-proton intermediate states to 8(81) by introducing the factor

$$\delta(q_0 - E_{\mathbf{q}}) = \delta(q_0 - (\mathbf{q}^2 + m^2)^{1/2})$$

in the summation over q. Upon integration the term in $\delta^{(3)}(\mathbf{q}-\mathbf{p}-\mathbf{k})$ gives rise to a pole at

$$E_{\mathbf{p}+\mathbf{k}} = E_{\mathbf{p}} + \omega_{\mathbf{k}} \tag{8(82a)}$$

multiplying a factor which is entirely determined in terms of the matrix elements 8(69). Similarly, the term in $\delta^{(3)}(\mathbf{q}-\mathbf{p}+\mathbf{k}')$ in 8(81) gives rise to a pole at

$$E_{\mathbf{p}-\mathbf{k}'} = E_{\mathbf{p}} - \omega_{\mathbf{k}'} \tag{8(82b)}$$

Both these poles lie outside the physical region. They correspond to the poles at

$$(p+k)^2 = -m^2 \tag{8(83a)}$$

and

$$(p-k')^2 = -m^2 \tag{8(83b)}$$

which arise from the lowest order $\pi - N$ scattering amplitudes 6(117a) and 6(117b) respectively. It is instructive to compare, say, 6(117a) with the exact expression 8(81) in the neighbourhood of the pole at $E_{\mathbf{p}+\mathbf{k}} = E_{\mathbf{p}} + \omega_{\mathbf{k}}$. Substituting 8(69) into 8(81) and enlisting the aid of the relation

$$\sum_{\tau=1}^{2} u_{\mathbf{q}\tau}\bar{u}_{\mathbf{q}\tau} = \frac{\gamma\,.\,q + im}{2im} \tag{8(84)}$$

to eliminate the summation over the spins of the intermediate proton, we find that the *residue* of the exact scattering amplitude at the pole 8(82a) is given by

$$-(2\pi)^4\delta^{(4)}(p+k-p'-k')\frac{1}{V^2}\left(\frac{m^2}{E_{\mathbf{p}'}E_{\mathbf{p}}2\omega_{\mathbf{k}}2\omega_{\mathbf{k}'}}\right)^{1/2} G_R^2\frac{1}{2E_{\mathbf{k}+\mathbf{p}}}$$

$$\times\,\bar{u}_{\mathbf{p}'\sigma'}\gamma_5(\gamma\,.\,p + \gamma\,.\,k + im)\gamma_5 u_{\mathbf{p}\sigma} \tag{8(85)}$$

since

$$\lim_{q\to p+k}\,(K[(p'-q)^2]K[(q-p)^2]) = K(k'^2)K(k^2)$$

$$= G_R^2 \tag{8(86)}$$

by virtue of 8(70). In deriving 8(85) we have replaced a factor V^{-1} by $(2\pi)^{-3}$ in accordance with the fact that q is treated as a continuous variable. The important point to note is that 8(85) is *identical* with the residue of 6(117a) at the pole 8(82a)*, except for the substitution

$$G_0 \to G_R$$

Similar remarks apply to the residue of the scattering amplitude at the pole 8(82b).

The lowest order scattering amplitudes 6(117a) and 6(117b) with G_0 replaced by G_R constitute the *renormalized Born approximation* for $\pi^0 - p$ scattering. Although the appearance of these renormalized Born terms in the scattering amplitude follows from renormalization theory, we stress that here they have been derived without recourse to perturbation theory and, indeed, without assuming the existence of a fundamental Yukawa coupling Lagrangian. The renormalized Born approximation plays an important role in dispersion theory†. In particular, the fact that the residues at the poles are proportional to G_R^2 has been exploited to *measure* the renormalized $\pi - N$ coupling constant experimentally. The technique consists essentially in extrapolating the experimental data for forward $\pi^{\pm} - p$ scattering into the unphysical region near the pole with the aim of determining the residue at the pole and, hence, G_R^2 [Haber-Schaim (1956)]. Measurement yields the value

$$\frac{G_R^2}{4\pi} = 14{\cdot}6 \pm 00{\cdot}2 \qquad 8(87)$$

8-2 Weak interaction currents and coupling constants

Introduction We now turn to the application of the techniques of Section 8.1 to the theory of weak interactions. We restrict our attention

* Comparison of the two expressions is aided by writing the Feynman propagator in 6(117a) as the sum of two terms

$$\frac{1}{(p+k)^2+m^2} = \frac{1}{2E_{\mathbf{p}+\mathbf{k}}}\left(\frac{1}{E_{\mathbf{p}+\mathbf{k}}-E_{\mathbf{p}}-\omega_{\mathbf{k}}}+\frac{1}{E_{\mathbf{p}+\mathbf{k}}+E_{\mathbf{p}}+\omega_{\mathbf{k}}}\right)$$

Although the accompanying pole at $E_{\mathbf{p}+\mathbf{k}} = -E_{\mathbf{p}}-\omega_{\mathbf{k}}$ has not appeared explicitly in the present treatment, it must exist on grounds of Lorentz covariance. To exhibit it explicitly we would need to go beyond the one-proton intermediate state approximation. The technique used here to exhibit the pole at (8.82a) recalls the treatment of $\pi - N$ scattering by means of 'old-fashioned' perturbation theory [S. S. Schweber, *Introduction to Relativistic Quantum Field Theory*, Row, Peterson and Co., 1961, Section 13b]. In this version of perturbation theory, momentum, *though not energy*, is conserved in intermediate states. In contrast, the fully relativistic Feynman approach of Chapter 6 conserves both energy and momentum at each vertex but the momenta of the particles in the intermediate state are taken off their physical mass shells.

† See, for example, the review article by S. Gasiorowicz (1960).

to the strangeness-conserving semileptonic processes for which the interaction Lagrangian is given by

$$\mathscr{L}_I^{wk} = G\cos\theta\frac{1}{\sqrt{2}}(J_+^\lambda + A_+^\lambda)l_\lambda + \text{herm. conj.} \quad 8(88)$$

Here, as indicated in Section 5-3, $\theta = 0{\cdot}26$ is the Cabibbo angle, G is the coupling constant as determined by μ-decay experiments,

$$G = (1{\cdot}03 \pm 0{\cdot}003)\times 10^{-5}\, m_p^{-2} \quad 8(89)$$

J_+^λ and A_+^λ are the $\Delta S = 0$, $\Delta I_3 = +1$ vector and axial vector hadronic currents, and

$$l_\lambda(x) = i\bar{e}(x)\gamma_\lambda(1+\gamma_5)\nu_e(x) + i\bar{\mu}(x)\gamma_\lambda(1+\gamma_5)\nu_\mu(x) \quad 8(90)$$

is the leptonic current. For the total Lagrangian of the system we write, neglecting electromagnetic interactions,

$$\mathscr{L}_I = \mathscr{L}_I^{st} + \mathscr{L}_I^{wk} \quad 8(91)$$

where $\mathscr{L}_I^{st}$ is the (unknown) interaction Lagrangian for the strong interactions. We make no assumptions about the form of $\mathscr{L}_I^{st}$ or about the explicit expression of the hadron currents $J_\pm^\lambda$ and $A_\pm^\lambda$ in terms of field operators. Nevertheless we assume, as in Section 5.3, that $\mathscr{L}_I^{st}$ conserves isotopic spin, that the currents $J_\pm^\lambda$ and $A_\pm^\lambda$ are the components

$$J_\pm^\lambda = J_1^\lambda \pm iJ_2^\lambda \quad 8(92a)$$

$$A_\pm^\lambda = A_1^\lambda \pm iA_2^\lambda \quad 8(92b)$$

of two isotopic vectors $\mathbf{J}^\lambda$ and $\mathbf{A}^\lambda$ and, furthermore, that $\mathbf{J}^\lambda$ is the conserved isotopic spin current

$$\partial_\lambda \mathbf{J}^\lambda(x) = 0 \quad 8(93)$$

whose existence is assured by the invariance of the strong interactions under SU_2 transformations. This last assumption is known as the *conserved vector current* (or CVC) hypothesis [Feynman (1958)]. In the absence of electromagnetic couplings, 8(93) is an exact conservation law; the violation of SU_2 by $\mathscr{L}_I^{wk}$ in 8(91) is essentially negligible.

To calculate S-matrix elements for transitions induced by $\mathscr{L}_I^{wk}$ we can proceed as in the opening paragraphs of Section 8.1 and transform to a picture in which the effects of the weak interaction bear only on the time-evolution of the state vectors but not on the field operators. Then the first order transition matrix element from an initial state $|i\rangle$ to a final state $|f\rangle$ is given by

$$-i\int \langle f|\mathscr{H}_I'^{wk}(x)|i\rangle\, d^3x \quad 8(94)$$

where in complete analogy with 8(9), the interaction Hamiltonian density

$$\mathscr{H}_I'^{wk} = -\mathscr{L}_I'^{wk} \tag{8(95)}$$

features hadron fields which are subject only to the effects of the strong interactions, and lepton fields $\bar{e}(x)$ and $\nu_e(x)$ which, being unaffected by the strong interactions are just free field operators.

β-Decay Let us apply 8(94) to the transition amplitude for β-decay

$$n \to p + e^- + \bar{\nu}_e \tag{8(96)}$$

Using 8(88) and 8(90) we see that the first order amplitude is given by

$$-\frac{G\cos\theta}{\sqrt{2}}\int d^3x\, {}^p\langle \mathbf{q}\tau|J_+^\lambda(x) + A_+^\lambda(x)|\mathbf{p}\sigma\rangle^n \frac{1}{\sqrt{V}}\sqrt{\frac{m_e}{E_{\mathbf{p}_e}}}\, \mathrm{e}^{-iq_e\cdot x}\,\mathrm{e}^{-iq_\nu\cdot x}$$

$$\times\, \bar{u}_{\mathbf{q}_e}\gamma_\lambda(1+\gamma_5)v_{\mathbf{q}_\nu} \tag{8(97)}$$

or

$$-\frac{G\cos\theta}{\sqrt{2}}(2\pi)^4\delta^{(4)}(p-q-q_e-q_\nu)\frac{1}{V}\sqrt{\frac{m_e}{E_{\mathbf{p}_e}}}\, {}^p\langle \mathbf{q}\tau|J_+^\lambda(0) + A_+^\lambda(0)|\mathbf{p}\sigma\rangle^n$$

$$\times\, \bar{u}_{\mathbf{q}_e}\gamma_\lambda(1+\gamma_5)v_{\mathbf{q}_\nu} \tag{8(98)}$$

where we have factored out the x-dependence of the current matrix elements in the usual way and integrated over x.

Now we have assumed that the weak vector current J_+^μ is the component $J_1^\mu + iJ_2^\mu$ of the conserved isotopic spin current $\mathbf{J}^\mu$. This has the important consequence that the matrix elements

$${}^p\langle \mathbf{q}\tau|J_+^\mu(0)|\mathbf{p}\sigma\rangle^n \tag{8(99)}$$

are completely determined in terms of the isovector *electromagnetic* form factors F_1^V and F_2^V, introduced in the preceding section. The point is that the isovector part J_μ^V of the hadronic electromagnetic current is just the third component of the isotopic spin current $\mathbf{J}_\mu$:

$$J_\mu^V = J_\mu^3 \tag{8(100)}$$

This can be seen explicitly by comparing, say, 5(114) with 5(98) for the simple case of interacting pions and nucleons. In the more general case, the statement 8(100) can be checked by comparing 5(149) with the corresponding expression for the electromagnetic current; this is left as an exercise. Armed with this knowledge, we can use isotopic spin invariance to extend 8(50b) to the isotopic vector $\mathbf{J}^{\mu}$, i.e.

$$^{N}\langle \mathbf{q}\tau|e_R\mathbf{J}^{\mu}(0)|\mathbf{p}\sigma\rangle^{N} = \left(\frac{m^2}{E_{\mathbf{q}}E_{\mathbf{p}}V^2}\right)^{1/2}\chi^{\dagger}\boldsymbol{\tau}\chi$$
$$\times\,\bar{u}_{\mathbf{q}\tau}[F_1^V(k^2)i\gamma_\mu + F_2^V(k^2)i\Sigma_{\mu\nu}k_\nu]u_{\mathbf{p}\sigma} \qquad 8(101)$$

where, since we are working in the limit of exact isotopic spin invariance, we have neglected electromagnetic interactions, setting (as in 8(56)),

$$e_R = e_0 \qquad 8(102)$$

or, equivalently, $Z_3 = 1$. The desired result

$$^{p}\langle \mathbf{q}\tau|J_+^{\mu}(0)|\mathbf{p}\sigma\rangle^{n} = \frac{2}{e_R}\left(\frac{m^2}{E_{\mathbf{q}}E_{\mathbf{p}}V^2}\right)^{1/2}\bar{u}_{\mathbf{q}\tau}[F_1^V(k^2)i\gamma_\mu + F_2^V(k^2)i\Sigma_{\mu\nu}k_\nu]u_{\mathbf{p}\sigma} \qquad 8(103)$$

now follows from 8(92a) and the fact that $\chi_p^{\dagger}(\tau_1 + i\tau_2)\chi_n = 2$.

Thus, for any momentum transfer k, the vector current contribution to the β-decay transition amplitude 8(98) is entirely determined in terms of the isovector electromagnetic form factors. We now define the renormalized weak vector coupling constant G_V by

$$\lim_{\mathbf{q}\to\mathbf{p}} {}^{p}\langle \mathbf{q}\tau|GJ_+^{\mu}(0)|\mathbf{p}\sigma\rangle^{n} = \frac{m}{E_{\mathbf{p}}V}G_V i\bar{u}_{\mathbf{q}\tau}\gamma_\mu u_{\mathbf{p}\sigma} \qquad 8(104)$$

in analogy with 8(37). G_V is the effective coupling constant which one measures in low momentum-transfer β-decay experiments. Referring to 8(103) and 8(51b) we see that

$$G_V = G \qquad 8(105)$$

The weak vector coupling constant is not renormalized by the strong interactions. This important result is a direct consequence of the CVC hypothesis. As in the case of the corresponding result 8(56) for hadron electrodynamics, the nonrenormalization of the coupling constant is ensured by current conservation. Experimentally, the equality 8(105) is verified to within 2%; this was in fact the observation which led Feynman and Gell-Mann to propose the CVC hypothesis. Further support for the CVC theory is provided by the fact that the existence of the term proportional to $\Sigma_{\mu\nu}k_\nu$ in 8(103)—the so-called *weak magnetism* term—has been confirmed experimentally, [Lee (1963)].

Turning to the matrix elements $^{p}\langle \mathbf{q}\tau|A_+^{\mu}|\mathbf{p}\sigma\rangle^{n}$ of the axial vector current we write, in analogy with 8(18) or 8(69),

$$^{N}\langle \mathbf{q}\tau|\mathbf{A}_{\mu}(0)|\mathbf{p}\sigma\rangle^{N} = \frac{1}{2}\left(\frac{m^2}{E_{\mathbf{q}}E_{\mathbf{p}}V^2}\right)^{1/2}\chi^{\dagger}\boldsymbol{\tau}\chi\bar{u}_{\mathbf{q}\tau}O_{\mu 5}(\mathbf{q},\mathbf{p})u_{\mathbf{p}\sigma} \qquad 8(106)$$

where we have extracted a factor $\frac{1}{2}$ for later convenience. Here $O_{\mu 5}(\mathbf{q}, \mathbf{p})$ is the pseudovector operator

$$O_{\mu 5}(\mathbf{q}, \mathbf{p}) = i\gamma_\mu\gamma_5 O'_1 + i\Sigma_{\mu\nu}k_\nu\gamma_5 O'_2 + ik_\mu\gamma_5 O'_3 + iP_\mu\gamma_5 O'_4 + i\Sigma_{\mu\nu}P_\nu\gamma_5 O'_5 \quad 8(107)$$

where $k = q-p$, $P = q+p$ and where, as in 8(19), the $O'_i (i = 1 \ldots 5)$ are functions of k^2. The easily verified relations

$$\bar{u}_{\mathbf{q}\tau}\Sigma_{\mu\nu}P_\nu\gamma_5 u_{\mathbf{p}\sigma} = -2m\bar{u}_{\mathbf{q}\tau}\gamma_\mu\gamma_5 u_{\mathbf{p}\sigma} - ik_\mu\bar{u}_{\mathbf{q}\tau}\gamma_5 u_{\mathbf{p}\sigma}$$

$$\bar{u}_{\mathbf{q}\tau}\Sigma_{\mu\nu}k_\nu\gamma_5 u_{\mathbf{p}\sigma} = -iP_\mu\bar{u}_{\mathbf{q}\tau}\gamma_5 u_{\mathbf{p}\sigma}$$

indicate that the terms in $\Sigma_{\mu\nu}\gamma_5 P_\nu$ and $\Sigma_{\mu\nu}\gamma_5 k_\nu$ in 8(107) can be amalgamated into the remaining three terms proportional to $i\gamma_\mu\gamma_5$, $ik_\mu\gamma_5$ and $iP_\mu\gamma_5$. Thus

$$O_{\mu 5}(\mathbf{q}, \mathbf{p}) = \mathscr{F}_1(k^2)i\gamma_\mu\gamma_5 + \mathscr{F}_2(k^2)ik_\mu\gamma_5 + \mathscr{F}_3(k^2)iP_\mu\gamma_5 \quad 8(108)$$

Finally, the term in iP_μ can be eliminated by the following argument, based on the invariance of the strong interactions under charge conjugation. In a simple model of the type considered in Section 5.3, the nucleon contribution to the axial vector current $\mathbf{A}_\mu$ is given by

$$\frac{i}{2}[\bar{N}, \gamma_\mu\gamma_5\boldsymbol{\tau}N]$$

and under charge conjugation the third component of the current transforms according to

$$\mathscr{C}\frac{i}{2}[\bar{N}, \gamma_\mu\gamma_5\tau_3 N]\mathscr{C}^{-1} = \frac{i}{2}[\bar{N}, \gamma_\mu\gamma_5\tau_3 N]$$

as can easily be verified, using 3(182a), 3(182b), 3(183), and the fact that $\tau_3^T = \tau_3$. We now make the explicit *assumption* that the same transformation law holds for A_3^μ, i.e.

$$\mathscr{C}A_3^\mu\mathscr{C}^{-1} = A_3^\mu \quad 8(109)$$

In that case we will have for, say, a one-proton state,

$$^p\langle\mathbf{q}\tau|A_3^\mu(0)|\mathbf{p}\sigma\rangle^p = {}^p\langle\mathbf{q}\tau|\mathscr{C}^{-1}\mathscr{C}A_3^\mu(0)\mathscr{C}^{-1}\mathscr{C}|\mathbf{p}\sigma\rangle^p$$

$$= {}^p\langle\mathbf{q}\tau|\mathscr{C}^{-1}A_3^\mu(0)\mathscr{C}|\mathbf{p}\sigma\rangle^p \quad 8(110)$$

Now the application of $\mathscr{C}$ to a one-proton state $|\mathbf{p}\sigma\rangle^p$ creates an *anti*proton state with the same momentum and spin*

$$\mathscr{C}|\mathbf{p}\sigma\rangle^p = |\mathbf{p}\bar{\sigma}\rangle^{\bar{p}} \quad 8(111)$$

* We adhere to the convention that for antinucleons $\bar{\sigma} = 1$ (i.e. $\sigma = 2$) means spin up, while $\bar{\sigma} = 2$ ($\sigma = 1$) means spin down. The opposite convention holds for nucleon states, which is why σ is replaced by $\bar{\sigma}$ under charge conjugation. [See the discussion following 1(67) and 3(171).]

For the free field theory 8(111) can be verified with the aid of 3(182a), 3(182b) and 1(67). For the interacting theory the same result follows, providing the theory is charge conjugation invariant, since $\mathscr{C}$ then satisfies $\mathscr{C}\phi_{\text{in}}^{A}\mathscr{C}^{-1} = \phi_{\text{in}}^{Ac}$ where the $\phi_{\text{in}}^{A}(A = 1 \ldots N)$ are the in-fields, and we can construct $\mathscr{C}$ in terms of the in-field creation and destruction operators in the same way as in the free field case. We now set

$$^{\bar{p}}\langle \mathbf{q}\bar{\tau}|A_3^{\mu}(0)|\mathbf{p}\bar{\sigma}\rangle^{\bar{p}} = -\frac{1}{2}\left(\frac{m^2}{E_{\mathbf{q}}E_{\mathbf{p}}V^2}\right)^{1/2} \bar{v}_{\mathbf{p}\bar{\sigma}}O_{\mu 5}(-\mathbf{p}, -\mathbf{q})v_{\mathbf{q}\bar{\tau}} \qquad 8(112)$$

where the form of the right-hand side and the appearance of an overall minus sign relative to 8(106) can be checked explicitly in perturbation theory*. Comparing with 8(106) we see that the condition 8(110) may be written in the form

$$\bar{u}_{\mathbf{q}\tau}O_{\mu 5}(\mathbf{q}, \mathbf{p})u_{\mathbf{p}\sigma} = -\bar{v}_{\mathbf{p}\bar{\sigma}}O_{\mu 5}(-\mathbf{p}, -\mathbf{q})v_{\mathbf{q}\bar{\tau}} \qquad 8(113)$$

Using 1(67) it is now straightforward to show that

$$\mathscr{F}_3(k^2) = 0 \qquad 8(114)$$

is both necessary and sufficient for consistency with 8(113). We have therefore established that the β-decay matrix element of the axial vector current must have the general form

$$^{p}\langle \mathbf{q}\tau|A_{+}^{\mu}(0)|\mathbf{p}\sigma\rangle^{n} = \left(\frac{m^2}{E_{\mathbf{q}}E_{\mathbf{p}}V^2}\right)^{1/2} \bar{u}_{\mathbf{q}\tau}[\mathscr{F}_1(k^2)i\gamma_{\mu}\gamma_5 + \mathscr{F}_2(k^2)ik_{\mu}\gamma_5]u_{\mathbf{p}\sigma} \qquad 8(115)$$

As in 8(104) we define the renormalized axial vector coupling constant G_A by

$$\lim_{q\to p} \langle \mathbf{q}\tau|GA_{+}^{\mu}(0)|\mathbf{p}\sigma\rangle = \frac{m}{E_{\mathbf{p}}V} G_A i\bar{u}_{\mathbf{q}\tau}\gamma_{\mu}\gamma_5 u_{\mathbf{p}\sigma} \qquad 8(116)$$

or, comparing with 8(115)

$$G_A = G\mathscr{F}_1(0) \qquad 8(117)$$

In this case, however, the form factors $\mathscr{F}_1$ and $\mathscr{F}_2$ are unknown a priori and, since $A_{+}^{\mu}(x)$ is not conserved, G_A and G can in principle differ widely. In fact, the experimental value for the ratio G_A/G_V is given by

$$G_A/G_V = 1{\cdot}18 \pm 0{\cdot}02 \qquad 8(118)$$

The fact that this ratio is quite close to unity suggests that the axial vector current may be *approximately* conserved. A successful theoretical calcula-

* See Problem 3.

tion of G_A/G based on this assumption has been performed by Adler (1965a, b) and Weisberger (1965) and will be discussed in Section 8.3.

Pion Decay We now turn to the pion decay process

$$\pi^- \to \mu^- + \bar{\nu}_\mu \tag{8(119)}$$

The first order transition amplitude for this decay, based on the interaction 8(88), is given by

$$-\frac{G\cos\theta}{\sqrt{2}}\int d^3x \langle 0|A_+^\lambda(x)|\mathbf{k}\rangle^{\pi^-}\frac{1}{V}\sqrt{\frac{m_\mu}{E_{\mathbf{p}_\mu}}}\,\mathrm{e}^{-iq_\mu . x}\,\mathrm{e}^{-iq_\nu . x}\bar{u}_{\mathbf{q}_\mu}\gamma_\lambda(1+\gamma_5)v_{\mathbf{q}_\nu} \tag{8(120)}$$

Note that the vector current does not contribute to π-decay since a nonzero matrix element

$$\langle 0|J_+^\lambda(x)|\mathbf{k}\rangle^{\pi^-}$$

would violate parity*.

We now appeal to the usual invariance arguments to write $\langle 0|A_+^\lambda(x)|\mathbf{k}\rangle^{\pi^-}$ in the form

$$\langle 0|A_+^\lambda(x)|\mathbf{k}\rangle^{\pi^-} = a\frac{1}{\sqrt{V}}\frac{1}{\sqrt{2\omega_{\mathbf{k}}}}ik^\lambda\,\mathrm{e}^{ik.x} \tag{8(121)}$$

where a is a constant. Inserting 8(121) into 8(120) and performing the x-integration, we obtain the following expression for the first order π-decay transition amplitude,

$$-(2\pi)^4\delta^{(4)}(k-q_\mu-q_\nu)\frac{G\cos\theta}{\sqrt{2}}a\frac{1}{V^{3/2}}\sqrt{\frac{m_\mu}{E_{\mathbf{q}_\mu}2\omega_{\mathbf{k}}}}ik^\lambda\,.\,\bar{u}_{\mathbf{q}_\mu}\gamma_\lambda(1+\gamma_5)v_{\mathbf{q}_\nu} \tag{8(122)}$$

This result is formally identical to 6(215), if we identify the phenomenological π-decay constant g_π as

$$g_\pi = \frac{G\cos\theta}{\sqrt{2}}a \tag{8(123)}$$

With this identification we are assured that, to first order in G, the general semileptonic weak interaction Lagrangian 8(88) yields the same π-decay transition amplitude as the phenomenological coupling 5(163). The pion decay lifetime τ has already been computed in Section 6-3 using the

* Here we have invoked the space-reflection invariance of the *strong* interactions!

phenomenological model; expressing the result, 6(219), in terms of a, we get

$$\tau^{-1} = \frac{G^2 \cos^2\theta \, a^2}{8\pi} m_\pi m_\mu^2 \left(1 - \frac{m_\mu^2}{m_\pi^2}\right)^2 \qquad 8(124)$$

A theoretical expression for the constant a, namely

$$a = \frac{\sqrt{2} m G_A}{G_R G} \qquad 8(125)$$

has been given by Goldberger and Treiman [Goldberger (1958)]. Here G_R and G_A are the renormalized $\pi - N$ and weak axial vector coupling constants respectively. Experimentally, the relation 8(125) is well satisfied. As against an experimental π-decay lifetime of $2{\cdot}56 \times 10^{-8}$ seconds, the formula 8(124) with 8(125) predicts a lifetime of $2{\cdot}7 \times 10^{-8}$ seconds. We turn now to the derivation of Eq. 8(125).

The Goldberger–Treiman Relation We have already remarked that the axial vector current cannot be completely conserved. Indeed, if

$$\partial_\lambda A_+^\lambda(x) = 0 \qquad 8(126)$$

then from 8(121) we have

$$0 = \langle 0|\partial_\lambda A_+^\lambda(x)|\mathbf{k}\rangle^{\pi^-} = ik_\lambda a \frac{1}{\sqrt{V}} \frac{1}{\sqrt{2\omega_\mathbf{k}}} ik^\lambda \, e^{ik.x}$$

$$= am_\pi^2 \frac{1}{\sqrt{V}} \frac{1}{\sqrt{2\omega_\mathbf{k}}} e^{ik.x} \qquad 8(127)$$

or $a = 0$, so that π-decay could not occur! Moreover from a theoretical point of view, we note that in most field theoretical models the conservation of the axial vector current is violated by the fact that the physical hadrons have finite mass. Thus the 4-divergence of the nucleon contribution

$$\partial_\mu(i\bar{N}\gamma_\mu\gamma_5\boldsymbol{\tau}N) = i(N\gamma_\mu\overleftarrow{\partial}_\mu)\gamma_5\boldsymbol{\tau}N - i\bar{N}\gamma_5\boldsymbol{\tau}(\gamma_\mu\partial_\mu N)$$

picks up a finite contribution

$$2mi\bar{N}\gamma_5\boldsymbol{\tau}N$$

from the nucleon mass terms in the field equations. There exist models in which this difficulty is circumvented [Nambu (1960a), Nishijima (1959)], but only at the price of introducing unphysical zero mass bosons into the theory. [See Section 9-5 and 10-3.]

Nevertheless, the nearness of the ratio G_A/G_V to unity suggests that a *partial* conservation law may hold true, as remarked in the paragraph following 8(118). Since $\partial_\lambda A^\lambda_+(x)$ has the same quantum numbers as the charged π^- field, let us make the simple assumption that the two operators are in fact proportional*.

$$\partial_\lambda A^\lambda_+(x) = c\phi_{\pi^-}(x) \tag{8(128)}$$

To determine c we compare 8(127), i.e.

$$\langle 0|\partial_\lambda A^\lambda_+(x)|\mathbf{k}\rangle^{\pi^-} = am_\pi^2 \frac{1}{\sqrt{V}}\frac{1}{\sqrt{2\omega_\mathbf{k}}}\mathrm{e}^{ik.x} \tag{8(129)}$$

with 7(99a)

$$\langle 0|\phi_{\pi^-}(x)|\mathbf{k}\rangle^{\pi^-} = Z_3^{1/2}\frac{1}{\sqrt{V}}\frac{1}{\sqrt{2\omega_\mathbf{k}}}\mathrm{e}^{ik.x}$$

This fixes the proportionality constant in 8(128) and we obtain the form

$$\partial_\lambda A^\lambda_+(x) = \frac{am_\pi^2}{Z_3^{1/2}}\phi_{\pi^-} \tag{8(130)}$$

Equation 8(130) is known as the partially conserved axial vector current (or PCAC) hypothesis [Gell-Mann (1960)]. We shall apply it to derive the Goldberger–Treiman relation and, in the following section, the Adler–Weisberger sum rule†.

From 8(130) and 8(71) we obtain

$$^p\langle \mathbf{q}\tau|\partial_\lambda A^\lambda_+(0)|\mathbf{p}\sigma\rangle^n = \left(\frac{m^2}{E_\mathbf{q}E_\mathbf{p}V^2}\right)^{1/2}\bar{u}_{\mathbf{q}\tau}i\gamma_5 u_{\mathbf{p}\sigma}\frac{\sqrt{2}am_\pi^2K(k^2)}{k^2+m_\pi^2} \tag{8(131)}$$

where the appearance of the factor $\sqrt{2}$ can be traced to the definitions 5(92a) and 5(92b). On the other hand

$$^p\langle \mathbf{q}\tau|\partial_\lambda A^\lambda_+(x)|\mathbf{p}\sigma\rangle^n = i(p_\lambda - q_\lambda)\,{}^p\langle \mathbf{q}\tau|A^\lambda_+(x)|\mathbf{p}\sigma\rangle^n$$

*In a sense which will be made precise in Section 9.4, 8(128) can be viewed as a *definition* of the pion field. The point is that, in present theory, there does not exist an 'absolute' definition of the pion field since there is no certitude as to the form of the fundamental Lagrangian and the nature of the fundamental fields. As will be shown in Section 9.4, any properly normalized local field operator can represent the pion field on the mass shell. The real content of 8(128) is therefore the assumption that $c^{-1}\partial_\lambda A^\lambda_+(x)$ is a good definition of $\phi_{\pi^-}(x)$ off the mass shell as well. One should also note that there exists a renormalizable field theory—the so called σ-model [Gell-Mann (1960)] in which 8(128) is satisfied as an operator identity.

† There exist alternative derivations of both the Goldberger–Treiman relation [Bernstein (1960)] and of the Adler–Weisberger sum rule [Weisberger (1965b)] based on dispersion relations in which the explicit assumption 8(130) is avoided.

whence, using 8(115),

$$
\begin{aligned}
&{}^{p}\langle \mathbf{q}\tau|\partial_\lambda A^\lambda_+(0)|\mathbf{p}\sigma\rangle^{n} \\
&= \left(\frac{m^2}{E_{\mathbf{q}}E_{\mathbf{p}}V^2}\right)^{1/2} \bar{u}_{\mathbf{q}\tau}[\mathscr{F}_1(k^2)\gamma\,.\,(q-p)\gamma_5+\mathscr{F}_2(k^2)(q-p)^2\gamma_5]u_{\mathbf{p}\sigma} \\
&= \left(\frac{m^2}{E_{\mathbf{q}}E_{\mathbf{p}}V^2}\right)^{1/2} \bar{u}_{\mathbf{q}\tau}\gamma_5 u_{\mathbf{p}\sigma}[2im\mathscr{F}_1(k^2)+k^2\mathscr{F}_2(k^2)] \qquad 8(132)
\end{aligned}
$$

where, in deriving the second line, we have used the Dirac equations 8(22a) and 8(22b). Comparing 8(131) and 8(132) at zero momentum transfer yields the relation

$$aK(0) = \sqrt{2}m\mathscr{F}_1(0)$$

or, by 8(117)

$$a = \frac{\sqrt{2}mG_A}{K(0)G} \qquad 8(133)$$

If we now assume that the form factor $K(k^2)$ varies slowly between $k^2 = -\mu^2$ and $k^2 = 0$, then by 8(70)

$$a = \frac{\sqrt{2}mG_A}{G_R G} \qquad 8(134)$$

which is the Goldberger–Treiman relation.

8-3 Renormalization effect for partially conserved currents

The Weak Vector Current In the preceding section we have seen that in the absence of electromagnetic interactions, the weak vector coupling constant is not renormalized by the strong interactions, i.e. we have

$$G_V = G \qquad 8(135)$$

owing to the fact that the weak vector current J^μ_+ is completely conserved. However, in the presence of electromagnetic interactions, the conservation law

$$\partial_\mu J^\mu_+ = 0 \qquad 8(136)$$

is only approximately true and we must therefore expect deviations from the exact equality 8(135). A general technique for computing such deviations or ‘renormalization effects’ has been developed by Fubini and Furlan [Fubini (1965)]. For the case of the vector current the method is based on the use of the commutation algebra of the SU_2 group.

Let us *assume** that the components of the total isotopic spin vector

$$\mathbf{I}(t) = -i\int \mathbf{J}_4(x)\,d^3x \qquad 8(137)$$

satisfy the commutation algebra

$$[I_i(t), I_j(t)] = i\varepsilon_{ijk}I_k(t) \qquad 8(138)$$

as in 5(102). Note that we have exhibited the time arguments explicitly: in the presence of electromagnetic interactions I_1 and I_2 are no longer constants of the motion, but vary with time according to the equation of motion

$$\begin{aligned}\dot{I}_i(t) &= \frac{1}{i}[I_i(t), H(t)]\\ &= \frac{1}{i}[I_i(t), H_{em}(t)] \qquad 8(139)\end{aligned}$$

the time variation being due only to the electromagnetic interaction Hamiltonian. From 8(138) we easily derive the commutation relation

$$[I_+(t), I_-(t)] = 2I_3 \qquad 8(140)$$

where

$$I_\pm(t) = -i\int J_\pm(x)\,d^3x = I_1 \pm iI_2 \qquad 8(141)$$

Taking the expectation value of 8(140) between one-proton states,

$$\begin{aligned}{}^p\langle\mathbf{p}'\sigma'|[I_+(t), I_-(t)]|\mathbf{p}\sigma\rangle^p &= 2\,{}^p\langle\mathbf{p}'\sigma'|I_3|\mathbf{p}\sigma\rangle\\ &= \delta_{\mathbf{pp}'}\delta_{\sigma\sigma'} \qquad 8(142)\end{aligned}$$

we introduce a sum over a complete set of physical states $|q\alpha\rangle$ on the left-hand side

$$\sum_{\mathbf{q}\alpha} {}^p\langle\mathbf{p}'\sigma'|I_+(t)|q\alpha\rangle\langle q\alpha|I_-(t)|\mathbf{p}\sigma\rangle^p - \sum_{k\beta} {}^p\langle\mathbf{p}'\sigma'|I_-(t)|k\beta\rangle\langle k\beta|I_+(t)|\mathbf{p}\sigma\rangle^p \qquad 8(143)$$

* Within the context of a particular *model* for the interacting hadrons, such as that considered in Section 5.3, Eq. 8(138) can be checked by applying the canonical equal-time commutation rules to the isotopic spin densities $\mathbf{J}_4(x)$. Thus we get

$$[J_4^i(x), J_4^j(x')]_{t=t'} = -\delta^{(3)}(\mathbf{x}-\mathbf{x}')\varepsilon^{ijk}J_4^k(x)$$

from which 8(138) follows. In general, however, the exact form of the strong interaction is unknown and 8(138) must be viewed as an explicit assumption. [See Gell-Mann (1962).]

where q and k range over the eigenvalues of the total energy-momentum operator and where α or β represent the eigenvalues of whatever other quantum numbers are required to complete the characterization of the physical states. The detailed nature of the sum over $q\alpha$, or $k\beta$, is determined by our assumptions regarding the spectrum of physical states*. Writing

$$\sum_{q\alpha} = \sum_{M_q} \sum_{\mathbf{q}} \sum_{\alpha} \tag{8(144)}$$

where

$$M_q = \sqrt{-q^2}$$

is the invariant mass parameter, and taking into account the selection rules, we see that the first term in 8(143) will pick up a contribution from the one-neutron state at

$$M_q = m$$

and that both terms in 8(143) will contain contributions from continuum states with a threshold

$$M_q = m + m_\pi$$

We begin by evaluating the one-neutron intermediate state contribution, writing

$$\begin{aligned}
{}^p\langle \mathbf{p}'\sigma'|I_+(t)|\mathbf{q}\tau\rangle^n &= -i\int d^3x\,{}^p\langle \mathbf{p}'\sigma'|J_+^4(x)|\mathbf{q}\tau\rangle^n \\
&= -i\int d^3x\, \mathrm{e}^{i(q-p')x}\,{}^p\langle \mathbf{p}'\sigma'|J_+^4(0)|\mathbf{q}\tau\rangle^n \\
&= \delta_{\mathbf{p}'\mathbf{q}}\frac{m}{E_{\mathbf{p}}}\frac{G_V}{G}\bar{u}_{\mathbf{p}'\sigma'}\gamma_4 u_{\mathbf{p}'\tau}
\end{aligned} \tag{8(145)}$$

where the last line follows from 8(104). Using 8(84) we obtain the one-neutron contribution to 8(143) in the form

$$\delta_{\mathbf{p}'\mathbf{p}}\left(\frac{m}{E_{\mathbf{p}}}\right)^2\left(\frac{G_V}{G}\right)^2 \bar{u}_{\mathbf{p}'\sigma'}\gamma_4\frac{\gamma\,.\,p+im}{2im}\gamma_4 u_{\mathbf{p}\sigma}$$

or after a few simple manipulations

$$\delta_{\mathbf{p}'\mathbf{p}}\delta_{\sigma'\sigma}\left(\frac{G_V}{G}\right)^2$$

* See Section 7.4.

Equation 8(142) is therefore of the form

$$\delta_{\mathbf{p}'\mathbf{p}}\delta_{\sigma'\sigma}\left(\frac{G_V}{G}\right)^2 + \text{continuum terms} = \delta_{\mathbf{p}'\mathbf{p}}\delta_{\sigma'\sigma}$$

Now if we suppose that the electromagnetic interactions are switched off, then, since

$$[I_{\pm}, H] = 0$$

the operators $I_{\pm}$ can only connect states with the same energy eigenvalue. Accordingly, only the one-neutron states can contribute to the sum 8(143), and we obtain

$$\left(\frac{G_V}{G}\right)^2 = 1$$

in agreement with the result 8(105) derived previously. In this case we say that the one-neutron state *saturates* the sum rule. The effect of the electromagnetic interaction is to generate additional contributions to the sum rule from matrix elements

$$^{P}\langle \mathbf{p}'\sigma'|I_{+}(t)|q\alpha\rangle \qquad 8(146)$$

where $|q\alpha\rangle$ is a continuum state with $M_q \geqslant (m+m_\pi)$. Writing

$$I_{+}(t) = \mathrm{e}^{-iHt} I_{+}(0)\, \mathrm{e}^{iHt}$$

and

$$^{P}\langle \mathbf{p}'\sigma'|I_{+}(t)|q\alpha\rangle = \mathrm{e}^{-i(q_0 - E_{\mathbf{p}'})t}\, {}^{P}\langle \mathbf{p}'\sigma'|I_{+}(0)|q\alpha\rangle$$

we can express the matrix elements 8(146) in terms of the matrix elements of the time derivative $\dot{I}_{+}(t)$:

$$^{P}\langle \mathbf{p}'\sigma'|I_{+}(t)|q\alpha\rangle = {}^{P}\langle \mathbf{p}'\sigma'|\dot{I}_{+}(t)|q\alpha\rangle \frac{i}{q_0 - E_{\mathbf{p}'}} \qquad 8(147)$$

Furthermore, by using the equation of motion 8(139), the above matrix elements can be related to the electromagnetic interaction, so as to yield a sum rule for $(G_V/G)^2$ in terms of electron–nucleon scattering amplitudes. We shall not pursue this development here [see Fubini (1965)].

Sum Rule for G_A/G We now turn to the important application of the Fubini–Furlan technique to the renormalization of the axial vector coupling constant G_A [Adler (1965a, b), Weisberger (1965)]. Two basic elements are necessary for the application of the Fubini–Furlan procedure.

The first—an expression for the time development of the integrated 'axial charge' or *chirality*

$$\chi_{\pm}(t) = -i\int d^3x A_{\pm}^4(x) \tag{8(148)}$$

is provided by the PCAC hypothesis. The latter may be written in the form

$$\partial_\lambda A_{\pm}^\lambda(x) = \frac{\sqrt{2}mm_\pi^2 G_A}{Z_3^{1/2}G_R G}\phi_{\pi\mp}(x) \tag{8(149)}$$

by combining 8(130) and 8(133). From 8(148) and 8(149), we get, assuming that the spatial components of $A_{\pm}^\mu(x)$ vanish at large distances

$$\dot{\chi}_{\pm}(t) = -i\int d^3x \partial_t A_{\pm}^4(x)$$

$$= \int d^3x \partial_\mu A_{\pm}^\mu(x)$$

$$= \frac{\sqrt{2}mm_\pi^2 G_A}{Z_3^{1/2}G_R G}\int d^3x \phi_{\pi\mp}(x) \tag{8(150)}$$

Thus the time derivative of the chirality operator is proportional to the space integral of the pion field. The second basic element in the Fubini–Furlan approach is the use of a commutation relation for the chirality operators $\chi_{\pm}(t)$ postulated by Gell-Mann (1962)*,

$$[\chi_+(t), \chi_-(t)] = 2I_3 \tag{8(151)}$$

We now follow the same procedure as in the vector current case. Taking the matrix element of 8(151) between one-proton states

$${}^p\langle \mathbf{p}'\sigma'|[\chi_+(t), \chi_-(t)]|\mathbf{p}\sigma\rangle^p = \delta_{\mathbf{p}'\mathbf{p}}\delta_{\sigma'\sigma} \tag{8(152)}$$

* This relation is suggested by the so-called *quark* model of the hadrons based on the triplet representation 5(139) of SU_3. In this model, the vector and axial vector currents are identified as $\mathbf{J}_\mu = i\bar{b}\gamma_\mu \boldsymbol{\tau} b$ and $\mathbf{A}_\mu = i\bar{b}\gamma_\mu\gamma_5\boldsymbol{\tau} b$ respectively, where $b = \binom{b_1}{b_2}$ is an isotopic doublet of Dirac fields similar to $N = \binom{n}{p}$. The equal-time commutation relations then yield

$$[A_4^i(x), A_4^j(x')]_{t=t'} = -\delta^{(3)}(\mathbf{x}-\mathbf{x}')\varepsilon^{ijk}J_4^k(x)$$

from which 8(151) follows.

we insert a sum over a complete set of states on the left-hand side:

$$\sum_{q,\alpha} {}^{p}\langle \mathbf{p}'\sigma'|\chi_+(t)|q\alpha\rangle \langle q\alpha|\chi_-(t)|\mathbf{p}\sigma\rangle^{p} - \sum_{k,\beta} {}^{p}\langle \mathbf{p}'\sigma'|\chi_-(t)|k\beta\rangle \langle k\beta|\chi_+(t)|\mathbf{p}\sigma\rangle^{p} \quad 8(153)$$

The one-neutron intermediate state contributes only to the first term in 8(153). Writing

$${}^{p}\langle \mathbf{p}'\sigma'|\chi_+(t)|\mathbf{q}\tau\rangle^{n} = -i\int d^3x\, {}^{p}\langle \mathbf{p}'\sigma'|A_+^4(x)|\mathbf{q}\tau\rangle^{n}$$

$$= \delta_{\mathbf{p}'\mathbf{q}}\frac{m}{E_{\mathbf{p}}}\frac{G_A}{G}\bar{u}_{\mathbf{p}'\sigma'}\gamma_4\gamma_5 u_{\mathbf{p}'\tau} \quad 8(154)$$

as in 8(145), the one-neutron intermediate state contribution to 8(153) is easily found to be

$$\delta_{\mathbf{p}'\mathbf{p}}\delta_{\sigma'\sigma}\left(\frac{G_A}{G}\right)^2\left(1-\frac{m^2}{E_{\mathbf{p}}^2}\right) \quad 8(155)$$

As in the vector current case, we observe that if the axial vector current were completely conserved, then the one-neutron state would saturate the sum rule and 8(152) would reduce to

$$\left(\frac{G_A}{G}\right)^2\left(1-\frac{m^2}{E_{\mathbf{p}}^2}\right) = 1 \quad 8(156)$$

or $G_A = G$ if we set $m = 0$ for consistency with $\partial_\lambda A_+^\lambda = 0$*. In fact, however, we have

$$\delta_{\mathbf{p}'\mathbf{p}}\delta_{\sigma'\sigma}\left(\frac{G_A}{G}\right)^2\left(1-\frac{m^2}{E_{\mathbf{p}}^2}\right)+\text{continuum terms} = \delta_{\mathbf{p}'\mathbf{p}}\delta_{\sigma'\sigma} \quad 8(157)$$

To evaluate the continuum contribution we write, as in 8(147)

$${}^{p}\langle \mathbf{p}'\sigma'|\chi_+(t)|q\alpha\rangle = {}^{p}\langle \mathbf{p}'\sigma'|\dot{\chi}_+(t)|q\alpha\rangle\frac{i}{q_0-E_{\mathbf{p}'}} \quad 8(158)$$

or, using 8.150)

$${}^{p}\langle \mathbf{p}'\sigma'|\chi_+(t)|q\alpha\rangle = \frac{i\sqrt{2}mm_\pi^2 G_A}{Z_3^{1/2}G_R G(q_0-E_{\mathbf{p}'})}V\delta_{\mathbf{p}'\mathbf{q}}\,\mathrm{e}^{-i(q_0-E_{\mathbf{p}'})t}\,{}^{p}\langle \mathbf{p}'\sigma'|\phi_{\pi^-}(0)|q\alpha\rangle$$

$$= -\frac{i\sqrt{2}mm_\pi^2 G_A}{Z_3^{1/2}G_R G(q_0-E_{\mathbf{p}'})}\frac{1}{(p'-q)^2+m_\pi^2}V\delta_{\mathbf{p}'\mathbf{q}}\,\mathrm{e}^{-i(q_0-E_{\mathbf{p}'})t}$$

$$\times\, {}^{p}\langle \mathbf{p}'\sigma'|j_{\pi^-}(0)|q\alpha\rangle \quad 8(159a)$$

* See the paragraph 8(127). If $m \neq 0$, the argument is inconclusive because 8(156) is incorrect as it stands. This is due to the fact that unphysical zero mass bosons must be introduced to make $\partial_\lambda A_+^\lambda(x) = 0$ consistent with $m \neq 0$ [Nambu (1960b)]. We would then get additional contributions to 8(155) from states featuring one neutron together with any number of zero energy bosons.

where $j_{\pi^-} = (\Box - \mu^2)\phi_{\pi^-}$. Similarly

$$^{p}\langle \mathbf{p}'\sigma'|\chi_{-}(t)|k\beta\rangle$$

$$= -\frac{i\sqrt{2}mm_\pi^2 G_A}{Z_3^{1/2}G_R G(q_0-E_{\mathbf{p}'})}\frac{1}{(p'-q)^2+m_\pi^2}V\delta_{\mathbf{p}'\mathbf{q}}\,e^{-i(q_0-E_{\mathbf{p}'})t}\,{}^{p}\langle \mathbf{p}'\sigma'|j_{\pi^-}(0)|k\beta\rangle \qquad 8(159b)$$

Thus we can write the continuum contribution to 8(157) in the form

$$\delta_{\mathbf{p}'\mathbf{p}}\delta_{\sigma'\sigma}\frac{2m^2m_\pi^4G_A^2}{Z_3G_R^2G^2}\sum_{\substack{q,\alpha\\(M_q\neq m)}}V^2\delta_{\mathbf{pq}}\frac{|^{p}\langle \mathbf{p}\sigma|j_{\pi^-}(0)|q\alpha\rangle|^2}{(q_0-E_{\mathbf{p}})^2[(p-q)^2+m_\pi^2]^2}-(\pi^-\to\pi^+) \qquad 8(160)$$

where we have set $t = 0$. We now define

$$\frac{1}{2}\sum_{\sigma=1}^{2}\sum_{\alpha}\left|\left(\frac{E_{\mathbf{p}}V}{m}\right)^{1/2}{}^{p}\langle \mathbf{p}\sigma|j_{\pi^\pm}(0)|q\alpha\rangle\left(\frac{q_0V}{M_q}\right)^{1/2}\right|^2 = Q_\pm[M_q,(p-q)^2] \qquad 8(161)$$

$Q_\pm$ must be a Lorentz scalar, since the factors $(E_{\mathbf{p}}V/m)^{1/2}$ and $(q_0V/M_q)^{1/2}$ cancel out the normalization factors in the matrix element $^{p}\langle \mathbf{p}\sigma|\phi_{\pi^\pm}(0)|q\alpha\rangle$. Moreover, as all internal variables are summed over, $Q_\pm$ can only be a function of the two independent invariants $q^2 = -M_q^2$ and $(p-q)^2$, as indicated. Averaging 8(157) over the proton spin we can now write the sum rule for $(G_A/G)^2$ as

$$1 = \left(\frac{G_A}{G}\right)^2\left(1-\frac{m^2}{E_{\mathbf{p}}^2}\right)+\frac{2m^2m_\pi^4G_A^2}{Z_3G_R^2G^2}\sum_{\substack{q\\M_q\neq m}}\delta_{\mathbf{pq}}\frac{1}{(q_0-E_{\mathbf{p}})^2E_{\mathbf{p}}q_0}$$

$$\times\frac{mM_q}{[(p-q)^2+m_\pi^2]^2}\{Q_-[M_q,(p-q)^2]-Q_+[M_q,(p-q)^2]\} \qquad 8(162)$$

As the next step we re-express the q_0-dependent factors in terms of M_q. Writing

$$q_0 = (\mathbf{q}^2+M_q^2)^{1/2} = (\mathbf{p}^2+M_q^2)^{1/2} = (E_{\mathbf{p}}^2+M_q^2-m^2)^{1/2} \qquad 8(163)$$

and

$$(q_0-E_{\mathbf{p}})^{-2} = (q_0+E_{\mathbf{p}})^2(q_0^2-E_{\mathbf{p}}^2)^{-2} = (q_0+E_{\mathbf{p}})^2(M_q^2-m^2)^{-2} \qquad 8(164)$$

the sum rule 8(162) becomes

$$1 = \left(\frac{G_A}{G}\right)^2\left(1-\frac{m^2}{E_{\mathbf{p}}^2}\right)+\frac{2m^2m_\pi^4G_A^2}{Z_3G_R^2G^2}\sum_{M_q\neq m}\sum_{\mathbf{q}}\delta_{\mathbf{pq}}\left\{\frac{[E_{\mathbf{p}}+(E_{\mathbf{p}}^2+M_q^2-m^2)^{1/2}]^2}{E_{\mathbf{p}}(E_{\mathbf{p}}^2+M_q^2-m^2)^{1/2}}\right\}$$

$$\times\frac{mM_q}{(M_q^2-m^2)^2[(p-q)^2+m_\pi^2]^2}\{Q_-[M_q,(p-q)^2]-Q_+[M_q,(p-q)^2]\} \qquad 8(165)$$

Equation 8(165) actually represents a family of sum rules for $(G_A/G)^2$—one for each value of $E_{\mathbf{p}}$. A useful sum rule—one in which the continuum contributions are related to experimentally measured quantities—is obtained by taking the limit $E_{\mathbf{p}} \to \infty$. We shall assume that this limit may be taken *inside* the summation over $\mathbf{q}$ and M_q*. Then noting that the limit of the quantity in curly brackets is 4 and that the limit of the momentum transfer

$$(p-q)^2 = -(E_{\mathbf{p}}-q_0)^2 = -[E_{\mathbf{p}}-(E_{\mathbf{p}}^2+M_q^2-m^2)^{1/2}]^2$$

is zero, we can write 8(165) in the simple form

$$1-\left(\frac{G}{G_A}\right)^2 = \frac{2m^2}{Z_3 G_R^2}\int_{m+m_\pi}^{\infty} dW \frac{4mW}{(W^2-m^2)^2}[K_+(W,0)-K_-(W,0)] \quad 8(166)$$

where we have set

$$\sum_{M_q} \equiv \int dW \sum_{M_q} \delta(W-M_q)$$

and defined

$$K_\pm[W,(p-q)^2] = \sum_{M_q \neq m} \delta(W-M_q)Q_\pm[M_q,(p-q)^2] \quad 8(167)$$

The crucial point of the Adler–Weisberger derivation is the recognition that for $(p-q)^2 = -m_\pi^2$, *the quantity* $K_\pm[W,(p-q)^2]$ *is directly related to the total unpolarized* $\pi^\pm - p$ *cross section* $\sigma_\pm(W)$ *at centre of mass energy* W. The exact relation, which will be established below is†

$$\sigma_\pm(W) = \frac{m}{|\mathbf{k}|}Z_3^{-1}\pi K_\pm(W,-m_\pi^2) \quad 8(168)$$

where $\mathbf{k}$ is the pion momentum. Thus the sum rule 8(166) relates $(G_A/G)^2$ to the $\pi^\pm - p$ cross sections for *zero* pion mass. For $m_\pi = 0$, the total centre of mass energy W is

$$W = E_{\mathbf{k}}+|\mathbf{k}| \quad 8(169)$$

* For a discussion of the significance of this assumption in terms of dispersion relations, see Adler (1965a, b).

† Note the key condition that $(p-q)^2$ should lie on the pion mass shell. This explains why it is necessary to take the limit $E_{\mathbf{p}} \to \infty$ in 8(165). Without this limiting process, the momentum transfer $(p-q)^2$ would vary with the summation over M_q and $\mathbf{q}$ and it would not be possible to relate the continuum terms to an integral over $\pi - p$ cross sections for a *fixed* pion mass.

so that

$$|\mathbf{k}| = \frac{W^2-m^2}{2W} \tag{8(170)}$$

and the $\pi^\pm - p$ cross section is given by

$$\sigma_\pm^{(m_\pi=0)}(W) = \frac{2m}{W^2-m^2} Z_3^{-1} \pi K_\pm(W, 0) \tag{8(171)}$$

Comparing with 8(166) we obtain the Adler–Weisberger sum rule

$$1 - \left(\frac{G}{G_A}\right)^2 = \frac{4m^2}{G_R^2 \pi} \int_{m+m_\pi}^{\infty} \frac{W\,dW}{W^2-m^2} [\sigma_+^{(m_\pi=0)}(W) - \sigma_-^{(m_\pi=0)}(W)] \tag{8(172)}$$

Experimentally this sum rule is very well satisfied. If the *physical* cross sections $\sigma_\pm(W)$ are inserted into the integral on the right-hand side, 8(172) gives the result

$$|G_A/G| = 1{\cdot}16$$

A more careful calculation, taking into account the corrections which arise in extrapolating $\sigma_\pm(W)$ to zero pion mass, yields

$$|G_A/G| = 1{\cdot}24 \tag{8(173)}$$

in good agreement with the experimental value $G_A/G = 1{\cdot}18$. We refer the reader to Adler and Weisberger's original papers for the details. To complete the derivation of 8(172) we now turn to the proof of Eq. 8(168).

The Optical Theorem The definition

$$S_{ab} = \langle a \text{ out} | b \text{ in} \rangle$$

of the S-matrix and the assumed completeness of the in and out states implies that S must be a unitary matrix. Hence we have the condition

$$S_{na}{}^* S_{nb} = \delta_{ab} \tag{8(174)}$$

or setting $S_{ab} = \delta_{ab} + iR_{ab}$,

$$i(R_{ab} - R_{ba}{}^*) = -\sum_n R_{na}{}^* R_{nb} \tag{8(175)}$$

Let us apply the above unitarity condition to $\pi - N$ scattering.

$$\mathbf{k} + (\mathbf{p}\sigma) \rightarrow \mathbf{k}' + (\mathbf{p}'\sigma') \tag{8(176)}$$

Then R_{ab} is of the form

$$R(\mathbf{k}', \mathbf{p}'\sigma'; \mathbf{k}, \mathbf{p}\sigma) = \frac{(2\pi)^4\delta^{(4)}(p'+k'-p-k)}{n_{\mathbf{p}'}n_{\mathbf{k}'}n_{\mathbf{p}}n_{\mathbf{k}}V^2}T(\mathbf{k}', \mathbf{p}'\sigma'; \mathbf{k}, \mathbf{p}\sigma) \quad 8(177)$$

where $n_{\mathbf{p}} = (E_{\mathbf{p}}/m)^{1/2}$ for nucleons and $n_{\mathbf{k}} = (2\omega_{\mathbf{k}})^{1/2}$ for pions. On the other hand, the matrix elements $R_{na}{}^*$ and R_{nb} which appear in the sum on the right-hand side of 8(175) are transition amplitudes from the initial $\pi - N$ state to a final n-particle state n*. R_{na} is then of the form

$$R(n; \mathbf{k}, \mathbf{p}\sigma) = \frac{(2\pi)^4\delta^{(4)}(P_n-p-k)}{N_n n_{\mathbf{p}}n_{\mathbf{k}}V^{1+n/2}}T(n; \mathbf{k}, \mathbf{p}\sigma) \quad 8(178)$$

where N_n is the product of the normalization factors n_i for the particles in the final state n and P_n is their total energy-momentum vector. Using 8(177) and 8(178) we write the unitarity condition in the form

$$\frac{i(2\pi)^4\delta^{(4)}(p'+k'-p-k)}{n_{\mathbf{p}'}n_{\mathbf{k}'}n_{\mathbf{p}}n_{\mathbf{k}}V^2}(T(\mathbf{k}', \mathbf{p}'\sigma'; \mathbf{k}, \mathbf{p}\sigma) - T^*(\mathbf{k}, \mathbf{p}\sigma; \mathbf{k}', \mathbf{p}'\sigma'))$$

$$= -\sum_n \frac{(2\pi)^8\delta^{(4)}(P_n-p-k)\delta^{(4)}(P_n-p'-k')}{N_n^2 n_{\mathbf{p}}n_{\mathbf{k}}n_{\mathbf{p}'}n_{\mathbf{k}'}V^{2+n}}T^*(n; \mathbf{k}', \mathbf{p}'\sigma')T(n; \mathbf{k}, \mathbf{p}\sigma)$$

or

$$i(T(\mathbf{k}', \mathbf{p}'\sigma'; \mathbf{k}, \mathbf{p}\sigma) - T^*(\mathbf{k}, \mathbf{p}\sigma; \mathbf{k}', \mathbf{p}'\sigma'))$$

$$= -(2\pi)^4 \sum_n \frac{\delta^{(4)}(P_n-p-k)}{N_n^2 V^n}T^*(n; \mathbf{k}', \mathbf{p}'\sigma')T(n; \mathbf{k}, \mathbf{p}\sigma)$$

which reduces to

$$2\,\mathrm{Im}\,T(\mathbf{k}, \mathbf{p}\sigma; \mathbf{k}, \mathbf{p}\sigma) = (2\pi)^4 \sum_n \frac{\delta^{(4)}(P_n-p-k)}{N_n^2 V^n}|T(n; \mathbf{k}, \mathbf{p}\sigma)|^2 \quad 8(179)$$

upon setting $\mathbf{k} = \mathbf{k}'$, $\mathbf{p} = \mathbf{p}'$ and $\sigma = \sigma'$. We now observe that the sum on the right-hand side of 8(179) is closely related to the total $\pi - N$ cross section in the centre of mass system. The latter is given in terms of the amplitude $T(n; \mathbf{k}, \mathbf{p}\sigma)$ by the expression†

$$\sigma_{\mathrm{pol}} = (2\pi)^4 \frac{1}{F}\frac{1}{n_{\mathbf{p}}^2 n_{\mathbf{k}}^2 V}\sum_n \frac{\delta^{(4)}(P_n-p-k)}{N_n^2 V^n}|T(n; \mathbf{k}, \mathbf{p}\sigma)|^2 \quad 8(180)$$

* The final state n will contain, in general, additional mesons and baryon pairs, as well as the scattered pion and nucleon, in accordance with the selection rules and energy-momentum conservation.

† See 6(136). σ_{pol} is the cross section for a given initial proton spin state.

where

$$F = \frac{1}{V}\left(\frac{|\mathbf{k}|}{\omega_{\mathbf{k}}} + \frac{|\mathbf{k}|}{E_{\mathbf{k}}}\right) \tag{8(181)}$$

is the incident flux in the centre of mass system. Combining 8(180) and 8(181) we have

$$\sigma_{\text{pol}} = (2\pi)^4 \frac{m}{2|\mathbf{k}|W} \sum_n \frac{\delta^{(4)}(P_n - p - k)}{N_n^2 V^n} |T(n; \mathbf{k}, \mathbf{p}\sigma)|^2 \tag{8(182)}$$

where W is the total centre of mass energy

$$W = E_{\mathbf{k}} + \omega_{\mathbf{k}} \tag{8(183)}$$

Comparison of 8(182) with 8(179) yields the so-called *optical theorem*

$$\operatorname{Im} T(\mathbf{k}, \mathbf{p}\sigma; \mathbf{k}, \mathbf{p}\sigma) = \frac{|\mathbf{k}|W}{m} \sigma_{\text{pol}} \tag{8(184)}$$

relating the imaginary part of the forward scattering amplitude to the total cross section. In the following, we shall apply the optical theorem for the *unpolarized* total cross section, averaging both sides of 8(184) over the initial proton spin states.

We now compute the imaginary part of $T(\mathbf{k}, \mathbf{p}\sigma; \mathbf{k}, \mathbf{p}\sigma)$ for the case of $\pi^\pm p$ scattering, using the reduction formalism. For $\pi^0 p$ scattering, the relevant term in the scattering amplitude iR_{ab} is given by 8(81); to describe $\pi^\pm p$ scattering we need only make the replacement*

$$\langle \mathbf{p}'\sigma' | j(0) | q\alpha \rangle \langle q\alpha | j(0) | \mathbf{p}\sigma \rangle \rightarrow {}^p\langle \mathbf{p}'\sigma' | j_{\pi\pm}(0) | q\alpha \rangle \langle q\alpha | j_{\pi\mp}(0) | \mathbf{p}\sigma \rangle^p$$

Then, taking into account the definition 8(177) of T_{ab}, we find

$$\begin{aligned} \operatorname{Im} T_\pm(\mathbf{k}, \mathbf{p}\sigma; \mathbf{k}, \mathbf{p}\sigma) = Z_3^{-1}(n_{\mathbf{p}}^2) V^2 \pi \sum_{M_q \neq m} \int d^3 q \, \delta(q_0 - E_{\mathbf{p}} - \omega_{\mathbf{k}}) \\ \times \delta^{(3)}(\mathbf{q} - \mathbf{p} - \mathbf{k}) \sum_\alpha |{}^p\langle \mathbf{p}\sigma | j_{\pi\pm}(0) | q\alpha \rangle|^2 \end{aligned} \tag{8(185)}$$

where we have used 8(144), replacing $\Sigma_{\mathbf{q}}$ by $V(2\pi)^{-3} \int d^3 q$. The one-neutron pole terms give no contribution to the imaginary part in the physical region. The same applies to the 'crossed' continuum contribution in 8(81): in the centre of mass system $\mathbf{p} + \mathbf{k} = 0$, the values of q^2 corresponding to a non-vanishing imaginary part are

$$\begin{aligned} q^2 &= (\mathbf{p} - \mathbf{k}')^2 - (E_{\mathbf{p}} - \omega_{\mathbf{k}'})^2 \\ &= 2\mathbf{k}^2(1 + \cos\theta) + (E_{\mathbf{p}} - \omega_{\mathbf{k}})^2 \end{aligned} \tag{8(186)}$$

* See Problem 4.

and for real momenta and physical scattering angles, 8(186) yields an invariant mass M_q which is always less than $m+m_\pi$. Averaging 8(185) over proton spins and recalling the definition 8(161) of $Q_\pm[M_q, (p-q)^2]$, we have

$$\frac{1}{2}\sum_\sigma \operatorname{Im} T_\pm(\mathbf{k}, \mathbf{p}\sigma; \mathbf{k}, \mathbf{p}\sigma) = Z_3^{-1}\pi \sum_{M_q \neq m} \int d^3q\delta(q_0 - E_\mathbf{p} - \omega_\mathbf{k})$$

$$\delta^{(3)}(\mathbf{q}-\mathbf{p}-\mathbf{k})\frac{M_q}{q_0}Q_\pm[M_q, (p-q)^2]$$

$$= Z_3^{-1}\pi \sum_{M_q \neq m} \int d^3q\delta(q_0 - E_\mathbf{p} - \omega_\mathbf{k})$$

$$\delta^{(3)}(\mathbf{q}-\mathbf{p}-\mathbf{k})\frac{M_q}{q_0}Q_\pm(M_q, -m_\pi^2) \qquad 8(187)$$

where we have set

$$(p-q)^2 = (\mathbf{p}-\mathbf{q})^2 - (E_\mathbf{p} - q_0)^2 = \mathbf{k}^2 - \omega_\mathbf{k}^2 = -m_\pi^2 \qquad 8(188)$$

by virtue of the delta function factors in 8(187). If we now apply the optical theorem and the centre of mass equations

$$\mathbf{p}+\mathbf{k} = 0 \qquad \mathbf{q} = 0$$

$$E_\mathbf{k} + \omega_\mathbf{k} = W \qquad q_0 = M_q \qquad 8(189)$$

we see that the unpolarized cross section is given by

$$\sigma_\pm(W) = \frac{m}{|\mathbf{k}|W} Z_3^{-1}\pi \sum_{M_q \neq m} \int d^3q\delta(q_0 - W)\delta^{(3)}(\mathbf{q})\frac{M_q}{q_0}Q_\pm(M_q, -m_\pi^2)$$

$$= \frac{m}{|\mathbf{k}|W} Z_3^{-1}\pi \sum_{M_q \neq m} \delta(M_q - W)Q_\pm(M_q, -m_\pi^2) \qquad 8(190)$$

Using 8(167) we obtain

$$\sigma_\pm(W) = \frac{m}{|\mathbf{k}|W} Z_3^{-1}\pi K_\pm(W, -m_\pi^2) \qquad 8(191)$$

as stated earlier.

8-4 Spectral representations of two point functions

Introduction In this section we derive certain general properties of vacuum expectation values of products of two Heisenberg field operators*

$$F(x_1 - x_2) = \langle 0|\phi_A(x_1)\phi_B(x_2)|0\rangle \qquad 8(192)$$

* As indicated in 8(192) and shown in Section 7.5 (cf. the paragraph following 7(95)), a vacuum expectation value of the type 8(192) can only depend on the difference $x_1 - x_2$.

Functions of the type 8(192) are known as 'two-point functions'. By inserting between ϕ_A and ϕ_B a summation

$$\sum_{q\alpha} |q\alpha\rangle\langle q\alpha|$$

over a complete set of physical states, we can exhibit these functions in the form of 'spectral' integrals over the corresponding free functions [Umezawa (1951), Källén (1952), Lehmann (1954)]. In turn, these spectral representation yields sum rules for the self masses and wave function renormalization constants.

Neutral Pseudoscalar Field We begin by treating the simple case in which ϕ_A represents a neutral pseudoscalar field. Let us consider the two-point function

$$i\Delta^{(+)}(x_1 - x_2) = \langle 0|\phi(x_1)\phi(x_2)|0\rangle \qquad 8(193)$$

and insert a sum over a complete set of physical states between $\phi(x_1)$ and $\phi(x_2)$:

$$i\Delta^{(+)\prime}(x_1 - x_2) = \sum_{k\alpha} \langle 0|\phi(x_1)|k\alpha\rangle\langle k\alpha|\phi(x_2)|0\rangle \qquad 8(194)$$

As in 8(144) we can break up the summation over k into a sum over $\mathbf{k}$ and a sum over the invariant mass parameter $M_k = \sqrt{-k^2}$. Substituting

$$\sum_{\mathbf{k}} \rightarrow V(2\pi)^{-3} \int d^3k$$

we can write 8(194) as

$$\begin{aligned} i\Delta^{(+)\prime}(x_1 - x_2) &= \sum_{M_k} \frac{V}{(2\pi)^3} \int d^3k \sum_{\alpha} \langle 0|\phi(x_1)|k\alpha\rangle\,\langle k\alpha|\phi(x_2)|0\rangle \\ &= \sum_{M_k} \frac{V}{(2\pi)^3} \int d^4k\,\theta(k_0)\delta(k_0 - \sqrt{\mathbf{k}^2 + M_k^2})\, e^{ik(x_1 - x_2)} \\ &\quad \times \sum_{\alpha} |\langle 0|\phi(0)|k\alpha\rangle|^2 \end{aligned} \qquad 8(195)$$

where in deriving the last line we have set

$$\langle 0|\phi(x_1)|k\alpha\rangle = e^{ik.x_1}\langle 0|\phi(0)|k\alpha\rangle \qquad 8(196a)$$

$$\langle k\alpha|\phi(x_2)|0\rangle = e^{-ik.x_2}\langle k\alpha|\phi(0)|0\rangle \qquad 8(196b)$$

and inserted a factor $\theta(k_0)$ in the integral over d^4k. The latter step is permissible, since all physical states must have positive-definite energy.

Finally we set

$$\sum_{M_k} = \int_0^\infty d\sigma^2 \sum_{M_k} \delta(\sigma^2 - M_k^2)$$

and define the function

$$\rho(\sigma^2) = \sum_{M_k} \delta(\sigma^2 - M_k^2) \sum_\alpha |\langle 0|\phi(0)|k\alpha\rangle (2k_0 V)^{1/2}|^2 \qquad 8(197)$$

Note that ρ is both *Lorentz invariant* and *positive-definite*; since all internal variables are summed over the right-hand side, ρ can only be a function of the invariant $k^2 = -\sigma^2$. With the above definition, we can write 8(195) in the form

$$i\Delta^{(+)\prime}(x_1 - x_2) = \frac{1}{(2\pi)^3} \int_0^\infty d\sigma^2 \rho(\sigma^2) \int d^4k \theta(k_0) \frac{1}{2k_0} \delta(k_0 - \sqrt{\mathbf{k}^2 + \sigma^2})$$
$$\times e^{ik.(x_1 - x_2)}$$

$$= \frac{1}{(2\pi)^3} \int_0^\infty d\sigma^2 \rho(\sigma^2) \int d^4k \theta(k_0) \delta(k^2 + \sigma^2) e^{ik.(x_1 - x_2)} \qquad 8(198)$$

The integral over d^4k is now readily identified as the free function $i\Delta^{(+)}(x;\sigma^2)$ for rest mass σ. Indeed, by 3(84a),

$$i\Delta^{(+)}(x;\sigma^2) = \frac{1}{(2\pi)^3} \int d^4k\, \theta(k_0) \delta(k^2 + \sigma^2) e^{ik.x}$$

Thus 8(198) may be written as a weighted integral over the invariant mass parameter of the free field function:

$$i\Delta^{(+)\prime}(x_1 - x_2) = \int_0^\infty d\sigma^2 \rho(\sigma^2) \Delta^{(+)}(x_1 - x_2; \sigma^2) \qquad 8(199)$$

Equation 8(199) is the Lehmann spectral representation for $\Delta^{(+)\prime}(x_1 - x_2)$. The invariant function $\rho(\sigma^2)$ is known as the *spectral function*. With the usual assumptions about the nature of the spectrum of physical states, we see that the integral over $d\sigma^2$ will contain a discrete contribution from the one-meson state at $\sigma = \mu$, μ being the meson mass, and a continuum contribution from scattering states with zero baryon number. The vacuum state, at $\sigma = 0$, will give no contribution; the invariance of the vacuum under a parity transformation and the transformation law

5(165b) implies that

$$\langle 0|\phi(0)|0\rangle = \langle 0|\mathscr{P}\phi(0)\mathscr{P}^{-1}|0\rangle = -\langle 0|\phi(0)|0\rangle = 0$$

so that $\rho(0) = 0$.

The spectral representation of the function

$$-i\Delta^{(-)\prime}(x_1 - x_2) = \langle 0|\phi(x_2)\phi(x_1)|0\rangle \qquad 8(200)$$

is derived in exactly the same way. We find

$$\Delta^{(-)\prime}(x_1 - x_2) = \int_0^\infty d\sigma^2 \rho(\sigma^2)\Delta^{(-)}(x_1 - x_2; \sigma^2) \qquad 8(201)$$

with the *same* spectral function $\rho(\sigma^2)$. Combining 8(199) and 8(201), we obtain the spectral representation for the boson propagator,

$$\Delta_F'(x_1 - x_2) = \int_0^\infty d\sigma^2 \rho(\sigma^2)\Delta_F(x_1 - x_2; \sigma^2) \qquad 8(202)$$

or, in momentum space,

$$\Delta_F'(k^2) = \int_0^\infty d\sigma^2 \rho(\sigma^2)\Delta_F(k^2; \sigma^2) = \int_0^\infty d\sigma^2 \rho(\sigma^2)\frac{-i}{k^2 + \sigma^2 - i\varepsilon} \qquad 8(203)$$

It is frequently useful to isolate the contribution of the one-meson state to the spectral integral. Comparing 8(203) with 7(96c) we see that

$$\rho(\sigma^2) = Z_3\delta(\sigma^2 - \mu^2) + \text{continuum contributions} \qquad 8(204)$$

and

$$i\Delta_F'(k^2) = \frac{Z_3}{k^2 + \mu^2 - i\varepsilon} + \int_{m_1^2}^\infty d\sigma^2 \frac{\rho(\sigma^2)}{k^2 + \sigma^2 - i\varepsilon} \qquad 8(205)$$

where m_1, the mass of the lightest continuum state, is effectively 3μ since $\langle 0|\phi|2\pi\rangle$ vanishes by space reflection invariance.

Constraint on $\rho(\sigma^2)$ The spectral function $\rho(\sigma^2)$ cannot, of course, be determined exactly. However, we can derive a constraint equation for ρ by enlisting the aid of the canonical equal-time commutation relations. From 8(200) and 8(201) we obtain the spectral representation of the exact commutator function

$$\Delta'(x_1 - x_2) = -i\langle 0|[\phi(x_1), \phi(x_2)]|0\rangle$$

in the form

$$\Delta'(x_1 - x_2) = \int_0^\infty d\sigma^2 \rho(\sigma^2) \Delta(x_1 - x_2; \sigma^2) \qquad 8(206)$$

Let us take the derivative of both sides with respect to x_1^0, and set $x_1^0 = x_2^0$:

$$\frac{\partial}{\partial t_1} \Delta'(x_1 - x_2)\bigg|_{t_1 = t_2} = \int_0^\infty d\sigma^2 \rho(\sigma^2) \frac{\partial}{\partial t_1} \Delta(x_1 - x_2; \sigma^2)\bigg|_{t_1 = t_2} \qquad 8(207)$$

On the other hand, the canonical equal-time commutation rules with $\pi = \dot{\phi}$ give

$$i\frac{\partial}{\partial t_1} \Delta'(x_1 - x_2)\bigg|_{t_1 = t_2} = i\frac{\partial}{\partial t_1} \Delta(x_1 - x_2)\bigg|_{t_1 = t_2} = -\delta^{(3)}(\mathbf{x} - \mathbf{x}') \qquad 8(208)$$

Comparing with 8(207) we get the important constraint

$$\int_0^\infty \rho(\sigma^2)\, d\sigma^2 = 1 \qquad 8(209)$$

Sum Rule for Z_3 Equation 8(209) provides a non-perturbative characterization of Z_3. Isolating the contribution of the one-particle states by means of 8(204), we get the sum rule

$$Z_3 + \int_{m_1^2}^\infty d\sigma^2 \rho(\sigma^2) = 1 \qquad 8(210)$$

expressing Z_3 in terms of a sum of squares of physical matrix elements of the form $\langle 0|\phi(0)|q\alpha\rangle$. From 8(210) and the fact that $\rho(\sigma^2)$ is positive-definite, we deduce that

$$0 \leqslant Z_3 < 1 \qquad 8(211)$$

This constraint was referred to in Section 7.5 in connection with the question of strong and weak operator convergence. Z_3 can only be unity if $\rho(\sigma^2)$ is zero for all $\sigma^2 \geqslant m_1^2$, i.e. if all matrix elements $\langle 0|\phi|n \text{ part.}\rangle$ vanish for $n > 1$. Obviously this possibility is realized only for free fields.

Mass Formula To make further headway we need more explicit dynamical assumptions. If we assume that ϕ obeys the Lagrangian equation of motion

$$(\Box - \mu_0^2)\phi(x) = iG_0 \tfrac{1}{2}[\bar{\psi}(x), \gamma_5 \psi(x)] \qquad 8(212)$$

then we can derive the following sum rule for the self-mass $\delta\mu^2$:

$$\delta\mu^2 = \int_0^\infty (\mu^2 - \sigma^2)\rho(\sigma^2)\, d\sigma^2 \qquad 8(213)$$

To show this, we operate with the Klein–Gordon operator on both sides of 8(206) and use 8(212):

$$(\Box_x - \mu_0^2)\langle 0|[\phi(x), \phi(y)]|0\rangle$$

$$= iG_0\frac{1}{2}\langle 0|[[\overline{\psi}(x), \gamma_5\psi(x)], \phi(y)]|0\rangle = i\int_0^\infty d\sigma^2\rho(\sigma^2)(\sigma^2 - \mu_0^2)\Delta(x-y\,;\sigma^2)$$

We now differentiate with respect to y_0 and set $x_0 = y_0$. Since

$$[\psi(x), \dot{\phi}(y)] = [\overline{\psi}(x), \dot{\phi}(y)] = 0 \qquad \text{for} \quad x_0 = y_0$$

we have

$$[[\overline{\psi}(x), \gamma_5\psi(x)], \dot{\phi}(y)] = 0 \qquad \text{for} \quad x_0 = y_0$$

Hence, using 8(208), we obtain

$$\int d\sigma^2\,\rho(\sigma^2)(\sigma^2 - \mu_0^2) = 0$$

which is easily seen to be equivalent to 8(213) by virtue of 8(209). Notice that the lower limit of integration in 8(213) is effectively $\sigma^2 = \mu^2$:

$$\delta\mu^2 = -\int_{\mu^2}^\infty (\sigma^2 - \mu^2)\rho(\sigma^2)\, d\sigma^2 \qquad 8(214)$$

We conclude that $\delta\mu^2$ is negative, i.e. that the bare mass is larger than the physical mass*.

Renormalized Propagator Frequently the spectral decomposition is applied to the renormalized propagator $\boldsymbol{\Delta}_{\mathrm{F}}'$. The latter is related to Δ_{F}' by

$$\boldsymbol{\Delta}_{\mathrm{F}}' = Z_3^{-1}\Delta_{\mathrm{F}}' \qquad 8(215)$$

and may be written in terms of the renormalized fields $\phi_R = Z_3^{-1/2}\phi$ as

$$\boldsymbol{\Delta}_{\mathrm{F}}'(x_1 - x_2) = \langle 0|T\phi_R(x_1)\phi_R(x_2)|0\rangle \qquad 8(216)$$

* The opposite is true for Dirac particles.

We have then, instead of 8(202) and 8(203),

$$\Delta'_F(x_1 - x_2) = \int_0^\infty d\sigma^2 \rho_R(\sigma^2) \Delta_F(x_1 - x_2; \sigma^2) \qquad 8(217)$$

and

$$\Delta'_F(k^2) = \int_0^\infty d\sigma^2 \rho_R(\sigma^2) \Delta_F(k^2) \qquad 8(218)$$

where $\rho_R = Z_3^{-1}\rho$ satisfies the constraint

$$Z_3^{-1} = \int_0^\infty \rho_R(\sigma^2)\, d\sigma^2 \qquad 8(219)$$

Instead of 8(213), the mass formula now reads

$$\delta\mu^2 = Z_3 \int_0^\infty (\mu^2 - \sigma^2) \rho_R(\sigma^2)\, d\sigma^2 \qquad 8(220)$$

Perturbative Calculation of ρ_R The advantage of working with the renormalized propagator and spectral function is that we can deal with renormalized quantities throughout. Thus we can use the equation of motion for $\phi_R = Z_3^{-1/2}\phi$ in terms of $\psi_R = Z_2^{-1/2}\psi$

$$(\Box - \mu^2)\phi_R = iG_R Z_1 Z_3^{-1} \tfrac{1}{2}[\bar{\psi}_R, \gamma_5 \psi_R] - \delta\mu^2 \phi_R \qquad 8(221)$$

to calculate ρ_R as a power series in the *renormalized* coupling constant [Lehmann (1954)]. Let us expand ρ_R according to

$$\rho_R(\sigma^2) = \rho_R^{(0)}(\sigma^2) + G_R^2 \rho_R^{(1)}(\sigma^2) + \ldots$$

and compute ρ_R to terms of order G_R^2. In zeroth order, clearly

$$\Delta'_F = \Delta'_F = \Delta_F$$

so that

$$\rho_R^{(0)}(\sigma^2) = \delta(\sigma^2 - \mu^2)$$

To determine $\rho_R^{(1)}$, we form the expression

$$(\Box_x - \mu^2)(\Box_y - \mu^2)\Delta^{(+)\prime}(x-y) = -i\langle 0|(\Box_x - \mu^2)\phi_R(x)(\Box_y - \mu^2)\phi_R(y)|0\rangle$$

and use the equation of motion 8(221) to evaluate the right-hand side to terms of order G_R^2. We easily find

$$\langle 0|(\square_x-\mu^2)\phi_R(x)(\square_y-\mu^2)\phi_R(y)|0\rangle$$

$$\cong -\tfrac{1}{4}G_R^2\langle 0|[\bar{\psi}_R(x),\gamma_5\psi_R(x)]\,[\bar{\psi}_R(y),\gamma_5\psi_R(y)]|0\rangle$$

$$\cong G_R^2\,\mathrm{Tr}\,[\gamma_5 S^{(+)}(x-y)\gamma_5 S^{(-)}(y-x)]$$

where we have dropped all higher order terms; in particular we have set $Z_3 = Z_1 = 1$, dropped the contribution from the $\delta\mu^2$ term in 8(121), and replaced the exact vacuum state and renormalized field operators by the corresponding unperturbed quantities. Thus,

$$(\square_x-\mu^2)(\square_y-\mu^2)\boldsymbol{\Delta}^{(+)\prime}(x-y)$$
$$= -iG_R^2\,\mathrm{Tr}\,[\gamma_5 S^{(+)}(x-y)\gamma_5 S^{(-)}(y-x)]$$
$$= 4iG_R^2[\partial_\mu\Delta^{(+)}(x-y)\partial_\mu\Delta^{(+)}(x-y)+\mu^2\Delta^{(+)}(x-y)\Delta^{(+)}(x-y)]$$

Using 3(84) and the spectral representation of $\boldsymbol{\Delta}^{(+)\prime}$, the above equality can be transformed to

$$-i\frac{1}{(2\pi)^3}\theta(k_0)(\sigma^2-\mu^2)^2\rho_R^{(1)}(\sigma^2)$$

$$= 4i\frac{1}{(2\pi)^6}\int d^4p\,\theta(p_0)\theta(k_0-p_0)\delta(p^2+m^2)\delta[(k-p)^2+m^2]$$

$$\times[-m^2-p^2+k\,.\,p] \qquad 8(222)$$

In the rest frame, $\mathbf{k} = 0$, $\mathbf{k}_0 = \sigma$, this reduces to

$$(\sigma^2-\mu^2)^2\rho_R^{(1)}(\sigma^2) = -\frac{4}{(2\pi)^3}\int d^4p\,\theta(p_0)\theta(\sigma-p_0)\delta(p_0^2-\mathbf{p}^2-m^2)$$

$$\times\,\delta[m^2-\sigma^2+\mathbf{p}^2-p_0^2+2p_0\sigma][-m^2-p^2+p_0^2-p_0\sigma]$$

The integral on the right-hand side is now easily evaluated; the final result for $\rho_R^{(1)}$ is

$$\rho_R^{(1)}(\sigma^2) = \frac{1}{8\pi^2}\theta(\sigma^2-4m^2)\frac{(\sigma^2-4m^2)^{1/2}}{(\sigma^2-\mu^2)^2}\sigma \qquad 8(223)$$

Thus we get, to order G_R^2

$$\rho_R(\sigma^2) = \delta(\sigma^2-\mu^2)+\frac{G_R^2}{8\pi^2}\theta(\sigma^2-4m^2)\frac{(\sigma^2-4m^2)^{1/2}}{(\sigma^2-\mu^2)^2}\sigma \qquad 8(224)$$

$$\Delta'_F(k^2) = \frac{-i}{k^2+\mu^2} - \frac{iG_R^2}{8\pi^2}\int_{4m^2}^{\infty} d\sigma^2 \frac{(\sigma^2-4m^2)^{1/2}\sigma}{(\sigma^2-\mu^2)^2(k^2+\sigma^2)} \qquad 8(225)$$

and

$$Z_3^{-1} = 1+\frac{G_R^2}{8\pi^2}\int_{4m^2}^{\infty} d\sigma^2 \frac{(\sigma^2-4m^2)^{1/2}\sigma}{(\sigma^2-\mu^2)^2} \qquad 8(226)$$

As an interesting sidelight, we observe that 8(226) also gives the expansion of Z_3 correctly to terms of order G_R^2. In other words, the expansion

$$Z_3 = \frac{1}{1+\dfrac{G_R^2}{8\pi^2}\displaystyle\int_{4m^2}^{\infty} d\sigma^2 \frac{(\sigma^2-4m^2)^{1/2}\sigma}{(\sigma^2-\mu^2)^2}+\ldots}$$

$$= 1-\frac{G_R^2}{8\pi^2}\int_{4m^2}^{\infty} d\sigma^2 \frac{(\sigma^2-4m^2)^{1/2}\sigma}{(\sigma^2-\mu^2)^2}+\ldots \qquad 8(227)$$

is correctly given to order G_R^2, since higher order corrections to Z_3^{-1} will not affect the term of order G_R^2 in Z_3. Observe that the spectral integral in 8(227) is logarithmically divergent; if we evaluate it in terms of a cut-off Λ^2 we find, in the limit of large Λ^2/m^2,

$$Z_3 = 1-\frac{G_R^2}{8\pi^2}\log\frac{\Lambda^2}{m^2} \qquad 8(228)$$

Vector Fields We now turn to the spectral decomposition of two-point functions for neutral vector fields. Treating first the case of a *massive* vector field, we consider the function

$$i\Delta_{\mu\nu}^{(+)\prime}(x_1-x_2) = \langle 0|A_\mu(x_1)A_\nu(x_2)|0\rangle$$

and insert a sum over a complete set of physical states between A_μ and A_ν. This yields

$$i\Delta_{\mu\nu}^{(+)\prime}(x_1-x_2) = \sum_{M_k}\frac{V}{(2\pi)^3}\int d^4k\,\theta(k_0)\delta(k_0-\sqrt{\mathbf{k}^2+M_k^2})\,e^{ik(x_1-x_2)}$$
$$\times\sum_\alpha \langle 0|A_\mu(0)|k\alpha\rangle\langle k\alpha|A_\nu(0)|0\rangle \qquad 8(229)$$

Applying the usual invariance arguments, we define two scalar functions

ρ_V and ρ'_V such that

$$\delta_{\mu\nu}\rho_V(\sigma^2)+k_\mu k_\nu\rho'_V(\sigma^2) = \sum_{M_k}\delta(\sigma^2-M_k^2) \times \sum_\alpha \langle 0|A_\mu(0)|k\alpha\rangle(2k_0V)\langle k\alpha|A_\nu(0)|0\rangle$$

in analogy with 8(197). Applying the subsidiary condition $\partial_\mu A_\mu(x) = 0$ to both sides of 8(229) we obtain the constraint

$$k_\mu(\delta_{\mu\nu}\rho_V+k_\mu k_\nu\rho'_V) = k_\nu(\rho_V-M_k^2\rho'_V) = 0$$

and, hence,

$$\left(\delta_{\mu\nu}+\frac{k_\mu k_\nu}{\sigma^2}\right)\rho_V(\sigma^2) = \sum_{M_k}\delta(\sigma^2-M_k^2)\sum_\alpha\langle 0|A_\mu(0)|k\alpha\rangle(2k_0V)\langle k\alpha|A_\nu(0)|0\rangle \qquad 8(230)$$

As in the pseudoscalar case, the spectral function $\rho_V(\sigma^2)$ is positive-definite. To show this, we set $\mu = \nu$ in 8(230), sum over μ

$$3\rho_V(\sigma^2) = \sum_{M_k}\delta(\sigma^2-M_k^2)\sum_\alpha\langle 0|A_\mu(0)|k\alpha\rangle(2k_0V)\langle k\alpha|A_\mu(0)|0\rangle$$

and eliminate the $\mu = 4$ term on the right-hand side by using the subsidiary condition

$$k_4\langle 0|A_4(0)|k\alpha\rangle = -\mathbf{k}\langle 0|\mathbf{A}(0)|k\alpha\rangle$$

Choosing the particular frame $\mathbf{k} = 0$*, we see that $\rho_V(\sigma^2)$ is given by the positive-definite expression

$$\rho_V(\sigma^2) = \tfrac{1}{3}\sum_{M_k}\delta(\sigma^2-M_k^2)\sum_\alpha(2\sigma V)\sum_{i=1}^{3}|\langle 0|A_i(0)|k\alpha\rangle|^2$$

Substituting 8(230) into 8(229) we obtain the Lehmann–Källén representation for $\Delta_{\mu\nu}^{(+)\prime}$

$$\Delta_{\mu\nu}^{(+)\prime}(x_1-x_2) = \frac{-i}{(2\pi)^3}\int_0^\infty d\sigma^2\rho_V(\sigma^2)\int d^4k\theta(k_0)\delta(k^2+\sigma^2)\,e^{ik.(x_1-x_2)} \times\left(\delta_{\mu\nu}+\frac{k_\mu k_\nu}{\sigma^2}\right)$$

$$= \int_0^\infty d\sigma^2\rho_V(\sigma^2)\left(\delta_{\mu\nu}-\frac{\partial_\mu\partial_\nu}{\sigma^2}\right)\Delta^{(+)}(x_1-x_2;\sigma^2) \qquad 8(231a)$$

* Note that $k_0 > 0$ throughout, since the vacuum state gives no contribution to the spectral integral.

Similarly, the spectral decomposition of the function

$$-i\Delta_{\mu\nu}^{(-)\prime}(x_1-x_2) = \langle 0|A_\nu(x_2)A_\mu(x_1)|0\rangle$$

is given by

$$\Delta_{\mu\nu}^{(-)\prime}(x_1-x_2) = \int_0^\infty d\sigma^2 \rho_V(\sigma^2)\left(\delta_{\mu\nu}-\frac{\partial_\mu\partial_\nu}{\sigma^2}\right)\Delta^{(-)}(x_1-x_2;\sigma^2) \quad 8(231b)$$

with the *same* spectral function $\rho_V(\sigma^2)$ [see Problem 6]. Thus the vector propagator

$$\Delta_{F\mu\nu}'(x_1-x_2) = \langle 0|TA_\mu(x_1)A_\nu(x_2)|0\rangle$$

has the spectral representation

$$\Delta_{F\mu\nu}'(x_1-x_2) = \int_0^\infty d\sigma^2 \rho_V(\sigma^2)\Delta_{F\mu\nu}(x_1-x_2;\sigma^2) \quad 8(232)$$

or, in momentum space

$$\Delta_{F\mu\nu}'(k) = \int_0^\infty d\sigma^2 \rho_V(\sigma^2)\left(\delta_{\mu\nu}+\frac{k_\mu k_\nu}{\sigma^2}\right)\frac{-i}{k^2+\sigma^2-i\varepsilon} \quad 8(233)$$

where we have dropped the physically uninteresting $\delta_{\mu4}\delta_{\nu4}$ term*.

As in the pseudoscalar case, it is frequently convenient to work with the renormalized spectral function ρ_{VR}. The following sum rule for the neutral vector wave function renormalization constant can then be derived

$$Z_3^{-1} = \int_0^\infty d\sigma^2 \rho_{VR}(\sigma^2) \quad 8(234)$$

[see Problem 7], and the perturbative calculation of ρ_{VR} can be carried out by following the same procedure as in the derivation of 8(223), assuming a trilinear coupling of the type 5(76). The result is

$$\rho_{VR}(\sigma^2) = \delta(\sigma^2-\mu_V^2)+\frac{G_{VR}^2}{12\pi^2}\theta(\sigma^2-4m^2)\left(1-\frac{4m^2}{\sigma^2}\right)^{1/2}\frac{(\sigma^2+2m^2)}{(\mu_V^2-\sigma^2)^2} \quad 8(235)$$

where μ_V and m are respectively the vector meson and nucleon rest masses [see Problem 8]. From 8(234) and 8(235) we get, as in 8(228)

$$Z_3 = 1-\frac{G_R^2}{12\pi^2}\log\frac{\Lambda^2}{m^2} \quad 8(236)$$

in the limit of large Λ/m.

* See the remarks concerning the cancellation of normal-dependent terms in Section 6.2.

For the *photon* propagator, the argument leading to 8(233) must be modified slightly, since we can no longer set $\partial_\mu A_\mu(x) = 0$ as an operator identity. In an arbitrary gauge, two spectral functions will appear in the Lehmann–Källén representation but we are at liberty to select the Landau gauge, in which

$$k_\mu D'_{F\mu\nu}(k) = 0$$

In that case, the spectral representation for $D'_{F\mu\nu}(k)$ has the form

$$D'_{F\mu\nu}(k) = \int_0^\infty d\sigma^2 \rho_\gamma(\sigma^2)\left(\delta_{\mu\nu} - \frac{k_\mu k_\nu}{k^2}\right)\frac{-i}{k^2 - i\varepsilon} \qquad 8(237)$$

where the renormalized photon spectral function $\rho_{\gamma R} = Z_3^{-1}\rho_\gamma$ is given, to first order in e_R^2, by 8(235) with $\mu_v = 0$ and $G_{VR} \to e_R$.

Dirac Fields We now outline the derivation of the Lehmann–Källén representation for Dirac two-point functions. Proceeding as before we insert a sum over a complete set of physical states $|k\tau\rangle$ in the function

$$-iS_{\alpha\beta}^{(+)}(x_1 - x_2) = \langle 0|\psi_\alpha(x_1)\bar{\psi}_\beta(x_2)|0\rangle \qquad 8(238)$$

to obtain

$$-iS_{\alpha\beta}^{(+)}(x_1 - x_2) = \sum_{M_k} \frac{V}{(2\pi)^3}\int d^4k\theta(k_0)\delta(k_0 - \sqrt{\mathbf{k}^2 + M_k^2})\, e^{ik.(x_1 - x_2)}$$
$$\times \sum_\tau \langle 0|\psi_\alpha(0)|k\tau\rangle\langle k\tau|\bar{\psi}_\beta(0)|0\rangle \qquad 8(239)$$

Defining

$$\rho_{\alpha\beta}(k) = \sum_{M_k} \delta(\sigma^2 - M_k^2)\sum_\tau \langle 0|\psi_\alpha(0)|k\tau\rangle(2k_0 V)\langle k\tau|\bar{\psi}_\beta(0)|0\rangle \qquad 8(240)$$

such that $\bar{u}\rho u$ is a Lorentz scalar, we can expand $\rho_{\alpha\beta}(k)$ in terms of the 16 independent Dirac matrices according to

$$-\rho_{\alpha\beta}(k) = S\delta_{\alpha\beta} + P(\gamma^5)_{\alpha\beta} + V_\mu(i\gamma^\mu)_{\alpha\beta} + A_\mu(i\gamma^\mu\gamma^5)_{\alpha\beta} + T_{\mu\nu}(\Sigma^{\mu\nu})_{\alpha\beta}$$

with coefficients S, P, V_μ, A_μ and $T_{\mu\nu}$ which are respectively scalar, pseudoscalar, vector, axial vector, and antisymmetric second rank tensor functions of k. Only the vector k_μ is available for the construction of these coefficients, so that the most general form for $\rho_{\alpha\beta}(k)$ must reduce to

$$-\rho_{\alpha\beta}(k) = k_\mu\rho_1(-k^2)(i\gamma^\mu)_{\alpha\beta} + \rho_2(-k^2)\delta_{\alpha\beta}$$

with only V and S contributions surviving. Here $\rho_1(-k^2)$ and $\rho_2(-k^2)$ denote scalar functions of $-k^2$. Equation 8(240) now becomes

$$\rho_1(\sigma^2)i\gamma\,.\,k+\rho_2(\sigma^2) = -\sum_{M_k}\delta(\sigma^2-M_k^2)\sum_{\tau}\langle 0|\psi(0)|k\tau\rangle(2k_0V)$$
$$\times\langle k\tau|\bar{\psi}(0)|0\rangle \qquad 8(241)$$

and the spectral decomposition of $S^{(+)\prime}(x_1-x_2)$ takes the form

$$S^{(+)\prime}(x_1-x_2) = \frac{-i}{(2\pi)^3}\int d\sigma^2\int d^4k\theta(k_0)\delta(k^2+\sigma^2)\,\mathrm{e}^{ik(x_1-x_2)}$$
$$\times[\rho_1(\sigma^2)i\gamma\,.\,k+\rho_2(\sigma^2)]$$
$$= \int_0^\infty d\sigma^2[\rho_1(\sigma^2)\gamma\,.\,\partial+\rho_2(\sigma^2)]\Delta^{(+)}(x_1-x_2\,;\sigma^2)$$

or, more conveniently, using 3(193),

$$S^{(+)\prime}(x_1-x_2) = \int_0^\infty d\sigma^2\{\rho_1(\sigma^2)S^{(+)}(x_1-x_2\,;\sigma^2)+[\sigma\rho_1(\sigma^2)+\rho_2(\sigma^2)]$$
$$\times\Delta^{(+)}(x_1-x_2\,;\sigma^2)\} \qquad 8(242)$$

The spectral representation of

$$-iS^{(-)\prime}_{\alpha\beta}(x_1-x_2) = \langle 0|\bar{\psi}_\beta(x_2)\psi_\alpha(x_1)|0\rangle$$

can be obtained from 8(242) by enlisting the aid of the charge conjugation transformation law

$$\mathscr{C}\psi\mathscr{C}^{-1} = \psi^C$$

and the C-invariance of the vacuum. We have

$$\langle 0|\bar{\psi}_\beta(x_2)\psi_\alpha(x_1)|0\rangle = \langle 0|\mathscr{C}\bar{\psi}_\beta(x_2)\mathscr{C}^{-1}\mathscr{C}\psi_\alpha(x_1)\mathscr{C}^{-1}|0\rangle$$
$$= -\langle 0|\psi_\gamma(x_2)\bar{\psi}_\delta(x_1)|0\rangle C^{-1}_{\gamma\beta}C_{\alpha\delta}$$

or, since $C^T = -C$,

$$S^{(-)\prime}_{\alpha\beta}(x_1-x_2) = -[C^{-1}S^{(+)\prime}(x_2-x_1)C]_{\beta\alpha} \qquad 8(243)$$

Obviously the same relation holds for the corresponding free functions, so that

$$S^{(-)\prime}(x_1-x_2) = \int d\sigma^2\{\rho_1(\sigma^2)S^{(-)}(x_1-x_2\,;\sigma^2)+[\sigma\rho_1(\sigma^2)+\rho_2(\sigma^2)]$$
$$\times\Delta^{(-)}(x_1-x_2\,;\sigma^2)\} \qquad 8(244)$$

where we have used 3(81). The spectral representation of the propagator

$$S'_F(x_1 - x_2) = \langle 0 | T\psi(x_1)\bar{\psi}(x_2) | 0 \rangle$$

can now be derived from 8(242) and 8(244) in the form

$$S'_F(x_1 - x_2) = \int_0^\infty d\sigma^2 \{\rho_1(\sigma^2) S_F(x_1 - x_2; \sigma^2) - [\sigma\rho_1(\sigma^2) + \rho_2(\sigma^2)] \times \Delta_F(x_1 - x_2; \sigma^2)\} \qquad 8(245)$$

or, in momentum space

$$S'_F(k) = \int_0^\infty d\sigma^2 \frac{-\gamma \,.\, k\rho_1(\sigma^2) + i\rho_2(\sigma^2)}{k^2 + \sigma^2} \qquad 8(246)$$

PROBLEMS

1. Derive the condition 8(10).
2. Establish the result 8(36) on the basis of the Feynman graph analysis of $\langle \mathbf{q}\tau | j_\mu(0) | \mathbf{p}\sigma \rangle$.
3. Verify that the right-hand side of 8(112) has the correct general form by applying perturbation theory to the model interaction 5(95) with

$$\mathbf{A}^\lambda = \frac{i}{2}[\bar{N}, \gamma^\lambda \gamma_5 \boldsymbol{\tau} N].$$

4. Apply the reduction formalism to $\pi^\pm - p$ scattering and derive the expression for the scattering amplitude corresponding to 8(81). Show that the imaginary part of the amplitude in the physical region is given by 8(185). Exhibit the renormalized Born terms for $\pi^+ p$ and $\pi^- p$ scattering.
5. Verify Eqs. 8(222) and 8(223).
6. Show that $\Delta_{\mu\nu}^{(+)\prime}$ and $\Delta_{\mu\nu}^{(-)\prime}$ must have the same spectral function.
7. Derive the sum rule 8(234) for the vector wave function renormalization constant. Derive a mass formula analogous to 8(213) for vector mesons.
8. Derive the result 8(235).
9. Establish the following properties of the Dirac spectral functions ρ_1 and ρ_2:

$$\rho_1(\sigma^2) \geqslant 0$$
$$\sigma\rho_1(\sigma^2) - \rho_2(\sigma^2) \geqslant 0$$
$$\int_0^\infty \rho_1(\sigma^2)\, d\sigma^2 = 1$$

10. Isolate the contribution of the one-particle states to the spectral decomposition of $S'_F(x_1 - x_2)$. Relate this contribution to the wave function renormalization constant Z_2.

REFERENCES

Adler, S. L. (1965a) *Phys. Rev. Letters* **14**, 1051

Adler, S. L. (1965b) *Phys. Rev.* **140**, B 736

Bernstein, J. (1960) (with S. Fubini, M. Gell-Mann and W. Thirring) *Nuovo Cimento* **17**, 757

Ernst, F. J. (1960) (with R. G. Sachs and K. C. Wali) *Phys. Rev.* **119**, 1105

Federbush, P. (1958) (with M. L. Goldberger and S. B. Treiman) *Phys. Rev.* **112**, 642

Feynman, R. P. (1958) (with M. Gell-Mann) *Phys. Rev.* **109**, 193

Fubini, S. (1965) (with G. Furlan) *Phys.* **1**, 229

Furry, W. H. (1951) *Phys. Rev.* **81**, 115

Gasiorowicz, S. (1960) *Fortschr. Physik* **8**, 665

Gell-Mann, M. (1960) (with M. Lévy) *Nuovo Cimento* **16**, 705

Gell-Mann, M. (1962) *Phys. Rev.* **125**, 1067

Goldberger, M. L. (1958) (with S. B. Treiman) *Phys. Rev.* **110**, 1178

Haber-Schaim, U. (1956) *Phys. Rev.* **104**, 1113

Källén, G. (1952) *Helv. Phys. Acta* **25**, 417

Lee, Y. K. (1963) (with L. W. Mo and C. S. Wu) *Phys. Rev. Letters* **10**, 253

Lehmann, H. (1954) *Nuovo Cimento* **11**, 342

Nambu, Y. (1960a) (with G. Jona-Lasinio) *Phys. Rev.* **122**, 345

Nambu, Y. (1960b) *Phys. Rev. Letters* **4**, 380

Nishijima, K. (1959) *Nuovo Cimento* **11**, 698

Umezawa, H. (1951) (with S. Kamefuchi) *Progr. Theoret. Phys.* **6**, 543

Ward, J. C. (1950) *Phys. Rev.* **78**, 1824

Weisberger, W. I. (1965) *Phys. Rev. Letters* **14**, 1047

9

Bound States

9-1 The Bethe–Salpeter equation

Introduction In nonrelativistic quantum mechanics, a bound state of two particles is described by a normalized wave function satisfying the two-body Schrödinger equation. To deal with bound state problems in relativistic quantum mechanics, on the other hand, it is natural to seek a covariant generalization of the Schrödinger wave function. This is the so-called *Bethe-Salpeter* wave function. In this section we shall derive the wave equation satisfied by Bethe-Salpeter amplitudes. Section 9.2 will be devoted to the normalization condition for these amplitudes.

We begin by reviewing the concept of the Schrödinger wave function from the point of view of nonrelativistic quantum field theory. This will serve as a springboard for our discussion of the Bethe-Salpeter wave functions, first in nonrelativistic theory and then in the relativistic case.

Nonrelativistic Theory To introduce the notion of the Bethe–Salpeter wave function, let us re-examine *non*relativistic quantum mechanics from the field point of view. The Schrödinger field in the presence of a central potential $V(\mathbf{x})$ was quantized in Section 3-4 and the resulting theory was shown to be fully equivalent to standard many-body Schrödinger theory for n identical particles. Here we shall be interested in the more general case in which the identical particles interact with one another through two-body forces.

Let us consider the quantum field theoretic Hamiltonian

$$H = -\frac{1}{2m}\int d^3x\psi^\dagger(\mathbf{x}, t)\nabla^2\psi(\mathbf{x}, t)+\tfrac{1}{2}\int d^3x\, d^3x'\psi^\dagger(\mathbf{x}', t) \times \psi^\dagger(\mathbf{x}, t)v(|\mathbf{x}-\mathbf{x}'|)\psi(\mathbf{x}, t)\psi(\mathbf{x}', t) \qquad 9(1)$$

where $v(|\mathbf{x}-\mathbf{x}'|)$ is a two-body potential and ψ is a field operator satisfying the commutation rules*

$$[\psi(\mathbf{x}, t), \psi^\dagger(\mathbf{x}', t)] = \delta^{(3)}(\mathbf{x}-\mathbf{x}')$$
$$[\psi(\mathbf{x}, t), \psi(\mathbf{x}', t)] = [\psi^\dagger(\mathbf{x}, t), \psi^\dagger(\mathbf{x}', t)] = 0 \qquad 9(2)$$

and the equation of motion

$$i\dot{\psi}(\mathbf{x}, t) = [\psi(\mathbf{x}, t), H]$$
$$= -\frac{1}{2m}\nabla^2\psi(\mathbf{x}, t) + \int d^3x'\psi^\dagger(\mathbf{x}', t)v(|\mathbf{x}-\mathbf{x}'|)\psi(\mathbf{x}', t)\psi(\mathbf{x}, t) \qquad 9(3)$$

Our first task is to recover the ordinary Schrödinger description of an n-particle system interacting through two-body forces. We shall do this by constructing the n-particle wave functions and showing that they satisfy the Schrödinger equation. From the commutation rules 9(2) it follows that

$$[N, H] = [\mathbf{P}, H] = 0 \qquad 9(4)$$

where N and $\mathbf{P}$ are the total particle-number and momentum operators

$$N = \int \psi^\dagger(x)\psi(x)\, d^3x \qquad 9(5)$$

$$\mathbf{P} = \int \psi^\dagger(x)(-i\nabla)\psi(x)\, d^3x \qquad 9(6)$$

Thus N and $\mathbf{P}$ are constants of the motion and we can select a basis $|n, \mathbf{k}, E, \alpha\rangle$ formed by the simultaneous eigenstates of N, $\mathbf{P}$ and H, with α denoting the eigenvalues of whatever other commuting observables are needed to form a complete set of states. From the commutation relations

$$[\psi^\dagger, N] = -\psi^\dagger \qquad 9(7a)$$
$$[\psi, N] = \psi \qquad 9(7b)$$

we infer that $\psi^\dagger$ and ψ are particle creation and destruction operators respectively. From 9(7a) for example, it follows that

$$N\psi^\dagger|n\rangle = \psi^\dagger N|n\rangle + \psi^\dagger|n\rangle = (n+1)\psi^\dagger|n\rangle$$

We define the vacuum state $|0\rangle$ by

$$\psi(\mathbf{x}, t)|0\rangle = 0 \qquad 9(8)$$

* We assume Bose–Einstein statistics. The discussion of the Fermi–Dirac case can be carried out in a parallel way.

for all $\mathbf{x}$ and t. It follows from 9(1), 9(5) and 9(6) that

$$H|0\rangle = \mathbf{P}|0\rangle = N|0\rangle = 0 \qquad 9(9)$$

We now proceed as in Section 3-4 to construct localized n-particle states

$$|\mathbf{x}_1 \ldots \mathbf{x}_n ; t\rangle = \frac{1}{\sqrt{n!}} \psi^\dagger(\mathbf{x}_1, t) \ldots \psi^\dagger(\mathbf{x}_n, t)|0\rangle \qquad 9(10)$$

and the corresponding normalized* probability amplitude

$$\begin{aligned} \Phi^n_{\mathbf{k}E\alpha}(\mathbf{x}_1 \ldots \mathbf{x}_n ; t) &= \langle \mathbf{x}_1 \ldots \mathbf{x}_n ; t | n, \mathbf{k}, E, \alpha\rangle \\ &= \frac{1}{\sqrt{n!}} \langle 0|\psi(\mathbf{x}_1, t) \ldots \psi(\mathbf{x}_n, t)|n, \mathbf{k}, E, \alpha\rangle \qquad 9(11) \end{aligned}$$

The amplitude 9(11) is the configuration space wave function for the n-particle system. Using 9(3), 9(2) and 9(8) it is straightforward to show that Φ satisfies the standard n-body Schrödinger equation in a two-body potential $v(|\mathbf{x}-\mathbf{x}'|)$:

$$\begin{aligned} i\partial_t \Phi^n_{\mathbf{k}E\alpha}(\mathbf{x}_1, \ldots \mathbf{x}_n ; t) &= -\frac{1}{2m} \sum_{i=1}^{n} \nabla_i^2 \Phi^n_{\mathbf{k}E\alpha}(\mathbf{x}_1, \ldots \mathbf{x}_n ; t) \\ &\quad + \tfrac{1}{2} \sum_{i \neq j}^{n} v(|\mathbf{x}_i - \mathbf{x}_j|) \Phi^n_{\mathbf{k}E\alpha}(\mathbf{x}_1, \ldots \mathbf{x}_n ; t) \qquad 9(12) \end{aligned}$$

[see Problem 2]. Moreover, Φ has the correct symmetry properties for a Bose–Einstein wave function since, by 9(2), it is symmetric under interchange of any pair of points $\mathbf{x}_i$ and $\mathbf{x}_j$.

Two-Particle Wave Functions We now restrict our discussion to two-particle systems. The Schrödinger wave function for such a system

$$\phi_{\mathbf{k}E\alpha}(\mathbf{x}_1, \mathbf{x}_2 ; t) = \frac{1}{\sqrt{2!}} \langle 0|\psi(\mathbf{x}_1, t)\psi(\mathbf{x}_2, t)|\mathbf{k}, E, \alpha\rangle \qquad 9(13)$$

satisfies the equation

$$\begin{aligned} i\partial_t \phi_{\mathbf{k}E\alpha}(\mathbf{x}_1, \mathbf{x}_2 ; t) &= -\frac{1}{2m} (\nabla_1^2 + \nabla_2^2) \phi_{\mathbf{k}E\alpha}(\mathbf{x}_1, \mathbf{x}_2 ; t) \\ &\quad + \tfrac{1}{2} \sum_{i \neq j}^{n} v(|\mathbf{x}_i - \mathbf{x}_j|) \phi_{\mathbf{k}E\alpha}(\mathbf{x}_1, \mathbf{x}_2 ; t) \qquad 9(14) \end{aligned}$$

* See Chapter 3, Problem 14.

obtained by setting $\Phi^{n=2} = \phi$ in 9(12). The dependence of $\phi_{\mathbf{k}E\alpha}$ on the centre of mass coordinate $(x_1+x_2)/2$ can be factored out either by rewriting 9(14) in terms of centre of mass and relative coordinates, or, alternatively, by setting

$$\begin{aligned}\phi_{\mathbf{k}E\alpha}(\mathbf{x}_1, \mathbf{x}_2; t) &= \frac{1}{\sqrt{2!}}\langle 0|\, e^{-i\mathbf{P}.\mathbf{a}}\psi(\mathbf{x}_1, t)\, e^{i\mathbf{P}.\mathbf{a}}\, e^{-i\mathbf{P}.\mathbf{a}}\psi(\mathbf{x}_2, t)\, e^{i\mathbf{P}.\mathbf{a}}\, e^{-i\mathbf{P}.\mathbf{a}}|\mathbf{k}, E, \alpha\rangle \\ &= \frac{1}{\sqrt{2!}}\langle 0|\psi(\mathbf{x}_1+\mathbf{a}, t)\psi(\mathbf{x}_2+\mathbf{a}, t)|\mathbf{k}, E, \alpha\rangle\, e^{-i\mathbf{k}.\mathbf{a}} \\ &= \frac{1}{\sqrt{2!}}\langle 0|\psi\left(\frac{\mathbf{x}_1-\mathbf{x}_2}{2}, t\right)\psi\left(-\frac{\mathbf{x}_1-\mathbf{x}_2}{2}, t\right)|\mathbf{k}, E, \alpha\rangle\, e^{i\mathbf{k}.(\mathbf{x}_1+\mathbf{x}_2)/2}\end{aligned} \qquad 9(15)$$

for $\mathbf{a} = -(\mathbf{x}_1+\mathbf{x}_2)/2$. In either case, of course, we rely on the invariance of the theory under spatial displacement. We shall write 9(15) in the form

$$\phi_{\mathbf{k}E\alpha}(\mathbf{x}_1, \mathbf{x}_2, t) = \frac{1}{\sqrt{V}}\, e^{i\mathbf{k}.\mathbf{X}-iEt}\phi_{\mathbf{k}E\alpha}(\mathbf{x}, 0) \qquad 9(16)$$

where $\mathbf{x} = \mathbf{x}_1 - \mathbf{x}_2$, $X = (\mathbf{x}_1+\mathbf{x}_2)/2$ and where we have factored out the time-dependence from

$$\phi_{\mathbf{k}E\alpha}(\mathbf{x}, t) = \frac{1}{\sqrt{2!}}\langle 0|\psi\left(\frac{\mathbf{x}}{2}, t\right)\psi\left(-\frac{\mathbf{x}}{2}, t\right)|\mathbf{k}, E, \alpha\rangle\sqrt{V} \qquad 9(17)$$

The amplitude $\phi_{\mathbf{k}E\alpha}(\mathbf{x}, 0)$ is the Schrödinger wave function for the relative motion of the two particles.

Let us lift the equal-time restriction in the definition 9(13) of the Schrödinger amplitude and define a new amplitude*

$$\chi_{\mathbf{k}E\alpha}(\mathbf{x}_1, t_1, \mathbf{x}_2, t_2) = \langle 0|T\psi(\mathbf{x}_1, t_1)\psi(\mathbf{x}_2, t_2)|\mathbf{k}, E, \alpha\rangle \qquad 9(18)$$

where T is the time-ordering symbol

$$T\psi(x_1)\psi(x_2) = \begin{cases}\psi(\mathbf{x}_1, t_1)\psi(\mathbf{x}_2, t_2) & t_1 > t_2 \\ \psi(\mathbf{x}_2, t_2)\psi(\mathbf{x}_1, t_1) & t_2 > t_1\end{cases} \qquad 9(19)$$

$\chi_{\mathbf{k}E\alpha}(x_1, x_2)$ is known as the *Bethe–Salpeter wave function.* Like the Schrödinger wave function, it is symmetric under interchange of x_1 and x_2. To derive the equation satisfied by $\chi_{\mathbf{k}E\alpha}$, we write

$$T\psi(x_1)\psi(x_2) = \tfrac{1}{2}\{\psi(x_1)\psi(x_2)\} + \tfrac{1}{2}\epsilon(t_1-t_2)[\psi(x_1), \psi(x_2)] \qquad 9(20)$$

* In accordance with the usual convention for Bethe–Salpeter wave functions, we omit the normalization factor $(2!)^{-1/2}$.

and apply the operator

$$\left(i\frac{\partial}{\partial t_1}+\frac{1}{2m}\nabla_1^2\right)\left(i\frac{\partial}{\partial t_2}+\frac{1}{2m}\nabla_2^2\right)$$

to both sides. Using the equation of motion 9(3) together with 9(2) and 9(8), we find [Schweber (1962)],

$$\left(i\frac{\partial}{\partial t_1}+\frac{1}{2m}\nabla_1^2\right)\left(i\frac{\partial}{\partial t_2}+\frac{1}{2m}\nabla_2^2\right)\chi_{\mathbf{k}E\alpha}(x_1, x_2)$$
$$= i\delta(t_1-t_2)v(|\mathbf{x}_1-\mathbf{x}_2|)\chi_{\mathbf{k}E\alpha}(x_1, x_2) \qquad 9(21)$$

[see Problem 4]. Equation 9(21) is the Bethe–Salpeter equation for the nonrelativistic problem, written in differential form. The factorization of the centre of mass dependence of $\chi_{\mathbf{k}E\alpha}(x_1, x_2)$ follows the same pattern as for the Schrödinger wave function. We have

$$\chi_{\mathbf{k}E\alpha}(x_1, x_2) = \frac{1}{\sqrt{V}}e^{ik.X}\chi_{\mathbf{k}E\alpha}(x) \qquad 9(22)$$

where $x = x_1-x_2$ and $X = (x_1+x_2)/2$. $\chi_{\mathbf{k}E\alpha}(x)$ is the Bethe–Salpeter amplitude for the relative motion.

It can be shown [Schweber (1962)] that the knowledge of the Schrödinger wave function uniquely determines the Bethe–Salpeter wave function. However, in *relativistic* quantum field theory, the Schrödinger wave function is not a useful quantity to work with, in view of its non-Lorentz-invariant character. Before turning to relativistic Bethe–Salpeter theory, let us examine the role of *bound states* in the present framework.

Bound States So far we have not distinguished between discrete bound states and continuum, or scattering, states. The distinction emerges upon converting the differential equation for the wave function into an *integral* equation.

To rewrite the Bethe–Salpeter equation 9(21) in integral form, we introduce the nonrelativistic propagator function

$$S_F(x_1-x_2) = \langle 0|T\psi(x_1)\psi^\dagger(x_2)|0\rangle \qquad 9(23)$$

which vanishes for $t_1 < t_2$. Using 9(20) to express the T-product and applying the equation of motion 9(3) we obtain the equation

$$\left(i\frac{\partial}{\partial t_1}+\frac{1}{2m}\nabla_1^2\right)S_F(x_1-x_2) = i\delta^{(4)}(x_1-x_2)+\langle 0|TJ(x_1)\psi^\dagger(x_2)|0\rangle \qquad 9(24)$$

where we have defined

$$J(x) = \left(i\frac{\partial}{\partial t}+\frac{1}{2m}\nabla^2\right)\psi(x) = \int d^4x'\psi^\dagger(x')\psi(x')\psi(x)\delta(t-t')v(|\mathbf{x}-\mathbf{x}'|) \quad 9(25)$$

Since by 9(8),

$$J(x)|0\rangle = \langle 0|J(x) = 0$$

we see that 9(24) reduces to the Green's function equation

$$\left(i\frac{\partial}{\partial t_1}+\frac{1}{2m}\nabla_1^2\right)S_F(x_1-x_2) = i\delta^{(4)}(x_1-x_2) \quad 9(26)$$

Using $S_F(x)$ to integrate 9(21) we obtain

$$\chi_{\mathbf{k}E\alpha}(x_1, x_2) = \chi^0_{\mathbf{k}E\alpha}(x_1, x_2) - \int d^4x\, d^4x' S_F(x_1-x)S_F(x_2-x')$$
$$i\delta(t-t')v(|\mathbf{x}-\mathbf{x}'|)\chi_{\mathbf{k}E\alpha}(x, x') \quad 9(27)$$

where the inhomogeneous term satisfies

$$\left(i\frac{\partial}{\partial t_1}+\frac{1}{2m}\nabla_1^2\right)\left(i\frac{\partial}{\partial t_2}+\frac{1}{2m}\nabla_2^2\right)\chi^0_{\mathbf{k}E\alpha}(x_1, x_2) = 0 \quad 9(28)$$

The crucial point is that $\chi^0_{\mathbf{k}E\alpha}$ *vanishes* in the case of a bound state having a discrete energy level $E < 0$. To see this we write the general solution to 9(28) in the form of a Fourier integral

$$\chi^0_{\mathbf{k}E\alpha}(x_1, x_2) = \int d^4k_1\, d^4k_2\, e^{ik_1.x_1+ik_2.x_2}\delta^{(4)}(k-k_1-k_2)$$
$$\times\left\{f(k_1, k_2)\delta\left(k_{10}-\frac{\mathbf{k}_1^2}{2m}\right)\delta\left(k_{20}-\frac{\mathbf{k}_2^2}{2m}\right)\right.$$
$$+g(k_1, k_2)\delta\left(k_{10}-\frac{\mathbf{k}_1^2}{2m}\right)$$
$$\left.+h(k_1, k_2)\delta\left(k_{20}-\frac{\mathbf{k}_2^2}{2m}\right)\right\} \quad 9(29)$$

Here the factor $\delta^{(4)}(k-k_1-k_2)$ guarantees that $\chi^0_{\mathbf{k}E\alpha}(x_1, x_2)$ has the correct dependence on the centre of mass coordinate $X = (x_1+x_2)/2$

$$\chi^0_{\mathbf{k}E\alpha}(x_1, x_2) \sim e^{ik.X}\chi^0_{\mathbf{k}E\alpha}(x_1-x_2)$$

as in 9(22) and the mass shell delta functions ensure that 9(29) satisfies 9(28). We now show that $\chi^0_{\mathbf{k}E\alpha}$ must also satisfy the conditions

$$\left(i\frac{\partial}{\partial t_2}+\frac{1}{2m}\nabla_2^2\right)\chi^0_{\mathbf{k}E\alpha}(x_1,x_2)=0 \qquad 9(30a)$$

$$\left(i\frac{\partial}{\partial t_1}+\frac{1}{2m}\nabla_1^2\right)\chi^0_{\mathbf{k}E\alpha}(x_1,x_2)=0 \qquad 9(30b)$$

and hence that

$$g(k_1,k_2)=0 \qquad 9(31a)$$

$$h(k_1,k_2)=0 \qquad 9(31b)$$

Applying the equation of motion 9(3) to 9(18) and using 9(20) to express the T-product, we get

$$\left(i\frac{\partial}{\partial t_2}+\frac{1}{2m}\nabla_2^2\right)\chi_{\mathbf{k}E\alpha}(x_1,x_2)=\left(i\frac{\partial}{\partial t_2}+\frac{1}{2m}\nabla_2^2\right)\langle 0|T\psi(x_1)\psi(x_2)|\mathbf{k},E,\alpha\rangle$$

$$=\langle 0|T\psi(x_1)J(x_2)|\mathbf{k},E,\alpha\rangle \qquad 9(32)$$

with $J(x)$ defined by 9(25). The right-hand side of 9(32) is equal to

$$\int d^4x\langle 0|T\psi(x_1)\psi^\dagger(x)|0\rangle\langle 0|\psi(x)\psi(x_2)|\mathbf{k},E,\alpha\rangle\delta(t_2-t)v(|\mathbf{x}_2-\mathbf{x}|)$$

so that

$$\left(i\frac{\partial}{\partial t_2}+\frac{1}{2m}\nabla_2^2\right)\chi_{\mathbf{k}E\alpha}(x_1,x_2)=\int d^4xS_F(x_1-x)\chi_{\mathbf{k}E\alpha}(x,x_2)\delta(t_2-t)v(|\mathbf{x}_2-\mathbf{x}|) \qquad 9(33)$$

Equation 9(30a) now follows by comparing 9(33) with the relation

$$\left(i\frac{\partial}{\partial t_2}+\frac{1}{2m}\nabla_2^2\right)\chi_{\mathbf{k}E\alpha}(x_1,x_2)=\left(i\frac{\partial}{\partial t_2}+\frac{1}{2m}\nabla_2^2\right)\chi^0_{\mathbf{k}E\alpha}(x_1,x_2)$$

$$+\int d^4xS_F(x_1-x)\chi_{\mathbf{k}E\alpha}(x,x_2)\delta(t_2-t)v(|\mathbf{x}_2-\mathbf{x}|)$$

derived from 9(27) with the aid of 9(26). A similar proof holds for 9(30b). Taking 9(31a) and 9(31b) into account, we see that the Fourier integral for $\chi^0_{\mathbf{k}E\alpha}$

$$\chi^0_{\mathbf{k}E\alpha}(x_1,x_2)=\int d^4k_1\,d^4k_2\,e^{ik_1.x_1+ik_2.x_2}\delta^{(4)}(k-k_1-k_2)$$

$$\times f(k_1,k_2)\delta\left(k_{10}-\frac{\mathbf{k}_1^2}{2m}\right)\delta\left(k_{20}-\frac{\mathbf{k}_2^2}{2m}\right)$$

must vanish for bound states, as the delta function restrictions are incompatible with a negative value of $E = k_0$. Thus in the case of a bound state the inhomogeneous term in the integral equation 9(27) vanishes and we are left with the *homogeneous* Bethe–Salpeter integral equation

$$\chi_{\mathbf{k}E\alpha}(x_1, x_2) = -\int d^4x\, d^4x' S_F(x_1 - x) S_F(x_2 - x') \times i\delta(t - t') v(|\mathbf{x} - \mathbf{x}'|) \chi_{\mathbf{k}E\alpha}(x, x') \qquad 9(34)$$

Relativistic Theory We now turn to the derivation of the two-particle Bethe–Salpeter equation in relativistic quantum field theory. As a convenient example, we treat the interaction of two spinor fields ψ_A and ψ_B—describing distinguishable particles of equal mass m—with a neutral scalar field ϕ

$$\mathscr{L}_{\text{int}}(x) = G_0 \tfrac{1}{2}[\bar{\psi}_A(x), \psi_A(x)]\phi(x) + (A \to B) \qquad 9(35)$$

Assuming the existence of a stable 2-fermion bound state $|k\rangle$ of mass M*

$$k^2 = -M^2 \qquad M < 2m$$

we shall derive the equation for its Bethe–Salpeter wave function $\chi_k(x_1, x_2)$ following the method of Gell-Mann and Low [Gell-Mann (1951)]. The essence of this method is to relate the Bethe–Salpeter wave function to the two-body propagator

$$K(x_1, x_2, x_3, x_4) = -\langle 0|T\psi_A(x_1)\psi_B(x_2)\bar{\psi}_A(x_3)\bar{\psi}_B(x_4)|0\rangle \qquad 9(36)$$

for which an integral equation can be derived from perturbation theory. This approach allows the integral equation for the wave function to be obtained *directly* without going through the differential form of the Bethe–Salpeter equation.

Let us select the time-ordering $t_1, t_2 > t_3, t_4$ in 9(36) and insert between ψ_B and $\bar{\psi}_A$ a sum over a complete set of states $|p, \alpha\rangle$ where p is the 4-momentum eigenvalue and α represents whatever other quantum numbers are needed to form a complete set. We have then

$$K(x_1, x_2; x_3, x_4) = -\sum_{p\alpha} \chi_{p\alpha}(x_1, x_2)\bar{\chi}_{p\alpha}(x_3, x_4) \qquad 9(37)$$

$$(t_1, t_2 > t_3, t_4)$$

* For simplicity we assume the bound state to be nondegenerate, so that it is uniquely characterized by its energy-momentum eigenvalue k_μ. The extension of the derivation to a set of linearly independent bound states $|k, \alpha\rangle$ of mass M is straightforward.

where $\chi_{p\alpha}$ and $\bar{\chi}_{p\alpha}$ are the Bethe–Salpeter amplitudes.

$$\chi_{p\alpha}(x_1, x_2) = \langle 0|T\psi_A(x_1)\psi_B(x_2)|p, \alpha\rangle \qquad 9(38a)$$

$$\bar{\chi}_{p\alpha}(x_1, x_2) = \langle p, \alpha|T\bar{\psi}_A(x_1)\bar{\psi}_B(x_2)|0\rangle \qquad 9(38b)$$

whose dependence on the centre of mass coordinate can be factored out in the same way as in 9(22):

$$\chi_{p\alpha}(x_1, x_2) = \frac{1}{\sqrt{V}} e^{ip.X} \chi_{p\alpha}(x) \qquad 9(39a)$$

$$\bar{\chi}_{p\alpha}(x_1, x_2) = \frac{1}{\sqrt{V}} e^{-ip.X} \bar{\chi}_{p\alpha}(x) \qquad 9(39b)$$

We remark that all states $|p, \alpha\rangle$ appearing in the sum on the right-hand side of 9(37) carry fermion number $+2$, since $\bar{\psi}_A(\psi_A)$ and $\bar{\psi}_B(\psi_B)$ each create (destroy) one fermion. In particular, the assumed existence of a 2-fermion bound state $|k\rangle$ with $k^2 = -M^2$ implies that for $t_1, t_2 > t_3, t_4$, $K(x_1, x_2; x_3, x_4)$ will pick up a contribution

$$-\sum_{k^2 = -M^2} \chi_k(x_1, x_2)\bar{\chi}_k(x_3, x_4)$$

from the bound state wave functions. To isolate the contribution of a single term $\chi_k\bar{\chi}_k$ we first form the inner product

$$\int d^3x_3\, d^3x_4 K(x_1, x_2; \mathbf{x}_3 t, \mathbf{x}_4 t)\gamma_4^A\gamma_4^B\chi_k(\mathbf{x}_3 t, \mathbf{x}_4 t)$$

for $t_1, t_2 > t$. For equal times we have

$$\begin{aligned}\bar{\chi}_{p\alpha}(\mathbf{x}_3 t, \mathbf{x}_4 t)\gamma_4^A\gamma_4^B &= \langle p, \alpha|\psi_A{}^\dagger(\mathbf{x}_3 t)\psi_B{}^\dagger(\mathbf{x}_4 t)|0\rangle \\ &= -\chi_{p\alpha}{}^*(\mathbf{x}_3 t, \mathbf{x}_4 t)\end{aligned}$$

so that

$$\begin{aligned}&\int d^3x_3\, d^3x_4 K(x_1, x_2; \mathbf{x}_3 t, \mathbf{x}_4 t)\gamma_4^A\gamma_4^B\chi_k(\mathbf{x}_3 t, \mathbf{x}_4 t) \\ &= \frac{1}{V}\sum_{p,\alpha} \chi_{p\alpha}(x_1, x_2) \int d\mathbf{X}\, e^{i(\mathbf{k}-\mathbf{p}).\mathbf{X}}\, e^{-i(k_0 - p_0)t} \int d^3x \chi_{p\alpha}^*(\mathbf{x})\chi_k(\mathbf{x}) \qquad 9(40)\end{aligned}$$

The integral over $\mathbf{X}$ projects out the momentum $\mathbf{k}$ from the sum over $\mathbf{p}$. To project out the discrete energy level k_0 from the sum over p_0 we perform the operation

$$L_{t\to-\infty} f(t) = \lim_{t\to-\infty} \frac{1}{|t|} \int_{2t}^{t} f(\tau)\, d\tau \qquad 9(41)$$

on both sides of 9(40). This eliminates all oscillating terms in t and yields the result

$$L_{t\to-\infty}\int d^3x_3\, d^3x_4 K(x_1, x_2; \mathbf{x}_3 t, \mathbf{x}_4 t)\gamma_4^A\gamma_4^B\chi_k(\mathbf{x}_3 t, \mathbf{x}_4 t) = \chi_k(x_1, x_2)\mathscr{P}_k \quad 9(42)$$

where

$$\mathscr{P}_k = \int d^3x \chi_k^*(\mathbf{x})\chi_k(\mathbf{x}) \quad 9(43)$$

Equation 9(42) expresses χ_k in terms of the two-body propagator K. Our next task is to derive an integral equation for K from perturbation theory.

Using the technique described in Section 7-3, we write $K(x_1, x_2; x_3, x_4)$ in the form

$$K(x_1, x_2; x_3, x_4) = \frac{(0|T\psi_A^{ip}(x_1)\psi_B^{ip}(x_2)\overline{\psi}_A^{ip}(x_3)\overline{\psi}_B^{ip}(x_4)U(\infty, -\infty)|0)}{(0|U(\infty, -\infty)|0)}$$

where the ψ^{ip} are interaction picture field operators and $|0)$ is the unperturbed vacuum state. From the above expression, one easily derives the perturbation expansion

$$K(1, 2; 3, 4) = S_F^A(1, 3)S_F^B(2, 4) - G_0^2\int d^4x_5\, d^4x_6 S_F^A(1, 5)S_F^B(2, 6)$$
$$\times \Delta_F(5, 6)S_F^A(5, 3)S_F^B(6, 4) + \ldots \quad 9(44)$$

as exhibited diagrammatically in Fig. 9.1 [see Problem 5]. Formally we may write this infinite series in the form

$$K(1, 2, 3, 4) = S_F'^A(1, 3)S_F'^B(2, 4) - \int d^4x_5\, d^4x_6\, d^4x_7\, d^4x_8$$
$$\times S_F'^A(1, 5)S_F'^B(2, 6)G(5, 6; 7, 8)S_F'^A(7, 3)S_F'^B(8, 4) \quad 9(45)$$

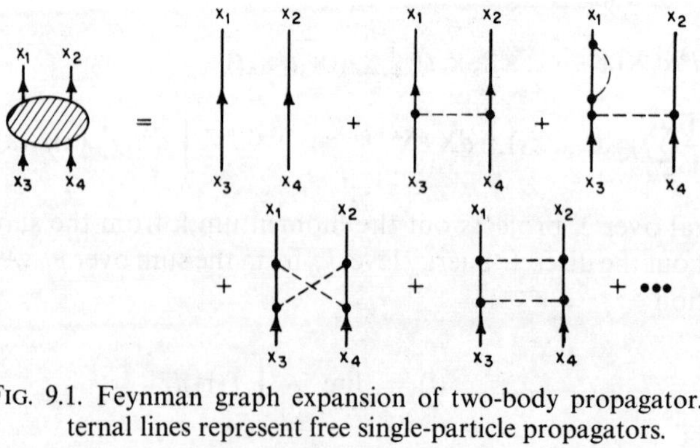

FIG. 9.1. Feynman graph expansion of two-body propagator. External lines represent free single-particle propagators.

where S'_F is the *exact* fermion propagator and where

$$G(5,6;7,8) = \sum_n G^{(n)}(5,6;7,8)$$

represents the sum of all graphs contributing to the *two-particle* interaction. The lowest order contribution to G is simply

$$G^{(0)}(5,6;7,8) = G_0^2\delta^{(4)}(x_5-x_7)\delta^{(4)}(x_6-x_8)\Delta_F(x_5-x_6) \qquad 9(46)$$

which, when substituted into 9(45) yields the G_0^2 term in 9(44) (with S_F replaced by S'_F).

As was noticed by Salpeter and Bethe [Salpeter (1951)], the expansion 9(45) can be recast in the form of an inhomogeneous integral equation. To do this, we introduce the notion of *Bethe–Salpeter irreducibility.* A fermion-fermion scattering graph will be termed *reducible* if one can draw a line through the graph which cuts no meson lines at all and each of the two fermion lines only once. If such a line cannot be drawn, the graph is *irreducible.* Examples of reducible and irreducible graphs are shown in Figs. 9.2 and 9.3. Denoting by $\bar{G}$ that part of G which refers to the sum of all

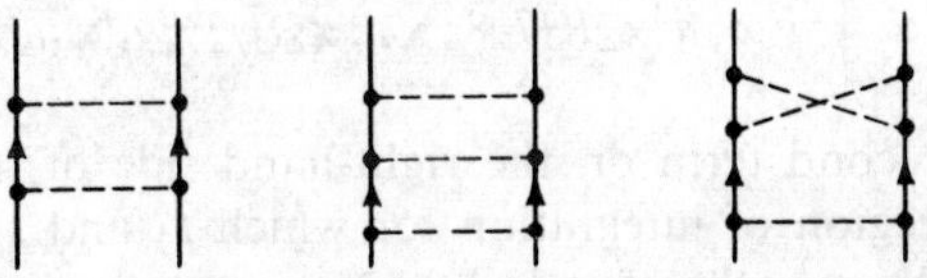

FIG. 9.2. Examples of Bethe–Salpeter reducible graphs.

irreducible graphs only, it is easy to see that 9(45) can be replaced by the integral equation

$$K(1,2;3,4) = S_F'^A(1,3)S_F'^B(2,4) - \int d^4x_5\, d^4x_6\, d^4x_7\, d^4x_8$$
$$S_F'^A(1,5)S_F'^B(2,6)\bar{G}(5,6;7,8)K(7,8;3,4) \qquad 9(47a)$$

or, equivalently,

$$K(1,2;3,4) = S_F'^A(1,3)S_F'^B(2,4) - \int d^4x_5\, d^4x_6\, d^4x_7\, d^4x_8$$
$$\times K(1,2;5,6)\bar{G}(5,6;7,8)S_F'^A(7,3)S_F'^B(8,4) \qquad 9(47b)$$

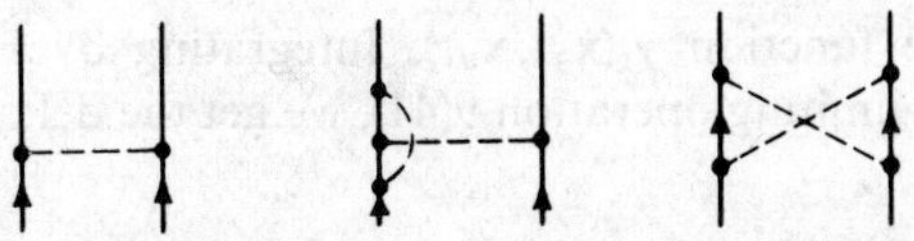

FIG. 9.3. Some Bethe–Salpeter irreducible graphs.

Indeed, the iterative solution of this equation,

$$K(1,2;3,4) = S_F'^A(1,3)S_F'^B(2,4) - \int d^4x_5\, d^4x_6\, d^4x_7\, d^4x_8$$

$$S_F'^A(1,5)S_F'^B(2,6)\bar{G}(5,6;7,8)S_F'^A(7,3)S_F'^B(8,4)$$

$$+\ldots$$

reproduces the perturbation expansion of K consisting of *all* graphs, reducible and irreducible. Armed with 9(42) and the inhomogeneous integral equation 9(47a) or 9(47b), we are now in a position to derive the equation for the Bethe–Salpeter amplitude $\chi_k(x_1, x_2)$.

Combining 9(42) and 9(47a), we have

$$\chi_k(x_1, x_2)\mathscr{P}_k = L_{t\to-\infty} \int d^3x_3\, d^3x_4 S_F'^A(x_1, \mathbf{x}_3 t)S_F'^B(x_2, \mathbf{x}_4 t)\gamma_4^A\gamma_4^B\chi_k(\mathbf{x}_3 t, \mathbf{x}_4 t)$$

$$-L_{t\to-\infty} \int d^4x_5\, d^4x_6\, d^4x_7\, d^4x_8 S_F'^A(1,5)S_F'^B(2,6)\bar{G}(5,6;7,8)$$

$$\times \int d^3x_3\, d^3x_4 K(7,8;\mathbf{x}_3 t, \mathbf{x}_4 t)\gamma_4^A\gamma_4^B\chi_k(\mathbf{x}_3 t, \mathbf{x}_4 t) \qquad 9(48)$$

Consider the second term on the right-hand side of this equation. As $t \to -\infty$, the region of integration for which t_7 and t_8 are less than t tends to zero. We may therefore use 9(42) to write this term in the form

$$-\mathscr{P}_k \int d^4x_5\, d^4x_6\, d^4x_7\, d^4x_8 S_F'^A(1,5)S_F'^B(2,6)\bar{G}(5,6;7,8)\chi_k(7,8)$$

We now turn to the first term on the right-hand side of 9(48). By the Lehmann–Källén representation for $S_F'(x)$, the coordinate dependence of $S_F'^A S_F'^B$ can be represented by products of wave functions

$$e^{i\mathbf{p}.(\mathbf{x}_1-\mathbf{x}_3)}\, e^{i\mathbf{p}'.(\mathbf{x}_2-\mathbf{x}_4)}\, e^{-ip_0(t_1-t)}\, e^{-ip_0'(t_2-t)}$$

integrated over $\mathbf{p}$, $\mathbf{p}'$ and over the invariant masses $M_p > m$ and $M_{p'} > m$. The above product is multiplied by a factor

$$e^{i\mathbf{k}(\mathbf{x}_3+\mathbf{x}_4)/2}\, e^{-ik_0 t}$$

from the wave function $\chi_k(\mathbf{x}_3 t, \mathbf{x}_4 t)$. Integrating over $\mathbf{x}_3$ and $\mathbf{x}_4$ and performing the limiting operation 9(41), we get the delta function restrictions

$$\mathbf{k} = \mathbf{p}+\mathbf{p}' \qquad k_0 = p_0 + p_0'$$

For M_p and $M_{p'}$ greater than m, these restrictions are incompatible with the mass shell constraint $k^2 = -M^2$ ($M < 2m$). Hence we conclude that the first term on the right-hand side of 9(48) vanishes and we are left with the *homogeneous* integral equation

$$\chi_k(1, 2) = -\int d^4x_5\, d^4x_6\, d^4x_7\, d^4x_8 S_F'^A(1, 5) S_F'^B(2, 6) \bar{G}(5, 6; 7, 8) \chi_k(7, 8) \qquad 9(49a)$$

for the bound state wave function. The Bethe–Salpeter equation for $\bar{\chi}_k(x_1, x_2)$ can be derived in the same way from 9(42) and 9(47b). The result is

$$\bar{\chi}_k(1, 2) = -\int d^4x_5\, d^4x_6\, d^4x_7\, d^4x_8 \bar{\chi}_k(5, 6) \bar{G}(5, 6; 7, 8) S_F'^A(7, 3) S_F'^B(8, 4) \qquad 9(49b)$$

Momentum Space It is frequently more convenient to apply the Bethe–Salpeter formalism in momentum space. To write the inhomogeneous equations 9(47a) and 9(47b) in momentum space, we note that by space-time translation invariance, $K(x_1, x_2; x_3, x_4)$ and $\bar{G}(x_1, x_2; x_3, x_4)$ can depend only on differences of coordinates, say, $x = x_1 - x_2$, $x' = x_3 - x_4$ and $X - X' = \frac{1}{2}(x_1 + x_2) - \frac{1}{2}(x_3 + x_4)$. Introducing Fourier transforms $K(p, q, k)$ and $\bar{G}(p, q, k)$ by

$$K(x_1, x_2; x_3, x_4) = \frac{1}{(2\pi)^8} \int d^4p\, d^4q\, d^4k\, e^{ik.(X-X')} e^{ip.x} e^{-iq.x'} K(p, q, k) \qquad 9(50)$$

and a similar equation for $\bar{G}$, we can write 9(47a) and 9(47b) as

$$\int d^4p' [I(p, p', k) + \bar{G}(p, p', k)] K(p', q, k) = \delta^{(4)}(p-q) \qquad 9(51a)$$

$$\int d^4q' K(p, q', k) [I(q', q, k) + \bar{G}(q', q, k)] = \delta^{(4)}(p-q) \qquad 9(51b)$$

where

$$I(p, q, k) = \delta^{(4)}(p-q) [S_F'^A(\tfrac{1}{2}k + p)]^{-1} [S_F'^B(\tfrac{1}{2}k - p)]^{-1} \qquad 9(52)$$

[see Problem 6].

We now calculate the contribution of the bound state to the Fourier transform $K(p, q, k)$. From 9(37) the bound state contribution to $K(x_1, x_2; x_3, x_4)$ is

$$-\frac{1}{(2\pi)^3} \int d^4k \chi_k(x) \bar{\chi}_k(x') e^{ik.(X-X')} \theta(k_0) \delta(k^2 + M^2) \theta(X_0 - X_0' - \tfrac{1}{2}|x_0| - \tfrac{1}{2}|x_0'|) \qquad 9(53)$$

where we have used 9(39a) and 9(39b). We have imposed the restriction $t_1, t_2 > t_3, t_4$ through the factor $\theta(X_0 - X'_0 - \frac{1}{2}|x_0| - \frac{1}{2}|x'_0|)$, which, as the reader can easily check, equals one for $t_1, t_2 > t_3, t_4$ and zero otherwise. Thus 9(53) represents the bound state contribution for *all* time orderings. Since

$$\theta(k_0)\delta(k^2 + M^2) = (2k_0)^{-1}\delta(k_0 - \omega_{\mathbf{k}})$$

with $\omega_{\mathbf{k}} = (\mathbf{k}^2 + M^2)^{1/2}$, we can write the bound state contribution as

$$-\frac{1}{(2\pi)^3}\int \frac{d^3k}{2\omega_{\mathbf{k}}} \chi_k(x)\bar{\chi}_k(x') e^{i\mathbf{k}.(\mathbf{X}-\mathbf{X}')} e^{-i\omega_{\mathbf{k}}(X_0 - X'_0)}\theta(X_0 - X'_0 - \tfrac{1}{2}|x_0| - \tfrac{1}{2}|x'_0|) \tag{9(54)}$$

Substituting the formula

$$\theta(y_0) = -\frac{1}{2\pi i}\int dp_0 \frac{1}{p_0 + i\varepsilon} e^{-ip_0 y_0}$$

into 9(54) and changing variables $p_0 \to k_0 - \omega_{\mathbf{k}}$, we obtain

$$-\frac{i}{(2\pi)^4}\int d^3k \frac{dk_0}{2\omega_{\mathbf{k}}} e^{i\mathbf{k}.(\mathbf{X}-\mathbf{X}')} e^{-ik_0(X_0 - X'_0)} \frac{1}{k_0 - \omega_{\mathbf{k}} + i\varepsilon} \chi'_k(x)\bar{\chi}'_k(x') \tag{9(55)}$$

where we have defined new amplitudes

$$\chi'_k(x) = e^{\frac{1}{2}i(k_0 - \omega_{\mathbf{k}})|x_0|}\chi_k(x) \tag{9(56a)}$$

$$\bar{\chi}'_k(x') = e^{\frac{1}{2}i(k_0 - \omega_{\mathbf{k}})|x'_0|}\bar{\chi}_k(x') \tag{9(56b)}$$

Setting

$$\chi'_k(x) = \frac{1}{(2\pi)^4}\int d^4p\, e^{ip.x}\chi'_k(p)$$

$$\bar{\chi}'_k(x') = \frac{1}{(2\pi)^4}\int d^4q\, e^{-iqx'}\bar{\chi}'_k(q)$$

the bound state contribution 9(55) may now be written in the form

$$-\frac{i}{(2\pi)^{12}}\int d^4p\, d^4q\, d^4k\, e^{ik.(X-X')} e^{ip.x} e^{-iq.x'}$$

$$\times \frac{1}{2\omega_{\mathbf{k}}} \frac{1}{k_0 - \omega_{\mathbf{k}} + i\varepsilon} \chi'_k(p)\bar{\chi}'_k(q)$$

We conclude that the bound state gives rise to a pole at $k_0 = \omega_{\mathbf{k}}$ in the Fourier transform $K(p, q, k)$. Since, by 9(56a) and 9(56b), χ'_k and $\bar{\chi}'_k$ reduce to the Bethe–Salpeter amplitudes χ_k and $\bar{\chi}_k$ at $k_0 = \omega_{\mathbf{k}}$, the *residue* of $K(p, q, k)$ at the pole is simply

$$-\frac{i}{(2\pi)^4}\frac{1}{2k_0}\chi_k(p)\bar{\chi}_k(q) \tag{9(57)}$$

where $\chi_k(p)$ and $\bar{\chi}_k(q)$ are the Fourier transforms of the Bethe–Salpeter amplitudes for relative motion

$$\chi_k(x) = \frac{1}{(2\pi)^4}\int d^4p\, e^{ip.x}\chi_k(p) \qquad (k_0 = \omega_{\mathbf{k}}) \tag{9(58a)}$$

$$\bar{\chi}_k(x') = \frac{1}{(2\pi)^4}\int d^4q\, e^{-iq.x'}\bar{\chi}_k(q) \qquad (k_0 = \omega_{\mathbf{k}}) \tag{9(58b)}$$

Thus we have shown that

$$K(p, q, k) = \frac{-i}{(2\pi)^4}\frac{1}{2k_0}\frac{1}{k_0-(\mathbf{k}^2+M^2)^{1/2}+i\varepsilon}\chi_k(p)\bar{\chi}_k(q) + \text{terms regular at } k_0 = \omega_{\mathbf{k}} \tag{9(59)}$$

Finally we show that the homogeneous Bethe–Salpeter equations 9(49a) and 9(49b) take the form

$$\int d^4p'[I(p, p', k)+\bar{G}(p, p', k)]\chi_k(p') = 0 \qquad (k_0 = \omega_{\mathbf{k}}) \tag{9(60a)}$$

$$\int d^4q'\bar{\chi}_p(q')[I(q', q, k)+\bar{G}(q', q, k)] = 0 \qquad (k_0 = \omega_{\mathbf{k}}) \tag{9(60b)}$$

in terms of $\chi_k(p)$ and $\bar{\chi}_k(q)$. Equations 9(60a) and 9(60b) can be derived by direct substitution of 9(39a), 9(39b) and 9(58a), 9(58b), using 9(50) and 9(52). A much simpler derivation makes use of the fact that 9(57) is the residue of $K(p, q, k)$ at $k_0 = \omega_{\mathbf{k}}$. To derive 9(60a) for example, we multiply 9(51a) by $k_0 - \omega_{\mathbf{k}}$ and take the limit $k_0 \to \omega_{\mathbf{k}}$. The inhomogeneous term drops out and, using 9(59), we immediately obtain 9(60a).

The Ladder Approximation In practice, the Bethe–Salpeter equation can only be treated in the so-called *ladder approximation.* This approximation consists in replacing the interaction function $\bar{G}(x_1, x_2; x_3, x_4)$ by its lowest order value 9(46) corresponding to the simple one-meson exchange graph. In addition we replace the dressed propagators $S_F'^A$ and $S_F'^B$ by their lowest order values S_F^A and S_F^B. The iterative solution to 9(47) will then generate the sequence of 'ladder' graphs shown in Fig. 9.4.

In the ladder approximation, the homogeneous Bethe–Salpeter equation 9(49a) reduces to

$$\chi_k(x_1, x_2) = -G_0^2 \int d^4x_5\, d^4x_6 S_F^A(x_1 - x_5) S_F^B(x_2 - x_6)$$
$$\times \Delta_F(x_5 - x_6)\chi_k(x_5, x_6) \quad (k_0 = \omega_{\mathbf{k}}) \qquad 9(61)$$

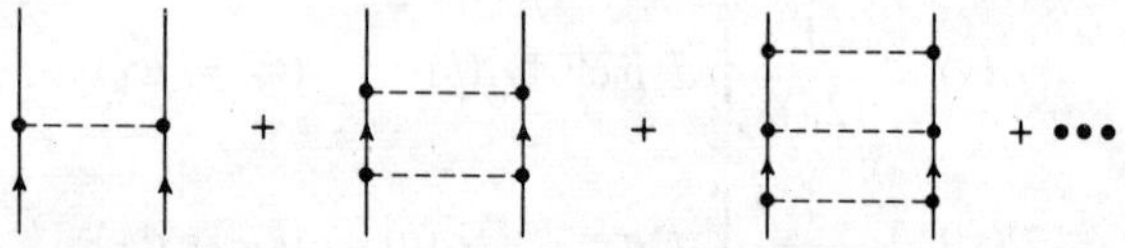

FIG. 9.4. Ladder approximation for the two-body propagator.

This equation is similar in structure to the nonrelativistic equation 9(34) and can be converted into the differential equation*

$$(\gamma^A . \partial + m)(\gamma^B . \partial + m)\chi_k(x_1, x_2) = G_0^2 \Delta_F(x_1 - x_2)\chi_k(x_1, x_2) \qquad 9(62a)$$

or, in momentum space,

$$S_F^A(\tfrac{1}{2}k + p)^{-1} S_F^B(\tfrac{1}{2}k - p)^{-1}\chi_k(p) = -\frac{G_0^2}{(2\pi)^4}\int d^4p' \Delta_F(p - p')\chi_k(p') \qquad 9(62b)$$

In their original paper, Salpeter and Bethe showed that, in the extreme nonrelativistic limit, neglecting the dependence of $\chi_k(p)$ on the relative energy k_0†, Equation 9(62b) reduces to the ordinary Schrödinger equation. An approximate solution to 9(62a) or 9(62b) was obtained by Salpeter and Bethe for the case of the deuteron ground state in scalar meson theory with scalar coupling [Salpeter (1951)]. Further progress was achieved by Wick (1954), who showed that one can define an analytic continuation of the Bethe–Salpeter amplitude to complex values of the relative time variable t. Correspondingly, the momentum space amplitude can be analytically continued to complex values of k_0. One can then transform 9(62b) by a 90° rotation of integration path in the complex k_0-plane, to obtain an integral equation which is considerably easier to treat analytically, [Cutkosky (1954)].

* This conversion is only possible in the ladder approximation. In the general case, 9(62a) is replaced by an integro-differential equation. Note that in converting an integral equation to differential form, information on the boundary conditions is lost and additional criteria are needed to distinguish bound states from scattering solutions.

† This is equivalent to replacing the Lorentz-invariant, retarded Yukawa interaction by one which is instantaneous in the centre of mass system.

9-2 Normalization of Bethe-Salpeter wave functions

Introduction We now turn to the normalization condition satisfied by Bethe–Salpeter wave functions. Although the homogeneous Bethe–Salpeter equation leaves the normalization of χ_k completely undetermined, it is clear from the definition 9(38) that the magnitude of χ is not arbitrary. The magnitude of the field operators $\psi_A(x_1)$ and $\psi_B(x_2)$ is fixed by the equal-time commutation rules, so that once we adopt a normalization convention, say $\langle \mathbf{k}|\mathbf{k}'\rangle = \delta_{\mathbf{k}\mathbf{k}'}$, for single-particle states, the magnitude of χ_k is essentially fixed.* Early approaches to the problem of determining the normalization condition related the amplitude χ_k to a conserved quantum number carried by the bound states, say baryon number or electric charge, [Nishijima (1953), (1954), (1955), Mandelstam (1955), Klein (1957)]. Here we shall follow a more general approach, [Allcock (1956), Cutkosky (1964), Lurié (1965a)], which is applicable even if the bound state carries no conserved quantum number.

Normalization Condition Our approach is based on the relation 9(59) between $\chi_k(p)$, $\bar{\chi}_k(q)$ and $K(p, q, k)$ in the neighbourhood of the pole at $k_0 = \omega_{\mathbf{k}}$. Since $K(p, q, k)$ obeys an inhomogeneous equation, its magnitude (and, hence, that of its residue at $k_0 = \omega_{\mathbf{k}}$) is essentially fixed and we should, in principle, be able to determine the normalization of χ_k. To extract the normalization condition we introduce the auxiliary quantity

$$Q(p, q, k) = \int d^4q'(k_0-\omega_{\mathbf{k}})K(p, q', k)\frac{\partial}{\partial k_0}[I(q', q, k)+\bar{G}(q', q, k)]$$

defined for $k_0 \neq \omega_{\mathbf{k}}$ as well as on the mass shell $k_0 = \omega_{\mathbf{k}}$. For convenience, we shall regard p and q as indices and write $Q(p, q, k)$ in the more compact operator form.

$$Q(k) = (k_0-\omega_{\mathbf{k}})K(k)\frac{\partial}{\partial k_0}[I(k)+\bar{G}(k)] \qquad 9(63)$$

In terms of this notation, 9(51b), 9(60a) and 9(59) appear as

$$K(k)[I(k)+\bar{G}(k)] = 1 \qquad 9(64a)$$

$$[I(k)+\bar{G}(k)]\chi_k = 0 \qquad (k_0 = \omega_{\mathbf{k}}) \qquad 9(64b)$$

$$\lim_{k_0\to\omega_{\mathbf{k}}} (k_0-\omega_{\mathbf{k}})K(k) = -\frac{i}{(2\pi)^4}\frac{1}{2k_0}\chi_k\bar{\chi}_k \qquad 9(64c)$$

* The same remark applies to the Schrödinger wave function, if defined as in 9(11) [see Chapter 3, Problem 14].

These three equations suffice to determine the normalization condition. Using 9(64a), we first obtain an alternative expression for $Q(k)$:

$$Q(k) = \frac{\partial}{\partial k_0}\left\{(k_0-\omega_{\mathbf{k}})K(k)[I(k)+\bar{G}(k)]\right\} - \frac{\partial[(k_0-\omega_{\mathbf{k}})K(k)]}{\partial k_0}[I(k)+\bar{G}(k)]$$

$$= 1 - \frac{\partial[(k_0-\omega_{\mathbf{k}})K(k)]}{\partial k_0}[I(k)+\bar{G}(k)] \qquad 9(65)$$

If we now operate on χ_k with Q in the form 9(65), and use 9(64b) for $k_0 = \omega_{\mathbf{k}}$, we get the simple equation

$$Q(k)\chi_k = \chi_k \qquad (k_0 = \omega_{\mathbf{k}})$$

On the other hand, if we operate with Q in the form 9(63), we obtain, for $k_0 = \omega_{\mathbf{k}}$

$$Q(k)\chi_k = -\frac{i}{(2\pi)^4}\frac{1}{2k_0}\chi_k\bar{\chi}_k\frac{\partial}{\partial k_0}[I(k)+\bar{G}(k)]\chi_k \qquad (k_0 = \omega_{\mathbf{k}})$$

by virtue of 9(64c). Comparing these two results we derive the condition

$$-\frac{i}{(2\pi)^4}\bar{\chi}_k\frac{\partial}{\partial k_0}[I(k)+\bar{G}(k)]\chi_k = 2k_0 \qquad (k_0 = \omega_{\mathbf{k}})$$

which, when written out in full, reads*

$$-\frac{i}{(2\pi)^4}\int d^4q\, d^4q'\,\bar{\chi}_k(q')\frac{\partial}{\partial k_0}[I(q',q,k)+\bar{G}(q',q,k)]\chi_k(q) = 2k_0$$

$$(k_0 = \omega_{\mathbf{k}}) \qquad 9(66)$$

Equation 9(66) is the normalization condition for Bethe–Salpeter wave functions. One can show that the corresponding condition for the non-relativistic Bethe–Salpeter wave function 9(18) reduces to the ordinary normalization condition for Schrödinger wave functions [see Problem 7].

Normalization of Elementary Particle Amplitudes The normalization condition 9(66) has a somewhat unfamiliar appearance. Nevertheless it is structurally very similar to the usual normalization condition for free elementary particle wave functions. To underline the correspondence, we shall ‘ derive ’ the normalization condition for elementary spinor wave functions by the same technique used in deriving 9(66).

Consider the free fermion propagator

$$S_F(x_1-x_2) = (0|T\psi(x_1)\bar{\psi}(x_2)|0)$$

* Note that in practical calculations with the ladder approximation, 9(66) is greatly simplified since $\partial\bar{G}/\partial k_0 = 0$ in the ladder approximation.

Its Fourier transform $S_F(p)$ satisfies

$$(\gamma . p - im)S_F(p) = S_F(p)(\gamma . p - im) = -1 \tag{9(67)}$$

We now consider $t_1 > t_2$ and obtain

$$S_F(x_1 - x_2) = \sum_{r=1}^{2} \int d^4p (0|\psi(x_1)|\mathbf{p}, r)(\mathbf{p}, r|\bar{\psi}(x_2)|0) \times \theta(p_0)\delta(p_0 - \sqrt{\mathbf{p}^2 + m^2}) \tag{9(68)}$$

since only one-fermion states contribute because of fermion number conservation. Using invariance under space-time translations, we write

$$(0|\psi(x_1)|\mathbf{p}, r) = \frac{1}{\sqrt{V}}\sqrt{\frac{m}{p_0}}\, e^{ip.x_1} u_{\mathbf{p}r} \tag{9(69a)}$$

$$(\mathbf{p}, r|\bar{\psi}(x_2)|0) = \frac{1}{\sqrt{V}}\sqrt{\frac{m}{p_0}}\, e^{-ip.x_2} \bar{u}_{\mathbf{p}r} \tag{9(69b)}$$

thereby defining spinors $u_{\mathbf{p}r}$ and $\bar{u}_{\mathbf{p}r}$ which satisfy

$$(\gamma . p - im)u_{\mathbf{p}r} = 0 \tag{9(70a)}$$

$$\bar{u}_{\mathbf{p}r}(\gamma . p - im) = 0 \tag{9(70b)}$$

as a consequence of the free field equations. From 9(68) and 9(69a), 9(69b), and the fact that one-fermion states do not contribute for $t_2 > t_1$, we deduce the result

$$S_F(p) = \frac{i}{p_0 - (\mathbf{p}^2 + m^2)^{1/2} + i\varepsilon} \frac{m}{p_0} \sum_{r=1}^{2} u_{\mathbf{p}r}\bar{u}_{\mathbf{p}r} + \text{terms regular at } p_0 = (\mathbf{p}^2 + m^2)^{1/2} \tag{9(71)}$$

Equations 9(67), 9(70a) and 9(71) correspond respectively to 9(64a), 9(64b) and 9(64c), and allow us to determine the normalization of the $u_{\mathbf{p}r}$. Proceeding exactly as in the Bethe–Salpeter case—introducing an auxiliary off-the-mass-shell quantity

$$[p_0 - (\mathbf{p}^2 + m^2)^{1/2}]S_F(p)\frac{\partial}{\partial p_0}(\gamma . p - im)$$

and using 9(67), 9(70a) and 9(71)—we deduce the orthonormality condition

$$i\bar{u}_{\mathbf{p}r}\frac{\partial}{\partial p_0}(\gamma . p - im)u_{\mathbf{p}s} = -\frac{p_0}{m}\delta_{rs} \tag{9(72)}$$

which reduces to the familiar relation*

$$u_{\mathbf{p}r}{}^{\dagger}u_{\mathbf{p}s} = \frac{p_0}{m}\delta_{rs} \qquad 9(73)$$

9-3 Bound state matrix elements

Introduction In this section we apply the Bethe–Salpeter techniques to the problem of evaluating the matrix element of a dynamical variable between two bound states, a problem first investigated by Mandelstam (1955) and by Nishijima (1953), (1954), (1955). We present a modified version of Mandelstam's approach, based on the Gell-Mann and Low technique of Section 9-2.

Formulation In field theory a dynamical variable is expressed in terms of a product of field operators and their derivatives. Our problem is to devise a technique for evaluating the general matrix element

$$M_{k'k}(x_i\ldots;y_i\ldots;z_i\ldots) = \langle k'|T\psi(x_i)\ldots\bar{\psi}(y_i)\ldots\phi(z_i)\ldots|k\rangle \qquad 9(74)$$

where $|k\rangle$ and $|k'\rangle$ are two bound states. We begin by considering the simpler quantity

$$M_k(x_i\ldots;y_i\ldots;z_i\ldots) = \langle 0|T\psi(x_i)\ldots\bar{\psi}(y_i)\ldots\phi(z_i)\ldots|k\rangle \qquad 9(75)$$

For simplicity we consider the same scalar coupling model as in Section 9-1, assuming $|k\rangle$ to be a nondegenerate two-fermion bound state with the Bethe–Salpeter wave function

$$\chi_k^{\alpha\beta}(x_1, x_2) = \langle 0|T\psi_A^{\alpha}(x_1)\psi_B^{\beta}(x_2)|k\rangle$$

Henceforth, the superscripts A and B will be suppressed to simplify the notation.

The matrix element 9(75) is a generalized Bethe–Salpeter amplitude and can be determined in terms of the Green's function

$$R(x_i\ldots;y_i\ldots;z_i\ldots;y_1,y_2) = \langle 0|T\psi(x_i)\ldots\bar{\psi}(y_i)\ldots\phi(z_i)\ldots\bar{\psi}(y_1)\bar{\psi}(y_2)|0\rangle \qquad 9(76)$$

by applying the technique of Gell-Mann and Low. For the time-ordering

$$x_{i0}\ldots, y_{i0}\ldots, z_{i0}\ldots > y_{10}, y_{20}$$

* Note the different distribution of p_0 factors between 9(68) and 9(69a), 9(69b) on the one hand, and 9(53) and 9(39a), 9(39b), on the other. Our convention for the latter is the one most commonly used in dealing with Bethe–Salpeter amplitudes. Equations 9(68) and 9(69a), 9(69b), on the other hand, are consistent with the practice followed throughout this book of normalizing one-particle states according to $\langle \mathbf{p}|\mathbf{p}'\rangle = \delta_{\mathbf{pp}'}$ and Dirac spinors according to the invariant prescription 9(73).

we write R in the form

$$\sum_{k,\alpha} \langle 0|T\psi(x_i)\ldots\bar{\psi}(y_i)\ldots\phi(z_i)\ldots|k,\alpha\rangle\langle k,\alpha|\bar{\psi}(y_1)\bar{\psi}(y_2)|0\rangle$$

Isolating the bound state contribution, we find, as in 9(42),

$$M_k(x_i\ldots;y_i\ldots;z_i\ldots)\mathscr{P}_k$$

$$= L_{t\to-\infty}\int d^3y_1\,d^3y_2 R(x_i\ldots;y_i\ldots;z_i\ldots;\mathbf{y}_1t,\mathbf{y}_2t)\gamma_4\gamma_4\chi_k(\mathbf{y}_1t,\mathbf{y}_2t) \quad 9(77)$$

where $\mathscr{P}_k$ is given by 9(43). We now define $T(x_i\ldots;y_i\ldots;z_i\ldots;y_1,y_2)$ to be the *truncated* Bethe–Salpeter irreducible part of $R(x_i\ldots;y_i\ldots;z_i\ldots;y_1,y_2)$, where by 'truncation' we mean the removal of the two S_F factors corresponding to the incoming lines, and where, as defined in Section 9.1, the Bethe–Salpeter-irreducible part of $R(x_i\ldots;y_i\ldots;z_i\ldots;y_1,y_2)$ is that part which *cannot* be split as in Fig. 9.5 into two pieces connected by two fermion lines. By definition,

$$R(x_i\ldots;y_i\ldots;z_i\ldots;y_1,y_2) = \int d^4y_3\,d^4y_4 T(x_i\ldots;y_i\ldots;z_i\ldots;y_3,y_4)$$
$$\times K(y_3,y_4;y_1,y_2) \quad 9(78)$$

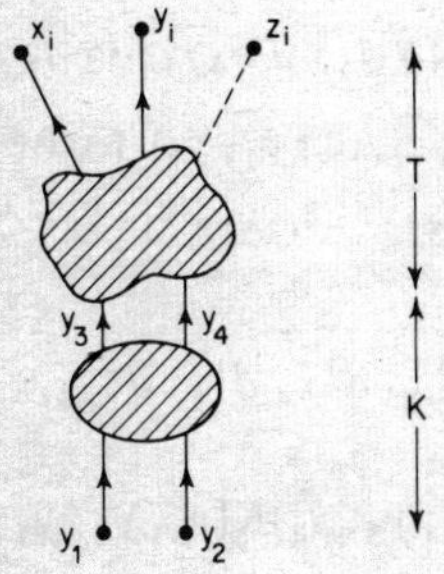

FIG. 9.5. A Bethe–Salpeter *reducible* part of $R(x_i\ldots;y_i\ldots;z_i\ldots;y_1,y_2)$ showing its decomposition into $T(x_i\ldots;y_i\ldots;z_i\ldots;y_3,y_4)$ and $K(y_3,y_4;y_1,y_2)$.

Inserting 9(78) into 9(77) and using 9(42) we obtain the formula

$$M_k(x_i\ldots;y_i\ldots;z_i\ldots) = \int d^4y_3\,d^4y_4 T(x_i\ldots;y_i\ldots;z_i\ldots;y_3,y_4)$$
$$\times \chi_k(y_3,y_4) \quad 9(79)$$

expressing M_k in terms of the Bethe–Salpeter amplitude χ_k and a function T which can, in principle, be evaluated by means of perturbation theory.

We are now in a position to treat the more general matrix element 9(74). Consider the generalized Bethe–Salpeter amplitude

$$\langle 0|T\psi(x_1)\psi(x_2)\psi(x_i)\ldots\bar{\psi}(y_i)\ldots\phi(z_i)\ldots|k\rangle \qquad 9(80)$$

for the time-ordering

$$x_{10}, x_{20} > x_{i0}, \ldots y_{i0}, \ldots z_{i0}, \ldots$$

and introduce a complete set of states between $\psi(x_2)$ and $\psi(x_i)$. Projecting out the contribution of the state $|k'\rangle$, we have

$$M_{kk'}(x_i \ldots ; y_i \ldots ; z_i \ldots)\mathscr{P}_{k'} = L_{t'\to\infty}\int d^3x_1\, d^3x_2 \bar{\chi}_{k'}(\mathbf{x}_1 t', \mathbf{x}_2 t')\gamma_4\gamma_4$$

$$\times \langle 0|T\psi(\mathbf{x}_1, t')\psi(\mathbf{x}_2, t')\psi(x_i)\ldots\bar{\psi}(y_i)\ldots\phi(z_i)\ldots|k\rangle \qquad 9(81)$$

Defining the Green's function

$$R(x_1, x_2; x_i \ldots ; y_i \ldots ; z_i \ldots ; y_1, y_2)$$
$$= \langle 0|T\psi(x_1)\psi(x_2)\ldots\psi(x_i)\ldots\bar{\psi}(y_i)\ldots\phi(z_i)\ldots\bar{\psi}(y_1)\bar{\psi}(y_2)|0\rangle$$

and applying 9(77) we obtain

$$M_{k'k}\mathscr{P}_{k'}\mathscr{P}_k = L_{t'\to+\infty}L_{t\to-\infty}\int d^3x_1\, d^3x_2\, d^3y_1\, d^3y_2$$

$$\times \bar{\chi}_{k'}(\mathbf{x}_1 t', \mathbf{x}_2 t')\gamma_4\gamma_4 R(\mathbf{x}_1 t', \mathbf{x}_2 t'; x_i \ldots ; y_i \ldots ; z_i \ldots ; \mathbf{y}_1 t, \mathbf{y}_2 t)\gamma_4\gamma_4\chi_k(\mathbf{y}_1 t, \mathbf{y}_2 t)$$

As a final step we define $T(x_1, x_2, x_i \ldots ; y_i \ldots ; z_i \ldots ; y_1, y_2)$ to be the truncated irreducible part of $R(x_1, x_2, x_i \ldots ; y_i \ldots ; z_i \ldots ; y_1, y_2)$ with respect to *both* x_1, x_2 and y_1, y_2:

$$R(x_1, x_2; x_i \ldots ; y_i \ldots ; z_i \ldots ; y_1, y_2) = \int d^4x_3\, d^4x_4\, d^4y_3\, d^4y_4$$

$$\times K(x_1, x_2; x_3, x_4)T(x_3, x_4; x_i \ldots ; y_i \ldots ; z_i \ldots ; y_3, y_4)K(y_3, y_4; y_1, y_2) \qquad 9(82)$$

[see Fig. 9.6]. Then, using 9(42) and the corresponding relation for $\bar{\chi}_k$:

$$L_{t\to\infty}\int d^3x_1\, d^3x_2 \bar{\chi}_k(\mathbf{x}_1 t, \mathbf{x}_2 t)\gamma_4\gamma_4 K(\mathbf{x}_1 t, \mathbf{x}_2 t; x_3, x_4) = \bar{\chi}_k(x_3, x_4)\mathscr{P}_k$$

we obtain the result

$$M_{k'k} = \int d^4x_3\, d^4x_4\, d^4y_3\, d^4y_4$$

$$\times \bar{\chi}_{k'}(x_3, x_4)T(x_3, x_4; x_i \ldots ; y_i \ldots ; z_i \ldots ; y_3, y_4)\chi_{k'}(y_3, y_4) \qquad 9(83)$$

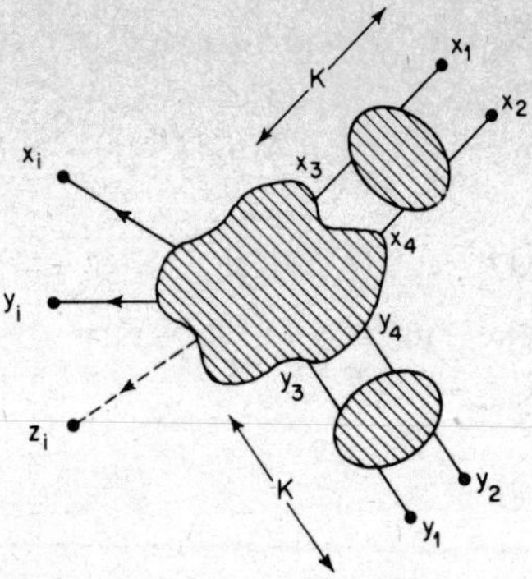

FIG. 9.6. A reducible part of $R(x_1, x_2; x_i \ldots; y_i \ldots; z_i \ldots; y_1, y_2)$ showing its decomposition into $T(x_3, x_4; x_i \ldots; y_i \ldots; z_i \ldots; y_3, y_4)$, $K(x_3, x_4; x_1, x_2)$, and $K(y_3, y_4; y_1, y_2)$.

Example As an illustration of the general technique, let us apply 9(83) to the bound state matrix elements of the charge-current density for particle A. We write the latter in the form

$$j_\mu(x) = \tfrac{1}{2}[\bar{\psi}_A(x), \gamma_\mu \psi_A(x)]$$

$$= -(\gamma_\mu)_{\alpha\beta} T\psi_A^\beta(x)\bar{\psi}_A^\alpha(x) \qquad 9(84)$$

where we have identified $T\psi^\beta(x)\bar{\psi}^\alpha(x)$ with the antisymmetrized product $\frac{1}{2}(\psi^\beta(x)\bar{\psi}^\alpha(x) - \bar{\psi}^\alpha(x)\psi^\beta(x))$ as in 3(202). Defining the Green's function

$$R(x_1, x_2; x_i; x_i; y_1, y_2) = \langle 0| T\psi^A(x_1)\psi^B(x_2)\psi^A(x_i)\bar{\psi}^A(x_i)\bar{\psi}^A(y_1)\bar{\psi}^B(y_2)|0\rangle$$

and retaining only the lowest order term in the expansion of the right-hand side, we find

$$R(x_1, x_2; x_i; x_i; y_1, y_2)$$

$$= S_F^A(x_1 - x_i)S_F^B(x_2 - y_2)S_F^A(x_i - y_1) + \ldots$$

$$= \int d^4x_3\, d^4x_4 S_F^A(x_1 - x_3)S_F^B(x_2 - x_4)\delta^{(4)}(x_3 - x_i)\delta^{(4)}(x_4 - y_2)$$

$$\times S_F^A(x_i - y_1) + \ldots$$

$$= \int d^4x_3\, d^4x_4\, d^4y_3\, d^4y_4 S_F^A(x_1 - x_3)S_F^B(x_2 - x_4)\delta^{(4)}(x_3 - x_i)\delta^{(4)}(y_3 - x_i)$$

$$\times S_F^B(x_4 - y_4)^{-1} S_F^A(y_3 - y_1)S_F^B(y_4 - y_2) + \ldots \qquad 9(85)$$

where we have defined the inverse propagator $S_F(x-y)^{-1}$ satisfying

$$\delta^{(4)}(x-x') = \int d^4y\, S_F(x-y)^{-1}S_F(y-x') \tag{9(86)}$$

or, since $(\gamma\,.\,\partial+m)S_F(y) = -i\delta^{(4)}(y)$,

$$S_F(x-y)^{-1} = i\delta^{(4)}(x-y)(\gamma\,.\,\partial_y+m) \tag{9(87)}$$

Writing 9(85) in the form

$$\begin{aligned}R(x_1, x_2; x_i; x_i; y_1, y_2) &= \int d^4x_3\, d^4x_4\, d^4y_3\, d^4y_4 K(x_1, x_2; x_3, x_4)\\ &\quad \times \delta^{(4)}(x_3-x_i)\delta^{(4)}(y_3-x_i)S_F^B(x_4-y_4)^{-1}K(y_3, y_4; y_1, y_2)+\ldots\end{aligned}$$

and comparing with 9(82) we identify the lowest order value of $T(x_3, x_4; x_i; x_i; y_3, y_4)$ as

$$T_0(x_1, x_2; x_i; x_i; x_3, x_4) = \delta^{(4)}(x_3-x_i)\delta^{(4)}(y_3-x_i)S_F^B(x_4-y_4)^{-1} \tag{9(88)}$$

Applying the general formula 9(83) and keeping track of spinor indices, we find

$$\begin{aligned}\langle k'|T\psi_A^\beta(x_i)\bar{\psi}_A^\alpha(x_i)|k\rangle &= \int d^4x_4\, d^4y_4 \bar{\chi}_{k'}^{\beta\rho}(x_i, x_4)S_{F\rho\sigma}^B(x_4-y_4)^{-1}\chi_k^{\sigma\alpha}(x_i, y_4)\\ &= i\int d^4x \bar{\chi}_{k'}^{\beta\rho}(x_i, x)(\gamma^B\,.\,\partial_x+m)_{\rho\sigma}\chi_k^{\sigma\alpha}(x_i, x)\end{aligned}$$

where we have used 9(87). Thus the bound state matrix elements of $j_\mu(x)$ are given in lowest order by

$$\langle k'|j_\mu(x_i)|k\rangle = -i\int d^4x \bar{\chi}_{k'}(x_i, x)\gamma_\mu^A(\gamma^B\,.\,\partial_x+m)\chi_k(x_i, x) \tag{9(89)}$$

χ_k and $\bar{\chi}_k$ being regarded as 16-component wave functions.

9-4 Field operators for composite particles

Introduction We have seen in Chapter 7 that the link between the S-matrix and the basic quantum field operators is contained in the *reduction formulae*. The first derivation of the reduction formulae given in Section 7-3 relied on perturbation theory and was necessarily restricted to the case of *elementary* particles. To free ourselves from the perturbative framework, a more abstract formulation was developed in Sections 7-4 and 7-5, in which the central role was played by the *asymptotic condition.* In this section we shall extend this formulation to the treatment of composite particles [Haag (1958), Nishijima (1958), (1961), (1964), (1965), Zimmermann (1958)].

Arbitrariness in the Choice of Field Operators Our aim is to extend the reduction formalism to scattering processes involving stable composite particles. The asymptotic in and out fields for such particles are constructed in exactly the same way as for elementary particles, using the procedure given in Section 7-4*. The chief problem we face is the definition of a suitable *interpolating* field which tends asymptotically to the in and out fields. Here we are aided by the realization that even for elementary particles there is an essential arbitrariness in the choice of the interpolating field.

Let us suppose that, asymptotically, a given elementary particle is represented by the incoming field $\phi^{\text{in}}(x)$ satisfying

$$(\square - \mu^2)\phi^{\text{in}}(x) = 0$$

In Section 7-5 it was understood that the field which tends asymptotically to $\phi^{\text{in}}(x)$ in the sense of 7(106a) is just the field operator $\phi(x)$ representing the particle in the Lagrangian. However, a closer examination of the proof of the asymptotic condition given at the end of Section 7-5 reveals that this need not be the case. In fact, the only properties of $\phi(x)$ used in proving the asymptotic condition were

(a) the normalization condition

$$\langle 0|\phi(x)|\mathbf{k}\rangle = Z_3^{1/2} \frac{1}{\sqrt{V}} \frac{1}{\sqrt{2\omega_{\mathbf{k}}}} e^{ik.x} \tag{9(90)}$$

where Z_3 is assumed to be nonzero, and

(b) causality, i.e. $\phi(x)$, must be a local field satisfying the principle of microscopic causality.

It follows that any local causal field $\varphi(x)$ *which satisfies* 9(90) *will tend asymptotically to* $\phi^{\text{in}}(x)$ *and* $\phi^{\text{out}}(x)$ *and is as good a candidate for the interpolating field as* $\phi(x)$†. It also follows that $\varphi(x)$ can be used in place of $\phi(x)$ in writing down reduction formulae. Thus, in practice, a wide variety of fields can be used to represent a given particle within a reduction formula. For a pseudoscalar meson, for example, if $\langle 0|T\bar{\psi}\gamma_5\psi|\mathbf{k}\rangle \neq 0$, we

* See the discussion leading to Equation 7(91).

† We must, of course, ensure that $\varphi(x)$ has the same Lorentz transformation properties as $\phi(x)$.

may take

$$\varphi(x) = Z_3^{1/2}\frac{1}{\sqrt{V}}\frac{1}{\sqrt{2\omega_{\mathbf{k}}}}\frac{1}{\langle 0|T\bar{\psi}(0)\gamma_5\psi(0)|\mathbf{k}\rangle}T\bar{\psi}(x)\gamma_5\psi(x) \qquad 9(91)$$

where the constant of proportionality has been chosen to ensure the normalization 9(90), i.e.

$$\langle 0|\varphi(x)|\mathbf{k}\rangle = Z_3^{1/2}\frac{1}{\sqrt{V}}\frac{1}{\sqrt{2\omega_{\mathbf{k}}}}\mathrm{e}^{ik.x}$$

Alternatively, we may take $\varphi(x)$ to be proportional to the divergence of the axial vector current:

$$\varphi(x) = -iZ_3^{1/2}\frac{1}{\sqrt{V}}\frac{1}{\sqrt{2\omega_{\mathbf{k}}}}\frac{1}{k_\mu\langle 0|T\bar{\psi}(0)\gamma_5\gamma_\mu\psi(0)|\mathbf{k}\rangle}\partial_\mu T\bar{\psi}(x)\gamma_5\gamma_\mu\psi(x) \qquad 9(92)$$

Actually, 9(91) and 9(92) should be defined in terms of a suitable limiting process. In place of 9(91) for example, Haag, Nishijima and Zimmermann write

$$\varphi(x) = \lim_{\substack{\xi^2>0\\ \xi\to 0}} Z_3^{1/2}\frac{1}{\sqrt{V}}\frac{1}{\sqrt{2\omega_{\mathbf{k}}}}\frac{T\bar{\psi}\left(x+\dfrac{\xi}{2}\right)\gamma_5\psi\left(x-\dfrac{\xi}{2}\right)}{\langle 0|T\bar{\psi}\left(\dfrac{\xi}{2}\right)\gamma_5\psi\left(-\dfrac{\xi}{2}\right)|\mathbf{k}\rangle} \qquad 9(93)$$

We shall not justify this here, but refer the reader to the original papers for detailed discussion.

We stress that the above arbitrariness in the choice of fields is only relative to the asymptotic condition and its consequences, i.e. the Yang–Feldman equations and the reduction formulae. A general Green's function

$$\langle 0|T\bar{\psi}(x_1)\ldots\psi(y_1)\ldots\phi(z_1)\ldots|0\rangle \qquad 9(94)$$

constructed with the aid of the fields, is defined both on and off the mass shell of the incoming and outgoing particles in momentum space. The S-matrix, on the other hand, is defined only *on* the mass shell and the function of the reduction formulae is, as indicated in Section 7-3, to single out the mass shell values of the Green's function to yield the corresponding S-matrix element. Accordingly, the arbitrariness in the choice of field applies only to the *mass shell* values of Green's functions of the type 9(94)*. We are not empowered to conclude that all matrix elements

* The class of field operators which give the same S-matrix is known as the Borchers' class [Borchers (1960)].

of ϕ and φ, as given by say 9(92), are equal, or that the replacement of ϕ by φ in 9(94) will leave the Green's function unaltered*. In particular, Equation 9(92) is a much weaker statement than the PCAC hypothesis, 8(130), used to derive the Goldberger–Treiman relation and the Adler–Weisberger sum rule in Chapter 8.

Composite Particles Having freed ourselves of the necessity of identifying the interpolating field with the basic field appearing in the Lagrangian, we are now in a position to treat both elementary and composite particles on an equal footing. For simplicity, we again refer to the scalar coupling model

$$\mathscr{L}_{\text{int}}(x) = G_0 \tfrac{1}{2}[\bar{\psi}_A(x), \psi_A(x)]\phi(x) + (A \to B) \qquad 9(95)$$

considered earlier in this chapter, and assume the existence of a spinless two-fermion bound state $|k\rangle$ with $k^2 = -M^2$. Asymptotically this particle will be represented by a scalar in-field ϕ_{in}^M satisfying

$$(\square - M^2)\phi_{\text{in}}^M = 0 \qquad 9(96)$$

and we seek an interpolating field ϕ^M which tends asymptotically to ϕ_{in}^M in the sense of weak operator convergence. Since

$$\langle 0|T\psi_A(x_1)\psi_B(x_2)|k\rangle \neq 0$$

the most obvious choice is

$$\phi^M(x) = \frac{1}{\sqrt{V}}\frac{1}{\sqrt{2k_0}}\frac{1}{\langle 0|T\psi_A(0)C^{-1}\psi_B(0)|k\rangle}T\psi_A(x)C^{-1}\psi_B(x) \qquad 9(97)$$

where we have formed the combination $-\bar{\psi}_A^c\psi_B = \psi_A C^{-1}\psi_B$ to obtain a Lorentz scalar, and normalized $\phi^M(x)$ according to

$$\langle 0|\phi^M(x)|k\rangle = \frac{1}{\sqrt{V}}\frac{1}{\sqrt{2k_0}}\mathrm{e}^{ik.x} \qquad 9(98)$$

The normalization is, of course, arbitrary, but once it is fixed it determines the proportionality coefficient which appears in the asymptotic condition. For the choice 9(98) we have

$$\lim_{t \to \mp\infty} \langle a|a_{\mathbf{k}}^M(t)|b\rangle = \langle a|a_{\mathbf{k}\,\substack{\text{in}\\\text{out}}}^M|b\rangle \qquad 9(99)$$

with a proportionality constant equal to 1 instead of $Z_3^{1/2}$. Here the operators $a_{\mathbf{k}}^M(t)$, $a_{\mathbf{k}\,\text{in}}^M$ and $a_{\mathbf{k}\,\text{out}}^M$ are defined in the usual way by relations of the

* The stronger *operator* equality $\phi = \varphi$ has been established within the framework of a simple model by Nishijima (1964). The proof is based on perturbation theory and relies crucially on the divergent character of the self-energy.

type 7(103) and 7(105) but with functions $f_{\mathbf{k}}^*$ which are solutions of the Klein–Gordon equation for mass M. The proof of 9(99) is identical to that given at the end of Section 7-5. One defines in and out fields with the aid of the retarded and advanced functions for mass M

$$\phi_{\text{in}}^{M}(x) = \phi^{M}(x) + \int \Delta_R(x-y; M)(\Box_y - M^2)\phi^{M}(y) \qquad 9(100)$$

and proceeds to show that ϕ_{in}^{M} and ϕ_{out}^{M} may be identified as the correct incoming and outgoing fields for the composite particle of mass M. The asymptotic condition 9(99) then follows by the same argument as in the paragraph following 7(117c).

We conclude that as far as the asymptotic condition and the reduction formulae are concerned, there is no clear distinction between elementary and composite particles. Both elementary and composite particles can be represented by a variety of local causal interpolating fields. Moreover, the same basic field appearing in the Lagrangian can be used to represent both elementary and composite particles within a reduction formula*.

9-5 Composite Bosons with $Z_3 = 0$

Introduction In this section we consider the following problem. Let us assume that in a self-coupled theory of the type 5(152), i.e.

$$\mathscr{L} = \mathscr{L}_0 + \mathscr{L}_f \qquad 9(101)$$

with

$$\mathscr{L}_0 = -\bar{\psi}(\gamma_\mu \partial_\mu + m)\psi \qquad 9(102)$$

$$\mathscr{L}_f = -g_0 \tfrac{1}{4}[\bar{\psi}, \gamma_5 \psi][\bar{\psi}, \gamma_5 \psi] \qquad 9(103)$$

the interaction is strong enough to produce a stable pseudoscalar bound state from a fermion-antifermion pair. Can the four-fermion interaction 9(103) then be replaced by an equivalent Yukawa coupling

$$\mathscr{L}_y = iG_0 \tfrac{1}{2}[\bar{\psi}, \gamma_5 \psi]\phi \qquad 9(104)$$

in which the bound state is represented by an elementary field? In various forms this problem has been considered by Jouvet (1956), Vaughn, Aaron and Amado [Vaughn (1961)], Rowe (1963), Lurié and Macfarlane [Lurié (1964)], and Zimmerman (1966). We shall show that, formally, the necessary and sufficient condition on the Yukawa theory is that Z_3, the boson

* If, for example, the model 9(95) gave rise to a bound state with the same quantum numbers as the field ϕ, the simplest choice for the interpolating field would be ϕ itself, multiplied by a suitable normalization factor.

wave function renormalization constant, be equal to zero. We shall restrict ourselves to the *chain approximation*, but the result is general and can be proved to all orders of perturbation theory.

A more difficult question is whether the $Z_3 = 0$ condition is a meaningful one. We shall see that in perturbation theory the $Z_3 = 0$ condition necessarily involves a high-momentum cutoff, which places it outside the framework of local relativistic quantum field theory.

Chain Approximation We seek to compare the fermion-fermion scattering amplitudes generated by the chain diagrams of Figs. 9.7 and 9.8. Some comment on the double-dot notation of Fig. 9.7 is required. We represent

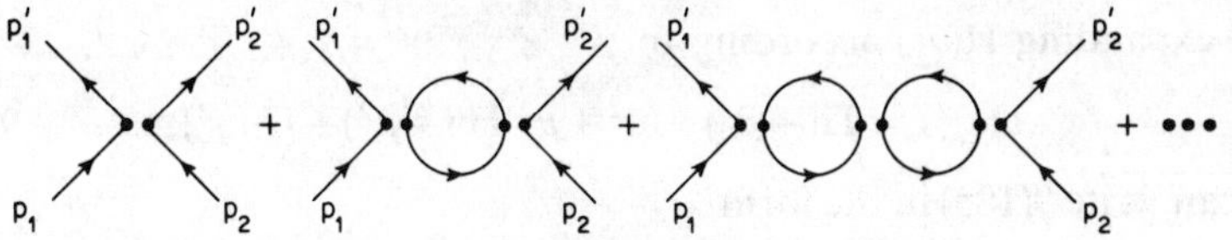

FIG. 9.7. Chain approximation for fermion–fermion scattering in the four-fermion theory.

a four-fermion vertex by an adjacent pair of dots at each of which the matrix γ_5 acts. This enables us to keep track of the spinor indices by simply following along fermion lines. We can thus distinguish the diagrams of

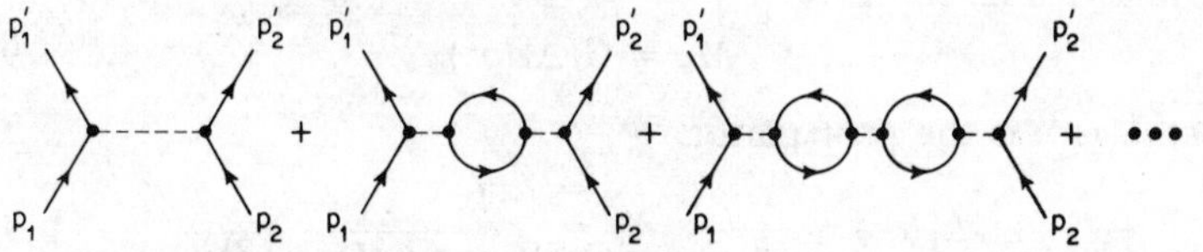

FIG. 9.8. Chain approximation for fermion–fermion scattering in the Yukawa theory.

Fig. 9.7 which involve closed fermion loops from, say, those of Fig. 9.9, which do not. We shall consider only the former in the chain approximation. The corresponding transition amplitude M_f can easily be calculated

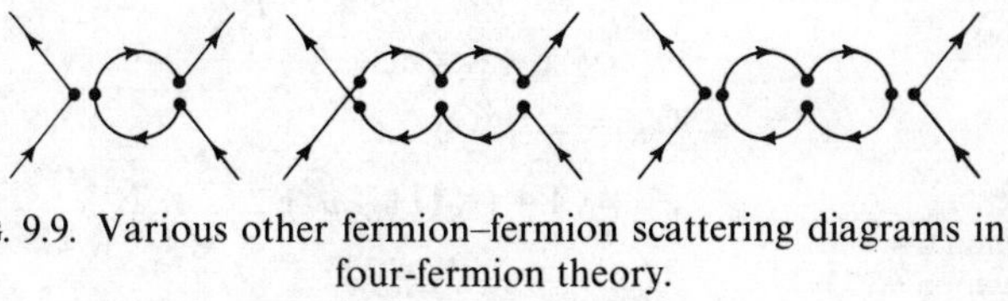

FIG. 9.9. Various other fermion–fermion scattering diagrams in the four-fermion theory.

by summing a geometric progression. We find [see Problem 10],

$$M_f = 2g_0 \frac{-i}{1-2g_0\Pi(q^2)} \tag{9(105)}$$

where $q = p_1' - p_1$ is the momentum transfer and $\Pi(q^2)$ refers to the basic fermion loop given by 6(222). We have omitted factors like $\bar{u}_{\mathbf{p}'}\gamma_5 u_{\mathbf{p}}$, referring to the external fermion lines. Let us assume that 9(105) has a pole at $q^2 = -\mu^2$ with $\mu < 2m$ corresponding to the exchange of a stable pseudoscalar boson of mass μ. The necessary and sufficient condition is

$$1 = 2g_0\Pi(-\mu^2) \tag{9(106)}$$

and expanding $\Pi(q^2)$ according to

$$\Pi(q^2) = \Pi(-\mu^2)+(q^2+\mu^2)\Pi'(-\mu^2)+\Pi_c(q^2) \tag{9(107)}$$

we can write 9(105) in the form

$$M_f = \frac{i}{(q^2+\mu^2)\Pi'(-\mu^2)+\Pi_c(q^2)} \tag{9(108)}$$

We note that $\Pi'(-\mu^2)$ diverges logarithmically and that $\Pi_c(q^2)$ is finite*.

We now compare 9(108) with the chain approximation for fermion-fermion scattering in the Yukawa theory. Recalling the work of Section 6-4 we can write the scattering amplitude corresponding to infinite set of diagrams in Fig. 9.8 as

$$M_y = G_0^2\Delta_F'(q^2) \tag{9(109)}$$

where $\Delta_F'(q^2)$ is the propagator

$$\Delta_F'(q^2) = \frac{-i}{q^2+\mu^2-G_0^2\Pi(q^2)+G_0^2\Pi(-\mu^2)} \tag{9(110)}$$

for a pseudoscalar boson of physical mass μ. Note that $\Pi(q^2)$ is given by the same expression 6(222) in both the four-fermion and Yukawa theories. Expanding $\Pi(q^2)$ as in 9(107) and renormalizing, we obtain the finite expression†

$$M_y = G_R^2 \frac{-i}{q^2+\mu^2-G_R^2\Pi_c(q^2)} \tag{9(111)}$$

where

$$G_R = Z_3^{1/2}G_0 \tag{9(112)}$$

$$Z_3 = 1+G_R^2\Pi'(-\mu^2) \tag{9(113)}$$

* See Section 6-4.

† See Equation 6(273).

Comparing 9(111) with 9(108) we see that the necessary and sufficient condition for equality is

$$G_R^2 = -\frac{1}{\Pi'(-\mu^2)} \tag{9(114)}$$

or, by 9(113)

$$Z_3 = 0 \tag{9(115)}$$

Identical considerations can obviously be applied to a scalar coupling model of the type 9(35).

Interpretation The argument presented above can be generalized to all orders of perturbation theory [Lurié (1964)] by comparing the Dyson equations for the basic Green's functions in the two theories. In place of 9(114) one finds the condition

$$G_R^2 = -\frac{1}{\mathbf{\Pi}^*{}_1{}'(-\mu^2)} \tag{9(116)}$$

where $\mathbf{\Pi}_1^{*\prime}(-\mu^2)$ is defined as in 6(371). By 6(372b), this is again the condition 9(115) generalized to all orders of perturbation theory. Thus the vanishing of the wave function renormalization constant appears as a condition characterizing a composite boson, when the latter is represented by an elementary field in Yukawa coupling theory. Heuristically, the condition $Z_3 = 0$ seems to be a natural one, since it implies that the bare, unrenormalized, field $\phi = Z_3^{1/2}\phi_R$ vanishes for a composite boson. Nevertheless, a closer examination of 9(114), or 9(116), reveals that the condition $Z_3 = 0$ is subject to a serious difficulty of interpretation. As indicated in Section 6-4, $\Pi'(-\mu^2)$, and more generally $\mathbf{\Pi}_1^{*\prime}(-\mu^2)$, are logarithmically divergent quantities in the limit of infinite cutoff. Thus 9(114) or 9(116) give $G_R = 0$ or no interaction! To get a nontrivial relation, it is therefore necessary to evaluate $\Pi'(-\mu^2)$ or $\mathbf{\Pi}_1^{*\prime}(-\mu^2)$ with the aid of a finite cutoff* and this immediately removes the $Z_3 = 0$ condition from the realm of local relativistic quantum field theory. Whether the $Z_3 = 0$ condition can be made meaningful within the framework of some non-local field theory remains at present an open question.†

The Nambu Model An interesting example of a model in which the bound state condition 9(106) is satisfied has been exhibited by Nambu and

* The same applies to the bound state condition 9(106) if $g_0 \neq 0$ since $\Pi(-\mu^2)$ diverges quadratically.

† See Lurié (1965b). Our argument has of course made crucial use of perturbation theory. It is conceivable that the $Z_3 = 0$ compositeness condition can be made consistent with locality if perturbation theory is abandoned.

Jona-Lasinio [Nambu (1961)]. This model is based on the Lagrangian density

$$\mathscr{L} = \mathscr{L}_0 + \mathscr{L}_I \tag{9(117)}$$

where $\mathscr{L}_0$ is the free Lagrangian

$$\mathscr{L}_0 = -\bar{\psi}\gamma_\mu \partial_\mu \psi \tag{9(118)}$$

and $\mathscr{L}_I$ is a four-fermion interaction of the $S+P$ type

$$\mathscr{L}_I = \tfrac{1}{4}g_0\{[\bar{\psi}, \psi][\bar{\psi}, \psi] - [\bar{\psi}, \gamma_5\psi][\bar{\psi}, \gamma_5\psi]\} \tag{9(119)}$$

A characteristic feature of this model is its invariance under the γ_5-transformation

$$\psi \to e^{i\alpha\gamma_5}\psi \tag{9(120)}$$

The invariance of the free part 9(118) follows from the absence of a bare mass term $-m_0\bar{\psi}\psi$. To check the invariance of $\mathscr{L}_I$ under 9(120) we observe that the bilinear densities $\bar{\psi}\psi$ and $i\bar{\psi}\gamma_5\psi$ transform according to

$$\begin{aligned} \bar{\psi}\psi &\to \bar{\psi}\psi \cos 2\alpha + i\bar{\psi}\gamma_5\psi \sin 2\alpha \\ i\bar{\psi}\gamma_5\psi &\to i\bar{\psi}\gamma_5\psi \cos 2\alpha - \bar{\psi}\psi \sin 2\alpha \end{aligned} \tag{9(121)}$$

which is just a rotation of the two-component vector $(\bar{\psi}\psi, i\bar{\psi}\gamma_5\psi)$.

The crucial assumption of the Nambu model is that despite the vanishing of the bare fermion mass, the *physical* mass m is nonzero. Then, in accordance with the Dyson mass renormalization prescription, we can adopt

$$\mathscr{L}'_0 = -\bar{\psi}(\gamma_\mu \partial_\mu + m)\psi \tag{9(122)}$$

as the free Lagrangian and

$$\mathscr{L}'_I = \tfrac{1}{4}g_0\{[\bar{\psi}, \psi][\bar{\psi}, \psi] - [\bar{\psi}, \gamma_5\psi][\bar{\psi}, \gamma_5\psi]\} + \delta m\bar{\psi}\psi \tag{9(123)}$$

as the new interaction Lagrangian. The self mass δm is given by

$$\delta m = -i\Sigma^*(p)|_{\gamma.p=im} \tag{9(124)}$$

where the right-hand side is calculated from the self-energy graphs of the theory, using the *physical* mass m. As discussed in Section 6-4, the identification 9(124) ensures that $\mathscr{L}'_I$ yields no additional self-mass effects. Since $m_0 = 0$, we have $m = \delta m$ and thus

$$m = -i\Sigma^*(p)|_{\gamma.p=im} \tag{9(125)}$$

Equation 9(125) is a self-consistency equation for m. Since the right-hand side diverges in any order of perturbation theory, a cutoff must necessarily be introduced to make 9(125) meaningful.

Using Wick's theorem, it is a simple matter to verify that to first order in g_0, the self-consistency condition 9(125) takes the form [see Problem 11]

$$m = 2g_0[\mathrm{Tr}\, S_F(0) - \gamma_5 \,\mathrm{Tr}\, \gamma_5 S_F(0) \\ -\tfrac{1}{2}\gamma_\mu \,\mathrm{Tr}\, \gamma_\mu S_F(0) + \tfrac{1}{2}\gamma_\mu\gamma_5 \,\mathrm{Tr}\, \gamma_\mu\gamma_5 S_F(0)] \qquad 9(126)$$

with

$$S_F(0) = \frac{1}{(2\pi)^4}\int \frac{-1}{\gamma \cdot p - im}\, d^4p \qquad 9(127)$$

The first two terms in 9(126) represent the contribution of the closed loop diagrams of Fig. 9.10. The last two terms represent the effect of the diagrams of Fig. 9.11, transformed with the aid of the Fierz identity

$$\delta^{\alpha\beta}\delta^{\gamma\delta} - \gamma_5^{\alpha\beta}\gamma_5^{\gamma\delta} = \tfrac{1}{2}\gamma_\mu^{\alpha\delta}\gamma_\mu^{\gamma\beta} - \tfrac{1}{2}(\gamma_\mu\gamma_5)^{\alpha\delta}(\gamma_\mu\gamma_5)^{\gamma\beta} \qquad 9(128)$$

FIG. 9.10. Closed-loop self-energy diagrams in four-fermion theory.

Only the first term on the right-hand side of 9(126) gives a finite contribution to the self-consistency condition and we find, using 9(127),

$$m = -\frac{8g_0 im}{(2\pi)^4}\int \frac{d^4p}{p^2 + m^2} \qquad 9(129)$$

or, assuming $m \neq 0$,

$$1 = -\frac{8g_0 i}{(2\pi)^4}\int \frac{d^4p}{p^2 + m^2} \qquad 9(130)$$

FIG. 9.11. Other lowest order self-energy diagrams in four-fermion theory.

Introducing an invariant cutoff at $p^2 = \Lambda^2$ and making the change of path $p_0 \to ip_0$, we easily obtain the condition

$$\frac{2\pi^2}{g_0\Lambda^2} = 1 - \frac{m^2}{\Lambda^2}\log\left(\frac{\Lambda^2}{m^2}+1\right) \qquad 9(131)$$

which is satisfied if

$$0 < \frac{2\pi^2}{g_0\Lambda^2} < 1 \qquad 9(132)$$

since the right-hand side of 9(131) is positive and $\leqslant 1$ for real Λ/m.

Consider now the fermion-fermion scattering amplitude generated by the chain diagrams of Fig. 9.7, in which all vertices are taken to be *pseudo-scalar**, i.e. equal to γ_5. The scattering amplitude is then just 9(105), i.e.

$$M_f^{ps} = 2g_0 \frac{-i}{1-2g_0\Pi(q^2)}$$

with $\Pi(q^2)$ given by 6(222). In accordance with our earlier discussion, the condition for the existence of a pseudoscalar bound state of mass μ is given by 9(106). The crucial point noted by Nambu and Jona-Lasinio is that at $\mu^2 = 0$, the bound state condition

$$\frac{1}{2g_0} = \Pi(0) = \frac{-4i}{(2\pi)^4}\int d^4p \frac{1}{p^2+m^2} \qquad 9(133)$$

is just identical to 9(130), so that, at least in the chain approximation, the existence of a massless pseudoscalar boson coupled to the fermion is *ensured* by the self-consistency requirement for the fermion mass. Identifying the residue of M_f^{ps} at $q^2 = 0$ with $-i$ times the square of the physical boson-fermion coupling constant, we obtain

$$G_R^2 = -\frac{1}{\Pi'(0)} \qquad 9(134)$$

which, as we have remarked earlier, is just the $Z_3 = 0$ condition for the Yukawa theory, in which the massless boson is introduced as a separate field†.

* They are either all pseudoscalar or all scalar. Mixed *S–P* loops vanish identically.

† Nambu has also examined the chain graphs generated by the scalar term in 9(119) and shown that, as a result of 9(130) the bound state condition for a *scalar* boson is satisfied for a scalar boson mass $\mu_s = 2m$. Thus in the chain approximation Nambu's model is equivalent to a Yukawa theory of the type $\mathscr{L}_y = \frac{1}{2}G_0[\bar{\psi}, \psi]\phi_s + \frac{1}{2}G_0[\bar{\psi}, \gamma_5\psi]\phi_p$ in which the wave function renormalization constants of both the scalar and pseudoscalar particles are equal to zero [Lurié (1964)].

To conclude, we remark that the masslessness of the boson bound state is closely connected to a peculiar degeneracy property of the vacuum state in Nambu's model. The connection is expressed by the *Goldstone's theorem*, which will be discussed in Section 10-3.

PROBLEMS

1. Verify that N and $\mathbf{P}$ commute with the nonrelativistic Hamiltonian 9(1).
2. Verify that $\Phi^n_{\mathbf{k}E\alpha}(\mathbf{x}_1 \ldots \mathbf{x}_n; t)$ satisfies the Schrödinger equation 9(12).
3. Use the invariance of the Hamiltonian 9(1) under space and time translations to separate out the centre of mass dependence of $\Phi^n_{\mathbf{k}E\alpha}(\mathbf{x}_1 \ldots \mathbf{x}_n; t)$.
4. Derive the nonrelativistic Bethe–Salpeter equation 9(21).
5. Check the G_0^2 term in the perturbation expansion 9(44).
6. Derive Eqs. 9(51a) and 9(51b).
7. Establish the normalization condition for the nonrelativistic Bethe–Salpeter amplitude. Prove that

$$\phi_{\mathbf{k}E\alpha}(\mathbf{q}) = \frac{1}{2\pi}\int dq_0 \chi_{\mathbf{k}E\alpha}(\mathbf{q}, q_0)$$

   where $\phi_{\mathbf{k}E\alpha}(\mathbf{q})$ and $\chi_{\mathbf{k}E\alpha}(q)$ are, respectively, the Fourier transforms of the Schrödinger and Bethe–Salpeter wave functions for the relative motion. Use the above result to establish the equivalence of the Bethe–Salpeter and Schrödinger normalization conditions in nonrelativistic theory.
8. Apply the technique of Section 9.3 to the *exact* fermion propagator and show that the normalization condition for $\langle 0|\psi(x)|\mathbf{p}, r\rangle$ reduces to the definition of Z_2.
9. Assuming the $\pi^0 NN$ interaction 5(62), show that if $|k\rangle$ is a one-π^0 state of 4-momentum k, the Fourier transform of

$$\chi_k(x_1, x_2) = \langle 0|T\psi(x_1)\bar{\psi}(x_2)|k\rangle$$

   is related to the $\pi^0 NN$ vertex function by

$$\chi_k(p) = Z_3^{1/2} G_0 S'_F(k+\tfrac{1}{2}p)\Gamma_5(k+\tfrac{1}{2}p, k-\tfrac{1}{2}p)S'_F(k-\tfrac{1}{2}p)$$

10. Derive the chain approximation result 9(105).
11. Show that 9(125) reduces to 9(126) to first order in g_0.

REFERENCES

Allcock, G. R. (1956) *Phys. Rev.* **104**, 1799
Borchers, H. J. (1960) *Nuovo Cimento* **15**, 784
Cutkosky, R. E. (1954) *Phys. Rev.* **96**, 1135
Cutkosky, R. E. (1964) (with M. Leon) *Phys. Rev.* **135 B**, 1445
Gell-Mann, M. (1951) (with F. E. Low) *Phys. Rev.* **84**, 350
Haag, R. (1958) *Phys. Rev.* **112**, 669
Jouvet, B. (1956) *Nuovo Cimento* **3**, 1133
Klein, A. (1957) (with C. Zemach) *Phys. Rev.* **108**, 126
Lurié, D. (1964) (with A. J. Macfarlane) *Phys. Rev.* **136 B**, 816
Lurié, D. (1965a) (with A. J. Macfarlane and Y. Takahashi) *Phys. Rev.* **140 B**, 1091

Lurié, D. (1965b) In *Proceedings Seminar on United Theories of Elementary Particles, Max Planck Institut, Munich, 1965*.
Mandelstam, S. (1955) *Proc. Roy. Soc.* **A 233**, 248
Nambu, Y. (1961) (with G. Jona-Lasinio) *Phys. Rev.* **122,** 345
Nishijima, K. (1953) *Prog. Theoret. Phys.* **10**, 549
Nishijima, K. (1954) *Prog. Theoret. Phys.* **12**, 279
Nishijima, K. (1955) *Prog. Theoret. Phys.* **13**, 305
Nishijima, K. (1958) *Phys. Rev.* **111**, 995
Nishijima, K. (1961) *Phys. Rev.* **122**, 298
Nishijima, K. (1964) *Phys. Rev.* **133 B**, 204; 1092
Nishijima, K. (1965) In *Proc. Trieste Seminar, 1965* (International Atomic Energy Agency)
Rowe, E. G. P. (1963) *Nucl. Phys.* **45**, 593
Salpeter, E. E. (1951) (with H. A. Bethe) *Phys. Rev.* **84**, 1232
Schweber, S. S. (1962) *Ann. Phys.*, **20**, 61
Vaughn, M. T. (1961) (with R. Aaron and R. D. Amado) *Phys. Rev.* **124**, 1258
Wick, G. C. (1954) *Phys. Rev.* **96**, 1124
Zimmerman, R. L. (1966) *Phys. Rev.* **141**, 1554
Zimmerman, W. (1958) *Nuovo Cimento* **10**, 597

10

The Functional Method

10-1 Schwinger's equations

Introduction In Section 6-4 we derived Dyson's equations for the Green's functions by summing infinite series of graphs in perturbation theory. In this section we present a powerful alternative approach to the Green's functions, originally pioneered by Schwinger (1951), on the basis of the quantum action principle [Section 3-3]. Our presentation follows the more conventional reformulation of Umezawa and Visconti [Umezawa (1955)] and Polivanov (1955).

Fermion Propagator in External Source We consider neutral pseudoscalar meson theory with the interaction Lagrangian density

$$\mathscr{L}_I(x) = iG_0\tfrac{1}{2}[\bar{\psi}(x), \gamma_5\psi(x)]\phi(x) \tag{10(1)}$$

to which we add the interaction term

$$\mathscr{L}_I^J(x) = J(x)\phi(x) \tag{10(2)}$$

representing the local interaction of the meson field with an external c-number source $J(x)$. Let us define the fermion Green's function in the presence of the external source by

$$S_F'(x, x')^J = \frac{\langle 0\ \text{out}|T\psi(x)\bar{\psi}(x')|0\ \text{in}\rangle_J}{\langle 0\ \text{out}|0\ \text{in}\rangle_J} \tag{10(3a)}$$

We have then, as in 7(71a)

$$S_F'(x, x')^J = \frac{(T\psi^{ip}(x)\bar{\psi}^{ip}(x')S^J)_0}{(S^J)_0} \tag{10(3b)}$$

where we have denoted the expectation value in the unperturbed vacuum $|0)$ simply by $(\ )_0$. Both the scattering operator

$$S^J = U(\infty, -\infty)^J$$

and the fermion Green's function $S_F'^J$ are now functionals of the external source. In particular we observe that

(a) the probability amplitude

$$\langle 0 \text{ out}|0 \text{ in}\rangle_J = (S^J)_0$$

for the vacuum to remain a vacuum is no longer just a phase factor;

(b) translation invariance arguments can no longer be applied when $J \neq 0$ so that $S_F'^J(x, x')$ is effectively a function of both $x-x'$ and $x+x'$.

We now apply the operator $\gamma \,.\, (\partial/\partial x)+m_0$ to both sides of 10(3a). Using the equation of motion

$$(\gamma \,.\, \partial + m_0)\psi(x) = iG_0\gamma_5\psi(x)\phi(x) \tag{10(4)}$$

and the identity

$$T\psi(x)\bar{\psi}(x') = \tfrac{1}{2}[\psi(x), \bar{\psi}(x')] + \tfrac{1}{2}\epsilon(x_0 - x_0')\{\psi(x), \bar{\psi}(x')\} \tag{10(5)}$$

valid for the Fermi–Dirac fields, we obtain*

$$\left(\gamma \,.\, \frac{\partial}{\partial x} + m_0\right) S_F'(x, x')^J$$

$$= -i\delta^{(4)}(x-x') + iG_0 \frac{\langle 0 \text{ out}|\, T\gamma_5\psi(x)\bar{\psi}(x')\phi(x)|0 \text{ in}\rangle_J}{\langle 0 \text{ out}|0 \text{ in}\rangle_J} \tag{10(6)}$$

The second term on the right-hand side can be expressed as a functional derivative of $S_F'^J$ with respect to J. To see this we compute the functional derivative of the scattering operator S^J appearing in 10(3b). We have

$$S^J = \sum_{n=0}^{\infty} \frac{i^n}{n!} \int d^4x_1 \ldots d^4x_n T(\mathscr{L}_I^{ip}(x_1) \ldots \mathscr{L}_I^{ip}(x_n))$$

with

$$\mathscr{L}_I^{ip} = iG_0 \tfrac{1}{2}[\bar{\psi}^{ip}, \gamma_5 \psi^{ip}] + J\phi^{ip}$$

and applying the elementary properties 2(7a), 2(7b) and 2(7c) of functional differentiation, we find

$$\frac{\delta S^J}{\delta J(x)} = \sum_{n=0}^{\infty} \frac{i^n}{n!} n \int d^4x_1 \ldots d^4x_{n-1} T\left(\frac{\partial \mathscr{L}_I^{ip}(x)}{\partial J(x)} \mathscr{L}_I^{ip}(x_1) \ldots \mathscr{L}_I^{ip}(x_{n-1})\right)$$

$$= i \sum_{n=0}^{\infty} \frac{i^{n-1}}{(n-1)!} \int d^4x_1 \ldots d^4x_{n-1} T(\phi^{ip}(x)\mathscr{L}_I^{ip}(x_1) \ldots \mathscr{L}_I^{ip}(x_{n-1}))$$

$$= iT\phi^{ip}(x)S^J \tag{10(7)}$$

* The analogous equation for $J = 0$ was derived in Section 7-3 [see the discussion preceding Eq. 7(74)].

We now use 10(7) to calculate the functional derivative of $S_F'^J$ as given by 10(3b)*. This yields the result

$$\frac{1}{i}\frac{\delta}{\delta J(y)} S_F'(x, x')^J$$

$$= \frac{1}{(S^J)_0}\frac{1}{i}\frac{\delta}{\delta J(y)}(T\psi^{ip}(x)\bar{\psi}^{ip}(x')S^J)_0 - S_F'(x, x')^J \frac{1}{(S^J)_0}\frac{1}{i}\frac{\delta}{\delta J(y)}(S^J)_0$$

$$= \frac{(T\psi^{ip}(x)\bar{\psi}^{ip}(x')\phi^{ip}(y)S^J)_0}{(S^J)_0} - S_F'(x, x')^J \frac{(T\phi^{ip}(y)S^J)_0}{(S^J)_0}$$

$$= \frac{\langle 0\,\text{out}|T\psi(x)\bar{\psi}(x')\phi(y)|0\,\text{in}\rangle_J}{\langle 0\,\text{out}|0\,\text{in}\rangle_J} - \varphi(y)^J S_F'(x, x')^J \qquad 10(8)$$

where we have defined φ^J to be the vacuum expectation value of ϕ

$$\varphi(y)^J = \frac{\langle 0\,\text{out}|\phi(y)|0\,\text{in}\rangle_J}{\langle 0\,\text{out}|0\,\text{in}\rangle_J} = \frac{(T\phi^{ip}(y)S^J)_0}{(S^J)_0} \qquad 10(9)$$

The 3-point function on the right-hand side of 10(8) is identical to that appearing in 10(6). Substituting, we obtain Schwinger's functional differential equation for $S_F'^J$:

$$\left(\gamma \cdot \frac{\partial}{\partial x} - G_0\gamma_5\frac{\delta}{\delta J(x)} - iG_0\gamma_5\varphi(x)^J + m_0\right) S_F'(x, x')^J = -i\delta^{(4)}(x - x') \qquad 10(10)$$

Equation for φ^J An accompanying equation for $\varphi(x)^J$ can be obtained by applying the Klein–Gordon operator $\Box - \mu_0^2$ to 10(9) and using the equation of motion

$$(\Box - \mu_0^2)\phi(x) = -\frac{i}{2}G_0[\bar{\psi}(x), \gamma_5\psi(x)] - J(x) \qquad 10(11)$$

This gives

$$(\Box - \mu_0^2)\varphi(x)^J = -J(x) - \tfrac{1}{2}iG_0\frac{\langle 0\,\text{out}|[\bar{\psi}(x), \gamma_5\psi(x)]|0\,\text{in}\rangle_J}{\langle 0\,\text{out}|0\,\text{in}\rangle_J} \qquad 10(12)$$

Making the identification

$$T\psi^\beta(x)\bar{\psi}^\alpha(x) = \tfrac{1}{2}(\psi^\beta(x)\bar{\psi}^\alpha(x) - \bar{\psi}^\alpha(x)\psi^\beta(x)) \qquad 10(13a)$$

*Equation 10(7) is the key relation in this entire approach. In Schwinger's original formulation [Schwinger (1951)], 10(7) is obtained directly from the action principle.

as in 9(84), we have

$$\frac{\langle 0\ \text{out}|\frac{1}{2}[\bar{\psi}(x), \gamma_5\psi(x)]|0\ \text{in}\rangle_J}{\langle 0\ \text{out}|0\ \text{in}\rangle_J} = -\gamma_5^{\alpha\beta}\frac{\langle 0\ \text{out}|\,T\psi^{\beta}(x)\bar{\psi}^{\alpha}(x)|0\ \text{in}\rangle_J}{\langle 0\ \text{out}|0\ \text{in}\rangle_J}$$

$$= -\text{Tr}\,\gamma_5 S'_F(x, x)^J \qquad 10(13b)$$

and 10(12) takes the form

$$(\Box - \mu_0^2)\varphi(x)^J = -J(x) + iG_0\,\text{Tr}\,\gamma_5 S'_F(x, x)^J \qquad 10(14)$$

Taken together, Equations 10(10) and 10(14) form a set of two coupled functional differential equations for $S_F'^J$ and φ^J.

Relation to Dyson's Equations Let us define the boson propagator

$$\Delta'_F(x, x')^J = \frac{1}{i}\frac{\delta\varphi(x)^J}{\delta J(x')} \qquad 10(15)$$

This reduces to the usual definition in the limit $J \to 0$: by 10(9) and 10(7) we have

$$\Delta'_F(x, x')^J = \frac{(T\phi^{ip}(x)\phi^{ip}(x')S^J)_0}{(S^J)_0} - \frac{(T\phi^{ip}(x)S^J)_0}{(S^J)_0}\frac{(T\phi^{ip}(x')S^J)_0}{(S^J)_0}$$

$$= \frac{\langle 0\ \text{out}|\,T\phi(x)\phi(x')|0\ \text{in}\rangle_J}{\langle 0\ \text{out}|0\ \text{in}\rangle_J} - \varphi(x)^J\varphi(x')^J \qquad 10(16)$$

and, since φ^J is the vacuum expectation value of a pesudoscalar field, it vanishes in the limit $J \to 0$. Combining 10(14) and 10(15) and noting that $\delta J(x)/\delta J(x') = \delta(x-x')$, we get the following equation for $\Delta'_F(x, x')^J$

$$(\Box - \mu_0^2)\Delta'_F(x, x')^J = i\delta^{(4)}(x-x') + G_0\,\text{Tr}\,\gamma_5\frac{\delta}{\delta J(x')}S'_F(x, x)^J \qquad 10(17)$$

Before proceeding further, we note that for $G_0 = 0$, the relations 10(14), 10(15) and 10(17) reduce to

$$(\Box - \mu_0^2)\varphi(x)^J = -J(x) \qquad 10(18a)$$

$$\Delta_F(x-x') = \frac{1}{i}\frac{\delta\varphi(x)^J}{\delta J(x')} \qquad 10(18b)$$

and

$$(\Box - \mu_0^2)\Delta_F(x-x') = i\delta^{(4)}(x-x') \qquad 10(18c)$$

respectively. In this case the Green's function $\Delta_F(x-x')$ is just the free Klein–Gordon propagator, independent of J, and the solution to 10(18a) with the boundary condition $\varphi^J = 0$ for $J = 0$ is

$$\varphi(x)^J = i\int \Delta_F(x-x')J(x')\,d^4x' \qquad 10(18d)$$

expressing φ^J as a linear functional of J. On the other hand, when $G_0 \neq 0$, the relation between φ^J and J becomes nonlinear and in place of 10(18d) we have the *Volterra expansion**

$$\varphi(x)^J = \int \left(\frac{\delta\varphi(x)^J}{\delta J(x')}\right)_{J=0} J(x')\,d^4x' + \frac{1}{2!}\int\int \left(\frac{\delta^2\varphi(x)^J}{\delta J(x')\delta J(x'')}\right)_{J=0} J(x')J(x'')\,d^4x'\,d^4x'' + \ldots \qquad 10(18e)$$

To exhibit 10(10) and 10(17) in the form of generalized Dyson equations for $S_F'^J$ and $\Delta_F'^J$ we use 2(7d) to write

$$\frac{1}{i}\frac{\delta S_F'(x,x')^J}{\delta J(z)} = \int \frac{\delta S_F'(x,x')^J}{\delta\varphi(y)^J}\frac{1}{i}\frac{\delta\varphi(y)^J}{\delta J(z)}\,d^4y = \int \frac{\delta S_F'(x,x')^J}{\delta\varphi(y)^J}\Delta_F'(y,z)^J\,d^4y \qquad 10(19)$$

Next we introduce the inverse fermion propagator $S_F'^{-1}(x,x')^J$ satisfying

$$\int S_F'^{-1}(x,\zeta)^J S_F'(\zeta,x')^J\,d^4\zeta = \int S_F'(x,\zeta)^J S_F'^{-1}(\zeta,x')^J\,d^4\zeta = \delta^{(4)}(x-x') \qquad 10(20a)$$

and differentiate 10(20a) with respect to $\varphi(y)^J$. This yields

$$\frac{\delta S_F'(x,x')^J}{\delta\varphi(y)^J} = -G_0\int d^4\xi\,d^4\zeta\,S_F'(x,\xi)^J\Gamma_5(\xi,\zeta;y)^J S_F'(\zeta,x')^J \qquad 10(20b)$$

where we have defined the generalized vertex function

$$\Gamma_5(x,x';y)^J = \frac{1}{G_0}\frac{\delta S_F'^{-1}(x,x')^J}{\delta\varphi(y)^J} \qquad 10(21)$$

Inserting 10(20b) into 10(19), we obtain the result†

$$\frac{1}{i}\frac{\delta S_F'(x,x')^J}{\delta J(z)} = -G_0\int d^4\eta\,d^4\xi\,d^4\zeta\,S_F'(x,\xi)^J\Gamma_5(\xi,\zeta;\eta)^J S_F'(\zeta,x')^J\Delta_F'(\eta,z)^J \qquad 10(22)$$

* The Volterra series is the functional analog of the Taylor series.

† From 10(22) we see that functional differentiation of $S_F'^J$ with respect to $J(z)$ can be interpreted graphically as the insertion of a boson vertex at z in the fermion propagator line. This is formally similar to Ward's identity in quantum electrodynamics [see Problem 1].

Finally we define, as in 6(399) and 6(395),

$$\Sigma^*(x, y)^J = -\int d^4x'\, d^4y' \gamma_5 S'_F(x, x')^J \Gamma_5(x', y; y')^J \Delta'_F(y', x)^J \quad 10(23)$$

$$\Pi^*(x, y)^J = i\int d^4x'\, d^4y' \operatorname{Tr}[\gamma_5 S'_F(x, x')^J \Gamma_5(x', y'; y)^J S'_F(y', x)^J] \quad 10(24)$$

Introducing 10(22) into 10(10) and 10(17), and using 10(23) and 10(24), we obtain the generalized Dyson equations

$$\left(\gamma\cdot\frac{\partial}{\partial x}+m_0-iG_0\gamma_5\varphi(x)^J\right)S'_F(x, x')^J - i\int G_0^2\Sigma^*(x, y)^J S'_F(y, x')^J\, d^4y$$
$$= -i\delta^{(4)}(x-x') \quad 10(25)$$

$$(\Box_x-\mu_0^2)\Delta'_F(x, x')^J + \int G_0^2\Pi^*(x, y)^J \Delta'_F(y, x')^J\, d^4y = i\delta^{(4)}(x-x') \quad 10(26)$$

or, in integrated form

$$S'_F(x, x')^J = S_F(x-x') - G_0\int S_F(x-y)\gamma_5\varphi(y)^J S'_F(y, x')^J\, d^4y$$
$$-\int S_F(x-y)G_0^2\Sigma^*(y, y')^J S'_F(y', x')^J\, d^4y\, d^4y' \quad 10(27)$$

$$\Delta'_F(x, x')^J = \Delta_F(x-x') + i\int \Delta_F(x-y)G_0^2\Pi^*(y, y')^J \Delta'_F(y', x')^J\, d^4y\, d^4y'$$
$$10(28)$$

Note the presence of the additional φ^J-dependent term on the right-hand side of 10(27). This term is represented graphically in Fig. 10.1. In the limit $J \to 0$, 10(27) and 10(28) go over into 6(399) and 6(394) respectively*.

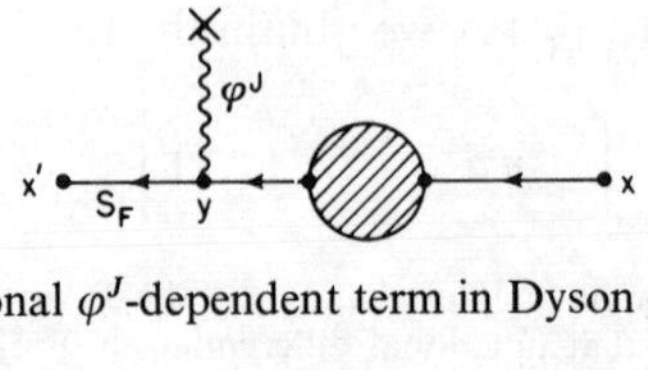

FIG. 10.1. Additional φ^J-dependent term in Dyson equation for $S_F'^J$.

* Note that $S_F'^J$ and $\Delta_F'^J$ can be viewed as functionals of φ^J since J never appears directly in 10(21), 10(23), 10(24), 10(27) or 10(28).

10-2 The Green's functional

Green's Functions and the Green's Functional Our starting point is the observation that the equality 10(7) can be applied to generate boson Green's functions of arbitrarily high order by functional differentiation of the vacuum–vacuum amplitude

$$\langle 0 \text{ out}|0 \text{ in}\rangle_J = (S^J)_0$$

Thus we have, for example

$$\langle 0 \text{ out}|\phi(x)|0 \text{ in}\rangle_J = \frac{1}{i}\frac{\delta}{\delta J(x)}\langle 0 \text{ out}|0 \text{ in}\rangle_J \tag{10(29)}$$

$$\langle 0 \text{ out}|T\phi(x)\phi(x')|0 \text{ in}\rangle_J = \frac{1}{i^2}\frac{\delta}{\delta J(x)}\frac{\delta}{\delta J(x')}\langle 0 \text{ out}|0 \text{ in}\rangle_J \tag{10(30)}$$

etc. . . .

Accordingly, the functional

$$Z[J] = \langle 0 \text{ out}|0 \text{ in}\rangle_J \tag{10(31)}$$

is known as the *boson Green's functional.* Boson Green's functions of any order can be generated by functional differentiation of $Z[J]$.

To extend this concept to fermion Green's functions we must introduce additional external sources of $\eta(x)$ and $\bar{\eta}(x)$ coupled to the fermion field operators $\bar{\psi}(x)$ and $\psi(x)$. We therefore add to the Lagrangian density the interaction

$$\mathscr{L}_I^{\eta\bar{\eta}}(x) = \bar{\eta}(x)\psi(x) + \bar{\psi}(x)\eta(x) \tag{10(32)}$$

The external sources η and $\bar{\eta}$ will be assumed to anticommute everywhere, both with each other and with ψ and $\bar{\psi}$. Following the same steps as in the derivation of 10(7), we obtain

$$\frac{\delta S^{\eta\bar{\eta}J}}{\delta\bar{\eta}(x)} = iT\psi^{ip}(x)S^{\eta\bar{\eta}J} \tag{10(33a)}$$

$$\frac{\delta S^{\eta\bar{\eta}J}}{\delta\eta(x)} = iT\bar{\psi}^{ip}(x)S^{\eta\bar{\eta}J} \tag{10(33b)}$$

and

$$\frac{\delta S^{\eta\bar{\eta}J}}{\delta J(x)} = iT\phi^{ip}(x)S^{\eta\bar{\eta}J} \tag{10(33c)}$$

where $S^{\eta\bar{\eta}J}$ is now the scattering operator in the presence of the external sources $\eta(x)$, $\bar{\eta}(x)$ and $J(x)$. From 10(33a), 10(33b) and 10(33c) we obtain the equalities

$$\langle 0\,\text{out}|\psi(x)|0\,\text{in}\rangle_{\eta\bar{\eta}J} = \frac{1}{i}\frac{\delta}{\delta\bar{\eta}(x)}\langle 0\,\text{out}|0\,\text{in}\rangle_{\eta\bar{\eta}J} \qquad 10(35a)$$

$$\langle 0\,\text{out}|\bar{\psi}(x)|0\,\text{in}\rangle_{\eta\bar{\eta}J} = \frac{1}{i}\frac{\delta}{\delta\eta(x)}\langle 0\,\text{out}|0\,\text{in}\rangle_{\eta\bar{\eta}J} \qquad 10(35b)$$

$$\langle 0\,\text{out}|\phi(x)|0\,\text{in}\rangle_{\eta\bar{\eta}J} = \frac{1}{i}\frac{\delta}{\delta J(x)}\langle 0\,\text{out}|0\,\text{in}\rangle_{\eta\bar{\eta}J} \qquad 10(35c)$$

$$\langle 0\,\text{out}|T\psi(x)\bar{\psi}(x')|0\,\text{in}\rangle_{\eta\bar{\eta}J} = \frac{1}{i^2}\frac{\delta}{\delta\bar{\eta}(x)}\frac{\delta}{\delta\eta(x')}\langle 0\,\text{out}|0\,\text{in}\rangle_{\eta\bar{\eta}J} \qquad 10(35d)$$

enabling us to express a general Green's function of arbitrarily high order in terms of functional derivatives of the general Green's functional

$$Z[\eta,\bar{\eta},J] = \langle 0\,\text{out}|0\,\text{in}\rangle_{\eta\bar{\eta}J} = (S^{\eta\bar{\eta}J})_0 \qquad 10(36)$$

In particular, to get the Green's functions in the *absence* of sources we simply set $\eta = \bar{\eta} = J = 0$ in 10(35); for example,

$$\langle 0|T\psi(x)\bar{\psi}(x')|0\rangle = \frac{1}{i^2}\left(\frac{\delta}{\delta\bar{\eta}(x)}\frac{\delta}{\delta\eta(x')}Z[\eta,\bar{\eta},J]\right)_{\eta=\bar{\eta}=J=0}$$

To summarize, once we know the vacuum–vacuum transition amplitude in the presence of the external sources η, $\bar{\eta}$ and J, we can calculate all the Green's functions. Hence, *the determinations of* $Z[\eta,\bar{\eta},J]$ *is equivalent to the solution of the field equations.* To turn this into an effective tool for treating quantum field theories, let us derive the equations satisfied by $Z[\eta,\bar{\eta},J]$.

Functional Equations for Z The field equations in the presence of the external sources η, $\bar{\eta}$ and J are given by

$$(\gamma\,.\,\partial + m_0)\psi(x) = iG_0\gamma_5\psi(x)\phi(x) + \eta(x) \qquad 10(37a)$$

$$\bar{\psi}(x)(\gamma\,.\,\overleftarrow{\partial} - m_0) = -iG_0\bar{\psi}(x)\gamma_5\phi(x) - \bar{\eta}(x) \qquad 10(37b)$$

$$(\Box - \mu_0^2)\phi(x) = -iG_0 T\bar{\psi}(x)\gamma_5\psi(x) - J(x) \qquad 10(37c)$$

where, in the last equation, we have replaced the antisymmetrized density $\frac{1}{2}[\bar{\psi}(x), \gamma_5\psi(x)]$ by $T\bar{\psi}(x)\gamma_5\psi(x)$ as in 10(13a). Taking the expectation value of these equations between $|0\,\text{in}\rangle$ and $|0\,\text{out}\rangle$ and using the equalities 10(35a), 10(35b), 10(35c) and 10(35d), we obtain the set of functional differential equations

$$\left(\gamma\,.\,\partial+m_0-G_0\gamma_5\frac{\delta}{\delta J(x)}\right)\frac{\delta}{\delta\bar{\eta}(x)}Z[\eta,\bar{\eta},J] = i\eta(x)Z[\eta,\bar{\eta},J] \qquad 10(38a)$$

$$\frac{\delta}{\delta\eta(x)}\left(\gamma\,.\,\overleftarrow{\partial}-m_0+G_0\gamma_5\frac{\delta}{\delta J(x)}\right)Z[\eta,\bar{\eta},J] = -i\bar{\eta}(x)Z[\eta,\bar{\eta},J] \qquad 10(38b)$$

$$(\Box-\mu_0^2)\frac{\delta}{\delta J(x)}Z[\eta,\bar{\eta},J] = -\left(G_0\frac{\delta}{\delta\eta(x)}\gamma_5\frac{\delta}{\delta\bar{\eta}(x)}+iJ(x)\right)Z[\eta,\bar{\eta},J] \qquad 10(38c)$$

satisfied by the Green's functional Z. We have thus replaced the operator field equations 10(37a), 10(37b) and 10(37c) by numerical functional differential equations*. As an illustration of how these equations can be applied, let us solve the system 10(38a), 10(38b) and 10(38c) in the case $G_0 = 0$.

Free Green's Functional Setting $G_0 = 0$ in 10(38a), 10(38b) and 10(38c) we obtain the set of equations

$$(\gamma\,.\,\partial+m_0)\frac{\delta}{\delta\bar{\eta}(x)}Z[\eta,\bar{\eta},J]_0 = i\eta(x)Z[\eta,\bar{\eta},J]_0 \qquad 10(39a)$$

$$\frac{\delta}{\delta\eta(x)}(\gamma\,.\,\overleftarrow{\partial}-m_0)Z[\eta,\bar{\eta},J]_0 = -i\bar{\eta}(x)Z[\eta,\bar{\eta},J]_0 \qquad 10(39b)$$

$$(\Box-\mu_0^2)\frac{\delta}{\delta J(x)}Z[\eta,\bar{\eta},J]_0 = -iJ(x)Z[\eta,\bar{\eta},J]_0 \qquad 10(39c)$$

whose solution $Z[\eta,\bar{\eta},J]_0$ represents the Green's functional for non-interacting fields. We begin by considering 10(39c). Using the free boson Green's function $\Delta_F(x-x')$ we can write this equation in the form

$$\frac{\delta}{\delta J(x)}Z[\eta,\bar{\eta},J]_0 = -\int d^4x'\Delta_F(x-x')J(x')Z[\eta,\bar{\eta},J]_0$$

The solution to this simple functional differential equation is

$$Z[\eta,\bar{\eta},J]_0 = Z[\eta,\bar{\eta}]_0\,e^{-\frac{1}{2}\int\int d^4x d^4x' J(x)\Delta_F(x-x')J(x')} \qquad 10(40)$$

where the ' constant of integration ' $Z[\eta,\bar{\eta}]_0$ is the free Green's functional for $J = 0$ but $\eta,\bar{\eta} \neq 0$. To determine $Z[\eta,\bar{\eta}]_0$ we insert 10(40) into 10(39a) and 10(39b), and integrate the resulting equations with the aid of the free fermion Green's function $S_F(x-x')$. This yields the functional differ-

* This recalls the passage from Heisenberg's matrix equations to Schrödinger's differential equation in ordinary quantum mechanics.

ential equations

$$\frac{\delta}{\delta\bar{\eta}(x)} Z[\eta, \bar{\eta}]_0 = -\int d^4x' S_F(x-x')\eta(x') Z[\eta, \bar{\eta}]_0 \qquad 10(41a)$$

$$\frac{\delta}{\delta\eta(x)} Z[\eta, \bar{\eta}]_0 = -\int d^4x' \bar{\eta}(x') S_F(x'-x) Z[\eta, \bar{\eta}]_0 \qquad 10(41b)$$

with the solution

$$Z[\eta, \bar{\eta}]_0 = e^{-\iint d^4x d^4x' \bar{\eta}(x) S_F(x-x')\eta(x')} \qquad 10(42)$$

Combining 10(40) and 10(42) we get the result

$$Z[\eta, \bar{\eta}, J]_0 = e^{-\iint d^4x d^4x' \bar{\eta}(x) S_F(x-x')\eta(x')} \times e^{-\frac{1}{2}\iint d^4x d^4x' J(x)\Delta_F(x-x') J(x')} \qquad 10(43)$$

for the Green's functional. The constant of integration is unity in this case, in accordance with the requirement

$$Z[\eta, \bar{\eta}, J]_0 = 1 \qquad \text{for} \quad \eta = \bar{\eta} = J = 0 \qquad 10(44)$$

It is a simple matter to check that 10(43) yields the correct free particle Green's functions. For the two-point fermion Green's function, for example, we find

$$\langle 0|T\psi(x)\bar{\psi}(x')|0\rangle = \frac{1}{i^2}\left(\frac{\delta}{\delta\bar{\eta}(x)}\frac{\delta}{\delta\eta(x')} Z[\eta, \bar{\eta}, J]_0\right)_{\eta=\bar{\eta}=J=0}$$
$$= S_F(x-x')$$

in agreement with the result of direct calculation.

Thus, for free fields, the Green's functional can be determined exactly and yields the familiar free-particle Green's functions by functional differentiation. We now give a brief account of the techniques developed by Schwinger (1954a, b) for the treatment of the interacting field case.

Formal Solution for Z Our goal is a formal expression for $Z[\eta, \bar{\eta}, J]$ which can be expanded in powers of the coupling G_0. As a preliminary, we derive a relation between $Z[\eta, \bar{\eta}, J]$ and the free Green's functional $Z[\eta, \bar{\eta}, J, B]_0$ where $B(x)$ is an additional external pseudoscalar field coupled to the pseudoscalar density $\bar{\psi}(x)\gamma_5\psi(x)$ according to

$$\mathscr{L}_I^B(x) = iG_0 T\bar{\psi}(x)\gamma_5\psi(x)B(x) \qquad 10(45)$$

As in 10(7), the functional derivatives of the scattering operator $S^{\eta\bar{\eta}JB}$

with respect to $J(x)$ and $B(x)$ are given by

$$\frac{1}{i}\frac{\delta S^{\eta\bar{\eta}JB}}{\delta J(x)} = T\phi^{ip}(x)S^{\eta\bar{\eta}JB}$$

$$\frac{1}{i}\frac{\delta S^{\eta\bar{\eta}JB}}{\delta B(x)} = iG_0 T\bar{\psi}^{ip}(x)\gamma_5\psi^{ip}(x)$$

and hence

$$\frac{1}{i^2}\frac{\delta}{\delta B(x)}\frac{\delta}{\delta J(x)}S^{\eta\bar{\eta}JB} = iG_0 T\bar{\psi}^{ip}(x)\gamma_5\psi^{ip}(x)\phi^{ip}(x)S^{\eta\bar{\eta}JB} \qquad 10(46)$$

On the other hand, if we replace the interaction 10(1) by

$$\mathscr{L}_I^{\lambda}(x) = i\lambda G_0\tfrac{1}{2}[\bar{\psi}(x), \gamma_5\psi(x)]\phi(x)$$

and vary λ we find, by the same procedure as in the derivation of 10(7),

$$\frac{1}{i}\frac{\partial}{\partial\lambda}S^{\eta\bar{\eta}JB} = iG_0\int d^4x T\bar{\psi}^{ip}(x)\gamma_5\psi^{ip}(x)\phi^{ip}(x)S^{\eta\bar{\eta}JB} \qquad 10(47)$$

Comparing 10(47) and 10(46) and taking the vacuum expectation value, we get the elementary differential equation

$$\frac{\partial}{\partial\lambda}Z[\eta,\bar{\eta},J,B] = \int d^4x\frac{\delta}{\delta B(x)}\frac{1}{i}\frac{\delta}{\delta J(x)}Z[\eta,\bar{\eta},J,B] \qquad 10(48)$$

for the Green's functional

$$Z[\eta,\bar{\eta},J,B] = \langle 0\text{ out}|0\text{ in}\rangle_{\eta\bar{\eta}JB} = (S^{\eta\bar{\eta}JB})_0$$

The solution to 10(48) is

$$Z[\eta,\bar{\eta},J,B] = \exp\left[\lambda\int d^4x\frac{\delta}{\delta B(x)}\frac{1}{i}\frac{\delta}{\delta J(x)}\right]Z[\eta,\bar{\eta},J,B]_0$$

where the 'constant of integration' is the free Green's functional $Z[\eta,\bar{\eta},J,B]_0$ in accordance with the requirement that for $\lambda = 0$, $Z[\eta,\bar{\eta},J,B]$ reduces to the vacuum–vacuum amplitude in the absence of quantum field theoretic interactions. Setting $\lambda = 1$, we get

$$Z[\eta,\bar{\eta},J,B] = \exp\left[\int d^4x\frac{\delta}{\delta B(x)}\frac{1}{i}\frac{\delta}{\delta J(x)}\right]Z[\eta,\bar{\eta},J,B]_0 \qquad 10(49)$$

To obtain a relation for $Z[\eta,\bar{\eta},J]$, we enlist the aid of the Volterra expansion 10(18e). We write the latter in the form

$$F[B+\alpha] = \sum_{n=0}^{\infty}\frac{1}{n!}\left[\int d^4x\alpha(x)\frac{\delta}{\delta B(x)}\right]^n F[B]$$

$$= \exp\left[\int d^4x\alpha(x)\frac{\delta}{\delta B(x)}\right]F[B] \qquad 10(50)$$

where B is some fixed function (taken equal to zero in 10(18e)). Applying 10(50) for $\alpha(x) = -i\delta/\delta J(x)$, i.e.

$$F\left[B+\frac{1}{i}\frac{\delta}{\delta J}\right] = \exp\left[\int d^4x \frac{\delta}{\delta B(x)}\frac{1}{i}\frac{\delta}{\delta J(x)}\right]F[B]$$

to 10(49), we obtain

$$Z[\eta, \bar{\eta}, J, B] = Z\left[\eta, \bar{\eta}, J, B+\frac{1}{i}\frac{\delta}{\delta J}\right]$$

or, setting $B = 0$,

$$Z[\eta, \bar{\eta}, J] = Z\left[\eta, \bar{\eta}, J, \frac{1}{i}\frac{\delta}{\delta J}\right]_0 \qquad 10(51)$$

Equation 10(51) relates the interacting Green's functional $Z[\eta, \bar{\eta}, J]$ to the free Green's functional $Z[\eta, \bar{\eta}, J, B]_0$ for $B = -i\delta/\delta J$. To apply 10(51) we must now compute $Z[\eta, \bar{\eta}, J, B]_0$.

The functional differential equations for $Z[\eta, \bar{\eta}, J, B]_0$

$$(\gamma . \partial + m_0 - iG_0\gamma_5 B(x))\frac{\delta}{\delta\bar{\eta}(x)}Z[\eta, \bar{\eta}, J, B]_0 = i\eta(x)Z[\eta, \bar{\eta}, J, B]_0 \qquad 10(52a)$$

$$\frac{\delta}{\delta\eta(x)}(\gamma . \overleftarrow{\partial} - m_0 + iG_0\gamma_5 B(x))Z[\eta, \bar{\eta}, J, B]_0 = -i\bar{\eta}(x)Z[\eta, \bar{\eta}, J, B]_0 \qquad 10(52b)$$

$$(\Box - \mu_0^2)\frac{\delta}{\delta J(x)}Z[\eta, \bar{\eta}, J, B]_0 = -iJ(x)Z[\eta, \bar{\eta}, J, B]_0 \qquad 10(52c)$$

are derived in the same way as the corresponding equations 10(39a), 10(39b) and 10(39c) for $B = 0$ and differ from the latter only by the replacement of m_0 by $m_0 - iG_0\gamma_5 B(x)$. Equation 10(52c) is identical to 10(39c) and we have, accordingly

$$Z[\eta, \bar{\eta}, J, B]_0 = Z[\eta, \bar{\eta}, B]_0\, e^{-\frac{1}{2}\iint d^4x d^4x' J(x)\Delta_F(x-x')J(x')} \qquad 10(53)$$

as in 10(40). To determine $Z[\eta, \bar{\eta}, B]_0$ we insert 10(53) into 10(52a) and 10(52b) and use the Green's function S_F^B satisfying

$$\left(\gamma . \frac{\partial}{\partial x} + m_0 - iG_0\gamma_5 B(x)\right)S_F(x, x')^B = -i\delta^{(4)}(x-x') \qquad 10(54a)$$

$$S_F(x, x')^B\left(\gamma . \frac{\overleftarrow{\partial}}{\partial x'} - m_0 + iG_0\gamma_5 B(x')\right) = i\delta^{(4)}(x-x') \qquad 10(54b)$$

to integrate the resulting equation. We find, as in 10(42)

$$Z[\eta, \bar{\eta}, B]_0 = Z[B]_0\, e^{-\iint d^4x d^4x' \bar{\eta}(x) S_F(x,x')^B \eta(x')} \qquad 10(55)$$

where the constant of integration is now $Z[B]_0 \neq 1$. Our final task is the determination of the free Green's functional $Z[B]_0$.

Applying the equality

$$\frac{1}{i}\frac{\delta}{\delta B(x)}S^B = iG_0 T\bar{\psi}^{ip}(x)\gamma_5\psi^{ip}(x)S^B$$

for the scattering operator S^B in the absence of quantum field theoretic couplings, we obtain the relation

$$\frac{1}{i}\frac{\delta}{\delta B(x)}Z[B]_0 = iG_0\langle 0\text{ out}|T\bar{\psi}(x)\gamma_5\psi(x)|0\text{ in}\rangle_B \tag{10(56)}$$

where $\psi(x)$ and $\bar{\psi}(x)$ interact only with $B(x)$:

$$[\gamma\,.\,\partial + m_0 - iG_0\gamma_5 B(x)]\psi(x) = 0 \tag{10(57a)}$$

$$\bar{\psi}(x)[\gamma\,.\,\overleftarrow{\partial} - m_0 + iG_0\gamma_5 B(x)] = 0 \tag{10(57b)}$$

Now the Green's function

$$\frac{\langle 0\text{ out}|T\psi(x)\bar{\psi}(x')|0\text{ in}\rangle_B}{\langle 0\text{ out}|0\text{ in}\rangle_B} \tag{10(58)}$$

can easily be seen to satisfy 10(54a) and 10(54b) and may be identified with $S_F(x,x')^B$. Accordingly, 10(56b) becomes

$$\frac{1}{i}\frac{1}{Z[B]_0}\frac{\delta Z[B]_0}{\delta B(x)} = -iG_0\,\mathrm{Tr}\,\gamma_5 S_F(x,x)^B \tag{10(59)}$$

or

$$\frac{1}{i}\frac{\delta Z[B]_0}{Z[B]_0} = -iG_0\int d^4x\delta B(x)\,\mathrm{Tr}\,[\gamma_5 S_F(x,x)^B] \tag{10(60)}$$

At this point it is convenient to adopt a more compact notation in which the space-time coordinates are regarded as matrix indices. Thus, the Green's function $S_F(x,x')^B$ is regarded as the matrix element

$$\langle x|S_F^B|x'\rangle \tag{10(61)}$$

of the operator S_F^B satisfying the equation

$$\begin{aligned}(\gamma\,.\,p - im_0 - G_0\gamma_5 B)S_F^B &= S_F^B(\gamma\,.\,p - im_0 - G_0\gamma_5 B)\\ &= -1\end{aligned} \tag{10(62)}$$

with

$$\langle x|p|x'\rangle = -i\partial_x\delta^{(4)}(x-x') \tag{10(63)}$$

$$\langle x|B|x'\rangle = \delta^{(4)}(x-x')B(x) \tag{10(64)}$$

In this notation the functional differential equation for $Z[B]_0$ becomes

$$\frac{1}{i}\frac{\delta Z[B]_0}{Z[B]_0} = -iG_0\,\mathrm{Tr}(\delta B\gamma_5 S_F^B) \tag{10(65)}$$

where the trace is taken over both spin and coordinate indices. We shall need the free Green's function S_F which obeys the equation

$$(\gamma . p - im_0)S_F = S_F(\gamma . p - im_0) = -1 \tag{10(66)}$$

in operator notation. Multiplying the two equations of 10(62) by S_F on the left and right respectively, we get the equations

$$(1 + G_0 S_F \gamma_5 B)S_F^B = S_F^B(1 + G_0 \gamma_5 B S_F) = S_F$$

with the symbolic solution

$$S_F^B = (1 + G_0 S_F \gamma_5 B)^{-1} S_F = S_F(1 + G_0 \gamma_5 B S_F)^{-1} \tag{10(67)}$$

We can therefore write 10(65) in the form

$$\begin{aligned}\delta \log Z[B]_0 &= G_0 \operatorname{Tr}(\delta B \gamma_5 S_F^B) = G_0 \operatorname{Tr}(\delta B \gamma_5 S_F (1 + G_0 \gamma_5 B S_F)^{-1}) \\ &= \operatorname{Tr}\{\delta(1 + G_0 \gamma_5 B S_F)(1 + G_0 \gamma_5 B S_F)^{-1}\}\end{aligned}$$

Schwinger (1954a) has shown that the above equation is sufficient to determine $Z[B]_0$ in the form*

$$Z[B]_0 = \det(1 + G_0 \gamma_5 B S_F) \tag{10(68)}$$

We refer the reader to Schwinger's paper for the proof of this statement. As the final step, we combine 10(51), 10(53), 10(55) and 10(68) to write $Z[\eta, \bar{\eta}, J]$ in the form

$$Z[\eta, \bar{\eta}, J] = \det\left(1 - iG_0 \gamma_5 \frac{\delta}{\delta J} S_F\right) e^{-\bar{\eta}\left(1 - iG_0 S_F \gamma_5 \frac{\delta}{\delta J}\right)^{-1} S_F \eta} \, e^{-\frac{1}{2} J \Delta_F J} \tag{10(69)}$$

where we have used operator notation and substituted 10(67) for S_F^B. Equation 10(69) is the desired result—a formal expression for $Z[\eta, \bar{\eta}, J]$ which can be expanded in powers of G_0.

Peturbation Expansion of $Z[\eta, \bar{\eta}, J]$ We now exhibit the first few terms in the perturbation expansion of $Z[\eta, \bar{\eta}, J]$. By calculating the appropriate functional derivatives, we should recover the lowest order terms in the perturbation expansion of the fermion and boson Green's functions.

We begin by neglecting the effect of the determinant in 10(69). Expanding the exponentials, we get, up to terms of order $\bar{\eta}\eta$

$$\left[1 - \bar{\eta}\left(1 - iG_0 S_F \gamma_5 \frac{\delta}{\delta J}\right)^{-1} \eta\right]\left[1 - \frac{1}{2} J \Delta_F J + \frac{1}{2!} \frac{1}{2} J \Delta_F J \frac{1}{2} J \Delta_F J - \dots\right] \tag{10(70)}$$

* This result is a consequence of the fact that the determinant of an operator X is completely defined by the equation $\delta(\log \det X) = \operatorname{Tr} \delta X . X^{-1}$ and the initial condition $\det 1 = 1$ [Schwinger (1954a)].

To calculate the exact boson Green's function in the absence of sources

$$\Delta_F'(x_1-x_2) = \langle 0|T\phi(x_1)\phi(x_2)|0\rangle = \frac{1}{i^2}\left(\frac{\delta}{\delta J(x_1)}\frac{\delta}{\delta J(x_2)}Z[\eta,\bar{\eta},J]\right)_{\eta=\bar{\eta}=J=0} \qquad 10(71)$$

we examine the term in J^2 in 10(70), i.e.

$$-\tfrac{1}{2}\int d^4x_1\, d^4x_2 J(x_1)\Delta_F(x_1-x_2)J(x_2)$$

Thus, in the approximation of neglecting the determinant, we find

$$\Delta_F'(x_1-x_2) = \Delta_F(x_1-x_2) \qquad 10(72)$$

To obtain the fermion Green's function in the absence of sources

$$S_F(x_1-x_2) = \langle 0|T\psi(x_1)\bar{\psi}(x_2)|0\rangle = \frac{1}{i^2}\left(\frac{\delta}{\delta\bar{\eta}(x)}\frac{\delta}{\delta\eta(x')}Z[\eta,\bar{\eta},J]\right)_{\eta=\bar{\eta}=J=0} \qquad 10(73)$$

we must examine the term in $\bar{\eta}\eta$ in 10(70). Expanding $\left(1-iG_0S_F\gamma_5\frac{\delta}{\delta J}\right)^{-1}$ we write 10(70) in the form

$$\left[1-\bar{\eta}\left[S_F+iG_0S_F\gamma_5\frac{\delta}{\delta J}S_F-G_0^2S_F\gamma_5\frac{\delta}{\delta J}S_F\gamma_5\frac{\delta}{\delta J}S_F+\ldots\right]\eta\right]$$

$$\times\left[1-\frac{1}{2}J\Delta_F J+\frac{1}{2!}\frac{1}{2}J\Delta_F J\frac{1}{2}J\Delta_F J+\ldots\right]$$

$$=\left[1-\int d^4x_1\, d^4x_2\bar{\eta}(1)\left[S_F(1,2)+\int d^4x_3 iG_0S_F(1,3)\gamma_5\frac{\delta}{\delta J(3)}S_F(3,2)\right.\right.$$

$$\left.\left.-G_0^2\int d^4x_3\, d^4x_4S_F(1,3)\gamma_5\frac{\delta}{\delta J(3)}S_F(3,4)\gamma_5\frac{\delta}{\delta J(4)}S_F(4,2)+\ldots\right]\eta(2)\right]$$

$$\times\left[1-\tfrac{1}{2}\int d^4x_1\, d^4x_2J(1)\Delta_F(1,2)J(2)+\ldots\right]$$

Carrying out the functional differentiations, and isolating the term in $\bar{\eta}\eta$

$$-\int d^4x_1\, d^4x_2\bar{\eta}(x_1)\left[S_F(x_1-x_2)+G_0^2\int d^4x_3\, d^4x_4S_F(x_1-x_3)\gamma_5S_F(x_3-x_4)\right.$$

$$\left.\times\gamma_5\Delta_F(x_3-x_4)S_F(x_4-x_2)+\ldots\right]\eta(x_2)$$

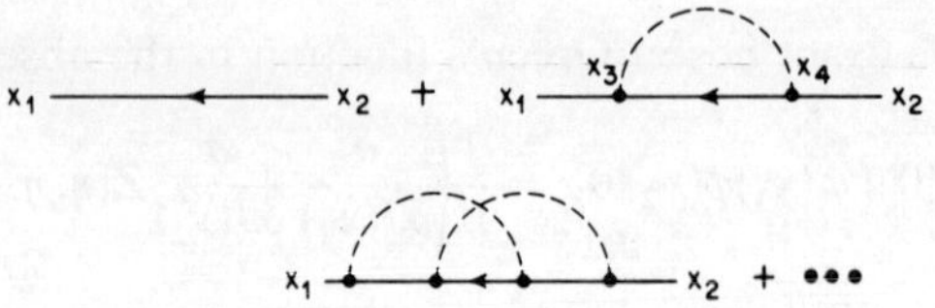

FIG. 10.2. Perturbative expansion of $S'_F(x_1 - x_2)$.

we can identify the terms in square brackets as the lowest order contributions to $S'_F(x_1 - x_2)$ shown in Fig. 10.2. In the approximation of neglecting the determinant, only the free boson propagator 10(72) will appear in the expansion of S'_F; closed fermion loops do not occur.

We now consider the effect of the determinant

$$\det\left(1 - iG_0\gamma_5 \frac{\delta}{\delta J} S_F\right) \tag{10(74)}$$

This is an infinite, continuous determinant and its expansion presents some difficulty. Mathematical methods for handling it have been developed by Schwinger (1954a, b). Here we shall restrict our attention to the lowest order terms in the expansion. Employing a discrete labelling for the space–time coordinates, we write the determinant in the form

$$\begin{vmatrix} 1 - iG_0\gamma_5 \dfrac{\delta}{\delta J(1)} S_F(1,1) & -iG_0\gamma_5 \dfrac{\delta}{\delta J(1)} S_F(1,2) & \cdots \\ -iG_0\gamma_5 \dfrac{\delta}{\delta J(2)} S_F(2,1) & 1 - iG_0\gamma_5 \dfrac{\delta}{\delta J(2)} S_F(2,2) & \cdots \\ \vdots & \vdots & \vdots \end{vmatrix} \tag{10(75)}$$

The determinant must of course also be evaluated with respect to the spin indices of the Dirac matrices: thus, for the single term

$$1 - iG_0\gamma_5 \frac{\delta}{\delta J(1)} S_F(1,1)$$

for example, we have the 4×4 determinant

$$\begin{vmatrix} 1 - iG_0 \dfrac{\delta}{\delta J(1)} [\gamma_5 S_F(1,1)]_{11} & -iG_0 \dfrac{\delta}{\delta J(1)} [\gamma_5 S_F(1,1)]_{12} & \cdots \\ -iG_0 \dfrac{\delta}{\delta J(1)} [\gamma_5 S_F(1,1)]_{21} & 1 - iG_0 \dfrac{\delta}{\delta J(1)} [\gamma_5 S_F(1,1)]_{22} & \cdots \\ \vdots & \vdots & \vdots \end{vmatrix}$$

which, when expanded up to terms of order G_0^2 gives

$$1-\mathrm{Tr}\, iG_0\gamma_5 S_F(1,1)\frac{\delta}{\delta J(1)} iG_0\gamma_5 S_F(1,1)\frac{\delta}{\delta J(2)} \qquad 10(76)$$

Here we have used the property

$$\mathrm{Tr}\,\gamma_5 S_F(1,1)=0 \qquad 10(77)$$

For the entire determinant 10(75) we get, to order G_0^2

$$\det\left(1-iG_0\gamma_5\frac{\delta}{\delta J}S_F\right)=1-\mathrm{Tr}\int d^4x_1\, d^4x_2 iG_0\gamma_5 S_F(x_1-x_2)\frac{\delta}{\delta J(x_1)}$$

$$\times\, iG_0\gamma_5 S_F(x_2-x_1)\frac{\delta}{\delta J(x_2)} \qquad 10(78)$$

Noting that the determinant commutes with the η-dependent exponential in 10(69), we compute the effect of the trace term in 10(78) on the factor $e^{-\frac{1}{2}J\Delta_F J}$:

$$G_0^2\,\mathrm{Tr}\int d^4x_1\, d^4x_2\gamma_5\frac{\delta}{\delta J(1)}S_F(1,2)\gamma_5 S_F(2,1)\frac{\delta}{\delta J(2)}$$

$$\times\left[1-\tfrac{1}{2}\int d^4x_1\, d^4x_2 J(1)\Delta_F(1,2)J(2)+\tfrac{1}{8}\int d^4x_1\ldots d^4x_4\right.$$

$$\left.\times\, J(1)\Delta_F(1,2)J(2)J(3)\Delta_F(3,4)J(4)-\ldots\right]$$

Evaluating the functional derivatives, we find

$$-G_0^2\,\mathrm{Tr}\int d^4x_1\, d^4x_2\gamma_5 S_F(1,2)\gamma_5 S_F(2,1)\Delta_F(1,2)$$

$$\times\left(1-\tfrac{1}{2}\int d^4x_1\, d^4x_2 J(1)\Delta_F(1,2)J(2)\right)$$

$$+\int d^4x_3\, d^4x_4\Delta_F(1,3)G_0^2\,\mathrm{Tr}\,[\gamma_5 S_F(3,4)\gamma_5 S_F(4,3)]\Delta_F(4,2)$$

$$\times\int d^4x_1\, d^4x_2 J(1)J(2) \qquad 10(79)$$

to terms of order J^2. Multiplying 10(79) by the η-dependent exponential in 10(69) and isolating the coefficients of the J^2 and $\bar{\eta}\eta$ terms respectively, we obtain the following corrections to Δ_F and S_F:

(a) for the boson propagator, the two graphs of Fig. 10.3,

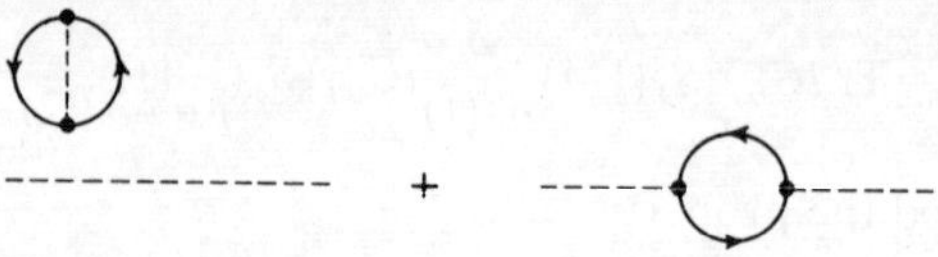

FIG. 10.3. Lowest order contributions to Δ'_F from $\det\left(1 - iG_0\gamma_5\frac{\delta}{\delta J}S_F\right)$.

(b) for the fermion propagator, the graph of Fig. 10.4.

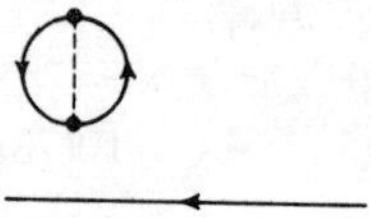

FIG. 10.4. Lowest order contribution to S'_F from $\det\left(1 - iG_0\gamma_5\frac{\delta}{\delta J}S_F\right)$.

Thus the effect of the determinant is to supply the fermion loops both in the boson propagator and the disconnected vacuum–vacuum diagrams*. This holds true in all orders of perturbation theory [Schwinger (1954a, b)].

In the following two sections we present two applications of the functional method. In Section 10-3, we apply Schwinger's equations for S'^J_F, Δ'^J_F and Γ^J_5 to give a proof of the Goldstone theorem [Goldstone (1961)] in the framework of a special model. In Section 10-4 we shall use the formalism of the Green's functional to construct the exact ' photon ' propagator in one-dimensional quantum electrodynamics [Schwinger (1962)].

10-3 The Goldstone theorem

Introduction In our discussion of the Nambu model in Section 9-5, the emergence of massless boson bound states in the chain approximation was seen to follow from the lowest order self-consistency condition for the fermion mass. We now investigate this phenomenon in detail with reference to the Yukawa coupling model,

$$\mathscr{L}_I = \tfrac{1}{2}G_0[\bar{\psi}, \psi]\phi_S + \tfrac{1}{2}G_0 i[\bar{\psi}, \gamma_5\psi]\phi_P \qquad 10(80)$$

* As indicated in Section 6-2, the vacuum–vacuum amplitude in the absence of external sources is just an infinite phase factor and may be disregarded. Note that the expansion of $Z[0, 0, 0] = (S)_0$ in terms of vacuum closed loops can be recovered from 10(69) by setting $\eta = \bar{\eta} = J = 0$, i.e.

$$Z[0, 0, 0] = \det\left(1 - iG_0\gamma_5\frac{\delta}{\delta J}S_F\right) e^{-(1/2)J\Delta_F J}\Big|_{J=0}$$

γ_5-Invariance We shall assume for the free Lagrangian that the bare fermion mass m_0 is zero and that the scalar and pseudoscalar bare masses are equal:

$$\mathscr{L}_0 = -\bar{\psi}\gamma_\mu\partial_\mu\psi - \tfrac{1}{2}\partial_\mu\phi_P\partial_\mu\phi_P - \tfrac{1}{2}\mu_0^2\phi_P^2 - \tfrac{1}{2}\partial_\mu\phi_S\partial_\mu\phi_S - \tfrac{1}{2}\mu_0^2\phi_S^2 \qquad 10(81)$$

The total Lagrangian $\mathscr{L} = \mathscr{L}_0 + \mathscr{L}_I$ is then invariant under the γ_5 transformation $\psi \to e^{i\alpha\gamma_5}\psi$ expressed as a rotation in 'parity space'

$$\begin{aligned} \bar{\psi}\psi &\to \bar{\psi}\psi \cos 2\alpha + i\bar{\psi}\gamma_5\psi \sin 2\alpha \\ i\bar{\psi}\gamma_5\psi &\to i\bar{\psi}\gamma_5\psi \cos 2\alpha - \bar{\psi}\psi \sin 2\alpha \end{aligned} \qquad 10(82)$$

combined with the rotation

$$\begin{aligned} \phi_S &\to \phi_S \cos 2\alpha + \phi_P \sin 2\alpha \\ \phi_P &\to \phi_P \cos 2\alpha - \phi_S \sin 2\alpha \end{aligned} \qquad 10(83)$$

From this invariance property, we easily infer that the axial vector current

$$j_\mu = i\bar{\psi}\gamma_\mu\gamma_5\psi + 2\phi_P\partial_\mu\phi_S - 2\phi_S\partial_\mu\phi_P \qquad 10(84)$$

obeys the conservation law [see Problem 2]

$$\partial_\mu j_\mu = 0 \qquad 10(85)$$

The chirality

$$\begin{aligned} \chi &= -i\int j_4(x)\, d^3x \\ &= \int (\bar{\psi}\gamma_4\gamma_5\psi - 2\phi_P\dot{\phi}_S + 2\phi_S\dot{\phi}_P)\, d^3x \end{aligned} \qquad 10(86)$$

is therefore conserved in time. Use of the equal time commutation rules yields the relations

$$[\psi, \chi] = \gamma_5\psi \qquad 10(87a)$$

$$[\phi_S, \chi] = -2i\phi_P \qquad [\phi_P, \chi] = 2i\phi_S \qquad 10(87b)$$

which ensure that $-\alpha\chi$ is the generator of infinitesimal γ_5-transformations

$$\delta\psi = i\alpha\gamma_5\psi = \frac{1}{i}[\psi, -\alpha\chi] \qquad 10(88a)$$

$$\delta\phi_S = 2\alpha\phi_P = \frac{1}{i}[\phi_S, -\alpha\chi] \qquad 10(88b)$$

$$\delta\phi_P = -2\alpha\phi_S = \frac{1}{i}[\phi_P, -\alpha\chi] \qquad 10(88c)$$

We can therefore identify $e^{i\alpha\chi}$ as the unitary operator representing the γ_5-transformation in Hilbert space. We have

$$e^{-i\alpha\chi}\psi\, e^{i\alpha\chi} = e^{i\alpha\gamma_5}\psi \qquad 10(89)$$

and

$$e^{-i\alpha\chi}\phi_S\, e^{i\alpha\chi} = \phi_S \cos 2\alpha + \phi_P \sin 2\alpha \qquad 10(90a)$$

$$e^{-i\alpha\chi}\phi_P\, e^{i\alpha\chi} = \phi_P \cos 2\alpha - \phi_S \sin 2\alpha \qquad 10(90b)$$

as is easily verified, using 10(88a), 10(88b) and 10(88c) and the identity 3(56).

Symmetry Breakdown and Degeneracy of the Vacuum State Our basic assumption is the nonvanishing of the vacuum expectation value of ϕ_S*

$$\varphi_S = \langle 0|\phi_S|0\rangle \neq 0 \qquad 10(91)$$

As we shall see presently, this assumption implies that the *physical* fermion mass m is nonzero. Before we examine the condition for the nonvanishing of φ_S in our model, we draw the reader's attention to the highly unconventional character of the assumption 10(91). We have assumed, as a general rule, that any symmetry of the Hamiltonian will also be exhibited by the vacuum state. In the case of charge conservation, for example, both the Hamiltonian *and* the vacuum state are invariant under an infinitesimal phase transformation generated by the charge Q. In other words, we have both

$$H = e^{-i\alpha Q} H\, e^{i\alpha Q} \qquad \text{or} \qquad [Q, H] = 0 \qquad 10(92)$$

and

$$|0\rangle = e^{i\alpha Q}|0\rangle \qquad \text{or} \qquad Q|0\rangle = 0 \qquad 10(93)$$

In the case of our γ_5-invariant model however, the invariance of the vacuum, as expressed by

$$|0\rangle = e^{-i\alpha\chi}|0\rangle \qquad \text{or} \qquad \chi|0\rangle = 0 \qquad 10(94)$$

contradicts the basic assumption 10(91). Indeed, if 10(94) holds, then we have

$$\begin{aligned}\varphi_S &= \langle 0|\phi_S|0\rangle \\ &= \langle 0|e^{-i\alpha\chi}\phi_S\, e^{i\alpha\chi}|0\rangle \\ &= \varphi_S \cos 2\alpha + \varphi_P \sin 2\alpha\end{aligned}$$

* Note that φ_S is a constant in the absence of sources, since by 7(84a)

$$\langle 0|\phi_S(x)|0\rangle = \langle 0|e^{-iP.x}\phi_S(0)\, e^{iP.x}|0\rangle = \langle 0|\phi_S(0)|0\rangle$$

where we have used 10(90a). Since $\varphi_P = \langle 0|\phi_P|0\rangle$ vanishes identically due to parity conservation, we conclude that for arbitrary α

$$\varphi_S = \varphi_S \cos 2\alpha$$

which is only possible for $\varphi_S = 0$. We conclude that a solution of the model with $\varphi_S \neq 0$ is necessarily associated with a nonsimple vacuum such that $\chi|0\rangle \neq 0$. In such a case we say that the vacuum state is in fact infinitely degenerate since the operator $e^{i\alpha\chi}$ carries a given vacuum state $|0\rangle$ into another vacuum* state $|\alpha\rangle$.

Goldstone, Salam and Weinberg [Goldstone (1962)] have shown under very general conditions† that if a continuous symmetry of the Lagrangian is broken by the vacuum state, then the theory must necessarily contain massless bosons. Here we shall restrict ourselves to the model 10(80) and show that to all orders of perturbation theory, the self-consistency condition for $\varphi_S \neq 0$ implies the existence of a massless pseudoscalar boson. We follow the approach of Bludman and Klein [Bludman (1963)] based on the functional techniques developed in Section 10-1.

Self-Consistency Condition for φ_S Let us introduce the external sources $J_S(x)$ and $J_P(x)$ coupled to $\phi_S(x)$ and $\phi_P(x)$ according to

$$\mathscr{L}_I^{J_S, J_P}(x) = \phi_S(x)J_S(x) + \phi_P(x)J_P(x) \qquad 10(95)$$

In the general case, the vacuum expectation values

$$\varphi_S(x)^J = \frac{\langle 0\text{ out}|\phi_S(x)|0\text{ in}\rangle_J}{\langle 0\text{ out}|0\text{ in}\rangle_J} \qquad 10(96)$$

$$\varphi_P(x)^J = \frac{\langle 0\text{ out}|\phi_P(x)|0\text{ in}\rangle_J}{\langle 0\text{ out}|0\text{ in}\rangle_J} \qquad 10(97)$$

with $J = (J_S, J_P)$, are functions of x, but for the purposes of deriving the self-consistency condition for φ_S we restrict ourselves to *constant* sources J_S and J_P. Translational invariance is then restored and φ_S^J and φ_P^J are again spacetime constants, as in the source-free case‡.

Taking the vacuum expectation value of the equations of motion

$$(\Box - \mu_0^2)\phi_S = -G_0\tfrac{1}{2}[\bar{\psi}, \psi] - J_S \qquad 10(98)$$

$$(\Box - \mu_0^2)\phi_P = -G_0\tfrac{1}{2}[\bar{\psi}, i\gamma_5\psi] - J_P \qquad 10(99)$$

* γ_5-invariance implies that $[H, \chi] = 0$, so that $e^{i\alpha\chi}$ must necessarily carry an eigenstate of H into another eigenstate with the same energy eigenvalue.

† See, however, Guralnik (1964).

‡ Note that $\varphi_P^J \neq 0$ for $J_P \neq 0$ [see below].

we obtain, as in 10(14), the equations

$$\mu_0^2\varphi_S^J = -G_0 \operatorname{Tr} S_F'(0)^J + J_S \tag{10(100)}$$

$$\mu_0^2\varphi_P^J = -G_0 \operatorname{Tr} i\gamma_5 S_F'(0)^J + J_P \tag{10(101)}$$

We shall treat $\varphi^J = (\varphi_S^J, \varphi_P^J)$ and $J = (J_S, J_P)$ as two-component vectors in parity space and write the above relations as

$$\mu_0^2\varphi_i^J = -G_0 \operatorname{Tr} O_i S_F'(x, x)^J + J_i \tag{10(102)}$$

$$(i = S, P)$$

where we have defined

$$O_S = 1 \qquad O_P = i\gamma_5 \tag{10(103)}$$

Equation 10(103) relates φ_i^J to the constant external sources J_S and J_P and to the fermion propagator $S_F'^J$.

We have remarked previously, in connection with 10(27), that $S_F'^J$ is effectively a function of φ^J. Thus 10(102) in the limit $J \to 0$

$$\mu_0^2\varphi_i = -G_0 \operatorname{Tr} O_i S_F'(0) \tag{10(104)}$$

becomes a self-consistency equation for φ_i. Our basic assumption is that 10(104) has a nontrivial solution $\varphi_i \neq 0$. Since 10(104) is a vector relation in the two-dimensional parity space, we must have

$$-G_0 \operatorname{Tr} O_i S_F'(0) = \varphi_i \Theta(\varphi^2) \tag{10(105)}$$

with Θ a γ_5-invariant function of the scalar product

$$\varphi^2 = \varphi_S\varphi_S + \varphi_P\varphi_P$$

Combining 10(104) and 10(105) we get the self-consistency equation

$$\mu_0^2 = \Theta(\varphi^2) \tag{10(106)}$$

which, we assume, has a nontrivial solution $\varphi^2 \neq 0$. Note that 10(106) determines only the magnitude of φ_i, not its direction. The latter is essentially arbitrary. If, as is convenient, we take φ_i to lie along the S-axis, then 10(106) becomes simply an equation for φ_S. Any particular choice of direction for φ_i breaks the original γ_5-symmetry. The arbitrariness in the direction of φ_i reflects the infinite degeneracy of the vacuum state. Thus, the frame in which $\varphi_P = \langle 0|\phi_P|0\rangle = 0$, is characterized by a vacuum state $|0\rangle$ which is an eigenstate of parity. In a different γ_5-frame, the new vacuum $|\alpha\rangle = e^{i\alpha\chi}|0\rangle$ is no longer an eigenstate of parity and the vector $\varphi = (\varphi_S', \varphi_P')$ will feature a nonvanishing pseudoscalar component. The transformation law for (φ_S, φ_P) is obtained from 10(90a) and 10(90b) by writing

$$\varphi_S = \langle 0|\phi_S|0\rangle = \langle 0|\mathrm{e}^{-i\alpha\chi}\,\mathrm{e}^{i\alpha\chi}\phi_S\,\mathrm{e}^{-i\alpha\chi}\,\mathrm{e}^{i\alpha\chi}|0\rangle$$
$$= \langle\alpha|(\phi_S \cos 2\alpha - \phi_P \sin 2\alpha)|\alpha\rangle$$
$$= \varphi'_S \cos 2\alpha - \varphi'_P \sin 2\alpha \qquad 10(107)$$

and similarly for φ_P.

We illustrate these remarks by reference to a simple approximation. Let us neglect, in the integral equation for $S_F'^J$ corresponding to 10(25), the term involving Σ^J. For constant sources J_S and J_P we have then [see Problem 3]

$$\left[\gamma\,.\,\frac{\partial}{\partial x} - G_0(\varphi_S^J + i\gamma_5\varphi_P^J)\right] S'_F(x-x')^J = -i\delta^{(4)}(x-x') \qquad 10(108)$$

or, in momentum space

$$S'_F(p)^J = \frac{-1}{\gamma\,.\,p + iG_0(\varphi_S^J + i\gamma_5\varphi_P^J)} \qquad 10(109)$$

In the limit $J \to 0$ 10(109) becomes

$$S'_F(p) = \frac{-1}{\gamma\,.\,p + iG_0(\varphi_S + i\gamma_5\varphi_P)} \qquad 10(110)$$

and if we assume that φ_i lies along S, 10(110) reduces to the standard form of the free fermion propagator with physical mass $m = G_0\varphi_S$:

$$S'_F(p) = \frac{-1}{\gamma\,.\,p + iG_0\varphi_S} \qquad 10(111)$$

The self-consistency condition 10(104) becomes

$$\mu_0^2\varphi_S = -G_0 \operatorname{Tr} S'_F(0)$$
$$= G_0\frac{1}{(2\pi)^4}\operatorname{Tr}\int d^4p\frac{1}{\gamma\,.\,p + iG_0\varphi_S} \qquad 10(112)$$

or, setting $m = G_0\varphi_S$ and evaluating the trace,

$$m = -\frac{4G_0^2 im}{\mu_0^2(2\pi)^4}\int\frac{d^4p}{p^2+m^2} \qquad 10(113)$$

This is identical to the self-consistency equation 9(129) of the Nambu model, if we identify $2g_0 = G_0^2/\mu_0^2$*.

* For a more detailed comparison of the Nambu model with the Yukawa coupling model 10(80), see Lurié (1964), especially Section 4A.

In an arbitrary γ_5-frame, the fermion mass is not m but

$$m_1 + i\gamma_5 m_2 = G_0\varphi_S + i\gamma_5 G_0 \varphi_P \qquad 10(114)$$

and the self-consistency condition 10(113) bears on the γ_5-invariant norm $m = (m_1^2 + m_2^2)^{1/2}$. The fermion propagator is 10(110) or

$$S_F'(p) = \frac{-1}{\gamma \, . \, p + im_1 - \gamma_5 m_2} \qquad 10(115)$$

which despite its odd form is perfectly consistent with parity conservation, since the vacuum state is now not an eigenstate of parity*.

Existence of the Goldstone Boson To exhibit the appearance of the massless boson in this model, we examine the boson propagator. We again apply the functional techniques of Section 10-1. For general sources $J_S(x)$ and $J_P(x)$ we define

$$\Delta'_{Fik}(x, x')^J = \frac{1}{i} \frac{\delta\varphi_i(x)^J}{\delta J_k(x')} \qquad 10(116)$$

$$i, k = (S, P)$$

The essential complication relative to the case treated in Section 10.1 is that the boson propagator is now a 2×2 matrix in parity space. The equations for Δ'^J_{Fik} can be derived by following the same procedure as in the derivation of 10(26). We find

$$(\Box_x - \mu_0^2)\Delta'_{Fik}(x, x')^J + \int G_0^2 \Pi_{ij}(x, y)^J \Delta'_{Fjk}(y, x')^J \, d^4y = i\delta^{(4)}(x - x')\delta_{ik} \qquad 10(117)$$

where

$$\Pi_{ij}(x, y)^J = \int d^4x' \, d^4y' \, \mathrm{Tr}[O_i S_F'(x, x')^J \Gamma_j(x', y'; y)^J S_F'(y', x)^J] \qquad 10(118)$$

with

$$\Gamma_j(x, x'; y) = \frac{1}{G_0} \frac{\delta S_F'^{-1}(x, x')^J}{\delta\varphi_i(y)^J} \qquad 10(119)$$

* We can express this by saying that the parity operator undergoes transformation together with the mass. For a general mass $m = (m_1, m_2)$ the matrix associated with the parity transformation is

$$\frac{m_1 - i\gamma_5 m_2}{(m_1^2 + m_2^2)^{1/2}} \gamma_4$$

rather than just γ_4.

[see Problem 4]. A simplified expression for $\Pi_{ij}(x, y)^J$, namely

$$\Pi_{ij}(x, y)^J = -\frac{1}{G_0}\frac{\delta}{\delta\varphi_j(y)^J}\operatorname{Tr} O_i S'_F(x, x)^J \tag{10(120)}$$

is obtained by combining 10(20b) with 10(118).

We now go over to the limit $J_S = J_P = 0$. Then, in accordance with our basic assumption, $\varphi_j(y)^J$ goes over into a finite constant vector φ_j. The Schwinger equation 10(117) becomes, in momentum space

$$[(k^2+\mu_0^2)\delta_{ij} - G_0^2\Pi_{ij}(k^2)]\Delta'_{Fjk}(k^2) = -i\delta_{ik} \tag{10(121)}$$

where $\Pi_{ij}(k^2)$ is given by the Fourier transform of 10(120) in the limit $J \to 0$. Let us examine the point $k^2 = 0$. Then

$$\Pi_{ij}(0) = -\frac{1}{G_0}\int d^4y \lim_{J\to 0}\frac{\delta}{\delta\varphi_j(y)^J}\operatorname{Tr} O_i S'_F(x, x)^J \tag{10(122)}$$

Since the Fourier transform for $k^2 = 0$ projects out the constant part of $\varphi_j(y)$, we can replace the functional differentiation in 10(122) by ordinary differentiation with respect to a constant external source and write

$$\Pi_{ij}(0) = -\frac{1}{G_0}\lim_{J\to 0}\frac{\delta}{\delta\varphi_j^J}\operatorname{Tr} O_i S'_F(0)^J \tag{10(123)}$$

Assuming that the covariant form 10(105) remains valid in the presence of a uniform external source, we obtain*

$$\begin{aligned} G_0^2\Pi_{ij}(0) &= -\frac{1}{G_0}\frac{\delta}{\delta\varphi_j}\operatorname{Tr} O_i S'_F(0) \\ &= \delta_{ij}\Theta(\varphi^2) + \varphi_i\varphi_j\Theta'(\varphi^2) \end{aligned} \tag{10(124)}$$

In the γ_5-frame for which $\varphi = (\varphi_S, 0)$, we see that the nondiagonal matrix elements Π_{12} and Π_{21} vanish and that Π_{22} is given by

$$G_0^2\Pi_{22}(0) = \Theta(\varphi^2) \tag{10(125)}$$

Thus, the boson propagator $\Delta'_{Fik}(0)$ is diagonal and its pseudoscalar matrix element

$$\Delta'_{F22}(0) = \frac{-i}{\mu_0^2 - G_0^2\Pi_{22}(0)} = \frac{-i}{\mu_0^2 - \Theta(\varphi^2)} \tag{10(126)}$$

* This will be the case if the vacuum expectation value φ_i^J points in the same direction as the external source J_i [see Problem 3 and the discussion by A. Klein in Proc. Seminar on Unified Theories of Elementary Particles, University of Rochester, July 1963 (unpublished)].

is infinite by virtue of 10(106), signalling the existence of the massless Goldstone boson. In an arbitrary γ_5-frame for which $\varphi = (\varphi_S, \varphi_P)$, the massless excitation will be associated with the direction perpendicular to φ.

10-4 One-dimensional quantum electrodynamics

Introduction As a second application of the functional method, we treat the model field theory of quantum electrodynamics in one space and one time dimension [Schwinger (1962)]. The model is completely soluble and we shall use the technique of the Green's functional to calculate the exact vector propagator. An interesting feature of the model is the appearance of a finite physical 'photon' mass, despite the fact that the electromagnetic field is minimally coupled to a conserved current*.

Field Equations and the Green's Functional The field equations for quantum electrodynamics in a world of one space and one time dimension are

$$\sigma_\mu \partial_\mu \psi = ie_0 \sigma_\mu A_\mu \psi \qquad 10(127a)$$

$$\Box A_\mu = -ie_0 \tfrac{1}{2}[\bar{\psi}, \sigma_\mu \psi] \qquad 10(127b)$$

where we note that the coupling constant e_0 now carries the dimension of mass. The Dirac operator $\sigma_\mu \partial_\mu$ stands for

$$\sigma_\mu \partial_\mu = \sigma_1 \frac{\partial}{\partial x_1} + \sigma_2 \frac{\partial}{\partial x_2} = \sigma_1 \frac{\partial}{\partial x_1} - i\sigma_2 \frac{\partial}{\partial x_0} \qquad 10(128)$$

where σ_1 and σ_2 are the Pauli matrices

$$\sigma_1 = \begin{pmatrix} 0 & 1 \\ 1 & 0 \end{pmatrix} \qquad \sigma_2 = \begin{pmatrix} 0 & -i \\ i & 0 \end{pmatrix} \qquad 10(129)$$

We have made the further simplification of setting the bare fermion mass m_0 equal to zero. Introducing the external sources J_μ, η and $\bar{\eta}$, the equations of motion become

$$\sigma_\mu \partial_\mu \psi = ie_0 \sigma_\mu A_\mu \psi + \eta \qquad 10(130a)$$

$$\Box A_\mu = -\frac{i}{2} e_0 [\bar{\psi}, \sigma_\mu \psi] - J_\mu \qquad 10(130b)$$

* See the remarks following 6(297). A similar phenomenon in the realistic case has been conjectured by Sakurai (1960) in an attempt to relate baryon number conservation to a gauge vector field in the same way as charge conservation is related to the electromagnetic field [Lee (1955)]. See also Johnson (1962).

From 10(130a) and 10(130b) we obtain, as in Section 10.2, the numerical functional differential equations for the Green's functional $Z[\eta, \bar{\eta}, J]$*

$$\sigma_\mu\left(\partial_\mu - e_0\frac{\delta}{\delta J_\mu(x)}\right)\frac{\delta}{\delta\bar{\eta}(x)}Z[\eta, \bar{\eta}, J] = i\eta(x)Z[\eta, \bar{\eta}, J] \qquad 10(131a)$$

$$\Box\frac{\delta}{\delta J_\mu(x)}Z[\eta, \bar{\eta}, J] = -\left(e_0\frac{\delta}{\delta\eta(x)}\sigma_\mu\frac{\delta}{\delta\bar{\eta}(x)}+iJ_\mu(x)\right)Z[\eta, \bar{\eta}, J] \qquad 10(131b)$$

Vector Propagator Proceeding as in Section 10.2, it is a simple matter to show that the exact vector propagator

$$\Delta'_{F\mu\nu}(x-x') = \langle 0|TA_\mu(x)A_\nu(x')|0\rangle$$

is given by

$$\Delta'_{F\mu\nu}(x-x') = \frac{1}{i^2}\left(\frac{\delta}{\delta J_\mu(x)}\frac{\delta}{\delta J_\nu(x')}Z[\eta, \bar{\eta}, J]\right)_{\eta=\bar{\eta}=J_\mu=0} \qquad 10(132)$$

Our task is therefore to exhibit the J_μ dependence of Z.

We follow the same procedure as in the derivation of the formal solution 10(69) for the pseudoscalar coupling case to write $Z[\eta, \bar{\eta}, J]$ in the form

$$Z[\eta, \bar{\eta}, J] = \mathrm{e}^{-\int d^2x d^2x' \bar{\eta}(x)S_F(x,x')^{\delta/i\delta J}\eta(x')}Z[J] \qquad 10(133)$$

in which only the functional dependence of Z on η and $\bar{\eta}$ is indicated explicitly. Here $S_F^{\delta/i\delta J}$ is the fermion Green's function in the presence of the external vector field $B_\mu = -i\delta/\delta J_\mu$ and in the absence of field theoretic couplings. Explicitly

$$S_F^{\delta/i\delta J} = \left(1 - ie_0 S_F\sigma_\mu\frac{\delta}{\delta J_\mu}\right)^{-1} S_F \qquad 10(134)$$

In four dimensions the evaluation of $Z[J]$ requires a perturbative expansion based on 10(134), as was carried out in Section 10.2. In the two dimensional case, however, $Z[J]$ can be determined exactly. Inserting 10(133) into 10(131b) and carrying out the functional differentiations with respect to η and $\bar{\eta}$, we obtain the functional differential equation

$$\Box\frac{\delta}{\delta J_\mu(x)}Z[J] = -[e_0 \operatorname{Tr}(\sigma_\mu S_F(x, x)^{\delta/i\delta J})+iJ_\mu]Z[J] \qquad 10(135)$$

* For convenience we shall denote the vector source J_μ by J whenever it appears as the argument of a functional. Actually the possibility of performing independent variations of the four components of J_μ requires careful analysis and is connected with the gauge variation of the propagator. The reader is referred to Zumino (1960) for a discussion of this point.

for $Z[J]$. We shall prove below that

$$\mathrm{Tr}\,(\sigma_\mu S_F(x, x)^{\delta/i\delta J}) = -\frac{e_0}{\pi}\left(\delta_{\mu\nu} - \partial_\mu \frac{1}{\Box}\partial_\nu\right)\frac{\delta}{\delta J_\nu(x)} \qquad 10(136)$$

where the symbol $\partial_\mu \Box^{-1}\partial_\nu$ stands for the integro-differential operator

$$-i\frac{\partial}{\partial x_\mu}\int d^2x' D_F(x-x')\frac{\partial}{\partial x'_\nu}$$

formed with the aid of the Green's function D_F satisfying

$$\Box D_F(x) = i\delta^{(2)}(x)$$

Inserting 10(136) into 10(135) we get the functional differential equation

$$\left[\Box\delta_{\mu\nu} - \frac{e_0^2}{\pi}\left(\delta_{\mu\nu} - \partial_\mu\frac{1}{\Box}\partial_\nu\right)\right]\frac{\delta}{\delta J_\nu(x)}Z[J] = -iJ_\mu Z[J] \qquad 10(137)$$

which, when integrated, yields the exact result

$$Z[J] = \mathrm{e}^{-\frac{1}{2}\int d^2x d^2x' J_\mu(x) G_{\mu\nu}(x-x') J_\nu(x')} \qquad 10(138)$$

where $G_{\mu\nu}$ is the Green's function satisfying

$$\left[\Box\delta_{\mu\lambda} - \frac{e_0^2}{\pi}\left(\delta_{\mu\lambda} - \partial_\mu\frac{1}{\Box}\partial_\lambda\right)\right]G_{\lambda\nu}(x) = i\delta_{\mu\nu}\delta^{(4)}(x) \qquad 10(139)$$

or, in momentum space

$$\left[k^2\delta_{\mu\lambda} + \frac{e_0^2}{\pi}\left(\delta_{\mu\lambda} - \frac{k_\mu k_\lambda}{k^2}\right)\right]G_{\lambda\nu}(k) = -i\delta_{\mu\nu} \qquad 10(140)$$

From 10(133) and 10(138) we immediately obtain the vector propagator $\Delta'_{F\mu\nu}$ by forming the right-hand side of 10(132). The result is simply

$$\Delta'_{F\mu\nu}(x-x') = G_{\mu\nu}(x-x')$$

Hence, solving 10(140)*, we obtain

$$\Delta'_{F\mu\nu}(k) = \frac{-i}{k^2 + \dfrac{e_0^2}{\pi}}\delta_{\mu\nu} \qquad 10(141)$$

which is just the momentum space propagator of a free massive vector boson of mass $e_0/\sqrt{\pi}$.

* Setting $G_{\mu\nu} = A\delta_{\mu\nu} + Bk_\mu k_\nu$ and solving 10(140) for A and B, we find $A = -i(k^2 + e_0^2/\pi)^{-1}$ and $B = e_0^2 A/\pi k^4$. The effect on the S-matrix of the $k_\mu k_\nu$ terms vanishes due to current conservation, and we effectively obtain 10(141).

The Induced Vacuum Current We now turn to the proof of 10(136). Our problem is to calculate the expression

$$\mathrm{Tr}\,[\sigma_\mu S_{\mathrm{F}}(x, x)^B] \tag{10(142)}$$

where $B_\mu(x)$ is an external electromagnetic field. Physically we are computing the current induced in the vacuum by the external field B_μ in the absence of quantum field theoretic couplings*.

Actually the expression 10(142) is highly singular and requires careful definition in terms of a suitable limiting process, [Schwinger (1959), Johnson (1961a, b)]. In particular, the definition

$$\mathrm{Tr}\,[\sigma_\mu S_{\mathrm{F}}(x, x)^B] = \lim_{x \to x'} [\sigma_\mu S_{\mathrm{F}}(x, x')^B]$$

is unsatisfactory, as it stands, as it yields an induced vacuum current which is not conserved, [see Chapter 7, Problem 5]. To obtain a conserved vacuum current, Schwinger defines

$$S_{\mathrm{F}}(x, x)^B = \lim_{x' \to x} S_{\mathrm{F}}(x, x')^B \, e^{-ie_0 \int_{x'}^{x} d\xi_\mu B_\mu(\xi)} \tag{10(143)}$$

with the additional instruction that we take symmetric limit—averaging the values obtained for $x = x' \pm 0$—along a spatial direction†. This definition yields the conserved vacuum current given by 10(136).

The propagator $S_{\mathrm{F}}(x, x')^B$ satisfies the equation

$$\sigma_\mu \left[\frac{\partial}{\partial x_\mu} - ie_0 B_\mu(x) \right] S_{\mathrm{F}}(x, x')^B = -i\delta^{(2)}(x - x') \tag{10(144)}$$

It is more convenient to work with

$$S_{\mathrm{F}d}(x, x')^B = \sigma_2 S_{\mathrm{F}}(x, x')^B \tag{10(145)}$$

which satisfies an equation involving only *diagonal* matrices,

$$[\sigma_3(\partial_1 - ie_0 B_1) - (\partial_0 + ie_0 B_0)] S_{\mathrm{F}d}(x, x')^B = -\delta^{(2)}(x - x') \tag{10(146)}$$

with

$$\sigma_3 = \begin{pmatrix} 1 & 0 \\ 0 & -1 \end{pmatrix}$$

* As in 10(13b), the trace 10(142) is just the negative of the vacuum expectation value of the current $\frac{1}{2}[\bar{\psi}, \sigma_\mu \psi]$.

† This serves the same purpose as the usual antisymmetrization prescription $\frac{1}{2}[\bar{\psi}, \sigma_\mu \psi]$ for the current operator, ensuring the subtraction of the infinite vacuum current when $B_\mu = 0$ [see below].

Equation 10(146) can be satisfied by setting

$$S_{Fd}(x, x')^B = S^0_{Fd}(x-x')\,e^{ie_0[\phi(x)-\phi(x')]} \qquad 10(147)$$

where $\phi(x)$ satisfies

$$(\sigma_3\partial_1 - \partial_0)\phi = \sigma_3 B_1 + B_0 \qquad 10(148)$$

and where S^0_{Fd} is the free Green's function satisfying

$$(\sigma_3\partial_1 - \partial_0)S^0_{Fd}(x-x') = -\delta^{(2)}(x-x') \qquad 10(149)$$

To evaluate S^0_{Fd} explicitly, we write

$$S^0_{Fd}(x) = i(\sigma_3\partial_1 + \partial_0)\Delta_F(x) \qquad 10(150)$$

where Δ_F is given by the analog in two dimensions of 3(94), i.e.

$$\Delta_F(x) = \theta(x_0)i\Delta^{(+)}(x) - \theta(-x_0)i\Delta^{(-)}(x)$$

with

$$\Delta^{(\pm)}(x) = \frac{\mp i}{2(2\pi)}\int \frac{dk}{|k|}\,e^{ikx - i|k|x_0}$$

A straightforward computation then gives [see Problem 5]

$$S^0_{Fd}(x) = \frac{1}{2\pi}\int_0^\infty dk\, e^{-i\sigma_3 k x_1 - ikx_0} \qquad \text{for} \quad x_0 > 0$$

$$= -\frac{1}{2\pi}\int_{-\infty}^0 dk\, e^{-i\sigma_3 k x_1 - ikx_0} \qquad \text{for} \quad x_0 < 0 \qquad 10(151)$$

and, taking the average of 10(151) for $x_0 = \pm 0$,

$$S^0_{Fd}(x) = \frac{\sigma_3}{2\pi i x} \qquad \text{for} \quad x_0 = 0 \qquad 10(152)$$

where we have used the formula

$$\int_0^\infty e^{-i\omega x}\,d\omega/\pi = \delta(x) + (\pi i x)^{-1}$$

Combining 10(145), 10(147) and 10(152) we now evaluate the limit along a spatial direction of

$$\frac{\sigma_1}{2\pi(x_1 - x_1')}\,e^{ie_0[\phi(x)-\phi(x')]}\,e^{-ie_0\int_{x_1'}^{x_1} d\xi_1 B_1(\xi)}$$

Expanding the exponentials for $x_1 \to x_1'$

$$\frac{\sigma_1}{2\pi(x_1 - x_1')}[1 + ie_0(x_1 - x_1')(\partial_1\phi - B_1)]$$

and taking the average for $x_1 \to x_1' \pm 0$ to remove the divergent first term, we find

$$S_F(x, x)^B = \frac{ie_0}{2\pi}\sigma_1(\partial_1\phi - B_1) \qquad 10(153a)$$

which by 10(148) also equals

$$\frac{e_0}{2\pi}\sigma_2(\partial_0\phi + B_0) = \frac{ie_0}{2\pi}\sigma_2(\partial_2\phi - B_2) \qquad 10(153b)$$

We thus obtain the covariant result

$$\text{Tr}\,[\sigma_\mu S_F(x, x)^B] = -\frac{ie_0}{\pi}\left(B_\mu - \partial_\mu \frac{1}{2}\,\text{Tr}\,\phi\right) \qquad 10(154)$$

Finally, we determine $\text{Tr}\,\phi$ by multiplying 10(148) by $\sigma_3\partial_1 + \partial_0$ and taking the trace of both sides. The result is

$$\Box\frac{1}{2}\,\text{Tr}\,\phi(x) = \partial_\nu B_\nu(x)$$

which, when integrated, yields

$$\frac{1}{2}\text{Tr}\,\phi(x) = -i\int D_F(x - x')\partial_\nu B_\nu(x')$$

$$\equiv \frac{1}{\Box}\partial_\nu B_\nu(x)$$

Thus we obtain

$$\text{Tr}\,[\sigma_\mu\, S_F(x, x)^B] = -\frac{ie_0}{\pi}\left(\delta_{\mu\nu} - \partial_\mu \frac{1}{\Box}\partial_\nu\right)B_\nu \qquad 10(155)$$

from which 10(136) follows by substituting $B_\nu \to -i\delta/\delta J_\nu$.

This completes our discussion of one-dimensional quantum electrodynamics. Functional techniques have also been applied to other one-dimensional models, such as the coupling of massless fermions to massive vector particles [Thirring (1964)]. Also, Sommerfield (1963) has investigated the model of a massless fermion field whose current is coupled both to itself and to a massive vector boson field. For wider applications of the functional method, see Umezawa (1955), Zumino (1960), Bialynicki-Birula (1960), Baker (1963), and for a review of the functional method in the theory of Green's functions see Kato, Kabayashi and Namiki [Kato (1960)].

10-5 Functional integration techniques

Introduction The functional approach to quantum field theory exploits the fact that a field is a system with an infinite number of degrees of freedom. These can be taken to be either the values of the field at different points of space, or, alternatively, the values of the coordinates corresponding to the 'normal modes' as in, say, Fourier analysis. As well as functional differentiation, the reverse operation of *functional integration* has important applications to field theory. In particular, we may expect that functional integration could be applied to integrate the functional differential equations for $Z(\eta, \bar{\eta}, J]$ derived in Section 10.2. In this section we present a brief nonrigorous introduction to this technique.

Mathematical Preliminaries We begin by considering a functional $F[\alpha]$ of a *real* function $\alpha(x)$ belonging to the Hilbert space of square-integrable functions. We are interested in attaching some sort of meaning to the *functional integral*

$$\int F[\alpha]\, d(\alpha) \qquad 10(156)$$

i.e. the integral of the functional $F[\alpha]$ over the *Hilbert space of functions* $\alpha(x)$. One conceivable way of defining such an integral is to subdivide the space on which the functions $\alpha(x)$ are defined into discrete cells and treat the average value of $\alpha(x)$ in each cell i as an independent variable α_i. The functional $F[\alpha]$ then becomes an ordinary *function* $F(\alpha_i)$ of a discrete set of variables α_i and the integral 10(156) may be defined as the limit of an ordinary multiple integral over the many variables α_i—the limit being taken for vanishing cell volume. In this limit, of course, the number of variables becomes continuously infinite. Instead of subdividing space into cells, a more convenient procedure is to decompose the function $\alpha(x)$ into 'normal modes'* by expanding it in terms of a complete orthonormal set of functions

$$\alpha(x) = \sum_{\lambda} \alpha_\lambda u_\lambda(x) \qquad 10(157)$$

* We have seen in Chapter 2 that either of these two ways of treating a field—subdivision into cells, or decomposition into normal modes—provides a suitable starting point for Lagrangian field theory.

One then regards the set of coefficients α_λ as the independent variables of integration. If we begin by restricting ourselves to the first n terms in the above expansion, then the functional $F[\alpha]$ becomes an ordinary function $F(\alpha_1 \ldots \alpha_n)$ of n variables; the functional integral 10(156) is then defined as the limit, for $n \to \infty$, of the multiple integral of this function over the variables $\alpha_1 \ldots \alpha_n$:

$$\int F[\alpha]\, d(\alpha) = \lim_{n\to\infty} \int F(\alpha_1 \ldots \alpha_n)\, d\alpha_1 \ldots d\alpha_n \qquad 10(158)$$

Whether the above limit exists in any but the most simple cases is at present an open mathematical problem. We can immediately exhibit a trivial instance where the limit *does* exist. Consider the Gaussian functional

$$e^{-\frac{1}{2}\alpha^2} \equiv e^{-\frac{1}{2}\int dx\alpha^2(x)} \qquad 10(159)$$

where, as a matter of notation, we shall frequently write the Hilbert space norm $\int dx\alpha^2(x)$ as α^2 and the scalar product of two functions $\alpha(x)$ and $\beta(x)$, i.e.,

$$\int dx\, \alpha(x)\beta(x)$$

as $\alpha \,.\, \beta$. Expanding $\alpha(x)$ in terms of a complete orthornormal set, as in 10(157), we have by our definition 10(158)

$$\begin{aligned}\int e^{-\frac{1}{2}\alpha^2}\, d\left(\frac{\alpha}{\sqrt{2\pi}}\right) &= \lim_{n\to\infty} \int e^{-\frac{1}{2}\sum_{\lambda=1}^{n}\alpha_\lambda^2}\, d\frac{\alpha_1}{\sqrt{2\pi}} \ldots d\frac{\alpha_n}{\sqrt{2\pi}} \\ &= \lim_{n\to\infty} \prod_{\lambda=1}^{n} \int e^{-\frac{1}{2}\alpha_\lambda^2}\, d\frac{\alpha_\lambda}{\sqrt{2\pi}} \\ &= 1 \end{aligned} \qquad 10(160)$$

where we have used the well-known formula

$$\int_{-\infty}^{\infty} e^{-\frac{1}{2}\alpha^2}\, d\frac{\alpha}{\sqrt{2\pi}} = 1$$

Note the crucial importance of the integration measure for this case. If we were to replace $d(\alpha/\sqrt{2\pi})$ by $d(m\alpha/\sqrt{2\pi})$ where m is some positive constant, then

$$\int_{-\infty}^{\infty} e^{-\frac{1}{2}\alpha^2} d\left(\frac{m\alpha}{\sqrt{2\pi}}\right) = \lim_{n\to\infty} m^n = \begin{cases} \infty & \text{if } m > 1 \\ 1 & \text{if } m = 1 \\ 0 & \text{if } m < 1 \end{cases} \tag{10(161)}$$

Only for the choice $m = 1$ do we get a finite result.

We have thus exhibited a trivial case—the Gaussian functional—for which the limit exists and the functional integral can be defined. As already stated, the existence of the limit in more complicated instances is still an open question. Nevertheless we can derive certain useful formulae [Friedrichs (1951), Symanzik (1954)] from the result 10(160), provided that we make certain assumptions. In particular we shall assume that

(i) Functional integration is translation invariant, i.e.

$$\int F[\alpha]\, d(\alpha) = \int F[\alpha+\phi]\, d(\alpha)$$

(ii) Functional integrals depend linearly on the integrand, so that we have the usual rules, familiar from calculus, for ordinary (and also functional) differentiation and integration under the functional integral sign.

Using assumption (i) and the result 10(160), we get

$$\int e^{-\frac{1}{2}(\alpha+\phi)^2}\, d\left(\frac{\alpha}{\sqrt{2\pi}}\right) = 1$$

or

$$\int e^{-\phi.\alpha-\frac{1}{2}\alpha^2}\, d\left(\frac{\alpha}{\sqrt{2\pi}}\right) = e^{\frac{1}{2}\phi^2}$$

As this is an identity in ϕ, we may replace ϕ by $-i\phi$ to obtain the formula

$$\int e^{i\phi.\alpha-\frac{1}{2}\alpha^2}\, d\left(\frac{\alpha}{\sqrt{2\pi}}\right) = e^{-\frac{1}{2}\phi^2} \tag{10(162)}$$

Next we evaluate the functional integral

$$\int e^{i\phi.\alpha-\frac{1}{2}\alpha^2} F[\alpha]\, d\left(\frac{\alpha}{\sqrt{2\pi}}\right)$$

By assumption (ii) it is equal to

$$F\left[\frac{1}{i}\frac{\delta}{\delta\phi}\right] \int e^{i\phi.\alpha-\frac{1}{2}\alpha^2}\, d\left(\frac{\alpha}{\sqrt{2\pi}}\right) \tag{10(163)}$$

so that, using 10(162) we get the important formula

$$\int e^{i\phi.\alpha-\frac{1}{2}\alpha^2}F[\alpha]\, d\left(\frac{\alpha}{\sqrt{2\pi}}\right) = F\left[\frac{1}{i}\frac{\delta}{\delta\phi}\right] e^{-\frac{1}{2}\phi^2} \qquad 10(164)$$

Next, we derive some useful formulae for functional integrals over *complex* functions. From 10(160) we infer that

$$\int e^{(-\frac{1}{2}\alpha^2-\frac{1}{2}\beta^2)}\, d\left(\frac{\alpha}{\sqrt{2\pi}}\right) d\left(\frac{\beta}{\sqrt{2\pi}}\right) = 1 \qquad 10(165)$$

for two real functions α and β. Setting

$$\varphi = \frac{1}{\sqrt{2}}(\alpha+i\beta) \qquad \varphi^* = \frac{1}{\sqrt{2}}(\alpha-i\beta) \qquad 10(166a)$$

and

$$d\left(\frac{\varphi}{\sqrt{\pi}}\right) = d\left(\frac{\alpha}{\sqrt{2\pi}}\right) d\left(\frac{\beta}{\sqrt{2\pi}}\right) \qquad 10(166b)$$

we can write 10(165) in the form

$$\int e^{-\varphi^*.\varphi}\, d\left(\frac{\varphi}{\sqrt{\pi}}\right) = 1 \qquad 10(167)$$

Moreover, 10(162) implies that

$$\int e^{-\frac{1}{2}\alpha^2-\frac{1}{2}\beta^2+i\gamma.\alpha+i\delta.\beta}\, d\left(\frac{\alpha}{\sqrt{2\pi}}\right) d\left(\frac{\beta}{\sqrt{2\pi}}\right) = e^{-\frac{1}{2}\gamma^2-\frac{1}{2}\delta^2} \qquad 10(168)$$

and setting

$$\chi = \frac{1}{\sqrt{2}}(\gamma+i\delta) \qquad \chi^* = \frac{1}{\sqrt{2}}(\gamma-i\delta) \qquad 10(169)$$

we get

$$\int e^{-\varphi^*.\varphi+i\chi^*.\varphi+i\varphi^*.\chi}\, d\left(\frac{\varphi}{\sqrt{\pi}}\right) = e^{-\chi^*.\chi} \qquad 10(170)$$

Equation 10(170) is the analog of 10(162) for functionals of complex functions. The analog of 10(164) is

$$\int F[\varphi^*, \varphi]\, e^{-\varphi^*.\varphi + i\chi^*.\varphi + i\varphi^*.\chi}\, d\left(\frac{\varphi}{\sqrt{\pi}}\right) = F\left[\frac{1}{i}\frac{\delta}{\delta\chi}, \frac{1}{i}\frac{\delta}{\delta\chi^*}\right] e^{-\chi^*.\chi} \qquad 10(171)$$

and is obtained directly from 10(170) and assumption (ii), in the same way as 10(164).

In our later discussion of the Green's functional we shall have recourse to *functional Fourier transformation.* For a functional $F[\alpha]$ of a real function, the functional Fourier transform $\hat{F}[\omega]$ is defined by

$$F[\alpha] = \int \hat{F}[\omega]\, e^{i\omega.\alpha}\, d\left(\frac{\omega}{\sqrt{2\pi}}\right) \qquad 10(172)$$

The inverse relation is

$$\hat{F}[\omega] = \int F[\alpha]\, e^{-i\omega.\alpha}\, d\left(\frac{\alpha}{\sqrt{2\pi}}\right) \qquad 10(173)$$

and in analogy with the usual integral representation of the delta function, we have here

$$\delta[\alpha - \beta] = \int e^{i(\alpha-\beta).\omega}\, d\left(\frac{\omega}{\sqrt{2\pi}}\right) \qquad 10(174)$$

expressing the 'delta-functional' as a functional integral. The delta-functional has the property that

$$\int F[\alpha]\delta[\alpha - \beta]\, d(\alpha) = F[\beta] \qquad 10(175)$$

To extend Fourier transformation to functionals of complex functions, we start, as before, with real functions α, β, γ, δ and go over to complex functions φ and χ by means of 10(166a), 10(166b) and 10(169). This yields the relation

$$F[\chi, \chi^*] = \int \hat{F}[\varphi^*, \varphi]\, e^{i\varphi^*.\chi + i\chi^*.\varphi}\, d\left(\frac{\varphi}{\sqrt{\pi}}\right) \qquad 10(176)$$

between the functional $F[\chi, \chi^*]$ and its Fourier transform $\hat{F}[\varphi^*, \varphi]$, and the inverse relation:

$$\hat{F}[\varphi^*, \varphi] = \int F[\chi, \chi^*]\, e^{-i\varphi^*.\chi - i\chi^*.\varphi}\, d\left(\frac{\chi}{\sqrt{\pi}}\right) \qquad 10(177)$$

Finally, we discuss changes in the functional integration 'volume element'. We seek a formula for the change in the real functional measure $d(\alpha)$ induced by the transformation

$$\alpha'(x) = \int K(x, y)\alpha(y)\, dy \qquad 10(178)$$

where the kernel $K(x, y)$ is assumed to be symmetric in x and y. Expanding the functions $\alpha(x)$ and $\alpha'(x)$ in terms of a complete set of real orthonormal functions $u_\lambda(x)$, as in 10(157), we can exhibit the transformation 10(178) in the form of a symmetric linear transformation

$$\alpha'_\lambda = \sum_{\mu=1}^{n} k_{\lambda\mu}\alpha_\mu \qquad 10(179)$$

on the expansion coefficients α_λ; the matrix $k_{\lambda\lambda'}$ is given by

$$k_{\lambda\lambda'} = \iint u_\lambda(x)K(x, x')u_{\lambda'}(x')\, dx\, dx' \qquad 10(180)$$

and n is kept finite for the time being. Now since $d(\alpha) = \Pi_\lambda\, d\alpha_\lambda$ and $d(\alpha') = \Pi_{\lambda'}\, d\alpha_{\lambda'}$ we have the relation

$$d(\alpha') = \det(k_{\lambda\lambda'})\, d(\alpha) \qquad 10(181)$$

where we now pass to the limit $n \to \infty$. Note that we have

$$\det(k_{\lambda\lambda'}) = \det(K_{xx'}) \qquad 10(182)$$

where we have written the kernel $K(x, x')$ in the form of a continuous matrix, as in 10(61)–10(64). The equality 10(182) follows from the definition 10(180) of $k_{\lambda\lambda'}$ written in the form

$$k_{\lambda\lambda'} = u_{\lambda x}K_{xx'}u_{\lambda' x'}$$

or

$$k = uKu^T$$

where u^T denotes the transposed matrix. Since $u^T = u^{-1}$, we get, upon taking the determinant of both sides,

$$\det k = \det\,(uKu^{-1}) = \det K$$

as stated. We may therefore write the relation 10(181) simply as

$$d(K\alpha) = \det(K)\, d(\alpha) \qquad 10(183)$$

In the case of functionals of a *complex* function $\varphi = (\alpha + i\beta)/\sqrt{2}$, if we make the change of variables

$$\alpha'(x) = \int K(x, y)\alpha(y)\, dy$$

$$\beta'(x) = \int K(x, y)\beta(y)\, dy \qquad 10(184)$$

with the same symmetric kernel K, we have, instead of 10(183)

$$d(K\varphi) = \det(K^2)\, d(\varphi) \qquad 10(185)$$

since $d(\varphi/\sqrt{\pi}) = d(\alpha/\sqrt{2\pi})\, d(\beta/\sqrt{2\pi})$.

As an application, consider the functional integral

$$\int e^{-\frac{1}{2}\alpha f \alpha}\, d\left(\frac{\alpha}{\sqrt{2\pi}}\right) = \int e^{-\frac{1}{2}\int dx\, \alpha(x) f(x) \alpha(x)}\, d\left(\frac{\alpha}{\sqrt{2\pi}}\right)$$

where f is some real function independent of the real function α. To evaluate the integral, we make the change of variables

$$\alpha' = f^{1/2}\alpha \qquad 10(186a)$$

$$d(\alpha) = \det(f^{-1/2})\, d(\alpha') \qquad 10(186b)$$

Note that the kernel $f^{1/2}$ is diagonal in this particular example. Using 10(186a) and 10(186b) we get the result

$$\int e^{-\frac{1}{2}\alpha f \alpha}\, d\left(\frac{\alpha}{\sqrt{2\pi}}\right) = \det(f^{-1/2}) \int e^{-\frac{1}{2}(\alpha')^2}\, d\left(\frac{\alpha'}{\sqrt{2\pi}}\right) = \det(f^{-1/2})$$

Similarly, for the integral

$$\int e^{-\chi^* f \chi}\, d\left(\frac{\chi}{\sqrt{\pi}}\right)$$

we get, upon making the change $\chi' = f^{1/2}\chi$ and

$$d\left(\frac{\chi}{\sqrt{\pi}}\right) = \det(f^{-1})\, d\left(\frac{\chi'}{\sqrt{\pi}}\right) \qquad 10(187)$$

the result

$$\int e^{-\chi^* f \chi}\, d\left(\frac{\chi}{\sqrt{\pi}}\right) = \det(f^{-1}) \int e^{-\chi'^* . \chi'}\, d\left(\frac{\chi'}{\sqrt{\pi}}\right) = \det(f^{-1}) \qquad 10(188)$$

In our work on the Green's functional we shall need to consider integrals over *anticommuting* spinoral functions ρ and $\rho^\dagger$ similar to the quantities η and $\eta^\dagger = \bar{\eta}\beta$ introduced in Section 10-2. The simplest such integral is

$$\int e^{-\rho^\dagger\rho}\, d\left(\frac{\rho}{\sqrt{\pi}}\right) \qquad 10(189)$$

There has been considerable discussion in the literature on the exact interpretation to be placed on these rather formal constructions. We shall not enter into this problem, but refer the reader to the papers by Matthews and Salam [Matthews (1955)], Burton and De Borde [Burton (1956)], Tobocman (1956), Candlin (1956), and Klauder (1960) in which this question is discussed. We shall assume that in analogy with 10(167) and 10(170) we have

$$\int e^{-\rho^*.\rho}\, d\left(\frac{\rho}{\sqrt{\pi}}\right) = 1 \qquad 10(190)$$

and

$$\int e^{-\rho^*.\rho + i\eta^*.\rho + i\rho^*.\eta} d\left(\frac{\rho}{\sqrt{\pi}}\right) = e^{-\eta^*.\eta} \qquad 10(191)$$

An important difference in the anticommuting case, and one which we state without proof, is that instead of 10(185) we have

$$d(K\rho) = \det(K^{-2})\, d(\rho) \qquad 10(192)$$

for the change in volume element under change of variables, [Symanzik (1954), Matthews (1955)]. The *inverse* kernel appears in the determinant in this case.

For our purposes, a more useful form of 10(191) is obtained by making the change of variables $\rho \to \beta^{1/2}\rho$ and noting that $\beta^{-1} = \beta$ and $\det(\beta) = 1$. This yields

$$\int e^{-\bar{\rho}.\rho + i\bar{\eta}\beta^{-1/2}\rho + i\bar{\rho}\beta^{-1/2}\eta}\, d\left(\frac{\rho}{\sqrt{\pi}}\right) = e^{-\bar{\eta}.\beta^{-1}\eta}$$

or, equivalently

$$\int e^{-\bar{\rho}.\rho + i\bar{\eta}.\rho + i\bar{\rho}.\eta}\, d\left(\frac{\rho}{\sqrt{\pi}}\right) = e^{-\bar{\eta}.\eta} \qquad 10(193)$$

upon substituting $\eta \to \beta^{-1/2}\eta$.

The Green's Functional We now apply the technique of functional integration to the formal solution of the functional differential equations for the Green's functional $Z[\eta, \bar{\eta}, J]$ [Symanzik (1954)]. We shall find that we get just the result 10(69).

The equations for $Z[\eta, \bar{\eta}, J]$, derived in Section 10-2, are

$$\left(\gamma . \partial + m_0 - G_0\gamma_5 \frac{\delta}{\delta J(x)}\right)\frac{\delta}{\delta\bar{\eta}(x)} Z[\eta, \bar{\eta}, J] = i\eta(x)Z[\eta, \bar{\eta}, J] \qquad 10(194a)$$

$$\frac{\delta}{\delta\eta(x)}\left(\gamma . \overleftarrow{\partial} - m_0 + G_0\gamma_5\frac{\delta}{\delta J(x)}\right)Z[\eta, \bar{\eta}, J] = -i\bar{\eta}(x)Z[\eta, \bar{\eta}, J] \qquad 10(194b)$$

$$(\Box - \mu_0^2)\frac{\delta}{\delta J(x)} Z[\eta, \bar{\eta}, J] = -\left(G_0 \frac{\delta}{\delta\eta(x)}\gamma_5\frac{\delta}{\delta\bar{\eta}(x)} + iJ(x)\right)Z[\eta, \bar{\eta}, J] \qquad 10(194c)$$

for the case of pseudoscalar coupling of neutral pseudoscalar bosons to fermions.

To solve these equations, we perform a functional Fourier transformation. We define the Fourier transformed functional $\hat{Z}[\bar{\rho}, \rho, \alpha]$ by

$$Z[\eta, \bar{\eta}, J] = \int \exp[i(\rho^\dagger\beta\eta + \bar{\eta} . \rho + J . \alpha)]\hat{Z}[\bar{\rho}, \rho, \alpha]\, d\left(\frac{\alpha}{\sqrt{2\pi}}\right) d\left(\frac{\rho}{\sqrt{\pi}}\right) \qquad 10(195)$$

where $\rho, \bar{\rho} = \rho^\dagger\beta$ and α are external source functions similar in character to η, $\bar{\eta}$ and J. We have commented earlier on the formal nature of functional integrals over anti-commuting functions; ignoring this difficulty, a short manipulation allows us to write 10(194a) 10(194b) and 10(194c) in the form

$$(\gamma . \partial + m_0 - iG_0\gamma_5\alpha(x))i\rho(x)\hat{Z}[\bar{\rho}, \rho, \alpha] = -\frac{\delta}{\delta\bar{\rho}(x)}\hat{Z}[\bar{\rho}, \rho, \alpha] \qquad 10(196a)$$

$$i\bar{\rho}(x)(\gamma . \overleftarrow{\partial} - m_0 + iG_0\gamma_5\alpha(x))\hat{Z}[\bar{\rho}, \rho, \alpha] = \frac{\delta}{\delta\rho(x)}\hat{Z}[\bar{\rho}, \rho, \alpha] \qquad 10(196b)$$

$$[(\Box - \mu_0^2)i\alpha(x) - G_0\bar{\rho}(x)\gamma_5\rho(x)]\hat{Z}[\bar{\rho}, \rho, \alpha] = \frac{\delta}{\delta\alpha(x)}\hat{Z}[\bar{\rho}, \rho, \alpha] \qquad 10(196c)$$

in terms of the transformed functional $\hat{Z}[\bar{\rho}, \rho, \alpha]$. For example to derive 10(196a) from 10(194a), we substitute 10(195), giving

$$\int \exp[i(\rho^\dagger . \beta\eta + \bar{\eta} . \rho + J . \alpha)](\gamma . \partial + m_0 - iG_0\gamma_5\alpha)i\rho\hat{Z}[\bar{\rho}, \rho, \alpha]\, d\left(\frac{\alpha}{\sqrt{2\pi}}\right) d\left(\frac{\rho}{\sqrt{\pi}}\right)$$

$$= \int i\eta \exp[i(\bar{\rho} . \eta + \bar{\eta} . \rho + J . \alpha)]\hat{Z}[\bar{\rho}, \rho, \alpha]\, d\left(\frac{\alpha}{\sqrt{2\pi}}\right) d\left(\frac{\rho}{\sqrt{\pi}}\right) \qquad 10(197)$$

and apply the inverse of 10(195), i.e.

$$\hat{Z}[\bar{\rho}', \rho', \alpha'] = \int e^{-i(\bar{\rho}'.\eta+\bar{\eta}.\rho'+J.\alpha')} Z[\eta, \bar{\eta}, J]\, d\left(\frac{J}{\sqrt{2\pi}}\right) d\left(\frac{\eta}{\sqrt{\pi}}\right) \qquad 10(198)$$

to both sides of 10(197). Writing the result for the right-hand side in the form

$$\int e^{-i(\bar{\rho}'.\eta+\bar{\eta}.\rho'+J.\alpha')}\left(-\frac{\overleftarrow{\delta}}{\delta\bar{\rho}'}\right) e^{i(\bar{\rho}.\eta+\bar{\eta}.\rho+J.\alpha)}$$
$$\times \hat{Z}[\bar{\rho}, \rho, \alpha]\, d\left(\frac{\alpha}{\sqrt{2\pi}}\right) d\left(\frac{\rho}{\sqrt{\pi}}\right) d\left(\frac{J}{\sqrt{2\pi}}\right) d\left(\frac{\eta}{\sqrt{\pi}}\right)$$

we easily get 10(196a).

The functional differential equations 10(196a), 10(196b) and 10(196c) for the Fourier transformed functional are elementary and the solution is easily seen to be

$$\hat{Z}[\bar{\rho}, \rho, \alpha] = C \exp\left\{i \int d^4x[\tfrac{1}{2}\alpha(x).(\Box-\mu_0^2)\alpha(x) - \bar{\rho}(x).(\gamma.\partial+m_0)\rho(x)\right.$$
$$\left. + iG_0\bar{\rho}(x).\gamma_5\rho(x)\alpha(x)]\right\} \qquad 10(199)$$

to within a constant C, which we shall fix later. When we insert the solution 10(199) back into 10(195), we obtain the Green's functional $Z[\eta, \bar{\eta}, J]$ in the form of a functional integral

$$Z[\eta, \bar{\eta}, J] = C \int d\left(\frac{\alpha}{\sqrt{2\pi}}\right) d\left(\frac{\rho}{\sqrt{\pi}}\right)$$
$$\times \exp \int d^4x[i\bar{\rho}.\eta + i\bar{\eta}.\rho + iJ.\alpha - \tfrac{1}{2}\alpha D_1\alpha - \bar{\rho}_2 D_2\rho - G_0\bar{\rho}\gamma_5\rho\alpha] \qquad 10(200)$$

where we have defined

$$D_1 = -i(\Box - \mu_0^2) \qquad 10(201)$$

$$D_2 = i(\gamma.\partial + m_0) \qquad 10(202)$$

Our task is now to evaluate the functional integral 10(200) by means of the formulae developed in the mathematical introduction. We first consider the integration over the space of real functions $\alpha(x)$. Writing 10(200) in the form

$$Z[\eta, \bar{\eta}, J] = C \int d\left(\frac{\alpha}{\sqrt{2\pi}}\right) \exp\left\{\int d^4x(iJ.\alpha - \tfrac{1}{2}\alpha D_1\alpha)\right\} f[\eta, \bar{\eta}, \alpha] \qquad 10(203)$$

we can evaluate the integral over α with the aid of the formula 10(164) by making the change of variable

$$\alpha' = D_1^{1/2}\alpha \qquad 10(204)$$

Applying 10(183) we get the relation

$$d(\alpha) = \det(D_1^{-1/2})\, d(\alpha')$$

between the 'volume elements'. Absorbing the factor $\det(D_1^{-1/2})$ in the constant C and using 10(164) we obtain

$$Z[\eta, \bar{\eta}, J] = Cf\left[\eta, \bar{\eta}, \frac{1}{i}\frac{\delta}{\delta J}\right] \exp\left[-\tfrac{1}{2}\int d^4x J(x) D_1^{-1}(x) J(x)\right]$$

or

$$Z[\eta, \bar{\eta}, J] = C\int d\left(\frac{\rho}{\sqrt{\pi}}\right) \exp\left\{\int d^4x\left[i\bar{\rho}\,.\,\eta + i\bar{\eta}\,.\,\rho - \bar{\rho}D_2\rho + iG_0\bar{\rho}\gamma_5\frac{\delta}{\delta J}\rho\right]\right\}$$
$$\times \exp\left\{-\tfrac{1}{2}\int d^4x\, d^4y J(x)\Delta_F(x-y)J(y)\right\} \qquad 10(205)$$

where we have inserted the explicit form of $f[\eta, \bar{\eta}, -i\delta/\delta J]$ from 10(200) and where we have used the identity

$$\frac{i}{\Box - \mu_0^2} J(x) = \int \Delta_F(x-y) J(y)\, d^4y \qquad 10(206)$$

To perform the remaining functional integration over ρ, we first write 10(205) in the form

$$Z[\eta, \bar{\eta}, J] = C\int d\left(\frac{\rho}{\sqrt{\pi}}\right) \exp\left\{\int d^4x[i\bar{\rho}\,.\,\eta + i\bar{\eta}\,.\,\rho - \bar{\rho}(S_F^{\delta/i\delta J})^{-1}\rho]\right\}$$
$$\times \exp\left\{-\tfrac{1}{2}\int d^4x\, d^4y J(x)\Delta_F(x-y)J(y)\right\} \qquad 10(207)$$

where

$$S_F^{\delta/i\delta J} = \left[D_2 + G_0\gamma_5\frac{1}{i}\frac{\delta}{\delta J}\right]^{-1} \qquad 10(208)$$

is the fermion Green's function* in the presence of the external field $\delta/i\delta J$. Next we make the change of variables

$$\rho' = (S_F^{\delta/i\delta J})^{-1/2}\rho \qquad 10(209)$$

and use 10(192) and 10(191). This yields

$$Z[\eta, \bar{\eta}, J] = C\det(S_F^{\delta/i\delta J})^{-1} \exp[-\bar{\eta}S_F^{\delta/i\delta J}\eta] \exp[-\tfrac{1}{2}J\Delta_F J]$$

* Compare 10(62).

and using 10(67) we recover the result 10(69)

$$Z[\eta,\bar\eta,J] = \det\left(1 - iG_0\gamma_5\frac{\delta}{\delta J}S_F\right)e^{-\bar\eta\left(1-iG_0S_F\gamma_5\frac{\delta}{\delta J}\right)^{-1}S_F\eta}$$

$$\times\, e^{-\frac{1}{2}J\Delta_F J} \qquad 10(210)$$

after setting $C = 1$ to ensure that in the absence of sources $Z[0, 0, 0]$ reduces to the infinite phase factor

$$\det\left(1 - iG_0\gamma_5\frac{\delta}{\delta J}S_F\right)e^{-\frac{1}{2}J\Delta_F J}\Big|_{J=0}$$

representing the sum of all disconnected vacuum–vacuum graphs.

Feynman's Action Principle The formula 10(200) or

$$\langle 0\text{ out}|0\text{ in}\rangle_{\eta\bar\eta J} = \int d\left(\frac{\alpha}{\sqrt{2\pi}}\right)d\left(\frac{\rho}{\sqrt{\pi}}\right)$$

$$\times \exp\int d^4xi\left[\bar\rho\eta + \bar\eta\rho + J\alpha + \frac{i}{2}\alpha D_1\alpha + i\bar\rho D_2\rho + iG_0\bar\rho\gamma_5\rho\alpha\right] \qquad 10(211)$$

has an important connection to Feynman's formulation of quantum mechanics [Feynman (1948)]. Let us perform an integration by parts on the $\frac{1}{2}\alpha D_1\alpha$ term in 10(211). The quantity in square brackets is thereby transformed to

$$-\tfrac{1}{2}\partial_\mu\alpha\partial_\mu\alpha - \frac{\mu_0^2}{2}\alpha^2 - \bar\rho(\gamma\,.\,\partial + m_0)\rho + iG_0\bar\rho\gamma_5\rho\alpha + \bar\rho\eta + \bar\eta\rho + J\alpha$$

which is just the Lagrangian density $\mathscr{L}(\rho, \alpha; \eta, J)$ for pseudoscalar meson theory with external fields η, $\bar\eta$ and J and with ψ and ϕ replaced by *classical* fields ρ and α respectively. Writing 10(211) as

$$\langle 0\text{ out}|0\text{ in}\rangle_{\eta\bar\eta J} = \int d\left(\frac{\alpha}{\sqrt{2\pi}}\right)d\left(\frac{\rho}{\sqrt{\pi}}\right)\exp\left(i\int d^4x\mathscr{L}(\rho\alpha;\eta J)\right) \qquad 10(212)$$

we see that the vacuum–vacuum transition amplitude is given as an integral over 'histories'—each history being a given choice of values for α and ρ over all space–time—with a weight factor e^{iS} where

$$S[\rho\alpha, \eta J] = \int d^4x\mathscr{L}(\rho\alpha, \eta J)$$

is the action integral for a given history $\rho\alpha$. Equation 10(212) is the extension to relativistic field theory of Feynman's action principle for non-relativistic quantum mechanics.*

* See R. P. Feynman and A. R. Hibbs, *Quantum Mechanics and Path Integrals*, McGraw Hill, 1965. Feynman's action principle is closely connected to Schwinger's action principle discussed in Section 3-3 [Matthews (1955), Burton (1955)].

PROBLEMS

1. Derive the Schwinger equations for the Green's functions in quantum electrodynamics. Apply the equation analogous to 10(21) to give a simple proof of Ward's identity.
2. Check the conservation law 10(85) for the axial vector current of the Yukawa coupling model 10(80).
3. Derive Schwinger's equation for $S_F'^J$ for the model discussed in Section 10-3. Show that in the approximation of Eq. 10(108), the vacuum expectation value φ_i^J $(i = S, P)$ points in the same direction as the external source vector J_i.
4. Derive the coupled Schwinger equations 10(117) for $\Delta_{Fik}'^J$.
5. Verify the result 10(151).

REFERENCES

Baker, M. (1963) (with K. Johnson and B. W. Lee) *Phys. Rev.* **133 B**, 209
Bialynicki-Birula, I. (1960) *Nuovo Cimento* **17**, 951
Bludman, S. A. (1963) (with A. Klein) *Phys. Rev.* **131**, 2364
Burton, W. K. (1955) *Nuovo Cimento* **1**, 355
Burton, W. K. (1956) (with A. H. de Borde) *Nuovo Cimento* **4**, 254
Candlin, D. J. (1956) *Nuovo Cimento* **4**, 231
Feynman, R. P. (1948) *Rev. Mod. Phys.* **20**, 367
Friedrichs, K. O. (1951) *Commun. Pure Appl. Math.* **4**, 161
Goldstone, J. (1961) *Nuovo Cimento* **19**, 154
Goldstone, J. (1962) (with A. Salam and S. Weinberg) *Phys. Rev.* **127**, 965
Guralnik, G. S. (1964) (with C. R. Hagen and T. W. B. Kibble) *Phys. Rev. Letters* **13**, 585
Johnson, K. (1961a) *Nucl. Phys.* **25**, 431
Johnson, K. (1961b) *Nuovo Cimento* **20**, 773
Johnson, K. (1962) In *Proc. Trieste Seminar 1962* (International Atomic Energy Agency)
Kato, T. (1960) (with T. Kobayashi and M. Namiki) *Supp. Prog. Theor. Phys.* No. 15
Klauder, J. R. (1960) *Ann. Phys.* **11**, 123
Lee, T. D. (1955) (with C. N. Yang) *Phys. Rev.* **98**, 1501
Lurié, D. (1964) (with A. J. Macfarlane) *Phys. Rev.* **136 B**, 816
Matthews, P. T. (1955) (with A. Salam) *Nuovo Cimento* **2**, 120
Polivanov, M. K. (1955) *Dokl. Akad. Nauk USSR* **100**, 1061
Sakurai, J. J. (1960) *Ann. Phys.* **11**, 1
Schwinger, J. (1951) *Proc. Nat. Acad. Sci.* **37**, 452
Schwinger, J. (1954a) *Phys. Rev.* **93**, 615
Schwinger, J. (1954b) *Phys. Rev.* **94**, 1362
Schwinger, J. (1959) *Phys. Rev. Letters* **3**, 296
Schwinger, J. (1962) *Phys. Rev.* **128**, 2425
Sommerfield, C. M. (1963) *Ann. Phys.* **26**, 1
Symanzik, K. (1954) *Z. Naturforschung* **9a**, 809
Thirring, W. E. (1964) (with J. E. Wess) *Ann. Phys.* **27**, 331
Tobocman, W. (1956) *Nuovo Cimento* **3**, 1213
Umezawa, H. (1955) (with A. Visconti) *Nuovo Cimento* **1**, 1079
Zumino, B. (1960) *J. Math. Phys.* **1**, 1

Index

(n indicates that the reference is found in the footnote)